Target Organ Toxicology Series

Toxicology of Skin

Target Organ Toxicology Series

Target Organ Toxicology Series

Toxicology of Skin

Edited by

Howard I. Maibach

Department of Dermatology
University of California
San Francisco, California, USA

USA	Publishing Office:	TAYLOR & FRANCIS 325 Chestnut Street Philadelphia, PA 19106 Tel: (215) 625-8900 Fax: (215) 625-2940
	Distribution Center:	TAYLOR & FRANCIS 7625 Empire Drive Florence, KY 41042 Tel: 1-800-624-7064 Fax: 1-800-248-4724
UK		TAYLOR & FRANCIS 27 Church Road Hove E. Sussex, BN3 2FA Tel: +44 (0) 1273 207411 Fax: +44 (0) 1273 205612

TOXICOLOGY OF SKIN

1 2 3 4 5 6 7 8 9 0

Printed by Sheridan Books, Ann Arbor, MI, 2000.
Cover design by Curt Tow.

A CIP catalog record for this book is available from the British Library.
   The paper in this publication meets the requirements of the ANSI Standard Z39.48-1984 (Permanence of Paper).

Library of Congress Cataloging-in-Publication Data

Toxicology of skin / edited by Howard I. Maibach.
       p. ; cm. —(Target organ toxicology series)
    Includes bibliographical references and index.
    ISBN 1-56032-802-9 (alk. paper)
    1. Dermatotoxicology.   2. Contact dermatitis.   3. Urticaria.   I. Maibach, Howard I.
II. Series.
    [DNLM: 1. Skin Diseases—metabolism.   2. Skin—metabolism.   3.
Skin—physiopathology.   4. Skin Absorption.   WR 140 T755 2000]
RL803.T69   2000
616.5—dc21

                                                                 00-037722

ISBN 1-56032-802-9 (case)

Contents

I. Irritant Dermatitis

II. Percutaneous Absorption

Contributors

S. Iris Ale *Department of Dermatology, University Hospital, Republic University, Montevideo, Uruguay*

Smita Amin *Department of Dermatology, University of California, San Francisco, CA, USA*

Angela N. Anigbogu *Department of Dermatology, University of California, San Francisco, California, USA*

Martin D. Barratt *Marlin Consultancy, Carlton, Bedford, United Kingdom*

E. Bloom *Department of Dermatology, University of California, San Francisco, California, USA and Department of Dermatology, University of Montreal, Montreal, Quebec, Canada*

Bruce A. Buchholz *Center for Accelerator Mass Spectrometry, Lawrence Livermore National Laboratory, Livermore, California, USA*

Veranne Charbonnier *Department of Dermatology, University of California, San Francisco, California, USA*

Tim Chartier *School of Medicine, University of Miami, Miami, Florida, USA*

Roger Cooke *Department of Biochemistry and Biophysics, University of California, San Francisco, California, USA*

Isaak Effendy *Department of Dermatology, University of Marburg, Germany*

Melanie S. Flint *Centers for Disease Control and Prevention, National Institute for Occupational Safety and Health, Morgantown, West Virginia, USA*

Søren Frankild *Department of Dermatology, Odense University Hospital, Denmark*

Shirley J. Gee *Department of Entomology, University of California, Davis, California, USA*

S. Douglas Gilman *Department of Entomology, University of California, Davis, California, USA (currently at Department of Chemistry, University of Tennessee, Knoxville, Tennessee, USA)*

Bruce D. Hammock *Department of Entomology, University of California, Davis, California, USA*

Whitney Hannon *School of Medicine, University of Miami, Miami, Florida, USA*

Kathryn L. Hatch *College of Agriculture, The University of Arizona, Tucson, Arizona, USA*

Rudolf A. Herbst *Department of Dermatology, University of California, San Francisco, California, USA and Department of Dermatology and Allergology, Hannover Medical University, Hannover, Germany*

David W. Hobson *DFB Pharmaceuticals, San Antonio Texas, USA*

B. Homey *Department of Dermatology, Heinrich-Heine University, Düsseldorf, Germany*

Jurij J. Hostýnek *Euromerican Technology Resources, Inc., Lafeyette, California, USA and Department of Dermatology, University of California, San Francisco, California, USA*

Xiaoying Hui *Department of Dermatology, University of California, San Francisco, California, USA*

Yoshiaki Kawasaki *Department of Dermatology, University of California, San Francisco, California, USA*

Mikyung Kwah *Department of Dermatology, University of California, San Francisco, California, USA*

Agis F. Kydonieus *Samos Pharmaceuticals LLC, Kendall Park, New Jersey, USA*

Antti I. Lauerma *Department of Dermatology, University of Helsinki, Helsinki, Finland*

P. Lehmann *Department of Dermatology, Heinrich-Heine University, Düsseldorf, Germany*

Yung-Hian Leow *National Skin Centre, Singapore*

Cheryl Levin *Department of Dermatology, University of California, San Francisco, California, USA*

Harald Loeffler *Department of Dermatology, University of Marburg, Germany*

Philip S. Magee *Department of Dermatology, University of California, San Francisco, California, USA*

Howard I. Maibach *Department of Dermatology, University of California, San Francisco, California, USA*

Graham A. Matthews *Imperial College at Silwood Park, Ascot, Berkshire, United Kingdom*

K. Milioni *Department of Dermatology, University of California, San Francisco, California, USA*

Jun-ichi Mizushima *Department of Dermatology, University of California, San Francisco, California, USA*

Boyce M. Morrison, Jr. *Colgate Palmolive Company, Piscataway, New Jersey, USA*

Marc Paye *Colgate Palmolive R& D, Milmort, Belgium*

Barbara R. Reed *University of Colorado Health Sciences Center, Denver, Colorado, USA*

Stephen D. Soileau *Gillette Medical Evaluation Laboratories, Needham, Massachusetts, USA*

Robert B. Strimling *Department of Dermatology, University of California, San Francisco, California, USA*

Hanafi Tanojo *Department of Dermatology, University of California, San Francisco, California, USA*

J. Richard Taylor *School of Medicine, University of Miami, Miami, Florida, USA*

Sally S. Tinkle *Centers for Disease Control and Prevention, National Institute for Occupational Safety and Health, Morgantown, West Virginia, USA*

Ethel Tur *Department of Dermatology, Tel Aviv Sourasky Medical Center and Sackler School of Medicine, Tel Aviv University, Tel Aviv, Israel*

John S. Vogel *Center for Accelerator Mass Spectrometry, Lawrence Livermore National Laboratory, Livermore, California, USA*

H. W. Vohr *Research Toxicology, Bayer AG, Wuppertal, Germany*

Ronald C. Wester *Department of Dermatology, University of California, San Francisco, California, USA*

John J. Wille *Bioderm Technologies Incorporated, Trenton, New Jersey, USA*

Hongbo Zhai *Department of Dermatology, University of California, School of Medicine, San Francisco, California, USA*

SECTION I
Irritant Dermatitis

Toxicology of Skin
Edited by Howard I. Maibach
Copyright © 2001 Taylor & Francis

1

Electron Paramagnetic Resonance Study for Defining the Mechanism of Irritant Dermatitis

Yoshiaki Kawasaki and Jun-ichi Mizushima

*Department of Dermatology, University of California,
San Francisco, California, USA*

Roger Cooke

*Department of Biochemistry and Biophysics, University of California,
San Francisco, California, USA*

Howard I. Maibach

*Department of Dermatology, University of California,
San Francisco, California, USA*

· **Introduction**
· **Materials and Methods**
    · Materials · Stratum Corneum Preparation, Spin Labeling, and Surfactant Treatment for Cadaver Skin · Stratum Corneum Preparation, Spin Labeling, and Surfactant Treatment for Stripped Stratum Corneum from Volunteers · EPR Spectral Measurement · Clinical Testing
· **Results and Discussion**
    · Effect of Surfactants on Intercellular Fluidity of Stratum Corneum Obtained from Cadaver Skin · Effect of Surfactant Mixtures (SLS/SLG) on Intercellular Lipid Fluidity of Stratum Corneum Obtained from Cadaver Skin · Effect of Incubation Time and Concentration of SLS on the Intercellular Lipid Fluidity of Cadaver Stratum Corneum · EPR Spectrum Measurement of Human-Stripped Stratum Corneum
· **Conclusion**

INTRODUCTION

The intercellular lipid lamellae in the stratum corneum constitute the main epidermal barrier to the diffusion of water and other solutes (Elias and Friend, 1975; Elias, 1981, 1983; Landman, 1986; Wertz and Downing, 1982). These lipids, arranged in multilayers between the corneocytes (Swartzendruber et al., 1989; Wertz et al., 1989), consist of ceramides (40%–50%), free fatty acids (15%–25%), cholesterol (15%–25%), and cholesterol sulfate (5%–10%) (Gray et al., 1982; Long et al., 1985; Swartzendruber et al., 1987). Information on the molecular structure of these lipids is important in elaborating a rational design of effective penetration enhancers for transdermal drug delivery (Woodford and Barry, 1986) and to understand the mechanism of irritant dermatitis and other stratum corneum disease. This information has been obtained by thermal analysis (Bouwstra et al., 1992; Golden et al., 1987; Van Duzee, 1975), X-ray diffraction study (Bouwstra et al., 1994; Bouwstra et al., 1991; Bouwstra et al., 1991; Garson et al., 1991; Vilkes et al., 1973; White et al., 1988), FT-IR spectroscopy (Bommannann et al., 1990; Krill et al., 1992), and electron paramagnetic resonance spectroscopy (EPR) (Rehfeld et al., 1988, 1990).

Several investigators (Barker et al., 1991; Faucher and Goddard, 1978; Froebe et al., 1990; Fulmer and Kramer, 1986; Giridhar and Acosta, 1993; Imokawa, 1980; Mokawa et al., 1975; Rhein et al., 1986, 1990; Wilmer et al., 1994) demonstrate that stratum corneum swelling, protein denaturation, lipid removal, inhibition of cellular proliferation, and chemical mediator release contribute to irritation reaction; however, the mechanism of irritant dermatitis has not been understood and defined completely yet.

On the other hand, permeability is increased by an increase in fluidity, both in biological and artificial membranes, suggesting the correlation between flux and fluidity (Golden et al., 1987; Knutson et al., 1985). The dynamic properties of intercellular lipid of stratum corneum are incompletely characterized; the effect of surfactants has not been studied in detail.

EPR-employing nitroxide spin probes, the spin labeling method, has been utilized as a valuable spectroscopic method to provide information about the dynamic structure of membranes (Curtain and Gorden, 1984; Sauerheber et al., 1977). Spin probes are specifically incorporated with the lipid or lipid part of biological membranes. Thus, each label reflects the properties of different membrane regions. EPR spectra of membrane-incorporated spin probes are sensitive to the rotational mobility, orientation of the probes, and polarity of the environment surrounding the probes.

This chapter investigates the influence of surfactants on the intercellular lipid structure of cadaver stratum corneum and the possibility of EPR spectra measurement on the stripped stratum corneum, which might reflect the actual skin lipid conditions.

MATERIALS AND METHODS

Materials

5-Doxyl stearic acid (5-DSA), purchased from Sigma Chemical Co. (St. Louis, MO), was used as the spin-labeling reagents without further purification.

TABLE 1.1. *Table of surfactants*

Category	Abbreviation	Chemical name	Purity	Supplier
Anionic	SLS	Sodium lauryl sulfate	>99%	Sigma Chemical Co. (USA)
	SL	Sodium laurate	>98%	Junsei Chemical Co. (Japan)
	SLES	Sodium lauryl POE(3) ether sulfate	Commercial grade	Kao Co. (Japan)
	SLEC	Sodium lauryl POE(3) ether carboxylate	Commercial grade	Sanyo Kasei Ltd. (Japan)
	SLG	Monosodium lauroyl glutamate	>98%	Ajinomoto Co., Inc. (Japan)
Cationic	MSAC	Monostearyl trimethyl ammonium chloride	>98%	Tokyo chemical Industry Co., Ltd. (Japan)
Amphoteric	HEA	N-[3-alkyl (Elias and Feingold, 1992; Bouwstra et al., 1992) oxy-2-hydroxypropyl]-L-arginine hydrochloride	>98%	Ajinomoto Co., Inc. (Japan)

The surfactants shown in Table 1.1 were used without further purification. Purity of all materials were as stated by the suppliers. Test solutions were prepared with deionized water (MILLI-Q reagent water system (Millipore Co., Bedford, MA)). Deionized water was the vehicle control. Chemical structures of surfactants and spin label used in this study are described in Figure 1.1.

Stratum Corneum Preparation, Spin Labeling, and Surfactant Treatment for Cadaver Skin

Human abdominal skin was obtained from fresh cadaver skin with a dermatome. Epidermis was separated from dermis by immersing the skin in 60°C water bath set for 2 minutes followed by mechanical removal. Then, the epidermis was placed stratum corneum side up on the filter paper and floated on 0.5% wt trypsin (type II; Sigma) in a Tris-HCl buffer solution (pH 7.4) for 2 hours at 37°C. After incubation, any softened epidermis was removed by mild agitation of the stratum corneum sheet. Stratum corneum was dried and stored in a desiccator at −70°C for 3–4 days. Details are described by Quan (Quan and Maibach, 1994; Quan et al., 1995).

5-DSA was used as a stearic acid spin-labeling agent. One slice of dry stratum corneum sheet (approximately 0.49 cm^2; ~0.7 cm × ~0.7cm) was incubated in 1.0 mg/dl 5-DSA aqueous solution (2.6×10^{-5} M; FW = 384.6) for 2 hours at 37°C and washed gently with deionized water to remove the excess of spin label.

Surfactant treatment was as follows: a spin-labeled stratum corneum was immersed in surfactant aqueous solutions and incubated at 37°C for hours. The stratum corneum was taken out of the surfactant solution at indicated times. After rinsing with deionized water and removing the excess water, stratum corneum was mounted on the flat-surface EPR cell and EPR spectra were recorded. The control EPR spectrum was recorded for the spin-labeled stratum corneum kept in the deionized water at 37°C instead of the surfactant solution.

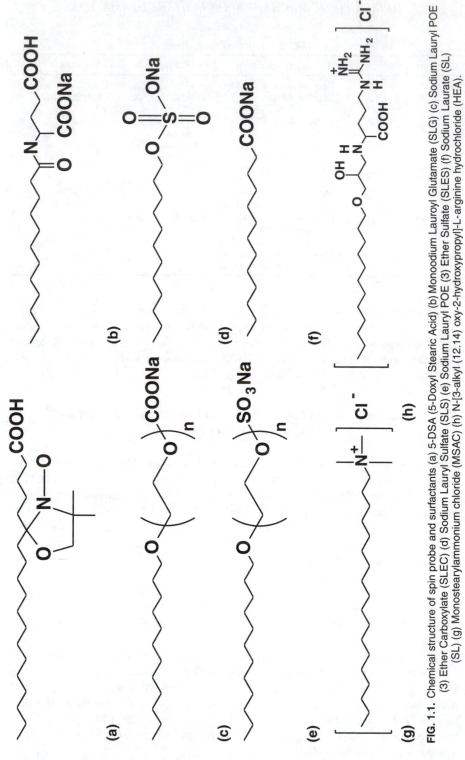

FIG. 1.1. Chemical structure of spin probe and surfactants (a) 5-DSA (5-Doxyl Stearic Acid) (b) Monoodium Lauroyl Glutamate (SLG) (c) Sodium Lauryl POE (3) Ether Carboxylate (SLEC) (d) Sodium Lauryl Sulfate (SLS) (e) Sodium Lauryl POE (3) Ether Sulfate (SLES) (f) Sodium Laurate (SL) (SL) (g) Monostearylammonium chloride (MSAC) (h) N-[3-alkyl (12.14) oxy-2-hydroxypropyl]-L-arginine hydrochloride (HEA).

Stratum Corneum Preparation, Spin Labeling, and Surfactant Treatment for Stripped Stratum Corneum from Volunteers

Stripped stratum corneum was obtained from the patch test sites of three healthy volunteers (3 male; mean age 34, ranging 30–38) after giving their informed consent. Stratum corneum was removed from volar side of forearm skin by a single stripping with one drop of cyanoacrylate resin onto a quartz glass (0.5 cm × 1.5 cm; Nihon Denshi CC, Tokyo, Japan) in accordance with the method of (Imokawa et al., 1991).

Stripped stratum corneum, which is attached on a quartz glass, was spin labeled with a drop (approximately 30 μl) of 1.0 mg/dl 5-DSA solution for 30 minutes at 37°C and then washed with deionized water to remove excess spin probe on the stripped skin surface.

EPR Spectral Measurement

One slice of stratum corneum previously labeled with 5-DSA was mounted on the flat surface EPR cell, whose active cell area is approximately 1 cm^2. EPR measurements were performed at approximately 23–25°C.

EPR spectra were obtained with a Bruker ER200D EPR spectrometer (Bruker Inc., Billierica, MA) with microwave power output of 20 mW; spectrum data were collected by an IBM PC using software written in PC/FORTH (ver. 3.2, Laboratory Microsystems, Marina del Rey, CA). The hyperfine splittings of labeled skin samples were determined with 100 gauss scan width, 2.0×10^5 receiver gain, and 100 msec

TABLE 1.2. *EPR spectral data and clinical data of SLS/SLG mixtures*

Sample name	Averaged S (mean ± S.D.) $n = 3$	Human patch (mean ± SD)	
		Visual	TEWL (g H$_2$O/m^2/h)
Control	0.856 ± 0.028	0.53 ± 0.08	13.0 ± 1.0
0.25wt % SLS	0.700 ± 0.021	0.73 ± 0.08	22.3 ± 1.7
0.50wt % SLS	0.662 ± 0.038	0.70 ± 0.10	22.3 ± 1.7
0.75wt % SLS	0.644 ± 0.027	0.87 ± 0.14	22.7 ± 1.5
1.00wt % SLS	0.560 ± 0.034	1.03 ± 0.15	25.4 ± 2.6
0.25wt % SLS + 0.75wt % SLG	0.809 ± 0.070	0.73 ± 0.08	20.0 ± 1.7
0.50wt % SLS + 0.50wt % SLG	0.714 ± 0.004	0.67 ± 0.08	20.7 ± 1.9
0.75wt % SLS + 0.25wt % SLG	0.656 ± 0.040	0.87 ± 0.11	21.2 ± 2.6
0.25wt % SLS + 1.00wt % SLG	0.808 ± 0.047	NA	NA
0.50wt % SLS + 1.00wt % SLG	0.790 ± 0.054	NA	NA
0.75wt % SLS + 1.00wt % SLG	0.744 ± 0.040	NA	NA
1.00wt % SLS + 1.00wt % SLG	0.663 ± 0.051	NA	NA
1.00wt % SLG	0.819 ± 0.023	0.67 ± 0.08	15.8 ± 1.1

Error bars = Mean ± SD, $n = 3$ for order parameters, Mean ± SD, $n = 15$ for clinical data. NA = Not available in reference (Lee et al., 1994).

time constant. Each sample was scanned several times; EPR signals were averaged to give a single estimate for the sample.

Triplet signals, which are sharp, can be observed when the spin probe (doxyl group) moves freely, as shown in Figure 1.2(a); however, the spectrum becomes broader (Figure 1.2[b]) when spin probe mobility is restricted by interaction with other components.

When the spin probe is incorporated in the highly oriented intercellular lipid structure in normal skin, the probe cannot move freely due to the rigidity of lipid structure, and its EPR spectrum represents the broad profile, such as in Figure 1.2(b). Once the normal structure is completely destroyed by chemical and/or physical stress, there is nothing to inhibit probe mobility and the EPR spectrum profiles become sharp as in Figure 1.2(a). EPR spectral profile represents the rigidity of the environment of the spin probe. To express the rigidity quantitatively, an order parameter is calculated from the EPR spectrum.

Order parameters (S) were calculated according to (Griffith and Jost, 1976), (Hubbel and McConnell, 1971), and (Marsh, 1981):

$$S = (A\| - A\perp)/[A_{ZZ} - 1/2(A_{XX} + A_{YY})](a_0/a_0'),$$

where $2A\|$ is identified with the outer maximum hyperfine splitting; A_{max} (Figure 1.2) and $A\perp$ is obtained from the inner minimum hyperfine splitting A_{min} (Figure 1.2).

a_0 is the isotropic hyperfine splitting constant for nitroxide molecule in the crystal state:

$$a_0 = (A_{XX} + A_{YY} + A_{ZZ})/3.$$

The values used to describe the rapid anisotropic motion of membrane-incorporated probes of fatty acid type are

$$(A_{XX}, A_{YY}, A_{ZZ}) = (6.1, 6.1, 32.4) \text{ Gauss.}$$

Similarly, the isotropic hyperfine coupling constant for the spin label in the membrane (a_0') is given by

$$a_0' = (A\| + 2A\perp)/3.$$

a_0' values are sensitive to the polarity of the environment of the spin labels, as increases in a_0' values reflect an increase in the polarity of the medium.

The order parameter (S) provides a measure of the flexibility of the spin labels in the membrane. It follows that $S = 1$ for highly oriented (rigid) and $S = 0$ for completely isotropic motion (liquid). Increases of order parameter reflect decreases in the segmental flexibility of the spin label, and, conversely, decreases in the order parameter reflect increases in the flexibility (Curtain and Gorden, 1984).

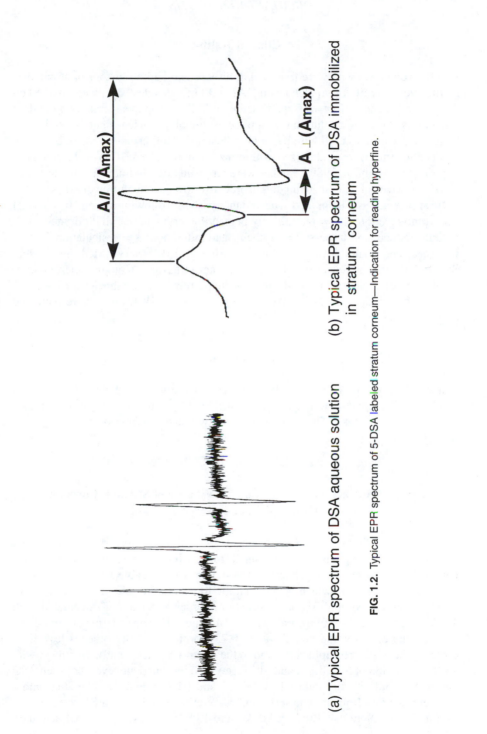

(a) Typical EPR spectrum of DSA aqueous solution

(b) Typical EPR spectrum of DSA immobilized in stratum corneum

FIG. 1.2. Typical EPR spectrum of 5-DSA labeled stratum corneum—Indication for reading hyperfine.

Clinical Testing

Healthy volunteers who were free of skin disease and had no history of atopic dermatitis were recruited for patch testing. After 30 minutes, test sites were marked on the subject's back or forearms and baseline of TEWL was measured. Two hundred microliters of each test solution was applied using polypropylene chambers (Hilltop; Cincinnati, Ohio) secured with paper tape (Scanpor, Norgesplaster, Oslo, Norway). Application sites were randomized to minimize anatomical bias (Cua et al., 1990; Van der Valk and Maibach, 1989). Patches were removed after 24 hours and the test sites were exposed to air at least 30 minutes in order to allow deconvolution of excess water.

Each site was visually graded in accordance with the following system: 0 = normal skin or no reaction; 0.5 = faint, barely perceptible erythema or slight dryness; 1 = definite erythema or dryness; 2 = erythema and induration; 3 = vesiculation. TEWL was measured quantitatively with an evaporimeter EP-1 (Servo Med, Stockholm, Sweden) or with a Tewameter TM-20 (Courage+Rhazaka, Cologne, Germany) at 30 minutes after patch removal. Readings were performed at a stable level 30 seconds or more after application of the TEWL probe on the skin. TEWL values were expressed as $g/m^2/h$ (Lee, 1994).

Statistical Analysis

Statistical analysis for EPR spectral data was conducted with using unpaired t-test with single tail. Clinical data were statistically analyzed by two-way analysis of variance (ANOVA, Fisher's test). The significance level was taken as $p < 0.05$.

RESULTS AND DISCUSSION

Effect of Surfactants on Intercellular Fluidity of Stratum Corneum Obtained from Cadaver Skin

The profile of EPR spectra of 5-DSA labeled depend on the surfactants treated. The corresponding order parameters S obtained from each EPR spectrum and the clinical data of 24-hour occlusive patch testing are summarized in Table 1.3.

Surfactant molecule, which is amphiphillic to water and lipid, might be incorporated into structured lipids (lamellar structure). Order parameter (S) calculated from 1.0% wt SLS-treated stratum corneum was 0.47, indicating disordering of lipid structure. On the contrary, the high S value (0.73) for 1.0% wt SLG means less of an effect on the structured lipid compared to the control S value (0.89). Treatment with 1.0% wt solution of SL, SLES, and SLEC revealed intermediate levels between SLG and SLS. Lipid disorder induced by MSAC and HEA, which are classified into a different category from anionic surfactant, also revealed intermediate levels between SLS and SLG. Note that 1.0% wt MSAC and 1.0% wt HEA, which are quaternary

TABLE 1.3. *Order parameters of stratum corneum treated with surfactant and clinical observation*

Category	Sample name	Concentration	Order parameter S	Visual scores (average ± SD)	TEWL values (average ± SD) g/m²/h
(Control)	Water		0.89 ± 0.04	0.00 ± 0.00	5.0 ± 1.1
Anionic	SLS	1.0% wt	0.47 ± 0.05	0.79 ± 0.30	13.6 ± 3.1
	SL	1.0% wt	0.65 ± 0.06	0.08 ± 0.20	7.1 ± 3.9
		5.0% wt	0.52 ± 0.04	0.67 ± 0.30	13.3 ± 3.7
	SLES	1.0% wt	0.62 ± 0.06	0.42 ± 0.30	7.6 ± 2.8
	SLEC	1.0% wt	0.62 ± 0.05	0.08 ± 0.20	7.4 ± 2.9
	SLG	1.0% wt	0.73 ± 0.07	0.04 ± 0.10	6.7 ± 3.5
		5.0% wt	0.77 ± 0.08	0.13 ± 0.20	5.5 ± 2.3
Cationic	MSAC	1.0% wt	0.68 ± 0.02	NA	10.2 ± 1.9
Amphoteric	HEA	1.0% wt	0.75 ± 0.02	NA	9.2 ± 0.8

NA = Data is not available at the same conditions as the anionics.

and amphoteric compounds, respectively, lead to less disorder in lipid structure than 1.0% wt SLS, although the irritation potential of surfactants is widely assumed to follow the pattern below, in which quaternaries are the most irritating: quaternaries > amphoterics > anionics > nonionics (Rieger, 1997). These two compounds have a plus charge. Their interaction with stratum corneum might be different from that of anionics, such as SLS. A plus charge might have more attractive interaction to proteins electrically because proteins are generally believed to be negatively charged.

The change of order parameter corresponds to the structural changes of lipid layers. Two phases can be speculated to increase fluidity of lipid structure (decreasing the order parameter). The first phase is an effect of surfactant incorporated into the lamellar structures. If surfactant interferes or decreases lateral interactions between lipids, mobility increases similar to the phase conversion from liquid crystal to gel in the lamellar layers. The second phase is the destruction of lamellar structure by micellization or solubilization of lipid by surfactant. In this case, lipids no longer have dimensional restrictions and gain much higher mobility.

The results shown in Table 1.1 indicate that mobility increase by SLG can be attributed to the phase one structural changes in the lipid layers, and SLS might cause further disruption of the structures of lipid layers. The role of water in the stratum corneum must also be considered for the effects of surfactant to lipid layers. Treatment with anionic surfactants might influence water penetration and/or skin swelling (Takino et al., 1996). (Rhein et al., 1986, 1990) examined the swelling of stratum corneum caused by surfactants and reported that swelling effect of surfactants suggest mechanism of action as the basis for in vivo irritation potential.

The correlation between order parameter S obtained from EPR spectrum and the clinical readings are summarized in Figure 1.3. The correlation coefficients (r^2) of visual score and TEWL values were 0.76 and 0.83, respectively. The order parameter

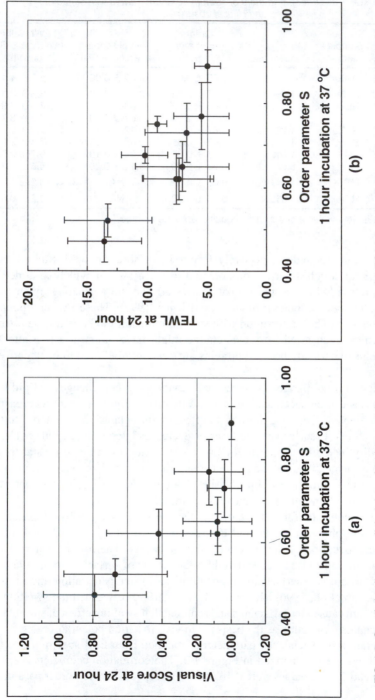

FIG. 1.3. Correlation between clinical data of 24 hour patch and order parameter S of 5-DSA labeled cadaver stratum corneum incubated in surfactants solution for 1 hour at 37°C. (a) Correlation between order parameter and visual score and (b) Correlation between order parameter and TEWL (Error bars: Mean ± SD, $n = 3$ for order parameter; Mean ± SD, $n = 14$ for clinical data).

S correlates better to TEWL values than to visual scores. This difference may be explainable in that TEWL is a direct measure of water barrier function, whereas visual scores represent total skin reactions, including physical or structural changes of skin tissue due to the physiological or biological reactions with surfactant. The order parameter might not predict the following skin reaction after the disorder of lipid structure, such as denaturation of proteins or mucosaccarides in dermis.

Order parameter measurement of stratum corneum may predict the minimal difference of irritating potential among various kinds of chemicals.

Effect of Surfactant Mixtures (SLS/SLG) on Intercellular Lipid Fluidity of Stratum Corneum Obtained from Cadaver Skin

Consumers are exposed daily to anionic surfactants in cleansing products, which may be irritating, especially when applied to sensitive skin sites. Although the properties of detergency and mildness seem to be contradictory, it is possible to reconcile these opposites by choice of surfactants that can reduce irritation potential. Some materials may reduce irritation potential of harsher surfactants, such as sodium lauryl sulfate (Kanari et al., 1992; Lee et al., 1994; Rhein et al., 1986, 1990).

SLS was the most severely irritating and SLG is the mildest amongst the anionic surfactants tested. The influence of surfactant mixtures (SLS/SLG) on intercellular lipid fluidity of stratum corneum obtained from cadaver skin is discussed in this section.

The profile of EPR spectra of 5-DSA labeled depends on the SLS concentration. The order parameters obtained from each EPR spectrum of stratum corneum treated with water, 0.25% wt, 0.50% wt, 0.75% wt, 1.00% wt SLS solutions, SLS/SLG mixtures (total concentration is constant at 1.00% wt) and 1.00% wt SLG solution were summarized in Table 1.3.

The order parameter of water-treated stratum corneum (vehicle control) was 0.86 ± 0.03. Anionic surfactants as an amphiphilic molecule might be incorporated into structured lipids (lamellar structure). Order parameter (*S*) calculated from 1.00% wt SLS-treated stratum corneum was 0.56 ± 0.03, indicating lipid structure disordering. On the contrary, the high *S* value (0.82 ± 0.02) for 1.00% wt SLG means less lipid structure disordered; 1.00% wt SLG almost equals to water. Treatment with 0.25% wt, 0.50% wt, and 0.75% wt SLS solutions revealed intermediate levels between 1.00% wt SLG and SLS.

Each order parameters of 5-DSA–labeled stratum corneum treated with SLS/SLG mixtures (total concentration is constant at 1.00% wt) showed higher values than that of 0.25% wt, 0.50% wt, 0.75% wt SLS, respectively. There were no statistically significant difference between 0.50% wt SLS and 0.50% wt SLS/0.50% wt SLG, and between 0.75% wt SLS and 0.75% wt SLS/0.25% wt SLG ($p > 0.05$). These results suggest that SLG-inhibited SLS induced lipid fluidization. To confirm the antifluidization of SLG, the SLS/SLG mixture solutions were prepared with making SLG concentration constant at 1.00% wt and measured EPR spectra of 5-DSA labeled

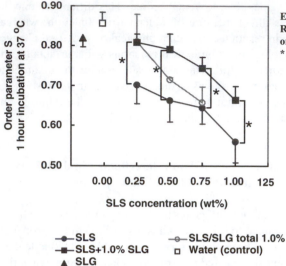

Error bar, mean ± SD for n=3;
Result for 1.00 wt% SLG is plotted
on the left for comparison;
* means p < 0.05

—●—SLS   —○—SLS/SLG total 1.0%
—■—SLS+1.0% SLG   □ Water (control)
▲ SLG

FIG. 1.4. Order parameter of 5-DSA labeled cadaver stratum corneum treated with water, SLS, SLG, and SLS/SLG mixtures (total concentration 1.00 wt%, 1.00 wt% SLG addition to the SLS solutions).

stratum corneum treated with them. The calculated order parameters S are plotted in Figure 1.4.

Order parameters at each SLS concentration (0.25, 0.50, 0.75, and 1.00% wt SLS) with 1.00% wt SLG showed higher values than those of SLS only solutions. There were statistically significant differences between with and without 1.00% wt SLG (p < 0.05), suggesting that the addition of 1.00% wt SLG inhibits the fluidization of intercellular lipid induced by SLS. It may be hypothesized that the direct interactions between SLS and intercellular lipids were interrupted; log P (partition coefficient; $\log\{[SLS]_{lipid}/[SLS]_{bulk}\}$) of SLS into the intercellular lipid might be decreased.

The role of water in the stratum corneum must also be considered for the effects of surfactant on lipid layers. (Alonso et al., 1995, 1996) reported that water increases the fluidity of intercellular lipid of rat stratum corneum at the region close to hydrophilic part, not at lipophillic part, deeply inside of intercellular lipid.

Order parameter correlated with the clinical readings (Figure 1.5). The correlation coefficients (r^2) of visual score and TEWL values were 0.73 and 0.83, respectively. The order parameter correlates to TEWL values better than to visual scores, which is the same result as the previous section.

The order parameters represent the disorder of stratum corneum induced by short-term surfactant contact; however, clinical data represent the skin irritation reaction induced by 24-hour occlusive contact with surfactants. Order parameter measurement of stratum corneum may predict the minimal difference of irritating potential among various chemicals.

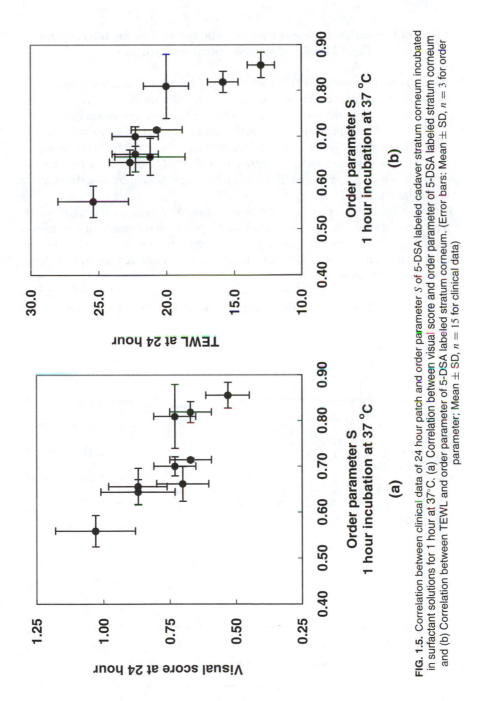

FIG. 1.5. Correlation between clinical data of 24 hour patch and order parameter S of 5-DSA labeled cadaver stratum corneum incubated in surfactant solutions for 1 hour at 37°C. (a) Correlation between visual score and order parameter of 5-DSA labeled stratum corneum and (b) Correlation between TEWL and order parameter of 5-DSA labeled stratum corneum. (Error bars: Mean ± SD, $n = 3$ for order parameter; Mean ± SD, $n = 15$ for clinical data)

Effect of Incubation Time and Concentration of SLS on the Intercellular Lipid Fluidity of Cadaver Stratum Corneum

As shown in the previous section, intercellular lipid fluidization induced by anionic surfactant depends on its chemical structure or pH; 1.0% wt SLS induces more fluidization than the other anionic surfactants. The order parameters shown in Table 1.1 were obtained at 1 hour incubation in surfactant solution kept at 37°C. We still question how long severe anionic surfactant SLS lead to fluidization in lipid ? How much alteration is induced by how much concentration of SLS? SLS solutions having various concentrations were prepared and EPR spectrum of their incubation time dependence is summarized in Figure 1.6.

With increasing incubation time, the order parameter S was decreased; however, each profile of incubation time dependence had a plateau at the region of 4 hours and thereafter. Skin lipid alteration induced by SLS might be completed within 4 hours at a given concentration; however, each alteration level in intercellular lipid depends on its SLS concentration.

As the concentration of SLS increases, the order parameter at 24-hour incubation decreases drastically in the range from 0 to 0.25% wt of SLS (Figure 1.7); however, the

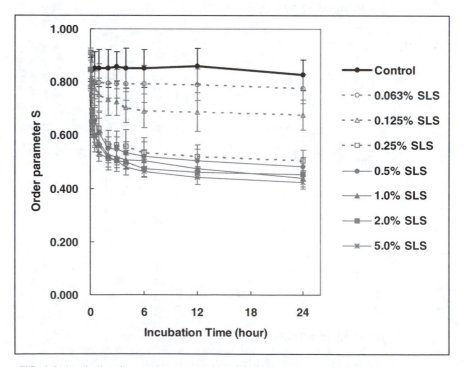

FIG. 1.6. Incubation time and concentration of SLS dependence of 5-DSA labeled cadaver stratum corneum. (Error bars; Mean ± SD for $n = 5$)

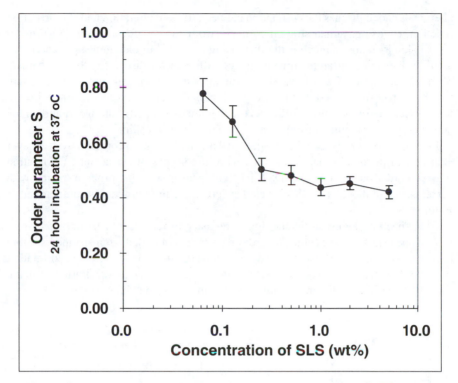

FIG. 1.7. Correlation between SLS concentration and order parameters of 5-DSA labeled cadaver stratum corneum incubated at 37°C for 24 hours. (Error bars; Mean ± SD for $n = 5$)

order parameters calculated from the stratum corneum treated with SLS more than 0.5% wt have no significant difference, showing around 0.45 ~ 0.49. This critical point between 0.25 and 0.5% wt (8.7 ~ 17.3 mM) may correspond to CMC for SLS at 37°C. Flockhalt and Elworthy (Rosen, 1978) reported that CMC of SLS is 8.6 mM at 40°C, 8.2 mM at 25°C.

This behavior is consistent with the following general agreement among experts: monomeric surfactants can penetrate the skin. Monomeric molecules are also the species that are initially adsorbed to the various surfaces within the skin; we cannot ignore the secondary bonding by hydrophobic effects. Thus, the concentration of monomeric species probably plays a principal role in skin and surfactant interactions (Rieger, 1995).

EPR Spectrum Measurement of Human-Stripped Stratum Corneum

The previous data are based on human stratum corneum obtained from cadaver skin. To define the structural changes in intercellular lipid induced by topical application of chemicals and to discuss the correlation between its lipid alteration and skin irritation reaction, choosing human stratum corneum from cadaver for a substrate as model site of skin irritation is much better than using animal skin like guinea pig and rat or lecithin liposomes. Stratum corneum is not sufficient for discussing the mechanism of irritant dermatitis. Cadaver stratum corneum is just a substrate, not a living system that has a recovering system induced by signals like chemical mediators. With the new procedure for measuring EPR spectra on human-stripped stratum corneum of the skin site, it may provide information of use conditions on skin.

In this section we discussed EPR spectral data on the stratum corneum from cadaver skin and stripped skin that was treated with three types of surfactants. Correlation between order parameters of 5-DSA-labeled cadaver stratum corneum treated with surfactants and those of 5-DSA-labeled stripped stratum corneum is summarized in Figure 1.8.

EPR spectra, having sufficient signal intensity to read, could be obtained from the stripped stratum corneum. The order parameters obtained from stripped stratum corneum are larger than those of cadaver stratum corneum; however, a high correlation between them were observed. Hence, the order parameters of cadaver stratum corneum reflects the fluidity of the intercellular lipid in the irritated skin.

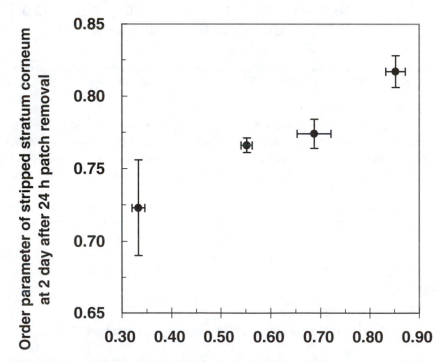

FIG. 1.8. Correlation of order parameters between cadaver stratum corneum and stripped stratum corneum. (Error bar; mean $\pm$ SD [$n = 5$])

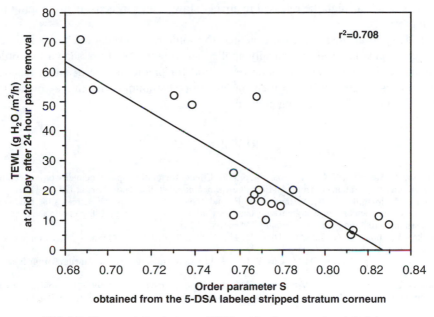

FIG. 1.9. The correlation between TEWL and order parameter at 2nd day after 24 hour patch removal.

The order parameter of SLS-treated cadaver stratum corneum is smaller than that of stripped stratum corneum. This difference might be explained by the barrier recovering property of skin itself. The epidermis can synthesize lipid immediately after barrier disruption (Grubauer et al., 1989). Skin barrier function is 80% repaired by 6–8 hours when skin is treated with acetone (Elias and Feingold, 1992).

The correlation between order parameter of stripped stratum corneum and clinical readings are follows: the correlation coefficient between order parameter and visual score, TEWL values on second day after patch removal were 0.526, 0.708, respectively. The correlation with TEWL is high, as shown in Figure 1.9. This result consists of in vitro data based on cadaver stratum corneum.

CONCLUSION

The toxic manifestation of topically applied substances may induce immediate phenomena (such as corrosion or primary irritation), delayed phenomena (such as sensitization), phenomena which require an additional vector (such as phototoxicity), and systemic phenomena (paraquat toxicity). Such reactions cannot occur unless the toxic agent reaches a viable part of the skin by going through the stratum corneum with the intercellular lipid structure disruption. If the toxicant can be stored in or sorbed by a skin layer without any alteration in lipid structure, it might not reach the viable tissues at all or might be released relatively slowly, thus effectively prolonging the symptoms.

EPR spin labeling is a robust method to monitor the structural change in intercellular lipids induced by topically applied surfactants. We have shown that order parameter is a facile and quantitative method for predicting the irritation reaction in skin. Especially, EPR measurement on stripped stratum corneum may reflect the actual skin condition focusing on lipid structure.

REFERENCES

Alonso, A., Meirelles, N. C., and Tabak, M. (1995): Effect of hydration upon the fluidity of intercellular membranes of stratum corneum: An EPR study, *Biochimica. et. Biophysica. Acta.*, 1237:6–115.

Alonso, A., Meirelles, N. C., Yushmanov, V. E., and Tabak, M. (1996): Water increases the fluidity of intercellular membranes of stratum corneum: Correlation with water permeability, elastic and electrical resistance properties. *J. Invest. Dermatol.*, 106:1058–1063.

Barker, J., Mitra, R., Griffiths, C., Dixit, V., and Nickoloff, B. (1991): Keratinocytes as initiators of inflammation, Lancet, 337:211–214.

Bommannann, D., Potts, R. O., and Guy, R. H. (1990): Examination of stratum corneum barrier function in vivo by infrared spectroscopy. *J. Invest. Dermatol.*, 95:403–408.

Bouwstra, J. A., Gooris, G. D., Van der Spek, J. A., Lavrijsen, S., and Bras, W. (1994): The lipid and protein structure of mouse stratum corneum: A wide and small angle diffraction study. *Biochim. Biophys. Acta.*, 1212:183–192.

Bouwstra, J. A., Gooris, G. S., De Vries, M. A., Van der Spek, J. A., and Bras, W. (1992): Structure of human stratum corneum as a function of temperature and hydration: A wide-angle X-ray diffraction study. *Intl. J. Phram.*, 84:205–216.

Bouwstra, J. A., De Vries, M. A., Doris, G. S., Bras, W., Brussee, J., and Ponec, M. (1991): Thermodynamic and structural aspects of the skin barrier. *J. Controlled Release*, 15:209–220.

Bouwstra, J. A., Gooris, G. S., Van der Spek, J. A., and Bras, W. (1991): Structural investigations of human stratum corneum by small-angle X-ray scattering. *J. Invest. Dermatol.*, 97:1005–1012.

Cua, A., Wilhelm, K., and Maibach, H. I. (1990): Cutaneous sodium lauryl sulfate irritation potential: Age and regional variability. *Br. J. Dermatol.*, 123:607–663.

Curtain, C. C., and Gorden, L. M. (1984): ESR spectroscopy of membranes. In: *Membranes, Detergents and Receptor Solubilization*. Edited by Venter, J. C., and Harrison L. C., pp 177–213 Alan R. Liss, New York.

Elias, P. M., and Friend, D. S. (1975): The permeability barrier in mammalian epidermis. *J. Cell. Biol.*, 65:180–191.

Elias, P. M. (1981): Epidermal lipids, membrane, and keratinization. *Int. J. Derm.*, 20:1–19.

Elias, P. M. (1983): Epidermal lipids, barrier function, and desquamation. *J. Invest. Dermatol.*, 80:44–49.

Elias, P. M., and Feingold, K. (1992): Lipids and the epidermal water barrier: Metabolism, regulation, and pathophysiology. *Semin. Dermatol.*, 11:176–82.

Faucher, J. A., and Goddard, E. D. (1978): Interaction of keratinous substrates with sodium lauryl sulfate. I: Sorption. *J. Soc. Cosmet. Chem.*, 29:323–337.

Froebe, C. L., Simon, F. A., Rhein, L. D., Cagan, R. H., and Kligman, A. (1990): Stratum corneum lipid removal by surfactants: Relation to in vivo irritation. *Dermatologica.*, 181:277–283.

Fulmer, A. W. and Kramer, G. J. (1986): Stratum corneum lipid abnormalities in surfactant-induced dry scaly skin. *J. Invest. Dermatol.*, 86:598–602.

Garson, J. C., Doucet, J., Leveuque, J. L., and Tsoucaris, G. (1991): Oriented structure in human stratum corneum revealed by X-ray diffraction. *J. Invest. Dermatol.*, 96:43–49.

Giridhar, J., and Acosta, D. (1993): Evaluation of cytotoxicity potential of surfactants using primary rat keratinocyte culture as an in vitro cutaneous model. *In Vitro Toxicol. A Journal of Molecular and Cellular Toxicology*, 6:33–46.

Golden, G. M., Guzek, D. B., Kennedy, A. H., McKie, J. E., and Potts, R. O. (1987): Stratum corneum lipid phase transitions and water barrier properties, *Biochemistry*, 26:2382–2388.

Golden, G. M., McKie, J. E., and Potts, R. O. (1987): Role of stratum corneum lipid fluidity in transdermal drug flux. *J. Pharm. Sci.*, 76:25–28.

Gray, G. M., White, R. J., and Yardley, H. J. (1982): Lipid composition of the superficial stratum corneum cells of pig epidermis. *Br. J. Dermatol.*, 106:59–63.

Griffith, O. H., and Jost, P. C. (1976): Lipid spin labels in biological membrane. In: *Spin Labeling Theory and Applications.* Edited by Berliner, L. J., pp. 453–523. Academic Press, New York.

Grubauer, G., Feingold, K., and Elias, P. M. (1989): Transepidermal water loss: The signal for recovery of barrier structure and function. *J. Lipid. Res.*, 30:232–333.

Hubbel, W. L., and McConnell, H. M. (1971): Molecular motion in spin-labeled phospholipids and membranes. *J. Am. Chem. Soc.*, 93:314–326.

Imokawa, G. (1980): Comparative study on the mechanism of irritation by sulfate and phosphate type anionic surfactants. *J. Soc. Cosmet. Chem.*, 31:45–66.

Imokawa, G., Abe, A., Jin, K., Higaki, Y., Kawashima, M., and Hidano A. (1991): Deceased level of ceramides in stratum corneum of atopic dermatitis: An etiologic factor in atopic dry skin? *J. Invest. Dermatol.*, 96:523–526.

Kanari, M., Kawasaki, Y., and Sakamoto, K. (1992): Acylglutamate as an anti-irritant for mild detergent system, (1993): *J. Soc. Cosmet. Chem. Jpn.*, 27:498–505.

Knutson, K., Potts, R. O., Guzek, D. B., Golden, G. M., Lambert, W. J., McKie, J. E., and Higuchi, W. I. (1985): Macro- and molecular physical-chemical considerations in understanding drug transport in the stratum corneum. *J. Controlled Release.*, 2:67–87.

Krill, S. L., Knutson, K., and Higuchi, W. I. (1992): The stratum corneum lipid thermotropic phase behavior. *Biochim. Biophys. Acta.*, 1112:281–286.

Landman, L. (1986): Epidermal permeability barrier: Transformation of lamellar granule-disks into intercellular sheets by a membrane-fusion process, a freeze-fracture study. *J. Invest. Dermatol.*, 87:202–209.

Lee, C. H., Kawasaki, Y., and Maibach, H. I. (1994): Effect of surfactant mixtures on irritant contact dermatitis potential in man: Sodium lauroyl glutamate and sodium lauryl sulfate, Contact Dermatitis, 30:205–209.

Long, S. A., Wertz, P. W., Strauss, J. S., and Downing, D. T. (1985): Human stratum corneum polar lipids and desquamation. *Arch. Dermatol. Res.*, 277:284–287.

Marsh, D. (1981): Electron paramagnetic resonance: Spin labels. In: *Membrane Spectroscopy.* Edited by Grell, E., pp. 51–142. Springer, Berline.

Mokawa, G., Sumura, K., and Katsumi, M. (1975): Study on skin roughness by surfactants. II: Correlation between protein denaturation and skin roughness. *J. Am. Oil Chem. Soc.*, 52:484–489.

Quan, D., and Maibach, H. I. (1994): An electron paramagnetic resonance study. I: Effect of Azone on 5-doxyl stearic acid-labeled human stratum corneum. *Int. J. Pharm.*, 104:61–72.

Quan, D., Cooke, R. A., and Maibach, H. I. (1995): An electron paramagnetic resonance study of human epidermal lipids using 5-doxyl stearic acid. *J. Controlled Release*, 36:235–241.

Rehfeld, S. J., Plachy, W. Z., William, W. I., and Elias, P. M. (1988): Calorimetric and electron spin resonance examination of lipid phase transitions in human stratum corneum: Molecular basis for normal cohesion and abnormal desquamation in recessive X-linked ichthyosis. *J. Invest. Dermatol.*, 91:499–505.

Rehfeld, S. J., Plachy, W. Z., Hou, S. Y. E., and Elias, P. M. (1990): Localization of lipid microdomains and thermal phenomena in murine stratum corneum and isolated membrane complexes: An electron spin resonance study. *J. Invest. Dermatol.*, 95:217–223.

Rhein, L. D., Robbins, C. R., Fernee, K., and Cantore, R. (1986): Surfactant structure effects on swelling of isolated human stratum corneum. *J. Soc. Cosmet. Chem.*, 37:125–139.

Rhein, L. D., Simion, F. A., Hill, R. L., Cagan, R. H., Mattai, J., and Maibach, H. I. (1990): Human cutaneous response to a mixed surfactant system: Role of solution phenomena in controlling surfactant irritation. *Dermatologica.*, 180:18–23.

Rieger, M. M. (1997): The skin irritation potential of quaternaries. *J. Soc. Cosmet. Chem.*, 48:307–317.

Rieger, M. M. (1995): *Surfactant interactions with skin, Cosmetics & Toiletries*, 110 (April):31–50.

Rosen, M. J. (1978): Micelle formation by surfactants. In: *Surfactants and Interfacial Phenomena*, pp. 83–122. John Wiley & Sons, New York.

Sauerheber, R. D., Gorden, L. M., Crosland, R. D., and Kuwahara, M. D. (1977):, Spin-label studies on rat liver and heart plasma membranes: Do probe interactions interfere with the measurement of membrane properties?. *J. Membr. Biol.*, 31:131–139.

Swartzendruber, D. C., Wertz, P. W., Kitko, D. J., Madison, K. C., and Downing, D. T. (1989): Molecular models of the intercellular lipid lamellae in mammalian stratum corneum. *J. Invest. Dermatol.*, 92:251–257.

Swartzendruber, D. C., Wertz, P. W., Madison, K. C., and Downing, D. T. (1987): Evidence that the corneocyte has a chemically bound lipid envelope. *J. Invest. Dermtol.*, 88:709–713.

Takino, Y., Kawasaki, Y., Sakamoto, K., and Higuchi, W. I. (1996): Influence of anionic surfactants to skin: The change of the water permeability and electric resistance. Abstract of 19th IFSCC International Congress, Sydney.

Van der Valk, P. G. M., and Maibach, H. I. (1989): Potential for irritation increases from the wrist to the cubital fossa. *Br. J. Dermatol.*, 121:709–712.

Van Duzee, B. F. (1975): Thermal analysis of human stratum corneum. *J. Invest. Dermatol.*, 65:404–408.

Vilkes, G. L., Nguyen, A. L., and Wildhauer, R. (1973): Structure-property relations of human and neonatal rat stratum corneum. I: Thermal stability of the crystalline lipid structure as studied by X-ray diffraction and differential thermal analysis. *Biochim. Biophys. Acta*, 304:267–275.

Wertz, P. W., and Downing, D. T. (1982): *Science*, 217:1261–1262.

Wertz, P. W., Swartzendruber, D. C., Kitko, D. J., Madison, K. C., and Downing, D. T. (1989): The role of the corneocyte lipid envelopes in cohesion of the stratum corneum. *J. Invest. Dermtol.*, 93:169–172.

White, S. H., Mirejovski, D., and King, G.I. (1988): Structure of lamellar lipid domains and corneocyte envelopes of murine stratum corneum: An X-ray diffraction study. *Biochemistry*, 27:3725–3732.

Wilmer, J., Burleson, F., Kayam, F., Kanno, J., and Luster, M. (1994): Cytokine induction in human epidermal keratinocytes exposed to contact irritants and its relation to chemical-induced inflammation in mouse skin. *J. Invest. Dermatol.*, 102:915–922.

Woodford, R. and Barry, B. W. (1986): Penetration enhancers and the percutaneous absorption of drugs: An update. *J. Toxicol. Cutaneous Ocul. Toxicol.*, 5:167–177.

Toxicology of Skin
Edited by Howard I. Maibach
Copyright © 2001 Taylor & Francis

2

Irritant Patch Test in Atopic Individuals

Harald Loeffler and Isaak Effendy

Department of Dermatology, University of Marburg, Germany

INTRODUCTION

Patients with an atopic history, particularly with an atopic dermatitis, suffer more frequently from irritant hand dermatitis than nonatopics (Kavli et al., 1987; Meding and Swanbeck, 1990; Nilsson and Bäck, 1986). It has been believed that atopic individuals might have an impaired skin barrier function, which is supported by higher transepidermal water loss (TEWL) values following irritant exposure in patients with an active atopic hand eczema (Cowley and Farr, 1992; Nassif et al., 1994; Tabata et al., 1998; Tupker et al., 1990). Consequently, irritants might penetrate easier through the epidermis and maintain the hand eczema. The mechanism might take place in the pathogenesis of hand dermatitis in atopic patients. So, it has been supposed that atopic individuals generally show an enhanced reaction to irritants; however, there are conflicting results concerning the skin susceptibility of patients with a history of atopic dermatitis but without any active lesions (Halkier-Sorensen, 1996; Stolz et al., 1997). It has been demonstrated that some atopic individuals showed normal reactions to irritants, whereas others showed intensively skin reactions (Gehring et al., 1998). Hence, the question is not *whether* atopic individuals do have stronger reaction in irritant tests, but *which* atopic individuals do reveal these stronger reactions.

Taking up the question which atopic individuals are more susceptible to irritants, we compared different groups of atopic individuals to healthy controls:

- Individuals with an active atopic dermatitis (AAD)

- Individuals with an inactive atopic dermatitis (IAD)
- Patients with a respiratory atopy (RA)

The purpose of our study was to determine different subgroups of atopic individuals, which might be discriminated by distinct reactions to irritant patch testing with SLS.

PATIENTS AND METHODS

Volunteers

Ninety-five volunteers (52 women, 43 men), recruited from the outpatient and inpatient clinic of the Department of Dermatology, University of Marburg, participated in this study; they were aged between 14 to 60 years. The study was approved by the ethical committee of the University of Marburg. Informed consent was obtained from all participants.

Four different groups of volunteers were defined as follows:

1. Active atopic dermatitis (AAD, 23 patients): The atopic dermatitis was diagnosed by typical clinical manifestations (e.g., flexural eczema), family atopic history, personal atopic history (without respiratory allergies), and signs of atopy (e.g., Hertoghe sign, dirty neck sign, xerosis, wool intolerance, white dermographism). Using these factors, the score of atopy (proposed by Diepgen et al., 1989, 1991) was calculated. According to this definition, the volunteers were defined as having AAD when their atopic score was higher than 10 *and* an acute eczema was present.
2. Inactive atopic dermatitis (IAD, 21 patients): Volunteers with an atopic score higher than 10 but *without* acute eczematous lesions at the time of testing and during the last 2 years were defined as patients with IAD.
3. Respiratory atopy (RA, 25 patients): Individuals with a typical history of rhinoconjuncitvitis or atopic asthma *and* showing at least one positive prick test to a relevant aeroallergen. The testing was performed when the patients had no rhinitic symptoms at the time of testing (e.g., in winter while the patient had rhinitic symptoms in summer). Volunteers with a coincidence of IAD and RA (or AAD and RA) were excluded.
4. Healthy controls (26 volunteers) without any history of atopy.

Test Procedure

Sixty microliters aqueous SLS 0,5% (SLS Sigma, 99% purity) were applied in Large Finn Chambers (inner diameter, 1,2 cm; Epitest Ltd., Hyrlä, Finland) for 48 hours on clinically unaffected skin at the flexor side of the forearm. TEWL was measured before (basal) and 1 hour after removal of the patch with the TEWAMETER TM210 (Courage & Khazaka, Cologne). The test evaluation was performed by two trained persons according to the guidelines of sodium lauryl sulphate exposure tests by the Standardization Group of the European Society of Contact Dermatitis (Tupker et al., 1997). The volunteers had rested at least one half hour in a room temperature between 20 and 22°C; the relative humidity varied between 35% and 62%. The tests were done within a period of 20 weeks during fall and winter 1997/98.

Statistical Methods

Data of measurements were calculated with SPSS software for Windows. After calculation of the Kolmogorov–Smirnov Test, TEWL values were shown as medium ± standard deviation. Differences in TEWL between the groups were calculated using ANOVA variance analysis and Duncan's Test.

RESULTS

Baseline Transepidermal Water Loss

Higher baseline TEWL values were seen at individuals with an AAD when compared with all other groups, but the differences were not statistically significant. A tendency to higher TEWL values of the control group (as compared to IAD and RA group) was observed (Table 2.1 and Figure 2.1).

Postexposure Transepidermal Water Loss

There were significant differences in TEWL values between basal and post-SLS tests in all three volunteer groups. Compared to healthy volunteers, significantly higher postexposure TEWL values have only been found in AAD group. No significant differences could be obtained between the groups of IAD, RA, and healthy individuals, neither in post-SLS values nor Δ TEWL. A tendency to slightly higher TEWL values in the IAD group only has been found when compared to the control group (Table 2.1 and Figure 2.1).

TABLE 2.1. *Basal, post-SLS, and Δ (post-basal) TEWL values*

	Active atopic dermatitis (AAD, n = 23)	Inactive atopic dermatitis (IAD, n = 21)	Respiratory allergy (RA, n = 25)	Controls (n = 26)
Basal-TEWL	8,6 ± 3,6	7,1 ± 2,1	7,2 ± 1,2	8,3 ± 3,3
Post-SLS-TEWL	25,4 ± 9,1*	19,6 ± 8,6	17,4 ± 7,2	17,9 ± 8,2
Δ-values	16,8 ± 9,0*	12,5 ± 8,9	10,2 ± 7,2	9,6 ± 7,7

TEWL = transepidermal water loss (g/m²h); mean ± standard error. * Significant difference to the control group (p ≤ 0, 05).

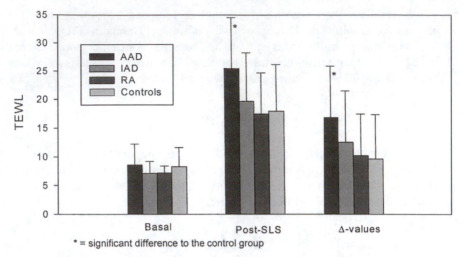

* = significant difference to the control group

FIG. 2.1. Patch testing with 0,5% SLS in distinct groups of atopic individuals.

DISCUSSION

When comparing studies in which patients with an atopic dermatitis were patch tested with irritants, remarkable differences can be found. When reviewing studies on irritant patch tests in atopic individuals, the problem of many contradictory results has been pronounced (Gallagher and Maibach, 1998). They mentioned the most important problems as follows (Table 2.2):

1. "Atopy" has been defined in different ways (RAST and IgE level, history or acute atopic dermatitis, history or acute symptoms of rhinoconjunctivitis); therefore, the investigated populations are not comparable. Mostly, individuals with atopy in one study are not like individuals with atopy in another one.
2. The acuity of the dermatitis is of importance. It has been demonstrated that a significantly higher response to sodium lauryl sulphate (SLS), as well as only a tendency to an increased skin susceptibility to SLS, is related to the severity of the dermatitis (Cowley and Farr, 1992; Nassif et al., 1994; Tupker et al., 1995a; 1990).
3. The method of evaluation is of importance. Although changes in the epidermal barrier might not be clinically visible (and not detectable by laser Doppler blood flow), distinct changes can be detected by the measurement of the TEWL. Hence, each method of test result evaluation (preferably bioengineering methods) has its own advantages and problems. They mostly are not comparable with each other.
4. The test procedure varies in most of the studies. Individual factors (gender, age, race), as well as different test procedures (test region, used irritants, open or occlusive test) and season of testing, makes most of the investigations incomparable.

TABLE 2.2. *Problems in compatibility of different studies on atopic patients*

Problem	Variable factors	Comments
Definition of atopy	Criterias of atopy: Only anamnestic features IgE level Criterias proposed by (Hanifin and Rajka, 1980; Hanifin et al., 1997; Diepgen et al., 1989; Diepgen et al., 1991; Sampson, 1990)	Apart from patient history or laboratory findings, use of a common atopy criteria (e.g., by Hanifin or Diepgen) is recommended
Definition of type and acuity of the dermatitis	Acuity of dermatitis influences irritant patch testings (Tupker et al., 1995a) Different types of dermatitis (atopic or nonatopic) influences the irritant patch test in a different manner	A clear definition and scores not only for the type, but also for the acuity of the dermatitis may be helpful
Individual variabilities	Age Race Body-side of testing	Matched volunteers are requested
Test procedures	Season of testing Different irritants Open or occlusive tests Time of irritant's exposure	Use of guidelines by the standardization group of the European Society of Contact Dermatitis (Pinnagoda et al., 1990; Tupker et al., 1997) is requested

Furthermore, the outcomes of irritant patch tests in patients with respiratory atopy (RA) like rhinitis or atopic asthma without a history of atopic dermatitis are contradictory as well. Whereas some authors found increased reactivity to SLS in patients with RA as compared to healthy controls (Nassif et al., 1994), other authors did not (Seidenari et al., 1996; Tupker et al., 1989). The season in which skin susceptibility was tested is of importance, which has recently been demonstrated (Tupker et al., 1995b): In individuals with AD, barrier function and skin hydration were reduced in winter.

To settle the question of which kind of atopic patients have an increased skin susceptibility, we investigated three distinct groups of atopic individuals: (1) patients with an AAD, (2) patients with an IAD, and (3) patients with a history of (RA).

The basal TEWL values did not differ significantly between atopic patients (neither AAD nor IAD) and healthy controls, confirming the findings of some previous studies (Agner, 1991; Iliev et al., 1997); however, individuals with AAD showed a slightly increased basal TEWL reflecting a subclinical disturbance of the epidermal barrier on clinical unaffected areas. This supports the studies, which have found (statistical significant) increased basal TEWL values in patients with AAD (van der Valk et al., 1985; Werner and Lindberg, 1985). Patients with an IAD showed normal values of the basal TEWL when compared to the control group. In most previous studies, however, basal TEWL tended to be higher in individuals with an IAD (Elsner and Burg, 1993; Stolz et al., 1997; Tupker et al., 1995a, 1990). The difference may be explained by the fact that the individuals in our study have not had any acute AD during the last 2 years, so their AD may have "burned out" for years and their skin is being stable; however, concerning increased skin susceptibility, a history of previous chronic irritant dermatitis is definitely of importance, as the basal TEWL in these individuals is significantly higher than that of healthy controls (Effendy et al., 1995).

Concerning post-SLS-TEWL, recent studies did not show any significant differences in skin susceptibility between patients with IAD and healthy controls (Elsner and Burg, 1993; Stolz et al., 1997). Also, a detailed study using ranges of concentrations and exposure times hardly showed differences between atopics (without differentiation of the intensity of their dermatitis) and healthy controls (Basketter et al., 1998). Generally, most previous studies (where mostly volunteers with an AAD were investigated) resulted in conflicting findings (Cowley and Farr, 1992; Nassif et al., 1994; Tupker et al., 1990). The discrepancies could likely be explained by various definitions of atopy or atopic dermatitis used. Concerning the definition of an atopic constitution, numerous conditions have been demanded. Mostly, lesions of atopic dermatitis must be visible (Tupker et al., 1995a; Uehara and Miyauchi, 1984); sometimes only a history of affected skin or a history of respiratory allergy is sufficient as a criterium (Klas et al., 1996; Tupker et al., 1990). In fact, even the two widely used criteria in Europe of atopy (or atopic dermatitis), proposed by (Hanifin and Rajka, 1980) and (Diepgen et al., 1989, 1991), differ from each other.

In this study, employing the definition of atopy by (Diepgen et al., 1989, 1991), only patients with an AAD showed an increased skin susceptibility to SLS. By contrast, individuals with an IAD that do not have any eczematous lesions did not show an increased skin susceptibility (Elsner and Burg, 1993; Stolz et al., 1997). There obviously is a difference in skin susceptibility between individuals without signs of dermatitis (IAD) and those with an active atopic dermatitis (AD).

Based on our results, the atopy score according to the definition of (Diepgen et al., 1989, 1991) might not be sufficient for studies on skin susceptibility, as individuals with an atopic constitution can be distinguished by the acuity of the dermatitis, namely, one group with an increased skin susceptibility to SLS (AAD) and another group without an increased skin susceptibility to SLS (IAD). It is likely that individuals with AAD may have a disturbed epidermal barrier, even at clinical unaffected areas. They are more susceptible to irritants like SLS, which enables an induction of hand dermatitis. The mentioned limitation of the atopy score is emphasized by (Gehring et al., 1998) as well. For investigations in the future, a clear distinction between these two groups of atopic individuals is required. In addition, for a comparison between clinical studies on atopic patients, the use of a clinical score system (e.g., SCORAD) would be helpful.

The influence of respiratory allergies on the skin susceptibility of atopic individuals has been discussed controversially, too. Whereas some authors reported an enhanced susceptibility to primary irritants (Hanifin et al., 1997; Tupker et al., 1990), others did not (Seidenari et al., 1996). Our study showed that patients with RA did not reveal an abnormal skin irritability. The reason for the discrepancies between our study and earlier studies could be a distinct seasonal time of testing. We performed the study in winter, and none of the volunteers had features of rhinoconjunctivitis, so the burden of the skin with aeroallergens may be relatively low. This seems to be important, as the epidermal barrier has been shown to be disrupted by aeroallergens during atopy patch test (Gfesser et al., 1996). Perhaps the skin of atopic patients would become more susceptible to primary irritants when it is simultaneously exposed to clinically relevant aeroallergens. (Nassif et al., 1994) showed an increased skin susceptibility in patients with RA suffering from rhinoconjunctivitis at the time of testing.

The role of the season in the skin susceptibility may be very complex. Apart from RA, (Tupker et al., 1995b) showed an increased skin susceptibility in atopic patients in winter season, suggesting the influence of dry air in winter. We consider, in addition, that the skin susceptibility in patients with RA may additionally be influenced by the degree of the aeroallergen burden. Thus, a pollen-allergy may provide an additional factor responsible for an increased skin susceptibility in spring and summer. Correspondingly, patients with an allergy to indoor allergens (e.g., mite dust) may have such a soreness over the whole year.

(Tupker et al., 1995a) demonstrated that the intensity of the skin response to irritants and bioactive agents is correlated to the disease severity. Hypothetically, the skin susceptibility of individuals with IAD and RA is therefore not primarily increased, but the susceptibility to primary irritants and allergens will increase as soon as the skin becomes eczematous and somewhat impaired. This could be induced by an endogenous activity of atopy and/or aeroallergens.

Concerning the pathogenetic hypothesis, the primary and secondary prevention of atopic dermatitis becomes more relevant. Preventing acute atopic dermatitis may result in a reduced prevalence of irritant contact dermatitis in atopic patients, too. Furthermore, an intact skin barrier can defense allergen penetration through the skin, therefore preventing the development of allergic contact dermatitis.

REFERENCES

Agner, T. (1991): Skin susceptibility in uninvolved skin of hand eczema patients and healthy controls. *Br. J. Dermatol.*, 125:140–146.

Basketter, D. A., Miettinen, J., and Lahti, A. (1998): Acute irritant reactivity to sodium lauryl sulfate in atopics and non-atopics. *Contact Dermatitis*, 38:253–257.

Cowley, N. C., and Farr, P. M. (1992): A dose-response study of irritant reactions to sodium lauryl sulphate in patients with seborrhoeic dermatitis and atopic eczema. *Acta Derm. Venereol.*, 72:432–435.

Diepgen, T. L., Fartasch, M., and Hornstein, O. P. (1989): Evaluation and relevance of atopic basic and minor features in patients with atopic dermatitis and in the general population. *Acta. Derm. Venereol.*, 144: (Suppl) 50–54.

Diepgen, T. L., Fartasch, M., and Hornstein, O. P. (1991): Kriterien zur Beurteilung der atopischen Haut-diathese. *Dermatosen*, 39:79–83.

Effendy, I., Löffler, H., and Maibach, H. I. (1995): Baseline transepidermal water loss in patients with acute and healed irritant contact dermatitis. *Contact Dermatitis*, 33:371–374.

Elsner, P., and Burg, G. (1993): Irritant reactivity is a better risk marker for nickel sensitization than atopy. *Acta. Derm. Venereol.*, 73:214–216.

Gehring, W., Gloor, M., and Kleesz, P. (1998): Predictive washing test for evaluation of individual eczema risk. *Contact Dermatitis*, 39:8–13.

Gfesser, M., Rakoski, J., and Ring, J. (1996): The disturbance of epidermal barrier function in atopy patch test reactions in atopic eczema. *Br. J. Dermatol.*, 135:560–565.

Hanifin, J. M., and Rajka, G. (1980): Diagnostic features of atopic dermatitis. *Acta. Derm. Venereol.*, 92: (Suppl) 44–47.

Hanifin, J. M., Storrs, F. J., Chan, S. C., and Nassif, A. (1997): Irritant reactivity in noncutaneous atopy. *J. Am. Acad. Dermatol.*, 37:139.

Iliev, D., Hinnen, U., and Elsner, P. (1997): Clinical atopy score and TEWL are not correlated in a cohort of metalworkers. *Contact Dermatitis*, 37:235–236.

Klas, P. A., Corey, G., Storrs, F. J., Chan, S. C., and Hanifin, J. M. (1996): Allergic and irritant patch test reactions and atopic disease. *Contact Dermatitis*, 34:121–124.

Nassif, A., Chan, S. C., Storrs, F. J., and Hanifin, J. M. (1994): Abnormal skin irritancy in atopic dermatitis and in atopy without dermatitis. *Arch. Dermatol.*, 130:1402–1407.

Pinnagoda, J., Tupker, R. A., Agner, T., and Serup, J. (1990): Guidelines for transepidermal water loss (TEWL) measurement: A report from the Standardization Group of the European Society of Contact Dermatitis. *Contact Dermatitis*, 22:164–178.

Sampson, H. I. (1990): Pathogenesis of eczema. *Clin. Exp. Allergy.*, 20:459–467.

Seidenari, S., Belletti, B., and Schiavi, M. E. (1996): Skin reactivity to sodium lauryl sulfate in patients with respiratory atopy. *J. Am. Acad. Dermatol.*, 35:47–52.

Stolz, R., Hinnen, U., and Elsner, P. (1997): An evaluation of the relationship between "atopic skin" and skin irritability in metalworker trainees. *Contact Dermatitis*, 36:281–284.

Tupker, R. A., Coenraads, P. J., Fidler, V., De Jong, M. C., Van der Meer, J. B., and De Monchy, J. G. (1995a): Irritant susceptibility and wheal and flare reactions to bioactive agents in atopic dermatitis. I: Influence of disease severity. *Br. J. Dermatol.*, 133:358–364.

Tupker, R. A., Coenraads, P. J., Fidler, V., De Jong, M. C., Van der Meer, J. B., and De Monchy, J. G. (1995b): Irritant susceptibility and wheal and flare reactions to bioactive agents in atopic dermatitis. II: Influence of season. *Br. J. Dermatol.*, 133:365–370.

Tupker, R. A., Coenraads, P. J., Pinnagoda, J., and Nater, J. P. (1989): Baseline transepidermal water loss (TEWL) as a prediction of susceptibility to sodium lauryl sulphate. *Contact Dermatitis*, 20:265–269.

Tupker, R. A., Pinnagoda, J., Coenraads, P. J., and Nater, J. P. (1990): Susceptibility to irritants: Role of barrier function, skin dryness and history of atopic dermatitis. *Br. J. Dermatol.*, 123:199–205.

Tupker, R. A., Willis, C., Berardesca, E., Lee, S. H., Fartasch, M., Agner, T., and Serup, J. (1997): Guidelines on sodium lauryl sulfate (SLS) exposure testing. *Contact Dermatitis*, 37:53–69.

Uehara, M., and Miyauchi, H. (1984): The morphologic characteristics of dry skin in atopic dermatitis. *Arch. Dermatol.*, 120:1186–1190.

van der Valk, P. G., Nater, J. P., and Bleumink, E. (1985): Vulnerability of the skin to surfactants in different groups of eczema patients and controls as measured by water vapour loss. *Clin. Exp. Dermatol.*, 10:98–103.

Werner, Y., and Lindberg, M. (1985): Transepidermal water loss in dry and clinically normal skin in patients with atopic dermatitis. *Acta. Derm. Venereol.*, 65:102–105.

Toxicology of Skin
Edited by Howard I. Maibacn
Copyright © 2001 Taylor & Francis

3

Quantification of Nonerythematous Irritant Dermatitis

Veranne Charbonnier and Howard I. Maibach

*Department of Dermatology, School of Medicine, University of California,
San Francisco, California, USA*

Boyce M. Morrison, Jr.

Colgate Palmolive Company, Piscataway, New Jersey, USA

Marc Paye

Colgate Palmolive R&D, Milmort, Belgium

INTRODUCTION

Human skin irritation is classically evaluated by visual and palpatory scoring. For more objective measurements, bioengineering methods for capacitance, transepidermal water loss (TEWL), and blood flow have been widely used (Agner and Serup, 1990; Berardesca and Maibach, 1988; De Boer and Bruynzeel, 1995; Tupker, et al., 1989; Tupker, 1994; Wihelm and Maibach, 1995; Wilhem et al., 1989), but the identification of substances of low potential or subclinical irritation remains problematic. Focusing on the stratum corneum, cellophane, and related tape methods to remove and analyze the stratum corneum via skin strippings have been developed and have numerous applications: quantifying stratum corneum (Dreher et al. 1998; Marttin

31

et al., 1996), barrier function perturbation (Schatz et al., 1993; Welzel et al., 1996), histological assessments (Gerritsen et al., 1994; van der Molen et al., 1997), percutaneous penetration (Coderch et al., 1996), and pharmacology (Letawe et al., 1996; Pershing et al., 1992). To evaluate nonerythematous irritant dermatitis, squamometry has recently appeared to be a sensitive complementary method to conventional skin color, TEWL, and hydration measurements (Charbonnier et al., 1998; Charbonnier et al.).

SQUAMOMETRY: METHODOLOGY

The use of the adhesive D-SQUAME disc as a harvesting method for the superficial desquamating layer of the stratum corneum has been discussed in detail (Miller, 1995). Minimal pigment in the loose clusters of corneocytes adhere to the adhesive disc. Image analysis of skin scales assumes that variations in the intensity of reflected or transmitted light from point to point in the image are governed by variations in the thickness and (condition) of the assembled stratum corneum flakes (Miller, 1996). In this chapter, we focus our attention on an alternative approach: staining the corneocytes to produce an image for further evaluation.

Squamometry is a noninvasive colorimetric evaluation of the level of alteration in the corneocyte layer collected by clear adhesive-coated discs. Xerotic and inflammatory changes in the stratum corneum can be so quantified (Pierard et al., 1992; Pierard and Pierard-Franchimont, 1996). The discs are applied onto the skin under controlled pressure. A short application time (15 seconds) enables the harvesting of the superficial corneocytes (superficial squamometry), and a long application time (1 hour) enables collection of a thicker layer of corneocytes (deep squamometry) (Pierard and Pierard-Franchimont, 1996; Pierard, 1996). The discs are stained for 30 seconds by dropping a solution of toluidine blue and basic fushsin in 30% alcohol in polyethylene, Polychrome Multiple Stain (PMS) (Delasco, IA) over the surface, followed by gentle rinsing in water. Measurements of the color of the samples in the $L^*a^*b^*$ mode are made using a reflectance colorimeter (Chromameter, Minolta). Calculation of the Chroma C^* values are done after $(a^*2+b^*2)1/2$. This parameter combines the values of the red and blue chromaticities, predominant colors of the PMS (Pierard et al., 1992). The Chroma C^* value has been shown to be proportional to the amount of stratum corneum harvested in xerotic situation (Pierard et al., 1992). The calculation of the Colorimetric Index of Mildness (CIM), where $CIM = L^*-C^*$, was performed (Pierard and Pierard-Franchimont, 1996) where L^* is the measure of luminance (Pierard et al., 1992). A trained person scored the discs with a microscope at x20 magnification (Paye and Morrison, 1996):

Intercorneocyte cohesion: $0 =$ large sheet; $1 =$ large clusters $+$ few isolated cells; $2 =$ small clusters $+$ many isolated cells; $3 =$ clusters in disruption, most cells isolated; $4 =$ all cells isolated, many cases of lysis.

Amount and distribution of dye found in cells: $0 =$ no staining; $1 =$ staining between cells or slight staining in cells; $2 =$ moderate staining in cells; $3 =$ large amount of dye in cells, but uniform; $4 =$ important staining in all cells, often with grains. This methodology has been used into the studies mentioned below.

SUBCLINICAL NONERYTHEMATOUS IRRITATION
WITH SURFACTANT

After a single occlusive application (24-hour patch test), low concentrations of sodium lauryl sulfate (SLS, 0.5% in water) can cause irritation, dryness, and tightness (Treffel and Gabard, 1996; Tupker et al., 1997; Wilhelm et al., 1994). These occlusive tests are too severe to observe subclinical damage, as erythema and skin barrier alterations predominate. In typical use, surfactant exposure usually is brief, of open application, and cumulative. The open application model becomes relevant when phenomena such as dryness and subclinical (i.e., nonvisible) irritation are induced. SLS can induce subclinical (i.e., nonvisible) skin damage in a repetitive open application test method (exaggerated model hand wash) as well as in a short-exposure patch test. Analysis of the skin surface via squamometry, following the methodology described above, offers a unique way of measuring skin changes when traditional methods do not and permits exploration of subclinical surfactant irritation (Charbonnier et al., 1998; Morrison et al., 1998) (Figure 3.1).

Squamometry has also been used for a better understanding of the mechanism of interaction between the surfactant and the stratum corneum: cellular damages by protein alterations or loss of intercellular cohesion (lipids alterations) and depth of adsorption of the surfactant into the stratum corneum (successive strippings) (Paye and Cartiaux, in press, 1999). A few assumptions may explain the fixation mechanism of the dye PMS on the cell: the dye should not penetrate the cell for a mild irritant, but should be fixed on the cell surface (desmosome? protein membrane?). In case the

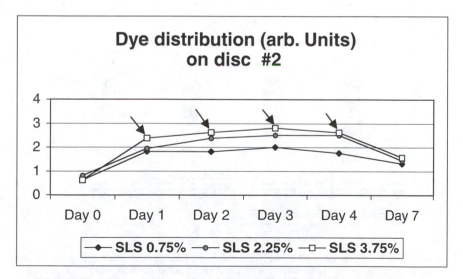

FIG. 3.1. Mean observer scored dye distribution data on disc #2 (out of 4) during an exaggerated hand wash procedure. Arrow indicates significant difference between SLS 0.75%/SLS 2.25% and SLS 0.75%/SLS 3.75% (p < 0.05). SLS = Sodium Lauryl Sulfate.

surfactant is more irritant, lipid alterations occur on the cell surface; the surfactant might penetrate the corneocyte and damage the intracorneocyte protein (keratin), allowing the PMS to be fixed in the cell where fixation sites are numerous. This hypothesis is consistent with the fact that the more the disc is colored, the greater the intercorneocyte cohesion loss and amount of dye found in cells scores are increased, the greater the skin damage.

To progress in the surfactant field, squamometry allows the move from exaggerated to more realistic test conditions without causing overt irritation (Paye and Cartiaux, in press, 1999). The advantages are obvious: no need to cause irritation to compare product mildness, more realistic test condition, direct study of the target for the surfactant, and it can be used as a quick screen.

Ranking surfactant solutions, as low as 3.75% in water, has been successful with squamometry (Pierard et al., 1992): SLS, sodium laureth sulfate (SLES), and sodium alpha olefin sulfonate (SAOS) have been tested under exaggerated hand-washing test. The results showed that bioengineering measurements (hydration parameter, TEWL, skin color) were not sufficiently sensitive to reveal significant difference between surfactants. Squamometry documented subclinical detergent stratum corneum effects and differentiated the surfactants.

Even when cohesion did not show a difference, chroma C*, CIM, and microscopic examination of dye fixation per cell were sufficiently sensitive to reveal differences between SLS, SLES, and SAOS (Figure 3.2). To refine the exaggerated hand-washing model, another study combined three daily washes at the laboratory and a typical

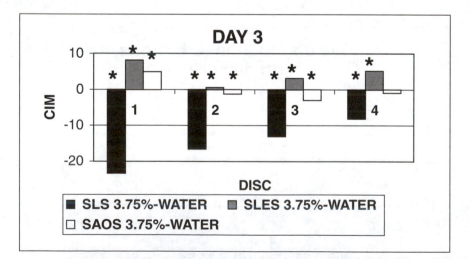

FIG. 3.2. Colorimetric Index of Mildness (=CIM) data on 4 successive discs during day #3 of an exaggerated hand wash procedure. (surfactant minus water control). Asterisk indicates significant difference between SLS, SLES, and SAOS at 3.75% (p < 0.05). SLS = Sodium Lauryl Sulfate, SLES = Sodium Laureth Sulfate, SAOS = Sodium Alpha Olefin Sulfate.

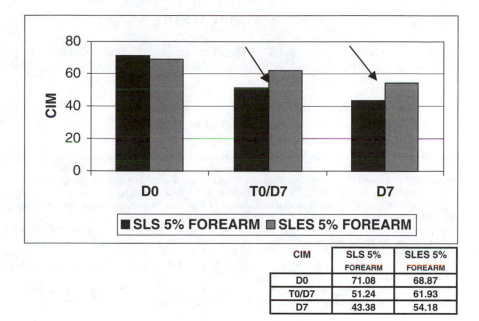

CIM	SLS 5% FOREARM	SLES 5% FOREARM
D0	71.08	68.87
T0/D7	51.24	61.93
D7	43.38	54.18

FIG. 3.3. Colorimetric Index of Mildness (=CIM) data during an open model assay. Arrow indicates significant difference between SLS and SLES 5% (p < 0.05). SLS = Sodium Lauryl Sulfate, SLES = Sodium Laureth Sulfate.

volunteer use at home for 1 week (Charbonnier et al., in press, 2000) and compared SLS and SLES. Once again, squamometry documented subclinical nonerythematous effects. Chroma C*, CIM, and microscopic examination of cell cohesion and dye fixation per cell were sufficiently sensitive to reveal differences between SLS at 5% and SLES at 5%. Most results were observed 1 hour after the day 7 wash, but in addition, the CIM (the higher, the milder) and the chroma C* (the lower, the milder) statistical analyses revealed a significant difference between SLS and SLES as early as T0/Day 7, which was after volunteers were self-dosed over the weekend (Figure 3.3). This suggests that squamometry was sufficiently sensitive to observe subclinical skin changes in an open application assay (use consumer test).

CONCLUSION

Squamometry appears to be

A facile and robust method to study and quantify nonerythematous irritant dermatitis
A tool to test products under nonexaggerated conditions
A sensitive and direct method to investigate the interaction of surfactants with the surface of skin
A promising technique

ACKNOWLEDGMENTS

The authors thank B.M. Morrison and M. Paye for their help. The presented surfactant studies were partially supported by Colgate Palmolive Company, Piscataway, New Jersey, USA.

REFERENCES

Agner, T., Serup, J. (1990): Sodium lauryl sulfate for irritant patch testing: A dose response study using bioengineering methods for determination of skin irritation., *J. invest. Dermatol.*, 95:543–547.

Berardesca, E., Maibach, H. I. (1988): Bioengineering and the patch test. *Contact Dermatitis*, 18:3–9.

Charbonnier, V., Morrison, B. M. Jr., Paye, M., Maibach, H. I. (1998): Open application assay in investigation of sub-clinical irritant dermatitis induced by SLS in man: Advantage of squamometry. *Skin Research and Technology*, 4:244–250.

Charbonnier V., Morrison, B. M. Jr., Paye, M., Maibach, H. I. Hand dryness by an exaggerated wash procedure in vivo with sodium lauryl sulfate, sodium laureth sulfate and sodium alpha olefin sulfonate. Manuscript in preparation.

Charbonnier, V., Morrison, B. M. Jr., Paye, M., Maibach, H. I. (in press, 2000): An Open Assay Model for induction of sub-clinical non-erythematous irritation. *Contact Dermatitis*.

Coderch, L., Oliva, M., Pons, A., de la Maza, A., Manich, A. M., Parra, J. L. (1996): Percutaneous penetration of liposome using the tape stripping technique. *International Journal of Pharmaceutics*, 139:197–203.

De Boer, E. M., Bruynzeel, D. P. (1995): Irritancy. In: *Bioengineering of the Skin: Cutaneous Blood Flow and Erythema*. Edited by Berardesca E., Elsner P., Maibach H. I., Boca Raton, CRC Press.

Dreher, F., Arens, A., Hostynek, J. J., Mudumba, S., Ademola, J., and Maibach, H. (1998): Colorimetric method for quantifying human stratum corneum removed by adhesive-tape-stripping. *Acta. Derm. Venereol. (Stockh)*, 78:1–4.

Gerritsen, J. P., van Erp, P. E. J., van Vlijmen-Willems, I. M. J. J., Lenders, L. T. M. and van de Kerkhof, C. M. (1994): Repeated tape stripping of normal skin: Histological assessment and comparison with events in psoriasis. *Arch. Dermatol. Res.*, 286:455–461.

Letawe, C., Pierard-Franchimont, C., Pierard, G. E. (1996): Squamometry in rating the efficacy of topical corticosteroids in atopic dermatitis. *Eur. J. Clin. Pharmacol.* 51:253–257.

Marttin, E., Neelissen-Subnel, M. T. A., De Haan, F. H. N. and Bodde, H. E., (1996): A critical comparison of methods to quantify stratum corneum removed by tape stripping. *Skin Pharmacol.* 9:69–77.

Miller, D. L. (1995): Sticky slides and tape techniques to harvest stratum corneum material. In: *Handbook of Non-invasive Methods and the Skin*. Edited by Serup, J., and Jemec, G. B. E. CRC Press, Boca Raton.

Miller, D. L. (1996): D-Square adhesive discs. In *Bioengineering of the skin: skin surface imaging and analysis*. Edited by K.P. Wilhelm, P. Elsner, E. Berardesca, and H. Maibach. CRC Press, Boca Raton.

Morrison, B. M., Jr., Cartiaux, Y., Paye, M., Charbonnier, V., Maibach, H. I. (1998): Demonstrating Invisible (Sub-clinical) Sodium Lauryl Sulfate Irritation with Squamometry. American Academy of Dermatolology, 56th Annual Meeting, Poster Session, Feb. 27–Mar. 4, Orlando, FL.

Paye, M. and Morrison, B. M. Jr., (1996): Non-visible skin irritation. *Proc. the Fourth World Surfactant Congress*, Barcelona, 3–7 June, 3:42–51.

Paye, M., Cartiaux, Y. Squamometry (in press, 1999): A tool to move from exaggerated to more and more realistic application conditions for comparing human skin compatibility of products. XXth IFSCC Congress, Oral Session, Sept. 1998, Cannes, France, and International Journal of Cosmetic Science.

Pershing, L. K., Lambert, L. D., Shah, V. P., and Lam, S. Y. (1992): Variability and correlation of chromameter and tape-stripping methods with visual skin blanching assay in the quantitative assessment of topical 0.05% betamethasone dipropionate bioavailability in humans. International Journal of Pharmaceutics, 86:201–210.

Pierard, G. E., Pierard-Franchimont, C., Saint-Leger, D, Kligman, A. M., (1992): Squamometry: The assessment of xerotic by colorimetry of D-squame adhesive discs. *J. Soc. Cosmet. Chem.*, 47:297–305.

Pierard, G. E., and Pierard-Franchimont, C. (1996): Drug and cosmetic evaluations with skin strippings. In: *Dermatology Research Technique*. Edited by H. I. Maibach. CRC Press, Boca Raton.

Pierard, G. E. (1996): EEMCO guidance to the assessment of dry skin (xerosis) and ichtyosis: Evaluation by stratum corneum strippings. *Skin Res. Technol.*, 2:3–11.

Schatz, H., Kligman, A. M., Manning, S. and Stoudemayer, T. (January/February 1993): Quantification of dry (xerotic) skin by image analysis of scales removed by adhesive discs (D-Squames). *J. Soc. Chem.*, 44:53–63.

Treffel, P. and Gabard, B. (1996): Measurements of Sodium Lauryl Sulfate-induced skin irritation, *Acta. Derm. Venereol.*, 76:341–343.

Tupker, R. A., Vermeulen, K., Fidler, V. and Coenraads, P. J. (1997): Irritancy testing of Sodium Lauryl Sulfate and other anionic detergents using an open exposure model. *Skin Research and Technology*, 3:133–136

Tupker, R. A., Pinnagoda, J., Coenraads, P. J., et al. (1989): The influence of repeated exposure to surfactants on the human skin as determined by transepidermal water loss and visual scoring. *Contact Dermatitis*, 18:3–9.

Tupker, R. A. Prediction of irritancy. In: *Bioengineering of the Skin: Water and the Stratum Corneum.* Edited by Elsner P., Berardesca E., Maibach H. I., Boca Raton, CRC Press.

van der Molen, R. G., Spies, F., van't Noordende, J. M., Boelsma, E., Mommaas, A. M. and Foerten, H. K. (1997): Tape stripping of human stratum corneum yield cell layers that originate from various depths because of furrows in the skin. *Arch. Dermatol. Res.*, 289:541–618.

Welzel, J., Wilhelm, P. and Wolff, H. H. (1996): Skin permeability barrier and occlusion: No delay of repair in irritated human skin. *Contact Dermatitis*, 35:163–168.

Wilhelm, K. P., Freitag, G. and Wolff, H. H. (1994): Surfactant-induced skin irritation and skin repair: Evaluation of the acute human irritation model by non invasive techniques., *Journal American Academy of Dermatology*, 30:944–949.

Wilhelm, K. P., Maibach, H. I. (1995): Evaluation of irritation tests by chromametric measurements. In: *Bioengineering of the Skin: Cutaneous Blood Flow and Erythema*. Edited by Berardesca E., Elsner P., Maibach H. I., Boca Raton, CRC Press.

Wilhelm, K. P., Surber, C., Maibach, H. I. (1989): Quantification of sodium lauryl sulfate irritant dermatitis in man: Comparison of four techniques-skin color reflectance, transepidermal water loss, laser Doppler flow measurement and visual scores. *Arch. Dermatol. Res.*, 281:293–295.

Toxicology of Skin
Edited by Howard I. Maibach
Copyright © 2001 Taylor & Francis

4

Detergents

Isaak Effendy

Department of Dermatology, University of Marburg, Germany

Howard I. Maibach

School of Medicine, Department of Dermatology, University of California, San Francisco, California, USA

INTRODUCTION

The term *deterge* derived from the Latin *detergere*, meaning to wipe off. The term *detergent*, which has existed at least since 1676, used to mean a cleansing agent. Until the late 19th century, the only man-made detergent was natural soap. Soap is chemically defined as the sodium or potassium (alkali) salts of fatty acids or similar products formed by the saponification or neutralization, by which triglycerides (fats and oils) or fatty acids are transformed with organic or inorganic bases into the corresponding alkali salt mixtures of fatty acids.

There are some reasons limiting the use of natural soaps. As the alkaline pH of the soap is induced by the hydrolysis of soap in aqueous solution, the pH value of the water raises to about 9 or 11, causing an increase in pH of the skin surface. This

39

provides a negative soap effect on the skin cleansing with soaps. Furthermore, soaps induce some irritation of the eyes and mucous membranes. The behavior of soap in hard water or saltwater seems somehow less convenient, as soap in such water, which is high in content of multivalent ions (e.g., calcium and magnesium), will hardly develop its foaming ability. Moreover, its critical shortage in Europe after World War I, particularly, provided an incentive for the development of synthetic surface-active agents (surfactants) as synthetic detergents (syndets). A synthetic process of sodium lauryl sulphate (anionic surfactant) was first described in Germany about 60 years ago (Löttermoser et al., 1933).

Nowadays, detergents particularly contain synthetic surfactants that concentrate at oil-water interfaces and hold cleansing as well as emulsifying properties. Furthermore, since the late 1940s, synthetic surfactants have been used in ever-growing proportions in consumer and industrial cleaning formulations. Among the various types, anionic surfactants have been used most frequently; they were reported to represent between 43% and 67% of the active ingredients in personal care, cosmetics, household, and industrial formulations in the United States. In 1992, total surfactant used in the United States was 2.3 billion kg, of which anionic surfactants made up 53% (Ainsworth, 1992).

CLASSIFICATION OF SURFACTANTS

A surfactant is defined as a compound that can reduce the interfacial tension between 2 immiscible phases. This is due to the molecule containing 2 localized regions, one being hydrophilic in nature and the other hydrophobic (Reynolds, 1999). The polar or hydrophilic region of the molecule may carry a positive or negative charge, giving rise to cationic or anionic surfactants, respectively. The presence in the same molecule of two moieties, in which one has affinity for the solvent and the other is antipathetic to it, is termed *amphipathy*. This dual nature is responsible for the phenomenon of hydrophobic (Reynolds, 1999).

The classification of surfactants is somewhat arbitrary. It is generally convenient, however, to categorize the chemicals according to their polar portion (hydrophilic head), as the nonpolar part usually is made up to alkyl or aryl groups (Zografi et al., 1990; Attwood et al., 1983; Rosen, 1978). The major polar groups of most synthetic surfactants are classified into four types (Table 4.1).

Anionic Surfactants

The most commonly used anionic agents are those containing alkyl carboxylates, sulphonates, and sulphate ions. Those containing carboxylate ions are known as *natural soaps*. Soaps, however, provide the oldest anionic surfactant: a *natural* surfactant made by a simple hydrolysis of natural materials.

Many alkyl sulphates are used as detergents, but by far the most popular member of this group is sodium lauryl sulphate (SLS). Unlike soaps, SLS is compatible with dilute

TABLE 4.1. *Classification of surfactants and their use*

Type of surfactant	Frequently used surfactants	Application
Anionic	Sodium lauryl sulphate, sodium lauryl ether sulphate, TEA-lauryl ether sulphate	Detergent, emulsifying, solubilizing, and wetting agent
Cationic	Quaternium-15, Quaternium-19, stearylalkoniumchloride	Preservatives (antimicrobial agent)
Amphoteric	Cocoamidopropyl betaine, coco betaine, disodium cocoamphodiacetate	Detergent, emulsifying agent, foam booster
Non-ionic	Polysorbat 20, cocamide DEA, lauramide DEA	Detergent, emulsifying agent, foam booster

Modified from Effendy and Maibach, 1995.

acid and with calcium and magnesium ions. The lower-chain-length compounds, around C12, have better wetting ability, whereas the higher members (C16–C20) have better detergent properties (Rosen, 1978). SLS has been reported to exhibit in vitro and in vivo antibacterial effects (Wade et al., 1992).

Cationic Surfactants

Many long-chain cations, such as amine salts and quaternary ammonium salts, are used as surfactants when dissolved in water; however, their use generally is limited to that of antimicrobial preservatives because of their bactericidal activity (Attwood et al., 1983; Zografi et al., 1990).

Amphoteric Surfactants

Amphoteric agents possess at least one anionic and one cationic group in its molecule. They have the detergent properties of anionic surfactants and the disinfectant properties of cationic surfactants. Their activity depends on the pH of the media in which they are used. Balanced amphoteric surfactants are reputed to be nonirritant to the eyes and skin and have therefore been used in so-called baby shampoos (Reynolds, 1999). The most ofetn used amphoterics are betaines, sulfobetaines, imidazolinium derivatives, and alkylamioacids (Bouillon, 1996)

Nonionic Surfactants

Nonionic surfactants have the advantage over ionic surfactants in that they are compatible with all other types of surfactants and their properties are minimally affected by pH. Moreover, they are generally less irritant than anionic or cationic surfactants. Nonionic surfactants are used as emulsifiers and solubilizing and wetting agents.

They have applications in the food, cosmetic, paint, pesticide, and textile industries (Reynolds, 1999).

CHOICE OF SURFACTANTS FOR DETERGENTS

In Europe a blend of alkyl sulfates and alkyl sulfosuccinates has mostly been employed. The pH value of the bar ranges between 5.5. and 7.0; however, in recent years, the use of sodium cocoyl isethionate has been increased in term of producing of mild bars following the American trend of skin cleanser.

In the United States the preferred mildness concept of skin cleanser makes the expensive sodium cocoyl isethionate the surfactant mainly used. To reduce the final cost of the formulation, the corresponding surfactant blend includes about 30% of fatty acid and fatty acid soap (Schoenberg, 1983). This is, in turn, responsible for the dull and somewhat slimy appearance of certain products. Other main surfactants used in the United States are sodium cocomonoglyceride sulphate and sodium cocoglyceryl ether sulphonate (Friedman et al., 1996).

In Japan, acyl glutamate provides a major surfactant used in one of the sophisticated expensive skin cleansers. In contrast, in other asian countries, natural sodium soaps still provide the main cleanser, as high-cost cleanser bars are hardly acceptable to the consumer.

The cleaning and lathering properties, plasticity, and skin compatibility will definitely depend on the used surfactants and their proportions. Alkyl sulphate and sulphosuccinate blends seem to have the highest cleansing properties, followed by acyl glutamate and triethanolamine soaps, whereas natural sodium soaps were ranked last (Spitz, 1990; Collwell, 1993; Friedman, 1996).

Certain surfactants have strong odor due to their origins, (e.g., coconut's fatty acid). To overcome such origin odor, highly concentrated fragrances are often required for the formulation. This, in turn, may create a risk for a contact sensitivity to fragrance for consumer.

In reality, the choice of the surfactants used as detergents may, indeed, not necessarily follow the basic aim of cleansing and washing, but rather the consumer's trend, which has been favorized, advertised by the manufacturers themselves in term of creating new products; however, at the least, a printed declaration of the used ingredients and hotline numbers for information regarding the product may be helpful for consumers.

IRRITANT PROPERTIES OF DETERGENTS

Surfactants used as detergents may cause skin irritation. The mechanisms of surfactant-induced irritant dermatitis are not yet fully understood (Lee et al., 1995). It has been reported that the effects of surfactants depend on both concentration and surfactant-lipid molar ratios. At low concentrations, surfactants can disrupt membranes that resulted in increased membrane permeability (Kalmanzon et al.,

1992), whereas at higher concentrations (above the critical micelle concentration), sur-factants cause cell lysis (Partearroyo et al., 1990). Thus, two opposing events, namely, interaction with the membrane and the permeant with the micelle, may be responsible for the overall effect of the surfactant on membrane permeability (Walters et al., 1993).

Anionic surfactants prove to be potent primary irritants to human and animal skin (Landsdown et al., 1972), and cationic surfactants are reputedly at least equally irri-tating (Scala et al., 1968; Landsdown et al., 1972), but more cytotoxic than anionics (Korting et al., 1994; Harvell et al., 1994). The irritation potential of nonionic sur-factants is believed to be the lowest (Landsdown et al., 1972; Imokawa et al., 1979; Willis et al., 1989; Zografi et al., 1990). Nevertheless, the irritancy ranking order of detergents cannot generally be made by the arbitrary classification of surfactants.

Mounting data suggest that change of epidermal lipid composition, protein denat-uration, epidermal cytokine release, epidermal barrier repair, and individual intrinsic factors can contribute to irritant responses (Wood et al., 1992; Nickoloff et al., 1994; Wilmer et al., 1994, Basketter et al., 1996). Interestingly, the pathogenesis of skin irritation seems to vary depending on the stimulus used (Willis et al., 1991; Yang et al., 1995; Fartasch et al., 1995, 1998). SLS, a widely used model irritant, has recently been shown to provoke damage of the nucleated parts of the epidermis and alterations of the lower parts of the stratum corneum (SC); however, the upper portions of the SC showed intact intercellular lipid layers that contradict the long-standing belief that surfactants damage the skin by delipidization (Fartasch et al., 1998). Other investiga-tors suggested that the epidermal response to detergent exposure is primarily directed at restoration of barrier function (Le et al., 1996).

Detergents are needed in everyday life; however, they provide a relevant risk factor in the development of irritant contact dermatitis. Hence, it is mandatory to search for less irritant detergents.

IRRITANCY RANKING OF DETERGENTS

In recent years numerous in vivo and in vitro studies on the irritant potential and the irritancy ranking order of detergents have been performed (Tables 4.2 and 4.3). In vivo data showed that SLS exhibited a higher irritancy than amphoteric surfactants (van der Falk et al., 1985; Willis et al., 1989; Korting et al., 1994); however, the detergent concentrations and the measurement methods employed may influence the test outcome, as only SLS in a high concentration was more, or at least equally, irritating than benzalkonium chloride (BAC) of low concentration (Willis et al., 1989; Berardesca et al., 1990). Although at a same concentration, BAC, clinically as well as in in vitro assay, demonstrated a higher irritant or cytotoxicity potential than SLS (Müller-Decker et al., 1994; Eun et al., 1994; Korting et al., 1994; Osborne et al., 1994; Harvell et al., 1994).

Tupker et al. (1999) have shown that different evaluation methods (visual scoring, bioengineering assessment) and exposure model (one-time occlusive test, repeated

TABLE 4.2. *In vivo irritancy ranking of frequently used surfactants*

Irritancy ranking	Irritancy test in human	Assessment	References
Soap $\geq$ SLS $\geqq$ ISE > SUC	One-time occlusive test	Visual scoring	Tupker et al., 1999
SLS > ISE $\geqq$ Soap > SUC	Repeated short-time occlusive	Visual scoring and TEWL	
SLS > ISE $\geqq$ Soap > SUC	Repeated short-time open	Visual scoring and TEWL	
SLS > SLES > CPAB > LESS > RMSS > PEG (each 1%)	2-day soap chamber test	TEWL, skin reflective color (SRC/chromameter)	Korting et al., 1994
0.5% SLS > 0.5% dodecyl trimethyl ammonium bromide > potassium soap	24-hour patch test	TEWL, capacitance	Wilhelm et al., 1994
N-alkyl-sulfate C_{12} > C_{8-10}, C_{14-16}	24-hour patch test	TEWL, SRC	Wilhelm et al., 1994
2% SLS > 2.9% LAS > 7.9% PEG-20 glyceryl monotallowate	5-day repeated occlusive application test (2 times daily)	Spectroscopic and visual scoring, TEWL, SRC, capacitance, skin replica	Zhou et al., 1991
7% SLS > 7% CAPB > 1% BAC > 10% sorbitan monolaurate	24-hour plastic occlusion stress test	Skin surface water loss (SSWL)	Berardesca et al., 1990
5% SLS > 0.5% BAC > 100% PG	48-hour patch test	Visual scoring	Willis et al., 1989
5% SLS = 0.5% BAC > 100% PG		Histology	
SLS > cocobetaine > CAPB (each 2%)	48-hour patch test	TEWL	van der Valk et al., 1985

AEOS-3E0 = alkyl (C_{12-14} average) ethoxy sulphate s, BAC = benzalkonium chloride; CAPB = cocoamidopropyl betaine; ISE = sodium cocoyl isethionate; LAS = linear alkyl (C_{12} average) benzene sulfonate; LESS = disodium laureth sulphate; PEG = polyethylene glycol; PG = propylene glycol; RMSS = disodium ricinoleamido monoethanolamido sulfosuccinate; SLES = sodium lauryl ether sulphate; SLS = sodium lauryl sulphate; SUC = disodium lauryl 3-ethoxysulfosuccinate.

Modified from Effendy and Maibach, 1995.

TABLE 4.3. *In vitro toxicity ranking of frequently used surfactants*

Toxicity ranking	In vitro test (cell culture)	Assessment	References
BAC > SLS > between 80	Human primary keratinocytes	Arachidonic acid and Interleukin-1α release, MTT (mitochondrial metabolic activity) assay	Müller-Decker et al., 1994
CTAB > SLS (at concentration: 3g/mg)	Normal human epidermal keratinocytes (NHEK)	MTT assay	Bigliardi et al., 1994
BAC > SLS (at concentration: 1×10^{-5}M)	Normal human oral and foreskin keratinocytes	MTT assay and lactat dehydrogenase (LDH) release	Eun et al., 1994
Cationic = amphoteric > anionic > non-ionic surfactants	NHEK, HaCaT cells, and 3T3 cells	Neutral red release and cell growth/protein	Korting et al., 1994
N-alkyl-sulfate $C_{12} > C_{14} > C_{10} > C_{16} > C_8$	HaCaT cells	Neutral red release	Wilhelm et al., 1994
BAC > SLS > between 20	Commercial human skin model* (Skin2)	MTT assay, LDH and PGE_2 release	Osborne et al., 1994
0.2% BAC > 0.5% SLS > 0.5% CAPB > 30% PG	Commercial human skin model* (Skin Equivalent)	MTT assay	Harvell et al., 1994

*Commercial human skin model = human dermal fibroblasts in a collagen-gel or a nylon-mesh matrix co-cultured with NHEK that have performed a stratified epidermis.

CTAB: cetyltrimethylammonium bromide.
Modified from Effendy and Maibach, 1995.

45

short-time occlusive, and repeated short-time open test) can vary the outcome of irritancy testing in mans (Table 4.3). The concordance among the different exposure methods has been found to be high when evaluated by transepidermal water loss (TEWL) but not by visual scoring, implying somewhat the superiority of the bioengineering assessment; however, visual scoring seems to be a "gold standard" in everyday use. This is one of the reasons when conducting irritancy test among the various methods that an exposure method which stimulates most in-use situations should be chosen.

To predict the irritant potential and the irritancy ranking order of detergents in men, certain aspects have to be considered (e.g., type of detergent, mode of exposure, in-use situation, choice of irritancy testing). It has been proposed that the repeated open test is the best way to imitate most real-life situations where the uncovered skin is exposed to detergents. The repeated occlusive test or the one-time patch test may be suitable to mimic situations in which the skin is occluded after irritation by detergents (Tupker et al., 1999). Finally, one should keep in mind that in vivo irritancy testing in humans remains crucial as long as in vitro tests do not provide a comparable predictor value.

REDUCED IRRITANT POTENTIAL OF MIXED SURFACTANT SYSTEMS

Blends of surfactants have been used in cosmetic and pharmaceutical formulas, particularly, in order to increase the acceptance of the product due to its reduced irritant potential, mildness, and comfort. For instance, there is antagonism or mutual inhibition in an acid-base neutralization and in anionic-cationic surfactant reaction.

SLS as well as linear C_{9-13} alkylbenzene sulphonate (LAS), when applied each alone at 20% to human skin, induced a noteable erythema. Nevertheless, a mixture of 20% SLS and 10% sodium lauryl ether-2E0 sulfate (SLES), or 10% cocoamidopropyl betaine (CAPB), or 10% cocodiethanolamine, caused significantly less erythema (Table 4.4). Similarly, a blend of 20% LAS + 10% SLES + 10% C_{9-11} alcohol 8 EQ (nonionic), a total surfactant level of 40%, was substantially less irritant than 20% LAS alone. Probably, irritant responses are not simply linked with the total concentration of surfactants used, but rather to the contents of the mixture (Dillarstone et al., 1993).

TABLE 4.4. *Reduced irritancy of mixed surfactant systems*

Mixture of surfactants vs. single surfactant	References
SLG + SLS < SLS	Kawasaki et al., 1999, Lee et al., 1994
SLS + DDAB < SLS	Hall-Manning et al., 1998
20% SLS + 10% SLES, or 10% CAPB, or 10% CDEA < 20% SLS 20% LAS + 10% SLES + 10% C_{9-11} alcohol 8EO < 20% LAS	Dillarstone et al., 1993

SLES = sodium lauryl ether 2EO sulphate; CAPB = cocoamidopropyl betaine; CDEA = cocodiethanolamine; DDAB = dimethyl dodecyl amido betaine; SLG = sodium lauroyl glutamate.

Likewise, the addition of sodium lauroyl glutamate (SLG), a mild surfactant, to an SLS solution induced less skin irritation than did SLS alone, as assessed by visual scoring and an evaporimeter (Lee et al., 1994). More recently, employing electron paramagnetic resonance, it was demonstrated that SLS at low concentration caused fluidization of intercellular lipids, perhaps due to interjection of SLS molecules into intercellular lipids; however, the addition of SLG to SLS could inhibit the intercellular lipid's impairment (Kawasaki et al., 1997, 1999).

Less irritant responses to a mixture of surfactants could basically be explained by competitive interactions between surfactants used. Initially, a reduced critical micelle concentration (CMC) of surfactants used may be responsible for the lowered irritation. Recent data indicate, however, that CMC may perhaps not be related to the reduced irritant reactions (Hall-Manning et al., 1998).

Effects of tandem applications of the same surfactants or different substances on human skin appeared incomparable to those of a mixed surfactant system (Effendy et al., 1996). The *overlap phenomenon* described higher TEWL values in the newly exposed human skin, perhaps due to SLS spread after prolonged treatment irritating the skin adjacent to the treated site. The authors have also shown intense irritant reactions in the partial overlapping region, implying a cumulative effect of a tandem application of SLS (Patil et al., 1994).

EFFECTS OF DETERGENTS ON DIFFERENT SKIN CONDITIONS

In general, children significantly show lower water content of horny layer when compared to adults. Use of detergents in children for 4 weeks in the winter remarkably decrease the hydration state of the skin surface, which could be countered by a regular use of emollient (Ohara et al., 1993; Yamamoto, 1996).

In the elderly, most substances take longer to penetrate normal skin (Roskos et al., 1992), but in dried skin, water-soluble substances may penetrate more easily (Stüttgen et al., 1990). The skin of the elderly generally seem to be more prone to "dry skin syndrome" dryness than young skin; presumably soaps and detergents rather than occupational irritants are responsible for this phenomenon. Excessive washing and in-adequate rinsing may lead to significant skin dryness (xerosis) and irritancy (Graham-Brown, 1996).

Atopics have been reported to be susceptible to the irritant effect of soaps and detergents, resulting in avoidance of washing (Lammintausta et al., 1981; Rystedt, 1985); however, washing with an alkaline soap improved the skin lesions in atopics (Uehara et al., 1985; Lechner 1990).

Skin cleansers have been postulated to be an important adjunct in the treatment for acne (Solomon et al., 1996); however, excessive cleansing may exacerbate the disease (Mills et al., 1975). Moreover, long-term use of neutral or alkaline surfactants was found to increase the amount of Propionibacteria on the skin (Korting et al., 1990).

Recent investigations showed that acidic syndets can be less irritant than a neutral or alkaline one, the pH being, respectively, 4.5 and 7.5 (Gehring et al., 1991; Korting et al., 1991). These data could be supported by the knowledge on the dependence of the bi-layer formation and thus water-retaining capacity of epidermal lipids controlled by the pH of the circumstances (Osborne et al., 1987; Friberg, 1990; Korting et al., 1996). The alkaline soaps induced a greater loss of fat from skin surface than did tap water or acidic detergents (Gfatter et al., 1997).

CONCLUSIONS

Because surfactants hold certain beneficial properties, their use in everyday life becomes nearly indispensable. They have applications not only in skin cleanser, but also in cosmetic, paint, pesticide, textile industries, and even food; however, the irritation potential of surfactants may relatively limit their employment. Therefore, development of less irritant, consumer-friendly surfactants or mixed surfactant detergent systems are of general interest.

There seem to be differences in the irritation potential between surfactants; however, the arbitrary classification of surfactants does not necessarily mirror the irritancy of each substance. Hence, illuminated assays to predict the irritation potential of surfactants remain required. Our theoretical and practical insights have significantly improved, yet the complexity of the interaction between skin and surfactants suggests that development will flourish with a multifactorial approach. A conjunction between the advancing techniques of biophysical chemistry and the more slowly evolving insights into animal and human skin biology should perhaps be the goal of the near future.

REFERENCES

Ainsworth, S. J. (1992): Soaps and detergents. *Chem. Engin. News.*, 70:27–63.

Attwood, D., Florence, A. T. (1983): *Surfactant System: Their Chemistry, Pharmacy and Biology*, pp. 1–11. Chapman and Hall, London, UK.

Basketter, D. A., Griffiths, H. A., Wang, X. M., Wilhelm, K. P., McFadden, J. (1996): Individual, ethnic and seasonal variability in irritant susceptibility of skin: the implications for a predictive human patch test. *Contact Dermatitis*, 35:208–213.

Berardesca, E., Fideli, D., Gabba, F., et al. (1990): Ranking of surfactant skin irritancy in vivo in man using the plastic occasion stress test (POST). *Contact Dermatitis*, 23:71–75

Bouillon, C. (1996): Shampoos. *Clinics Dermatol*, 14:113–121.

Colwell, S. M. (1993): Soap wars. *Soaps Cosm. Chem. Spec.*, 69:22–28.

Dillarstone, A., Paye, M. (1993): Antagonism in concentrated surfactant systems. *Contact Dermatitis*, 28:198.

Effendy, I., Maibach, H. I. (1995): Surfactants and experimental irritant contact dermatitis. *Contact Dermatitis*, 33:217–225.

Effendy, I., Weltfriend, S., Patil, S., Maibach, H. I. (1996): Differential irritant skin responses to topical retinoic acid and sodium lauryl sulphate: Alone and in crossover design. *Br. J. Dermatol.*, 134:424–430.

Eun, H. C., Chung, J. H., Jung, S. Y., Cho, K. H., Kim, K. H. (1994): A comparative study of the cytotoxicity of skin irritants on cultured human o and skin keratinocytes. *Br. J. Dermatol.* 130:24–28.

Fartasch, M., Schnetz, E., Diepgen, T. L. (1998): Characterization of detergent-induced barrier alterations: Effects of barrier cream on irritation. *J. Invest. Dermatol.*, 21–27.

Fartasch, M. (1995): Human barrier information and reaction to irritation. *Curr. Probl. Dermatol*, 23:95–103.

Friberg, S. E. (1990): Micelles, microemulsions, liquid crystals and the structure of stratum corneum lipids. *J. Soc. Cosmet Chem.*, 41:155–171.

Friedman, M., Wolf, R. (1996): Chemistry of soaps and detergents: Various types of commercial products and their ingredients. *Clinics Dermatol.*, 14:7–13.

Gehring, W., Gehse, M., Zimmermann, V. et al. (1991): Effects of pH changes in a specific detergent multicomponent emulsion on the water content of stratum corneum. *J. Soc. Cosmet. Chem.*, 42:327–333.

Gfatter, R., Hackl, P., Braun, F. (1997): Effects of soap and detergents on skin surface pH, stratum corneum hydration and fat content in infants. *Dermatology*, 195:258–262.

Graham-Brown, R. (1996): Soaps and detergents in the elderly. *Clinics Dermatol.* 14:85–87.

Hall-Manning, T. J., Holland, G. H., Rennie, G., Revell, P., Hines, J., Barratt, M. D., Basketter, D. A. (1998): Skin irritation potential of mixed surfactants systems. *Food Chem. Toxicol.*, 36:233–238.

Harvell, J., Tsai, Y. C., Maibach, H. I., et al. (1994): An in vivo correlation with three in vitro assays to assess skin irritation potential. *J. Toxic Cut. Ocular. Toxic.*, 13:171–183.

Harvell, J., Bason, M. M., Maibach, H. I. (1992): In vitro skin irritation assays: Relevance to human skin. *Clin. Toxicol.*, 30: 359–369.

Imokawa, G., Mishima, Y. (1979): Cumulative effect of surfactants on cutaneous horny layers: Lysosome labilizing action. *Contact Dermatitis*, 5:151–162.

Kalmanzon, E., Zlotkin, E., Cohen, R., Barenholz, Y. (1992): Liposomes as a model for the study of the mechanism of fish toxicity of sodium dodecyl sulfate in sea water. *Biochem. Biophys. Acta.*, 1103:148–156.

Kawasaki, Y., Quan, D., Sakamoto, K., Maibach, H. I. (1997): Electron paramagnetic resonance studies on the influence of anionic surfactants on human skin. *Dermatology*, 194:238–242.

Kawasaki, Y., Quan, D., Sakamoto, K., et al. (1999): Influence of surfactant mixtures on intercellular lipid fluidity and skin barrier function. *Skin Res. Tech.*, 5:96–101.

Korting, H. C., Braun-Falco, O. (1996): The effect of detergents on skin pH and its consequences. *Clinics Dermatol.*, 14:23–27.

Korting, H. C., Herzinger, T., Hartinger, A., Kerscher, M., Angerpointner, T., Maibach, H. I. (1994): Discrimination of the irritancy potential of surfactants in vitro by two cytotoxicity assays using normal human keratinocytes, HaCaT cells and 3T3 mouse fibroblast: Correlation with in vivo data from a soap chamber assay. *J. Dermatol. Sci.*, 7:119–129.

Korting, H. C., Huebner, K., Greiner, K., Hamm, G., Broun-Falco, O. (1990): Differences in the skin surface pH and bacterial microflora due to the long-term application of synthetic detergent preparations of pH 5.5 and 7.0. *Acta Derm. Venereol (Stockh)*, 70:429–431.

Korting, H. C., Megele, M., Mehringer, L., et al. (1991): Influence of skin cleansing preparation acidity on skin surface properties. *Int. J. Cosmet Sci.*, 13:91–112.

Lammintausta, K., Kalimo, K. (1981): Atopy and hand dermatitis in hospital wet work. *Contact Dermatitis*, 7:301–308.

Lansdown, A. B. C., Grasso, P. (1972): Physico-chemical factors influencing epidermal damage by surface active agents. *Br. J. Dermatol*, 86:361–373.

Le, M., Schalkwijk, J., Siegenthaler, G., van de Kerkhof, P. C., Veerkamp, J. H., van der Valk, P. G. (1996): Changes in keratinocyte differentiation following mild irritation by sodium dodecyl sulphate. *Arch. Dermatol Res.*, 288:684–690.

Lechner, W. (1990): Adjuvant treatment with synthetic detergent preparation in atopic dermatitis. In: *Skin Cleansing with Synthetic Detergents: Chemical Ecological and Clinical Aspects.* Edited by Braun-Falco O., Korting H. C., pp. 160–163. Springer-Verlag, Berlin.

Lee, C. H., Maibach, H. I. (1995): The sodium lauryl sulfate in contact dermatitis model. *Contact Dermatitis*, 33:1–7.

Lee, H. L., Kawasaki, Y., Maibach, H. I. (1994): Effect of surfactant mixtures on irritant contact dermatitis potential in man: Sodium lauroyl glutamate and sodium luryl sulphate. *Contact Dermatitis*, 30:205–209.

Löttermoser, A., Stoll, F. (1933): The surface and interfacial activities of salts of the fatty acid alcohol sulfuric acid esters. *Kolloid-Z*, 63:49–61.

Mills, O. H., Kligman, A. M. (1975): Acne detergicans. *Arch. Dermatol.*, 111:65–68.

Mueller-Decker, K., Fürstenberger, C., Marks, F. (1994): Keratinocyte-derived proinflammatory key mediators and cell viability as in vitro parameters of irritancy: A possible alternative to the Draize skin irritation test. *Toxicol Appl. Pharmacol.*, 127:99–108.

Nickoloff, B. J., Naidu, Y. (1994): Perturbation of epidermal barrier function correlates with initiation of cytokine cascade in human skin. *J. Am. Acad. Dermatol.*, 30:535–546.

Ohara, Y., Noda, A., Fujiwara, Y., et al. (1993): On improvement effects of skin-care-products in infants. *J. Pediatr. Dermatol*, 12:67–74.

Osborne, D. W., Friberg, S. E. (1987): Role of stratum corneum lipids as moisture retaining agent. *J. Dispers. Sci. Technol.*, 8:173–179.

Osborne, R., Perkins, M. A. (1994): An approach for development of alternative test methods based on mechanisms of skin irritation. *Food Chem. Toxicol.*, 32:133–142.

Partearroyo, M. A., Ostolaza, H., Goni, F. M., Barbera-Guillem, M. (1990): Surfactant-induced cell toxicity and cell lysis. *Biochem. Pharmacol.*, 40:1323–1328.

Patil, S. M., Singh, F., Maibach, H. I. (1994): Cumulative irritancy in man to sodium lauryl sulfate: The overlap phenomenon. *Int. J. Pharmacol.*, 110:147–155.

Reynolds, J. E. F. (1999): *Martindale: The Extra Pharmacopoeia.* 32nd ed. Royal Pharmaceutical Society, London, UK.

Rosen, M. J. 1978. *Surfactants and Interfacial Phenomena*, pp. 1–25. John Wiley & Sons, New York.

Roskos, K., Maibach, H. I. (1992): Percutaneous absorption and age: Implications for therapy. *Drugs & Aging*, 2:432–449.

Rystedt, I. (1985): Hand eczema and long-term prognosis in atopic dermatitis (dissertation). *Acta Derm Venereol* (Stockh) 65 (Suppl 117):1–59.

Scala, I., McOsker, D. E., Reller, H. H. (1968): The percutaneous absorption of ionic detergents. *J. Invest. Dermatol.*, 50:370–377.

Schoenberg, T. (1983): Formulating of mild skin cleansers. *Soaps Cosm. Chem. Spec.*, 59:33–38.

Solomon, B. A., Shalita, A. R. (1996): Effects of detergents on acne. *Clinics Dermatol.*, 14:95–99.

Spitz, L. (1990): *Soap Technology for the 1990's.* American Oil Chemists' Society, Campaign, IL.

Stuettgen, G., Ott, A. (1990): Senescence in the skin. *Br. J. Dermatol.*, 122(Suppl 35):43–48.

Tupker, R. A., Bunte, E. E., Fidler, V., et al. (1999): Irritancy ranking of anionic detergents using one-time occlusive, repeated occlusive and repeated open tests. *Contact Dermatitis*, 40:316–322.

Uehara, M., Takada, K. (1985): Use of soap in the management of atopic dermatitis. *Clin. Exp. Dermatol.*, 10:419–425.

Van den Valk, P. G. M., Nater, J. P., Bleumink, E. (1985): Vulnerability of the skin to surfactants in different groups of eczema patients and controls as measured by water vapour loss. *Clin. Exp. Dermatol.*, 10:98–103.

Wade, W. G., Addy, M. (1992): Antibacterial activity of some triclosan-containing toothpastes and their ingredients. *J. Periodontol*, 63:280–282.

Walters, K. A., Bialik, W., Brain, K. R. (1993): The effects of surfactants on penetration across the skin. *Int. J. Cosmet Sci.*, 15:260–270.

Wilhelm, K. P., Freitag, C., Wolff, H. H. (1994): Surfactant-induced skin irritation and skin repair. *J. Am. Acad. Dermatol.*, 30:944–949.

Wilhelm, K. P., Samblebe, M., Siegers, C. F. (1994): Quantitative in vitro assessment of N-alkyl sulphate-induced cytotoxicity in human keratinocytes (HaCaT): Comparison with in vivo human irritation tests. *Br. J. Dermatol.*, 130:18–23.

Willis, C. M., Stephens, C. J. M., Wilkinson, J. D. (1989): Epidermal damage induced by irritants in man: A light and electron microscopic study. *J. Invest. Dermatol.*, 93:695–699.

Willis, C. M., Stephens, C. J. M., Wilkinson, J. D. (1991): Selective expression of immune-associated surface antigens by keratinocytes in irritant contact dermatitis. *J. Invest. Dermatol.*, 96:505–511.

Wilmer, J. L., Burleson, F. G., Kayama, F., Kanno, J., Luster, M. I. (1994): Cytokine induction in human epidermal keratinocytes exposed to contact irritants and its relation to chemical-induced inflammation in mouse skin. *J. Invest. Dermatol.*, 102:915–922.

Wood, L. C., Jackson, S. M., Elias, P. M., Grunfeld, C., Feingold, K. R. (1992): Cutaneous barrier perturbation stimulates cytokine production in the epidermis of mice. *J. Clin. Invest.*, 90:482–487.

Yamamoto, K. (1996): Soaps and detergents in children. *Clinics Dermatol.*, 14:77–80.

Yang, L., Mao-Qiang, M., Taljebini, M., et al. (1995): Topical stratum corneum lipids accelerate barrier repair after tape stripping, solvent treatment and some but not all types of detergent treatment. *Br. J. Dermatol.*, 133:679–685.

Zhou, J., Mark, R., Stoudemayer, T., et al. (1991): The value of multiple instrumental and clinical methods, repeated patch applications, and daily evaluations for assessing stratum corneum changes induced by surfactants. *J. Soc. Cosmet Chem.*, 42:105–128.

Zografi, G., Schott, H., Swarbrick, J. (1990): Interfacial phenomena. In: *Remington's Pharmaceutical Sciences.* Edited by Gennaro, A. R., pp. 257–272. Mack, Easton, PA.

Toxicology of Skin
Edited by Howard I. Maibach
Copyright © 2001 Taylor & Francis

5

Permeability of Skin for Metal Compounds

Jurij J. Hostýnek

Euromerican Technology Resources, Inc., Lafeyette, California, USA and Department of Dermatology, School of Medicine, University of California, San Francisco, California, USA

INTRODUCTION

The largest human organ, the skin, is an envelope that almost totally shields the body from the external environment and helps maintain the integrity and function of its complex inner organism. When intact, the skin is a multilayered barrier that is impenetrable to microorganisms but is a membrane semipermeable to chemicals, selectively allowing their passage in both directions. It usually is disregarded as port of entry for xenobiotics, and this is a particularly serious risk in the work environment, where exposure and prevention still focuses primarily on inhalation and ingestion. In our present understanding of structure and function of the skin, its permeation is determined by the physicochemical parameters of permeants. Intuitively, and also based on the therapeutic action of dermatologicals and cosmeceuticals, we anticipate easy penetration by lipophilic compounds there, as is commonly experienced in the process of inunction, but it is rather counter-intuitive that polar structures such as water or electrolytes (e.g., salts in aqueous solution) should also be able to penetrate the skin to any significant degree. For water and a number of hydrophilic-charged molecules, however, including a great number of metal compounds, this could be demonstrated also, albeit the process proceeds at an average of two to three orders of magnitude slower than is the penetration of small-molecular-weight, lipophilic nonelectrolytes (Hostýnek et al., 1993). Prompt onset of adverse reactions observed in the skin of hypersensitive individuals following even casual contact with allergenic metals, such as nickel (Gilboa et al., 1988; Kanerva et al., 1998; Samitz and Katz, 1976) or chromium (Burrows, 1972) are evidence of their facile diffusion through the Stratum Corneum (SC).

As a major interface between man and his environment, the skin thus is an important route of entry as well as a line of defense against potentially hazardous agents. Metals and their compounds, ubiquitous in man's environment, particularly a number of highly toxic ones, are of major concern in occupational medicine and for regulatory agencies as, particularly on heavy and sustained exposure in the work place, they may find entrance into the human organism through the integument. This fact is clearly acknowledged in the 1988 NIOSH publication *Proposed National Strategies for the Prevention of Leading Work-Related Diseases and Injuries, Part 2*, a document that has been reinforced by the comprehensive position statement resulting from the American Academy of Dermatology–sponsored "National Conference on Environmental Hazards to the Skin" in 1992. These reports highlight a number of issues that directly point to and address the hazards of skin exposure to xenobiotics in general, and to toxic metals in particular. A number of issues raised there are excerpted here, as they appear germane to this discussion:

As new regulatory requirements have gone into effect to reduce permissible airborne exposure levels of potential carcinogens and toxins, percutaneous absorption has become a relatively more important route of exposure in terms of total body chemical burden. Of the chemical substances currently listed in RTECS, less than 5% have dermal LD_{50} data recorded, and even fewer have any cutaneous irritant effects reported; specific quantitative dose-response data, particularly as they address heavy-metal chemicals, are virtually nonexistent.

Since 1982, NIOSH has included occupational skin disorders on its list of the 10 leading work-related diseases and injuries.

Because large surface areas of the skin are often directly exposed to the environment, this organ is particularly vulnerable to occupational and environmental diseases and injuries.

Skin diseases account for >30% of all cases of chronic occupational disease.

Occupational exposure to heavy metals and their compounds is a major cause of dermatological disease due to irritation and allergic sensitization (Adams, 1983a, 1983b).

With respect to specific (allergic) and nonspecific (irritant) contact dermatitis (which may constitute up to 90% of workers' compensation claims for skin diseases), percutaneous absorption of potential allergens and irritants through the skin is a key determinant of the risk of skin sensitization and irritation, and may be enhanced if protective clothing entraps or occludes the irritant against the skin, by increased hydration of the SC, and by contact with anatomical sites where skin permeability is greater. (Exposure to nickel, chromium, cobalt, and mercury compounds are premier causes for allergic and irritant reactions in industrialized countries (Schnuch et al., 1997)).

For purposes of this review, literature has been scoured for quantitative data and for experimental techniques which have been used in vivo and in vitro to measure dermal absorption of metal compounds. Whenever data permits, the permeability coefficient Kp or the percent absorbed values are presented.

As evident from the data tabulated, the limited number of metals that have been investigated for their ability to penetrate the skin are mostly heavy metals ubiquitous in man's environment. Recognized to pose outstanding risks from exposure, they have come to special public attention, either for scientific or political reasons. Included in this review are only the quantitative aspects of passive diffusion of metal compounds through healthy, intact skin, and not their penetration promoted by mechanical or chemical methods.

Finally, the chapter also reviews recent results obtained in our laboratory with compounds of nickel, lead, and cadmium.

SKIN—A SEMIPERMEABLE BARRIER FOR ABSORPTION, STORAGE, AND ELIMINATION OF METALS

Although the effects from skin exposure to xenobiotics are substantially less than those from other routes, the large surface area of the body as a whole offers the opportunity for substantial absorption and concomitant systemic effects, as well as for excretion. Absorption is potentially beneficial as it allows transdermal delivery of therapeutics, but that process also provides access to toxics, due to accidental or unwitting exposure to the myriad chemical agents, natural or anthropogenic, that are present in man's environment (Table 5.1). Only larger structures (e.g., microorganisms, or proteins,

TABLE 5.1. *Benefits and risks from skin permeation by metals*

	Benefits	Risks
Ingress	Therapy (drug delivery) (Cu, Zn)	Toxic covalent organometallics penetrate with ease (Pb, Tl)
Egress	Detoxification through desquamation and sweat (Cd, Pb, Hg, Ni)	Loss of essential nutrients through desquamation and sweat (Cu, Fe, Zn)

TABLE 5.2. *Permeability coefficients Kp of uncharged molecules through human skin*

Compound	Kp x 10^4	Reference
Sucrose	0.052	Anderson et al., 1988
Hydrocortisone	1.2	Hadgraft and Rideout, 1987
Ethanol	8	Scheuplein and Blank, 1971
Water	10	Bronaugh et al., 1986
Phenol	82	Roberts et al., 1978
Benzene	100	Blank and McAuliffe, 1985
Ethyl ether	160	Scheuplein and Blank, 1971
Nicotine	190	Hadgraft and Rideout, 1987
n-Octanol	520	Scheuplein and Blank, 1971

peptides and synthetic polymers (as a rule, greater than 5000 daltons (Sage et al., 1993)) or highly polar compounds, such as sugars, will not diffuse through this barrier as long as it is healthy and intact. Limits in polarity, expressed as the logarithm of n-octanol/water distribution or $logP_{oct}$, are relevant to the penetration of the SC as well as other biological membranes; defined in structure-activity studies developed to predict biological processes, the optimal range is bracketed at $-0.5 > logP_{oct} > +3.5$ (Briggs et al., 1977). For perspective, the permeability coefficients for a selection of compounds of diverse size and polarity are listed in Table 5.2, with the permeability coefficient for water as a point of reference.

Several heavy metals, have a particular affinity for keratinizing structures and either are retained in the epidermis upon exposure, forming temporary or permanent reservoirs, or exhibit extended delays in the diffusion process. Even following systemic exposure, a number of these will concentrate in epidermal tissue, forming permanent deposits in their elemental state and imparting a characteristic coloration to the skin.

Finally, the skin and its appendages also function as an important secretory organ which plays a vital part in the natural process of detoxification and the maintenance of homeostatic balance. This selective process represents a particularly important safeguard against overload by toxics, as well as against loss of vitally important trace nutrients.

ELECTROLYTE TRANSPORT THROUGH THE SKIN

A significant number of compounds used in consumer and personal care products bear an electron charge: quats (antimicrobials), betaines (surfactants) or salts of aluminum, zirconium, lead, zinc, or mercury (antiperspirants, hair colorants, disinfectants, etc.). The suggestion that such charged molecules may penetrate the skin was formerly rejected off-hand due to lack of radioisotopes or microanalytical methods sensitive enough to detect low levels of permeants, and it still meets with skepticism now, but metal isotopes and modern microanalytical methods provide ample evidence which illustrates that process of permeation through the structures of the skin and its appendages. (Potts et al., 1992). The uptake of zwitterionic surfactants, for instance, has been measured in vivo (Bucks et al., 1993) and in vitro (Ridout et al., 1991), and the epidemiologically significant occurrence of allergic reactions on contact with metals or their compounds also provides ample evidence for the facile violation of the SC barrier.

MEASURING THE ABSORPTION OF METAL COMPOUNDS

In Vitro—Fick's First Law of Membrane Diffusion and the Permeation Coefficient Kp

The most commonly used in vitro technique for measuring percutaneous absorption involves placing a piece of excised skin in a two-chamber diffusion cell. It consists of a top chamber to receive an adequate volume of penetrant in solution, an O-ring to secure the skin in place, a temperature-controlled bottom chamber with continually circulating solution removing the penetrating amounts on the receptor side of the membrane and a sampling port to withdraw fractions at specific time intervals for analysis (Gummer et al., 1987). The solute diffuses from the fixed higher concentration medium in the donor chamber into the less concentrated solution in the receptor chamber. The advantage of diffusion experiments thus conducted in vitro lies in the applicability of Fick's first law of diffusion (Fick, 1855). Formulated to characterize passive diffusion of compounds across membranes in general, Fick's law has been shown to also apply to passive diffusion of xenobiotics through the SC of the skin. The law states that if the permeation process from an "infinite" reservoir reaches the point of steady state equilibrium (i.e., the concentrations in the donor and receptor phases remain constant over time), the steady-state penetration flux Jss per unit path length is proportional to the concentration gradient dC and to the penetrant's permeation constant Kp: $Jss = Kpd \times C$. The technique is easily standardized and it allows the determination of Kp through skin or other membranes as long as the barrier properties are not affected by either permeant or carrier solvent. That implies that the permeant may not react with barrier material (i.e., change barrier permeability with time). Because a number of heavy metals are known to be reactive with epidermal protein to some degree, however, this somewhat prejudices the general validity of Kp values. Still, determined under conditions of steady state and normalized for concentration, the Kp serves as the most useful descriptor of diffusion through skin. Expressing depth of permeation per unit of time (here in cm/h), human and animal penetration data currently available in the literature are presented in Tables 5.4–5.7, respectively.

In Vivo—Percentage of Dose Absorbed

It appears important that, whenever possible, absorption data be determined in vivo, in man or primates, for as accurate a risk assessment as possible; avoiding the uncertainties of cadaver skin or adjustments for species variation. One principal in vivo methodology, using human volunteers or monkeys, is the one developed by Feldmann and Maibach (1970; 1974, 1969), which is the most applied in vivo method for the

TABLE 5.3. *Deposition of metals in the skin*

Metal	Clinical term	Comment
Calcium	Calcinosis cutis	Long-term occupational exposure to high levels of dissolved calcium (water hardness), as experienced by oil field workers (Wheeland and Roundtree, 1985), miners (Sneddon and Archibald, 1958), and agricultural workers (Christensen, 1978). Benign and reversible hardening of the exposed skin.
Gold	Chrysiasis	Gold particles deposited cause bluish discoloration, remain in place permanently, and are located around blood vessels and in dermal macrophages (Cox and Marich, 1973; Everett, 1979; Granstein and Sober, 1981).
Silver	Argyria	Silver particles accumulate in the mucosa, nail lunulae, hair, and skin, resulting in permanent graying of the skin. Pigmentary changes are most pronounced in sun-exposed areas, hands, face, and neck, basal membrane of the epidermis, sweat glands, hair follicles, small cutaneous blood vessels, and the connective tissue surrounding the sebaceous glands (Pariser, 1978; Requena et al., 1985). Discoloration of the oral mucosa, tongue, sclera, and conjunctiva has also been described (Lee and Lee, 1994; Matsumura et al., 1992).
Arsenic	Hyperpigmentation	Systemic exposure to inorganic salts of arsenic (tri- and pentavalent) in therapy, drinking water, or occupational exposure in the mining industry (Levantine and Almeyda, 1973; Vente et al., 1993).
Bismuth	Bismuthia	Blue-black gingival line (Granstein and Sober, 1981). Histological examination reveals the presence of metallic granules in the papillary and reticular layers of the dermis.
Iron	Haemochromatosis	Permanent dark brown discoloration due to increased melanization and iron deposition in the skin (deeper dermis, especially in the macrophages of capillary endothelial cells) (Pollycove, 1966).
Lead	Saturnism	Pallor and lividity of the skin, dark lining of the gingivae due to subepithelial deposits of metallic lead, which is converted to lead sulfide (Birmingham, 1979).
Mercury	Hydrargyrosis cutis	Chronic dermal application of mercurials used as skin bleaches can lead to tissue accumulation of metallic mercury. Metallic mercury concentrates in the upper dermis (Burge and Winkelmann, 1970; Calvery et al., 1946; Granstein and Sober, 1981; Lamar and Bliss, 1966; Luders et al., 1968).
Selenium	Icteroid discoloration	High levels of selenium in soil and water result in elevated dietary intake, leading to dermatotoxic pathologies: chronic dermatitis, changes in nail structure, discoloration, and partial or total loss of hair (Harr and Muth, 1972).

TABLE 5.4. *In vitro permeability of human skin to metal compounds*

Compound	% Dose	$Kp \times 10^{4*}$ (cm/h)	Reference
$H_3{}^{73}AsO_4$	1.9		Wester et al., 1993
$H_3{}^{10}BO_3$	0.7–1.2	2.9–5.0	Wester et al., 1998
$Na_2B_4O_7$	0.14	1.7	Wester et al., 1998
DOT[†]	0.19	0.8	Wester et al., 1998
$^{109}CdCl_2$	0.1–0.6		Wester et al., 1992
$^{109}CdCl_2$ (in skin tissue)	2.7–12.7		Wester et al., 1992
$^{58}CoCl_2$		0.19–4.4	Wahlberg, 1965
$Na_2{}^{51}CrO_4$[‡]		0.19–15	Wahlberg, 1965; Wahlberg, 1970
$K_2Cr_2O_7$[‡]		0.54	Mali et al., 1964
$K_2Cr_2O_7$		2.3–27	Fitzerald and Brooks, 1979, Gammelgaard et al., 1992
$^{51}CrCl_3$[‡]		10–14	Wahlberg, 1970
$CrCl_3$		0.13–0.41	Fitzgerald and Brooks, 1979; Gammelgaard et al., 1992
$CrCl_3$ in skin		0.064–2.4	Gammelgaard et al., 1992
$Cr(NO_3)_3$		0.30	Gammelgaard et al., 1992
$Cr(NO_3)_3$ in skin		0.013	Gammelgaard et al., 1992
$K_2Cr_2O_7$		0.15–0.46	Gammelgaard et al., 1992
$HgCl_2$[†]		0.2–9.3	Wahlberg, 1965
Ammoniated ^{203}Hg (in stratum corneum)		0.002–0.03	Marzulli and Brown, 1972
^{203}Hg acetate (in stratum corneum)		0.001–0.03	Marzulli and Brown, 1972
LiCl		0.3	Phipps et al., 1989
$^{24}NaCl$		0.35	Tregear, 1966
$^{22}NaCl$		0.06–11.9	Kasting and Bowman, 1990; Kasting and Bowman, 1990
$^{63}NiSO_4$ (epidermis)	0.0–0.066	0–0.07	Samitz and Katz, 1976
$NiCl_2$		0–0.037	Fullerton et al., 1988; Fullerton et al., 1986
$NiCl_2$ (occluded)	0.2–15.5	0.03–2.3	Fullerton et al., 1986
$^{63}NiCl_2$	0.29–1.6		Emilson et al., 1993
$NiSO_4$	1	0.073	Fullerton et al., 1986
$NiSO_4$ (occluded)	0.62	0.09	Fullerton et al., 1986
Pb naphthenate		23	Rasetti et al., 1961

[*]Apparent permeability coefficient.
[†]Disodium octaborate decahydrate.
[‡]Disappearance method.

TABLE 5.5. *In vitro permeability of animal skin to metal compounds*

Compound	Species	$Kp \times 10^{4*}$ (cm/h)	Reference
[58]CoCl$_2$[†]	Guinea pig	0.38–12.2	Wahlberg, 1965a
Na$_2$[51]CrO$_4$[†]	Guinea pig	11–19	Wahlberg, 1970
[51]CrCl$_3$[†]	Guinea pig	14–21	Wahlberg, 1970
bis(Glycinato)-Cu	Cat	24	Walker et al., 1977
[203]HgCl$_2$[†]	Guinea pig	14.7–25.3	Wahlberg, 1965a,c,e
[42]KCl	Rabbit	22	Tregear, 1966, Tregear, 1966
[42]KCl	Pig	1.9	Tregear, 1966, Tregear, 1966
LiCl	Pig	9.9	Phipps et al., 1989
LiCl	Rabbit	19	Phipps et al., 1989
[24]NaCl[†]	Rabbit	23.6	Tregear, 1966, Tregear, 1966
[24]NaCl[†]	Pig	7.4	Tregear, 1966, Tregear, 1966
[22]NaCl[†]	Guinea Pig	8–16	Tregear, 1966, Tregear, 1966
[22]NaCl	Mouse	0.7–18	Bagniefski and Burnette, 1990 and Bagniefski, 1988
[65]Zn pyridine thione	Rabbit	0.033	Klaassen, 1976

[*]Apparent permeability coefficient.
[†]Disappearance method.

determination of skin absorption yielding a large part of the data so far available on skin penetration by drugs and pesticides avaible so far.

In human studies, following topical application, plasma levels of test compounds are low, usually falling below assay detection, and so it becomes necessary to use radiolabelled chemicals. The compound labeled with carbon-14 or tritium, or a metal isotope, is applied to the skin in a minimal volume of a volatile solvent that is left to evaporate, and the total amount of radioactivity excreted is then determined. The

TABLE 5.6. *In vivo permeability of human skin to metal compounds*

Compound	% Dose	$Kp \times 10^{4*}$ (cm/h)	Reference
[111]AgNO$_3$[†]	Max. 4		Nørgaard, 1954
H$_3$[10]BO$_3$		0.0019	Wester et al., 1998
Na$_2$B$_4$O$_7$		0.0018	Wester et al., 1998
Disodium octaborate tetrahydrate		0.0010	Wester et al., 1998
Na$_2$CrO$_4$		9.6–35	Baranowska-Dutkiewicz, 1981
K$_2$[51]Cr$_2$O$_7$ (in skin)		0.11–13	Mali et al., 1964
K$_2$Cr$_2$O$_7$		0.015–0.18	Corbett et al., 1997
[203]Hg	3.0–10.6	6060–2412	Hursh et al., 1989
[203]Hg (in stratum corneum)	0.3–1.3		Hursh et al., 1989
[24]NaCl		0.62	Tregear, 1966
[57]NiSO$_4$[†]	76–55		Nørgaard, 1955
[203]Pb acetate	0.058	0.005	Moore et al., 1980
[204]Pb(NO$_3$)$_2$ (occluded)		1.3	Stauber et al., 1994

[*]Apparent permeability coefficient.
[†]Disappearance method.

TABLE 5.7. *In vivo permeability of animal skin to metal compounds*

Compound	Species	% Dose	Kp x 10^{4*} (cm/h)	Reference
$^{110m}AgNO_3$	Guinea pig	1–3.9		Skog and Wahlberg, 1964
$Na_2H^{74}AsO_4$	Rat tail		0.027	Dutkiewicz, 1977
$H_3^{73}AsO_4$[‡]	Monkey	2.0–6.4		Wester et al., 1993
$^{115m}CdCl_2$[†]	Guinea pig	1.8	11	Skog and Wahlberg, 1964
$^{58}CoCl_2$[†]	Guinea pig	$\leq$1–4%	$\leq$6.4	Skog and Wahlberg, 1964
$^{58}CoCl_2$[†]	Guinea pig		2.1–10.1	Wahlberg, 1971
$Na_2^{51}CrO_4$[†]	Guinea pig		2–26	Wahlberg, 1968, Wahlberg, 1965a, Wahlberg, 1965b, Wahlberg, 1971
$^{51}CrCl_3$[†]	Guinea pig	1.6–2.2	12–14	Wahlberg, 1965b
$Na_2^{51}CrO_4$	Rat tail		8–15	Dutkiewicz and Przechera, 1966
$^{137}CsCl$	Rat	1.33 ± 0.21	$\leq$2.7	Stojanovic and Milivojevic, 1971
^{59}Fe chelates	Mouse skin	55–80		Minato et al., 1967
$^{203}HgCl_2$	Guinea pig	2.0–3.2	13–21	Skog and Wahlberg, 1964
$K_2^{203}HgI_4$	Guina pig	1.7–4.1	11–26	Skog and Wahlberg, 1964
$CH_3^{203}Hg$ dicyandiamide[§]	Guinea pig	3.4–4.5	22–29	Skog and Wahlberg, 1964
$^{24}NaCl$[†]	Pig		19.7	Tregear, 1966
$^{24}NaCl$[†]	Rabbit		30.2	Tregear, 1966
$Na_2^{75}SeO_3$	Rat tail		8.8	Dutkiewicz et al., 1971
$^{89}SrCl_2$[†]	Guinea pig		16–20	Wahlberg, 1968

[*]Apparent permeability coefficient.
[†]Disappearance method.
[‡]Corrected for urinary excretion.
[§]Methyl mercury dicyandiamide.

amount retained in the body is corrected for by determining the amount of radioactivity excreted following parenteral administration. The resulting radioactivity value is then expressed as the percent of the applied dose absorbed. Fick's postulates for membrane diffusion are not met there because a concentration of the penetrant cannot be defined, and neither is a steady state equilibrium reached with this method; a permeability coefficient characterizing the compound cannot be calculated. Nevertheless, together with the toxicity of a chemical, the percentage of a substance absorbed is the second factor critical for risk assessment. It gives a measure for human exposure and an indication of the amount potentially absorbed in (the worst) case of total body immersion. For a chemical applied at 100 mg/cm^2, at 1% absorption, for instance, the extent of systemic exposure to the compound applied on the entire body area of an average human adult (18,000 cm^2) would be in the milligram range in the course of 24 hours.

Data acquired by this method for metal compounds are included in Tables 5.3, 5.4 and 5.6, as alternatives to Kp values. Determined under most disparate experimental conditions, by different investigators, these percentage values are not readily correlated for comparison, and therefore are listed there under a separate heading.

Another method for determining percentage of dose absorbed is the "disappearance method," extensively used with radiolabelled metal salts, primarily by Wahlberg. Measuring the loss of material from the skin surface over time, the guinea pig is the model mostly used in vitro and in vivo, with a limited number of experiments conducted on human skin in vitro (Wahlberg, 1968a, 1968b). Consistently using that technique, Wahlberg studied a dozen metal salts for their skin diffusivity, and although the data on skin disappearance were obtained over relatively short exposure times, the results are valuable as benchmarks, allowing at least partial validation of other measurements. To make up for the overall paucity of data on skin absorption by metals, those values are included in the tables under the % absorbed heading, bearing a special notation.

Limitations in Measuring Percutaneous Absorption of Metal Compounds

For those metals where skin penetration was investigated, the data rarely are adequate for the calculation of flux or permeability coefficient; those that are amenable to quantitative analysis often follow diverse experimental protocols, which lead to results that are not necessarily comparable. While a permeability coefficient is ideally determined under steady-state conditions, the percutaneous absorption of metals has rarely met this criterion.

Another problem inherent in the in vitro approach is the lack of total-dose accountability, as results are based on the quantity of permeant found in the receptor phase of the in vitro system. The amount of chemical expressed as percentage of dose is taken to be the amount becoming available systemically, and material retained in the SC that may subsequently diffuse further, significant for risk assessment purposes, is not determined. Retention there can be reversible, as was shown to be the case for nickel (Samitz and Katz, 1976) or platinum (Fairlie and Whitehouse, 1991) or dynamic, as with essential elements subject to homeostatic regulation.

Nevertheless, whenever possible, literature data have been abstracted for this review and, as necessary, transformed so as to make them adequate for purposes of risk assessment. Because K_p values, are the best parameter available for comparison of percutaneous flux and estimation of skin absorption, they are presented in Tables 5.4–5.7.

PREDICTION OF PERCUTANEOUS ABSORPTION THROUGH STRUCTURE-PERMEABILITY RELATIONSHIPS

The experimental determination of permeability data through mammalian epidermis or, even more critical for risk assessment, through human skin is problematic. Aside from biological variability and the choice from among a variety of experimental parameters, there exist constraints of an ethical, legal, and financial nature which account for the paucity of reliable skin absorption data for chemicals. In the present environment of heightened public and official awareness of chemical risk, combined with the efforts to reduce the number of animals used in biological experimentation, it becomes increasingly important to construct and validate predictive mathematical models that are mechanistically and biologically founded and which can take the place of traditional in vivo testing. The development of quantitative structure-activity relationships (QSAR) as a predictive tool is thus gaining in importance in the emerging field of computational toxicology and in estimating percutaneous absorption in particular. Several algorithms have been developed which describe passive diffusion of small nonelectrolytes, correlating structural characteristics of permeant (bulk, polarity, hydrogen bonding) with rate of penetration (Anderson et al., 1988, 1989; Bunge et al., 1994; Flynn, 1990; Kasting and Robinson, 1993; Kasting et al., 1987; Potts and Guy, 1995; Pugh and Hadgraft, 1994). The laws governing passive diffusion of small nonelectrolytes through the epidermis have thus been systematically investigated and adequately defined in mathematical terms. As the skin is a typical example of a biological membrane, the postulates of Fick's law of diffusion are applicable for that purpose, especially when using in vitro models. Thus, the body of skin penetration data for nonelectrolytes is rapidly growing, particularly because the transdermal route as an alternative mode of dosing xenobiotics comports certain pharmacodynamic advantages increasingly being realized in medical devices designed for controlled release. QSAR models now make it possible to predict in vivo skin penetration and dosing of a variety of structures with adequate accuracy, thus significantly reducing the need for in vitro or in vivo experimentation.

Although size, polarity, and hydrogen bonding ability appear as the principal factors governing diffusion of nonelectrolytes through the SC (Potts and Guy, 1995), presumably through the intercellular pathway, a much larger number of criteria appears to apply to the diffusion of metal ions into and through the epidermis. For one, they seem to proceed by various routes: transcellular, extracellular, as well as through the skin's appendages (Potts et al., 1992); once they reach the epidermis, depending on their electropositivity, they can show pronounced idiosyncratic behavior vis-à-vis nucleophilic protein side chains (e.g., as they come in contact with cysteine, lysine, histidine or methionine residues). A number of metals, such as nickel and chromium, were shown to form deposits in the epidermis; chromium is also known to change its valence as it transits through the skin (Samitz and Katz, 1963, 1964; Samitz et al., 1964); cadmium, zinc or copper form reversible complexes with metallothioneins (Cousins, 1985; Goering and Klaassen, 1984); others are subject to homeostatic control (zinc (Burch et al., 1975; Derry et al., 1983), calcium (Pillai et al., 1993; i.e., they are either retained or redistributed throughout the organism in function of prevailing body burdens).

More subtle parameters, such as oxidation state, ionic charge and radius, electropositivity, redox potential, hydration state, counter ion, pH, polarity, and polarizability, appear to play a role in determining the rate at which metal-based compounds cross the skin. With few exceptions, the permeation route of these compounds remains undefined. It thus appears unlikely that analysis of the scarce data so far available on these compounds will be encompassing enough to allow expression of the involved processes by a single algorithm soon, by which to predict metal diffusion through the skin.

The development of a QSAR model to predict the diffusion potential of electrolytes in general and metal compounds in particular thus still appears to lie beyond present possibilities due to the number of factors that control such a process and that are not yet fully understood (Table 5.8).

TABLE 5.8. *Factors determining rate and route of skin penetration by metal compounds*

Ionic charge	Interacts with counter-ions in salts and co-atoms in oxo complexes (CrO_4, MnO_4, etc), with fixed charges in tissue. Determines ion radius and thus charge density.
Valency	Mutable in transit through tissue (Fe, Cr, Ni, Au), comporting changes in solubility and membrane diffusivity.
Hydration sphere	Stable primary and secondary h. s. make the ion equivalent to a polyhydroxy structure of corresponding bulk, retarding transit.
pH	Variable degree of salt dissociation, hydration of amphoteric metals (Al, B) (Freimuth and Fisher, 1958).
Counter ion	Impacts on compound solubility and polarity.
Concentration	Rate of diffusion of a number of heavy metals is not commensurate with applied concentration. Absolute absorption of some can reach a plateau value, decrease with increasing concentration (Cr, Hg (Wahlberg, 1968)) or be entirely independent of concentration (Sr (Loeffler, 1951)).
Electron shell polarizability	Impacts on nature of the bond (ionic-covalent).
Solubility	Determines route of diffusion (intracellular–intercellular-appendageal)
Protein precipitation	Protein binding and depot formation in SC can result in formation of a secondary barrier (Ni, Cr).
Depot formation	Reversible (Zn) or irreversible (Hg, Ag, Au).
Homeostasis	Endogenous factors determine equilibrium values (rates of uptake and transit) of essential elements (Na, Zn, Ca).
Anatomical site	Variable SC thickness and shunt density (Maibach et al., 1971; Wester and Maibach, 1980).
Transcellular/shunt pathway	Preferred route determines rate of diffusion.
Vehicle	Thermodynamic activity or release of the permeant depends on polarity of carrier.

FORMATION OF METAL DEPOTS IN THE SKIN

Long-term exposure to several heavy metals through occupation, therapy, or lifestyle can result in their accumulation in the organism, and their preferential deposition in dermal tissues can become visible as a characteristic discoloration imparted to the skin. As these metals are not subject to significant natural processes of elimination, they accumulate over a lifetime and form permanent deposits in their elemental state, visualized histologically as metal particles. With the exception of arsenic, however, they don't seem to have any untoward effects on the organism other than to alter the natural coloring of the skin. Metals known to form significant deposits and thus affect the integument are listed in Table 5.3.

SKIN AND ITS APPENDAGES AS SECRETORY ORGAN

Besides urinary and biliary excretion, the skin and its appendages are also important routes of elimination of toxic agents, such as metals and their compounds. Thus, particularly electropositive (transition) metals such as Cr, Ni, Hg, Pb, with affinity for sulfhydryl groups, and thereby keratin-rich tissues, find significant excretion via skin through exfoliation, hair, sweat, nails, and ear wax. The skin thus plays a role in detoxification as well as in maintaining an appropriate balance of certain essential elements. Copper or iron, for instance, eliminated excessively under stress conditions or psoriasis, through sweat or desquamation, respectively, elude the kidney's homeostatic control, and losses can result in untoward effects, such as heat stroke or a state of iron deficiency (Cohn and Emmett, 1978; Morris, 1987).

Hair

With modern analytical methods, which now include atomic absorption and emission, neutron activation, energy dispersive x-ray fluorescence and emission, and particle-induced x-ray emission (PIXE), analysis of hair for some 30 elements, including metals, are now routinely possible, yielding information on nutritional status as well as environmental exposure (Lal et al., 1987; Moon et al., 1986). Predominant elemental levels in the hair of adult humans are presented in Table 5.9 (Iyengar, 1987). Elemental hair analysis has not been found useful in establishing a longitudinal record of the essential trace elements present in the organism, and thus its nutritional status, as many among them are subject to short-term fluctuations due to homeostatic control (Rivlin, 1983).

Sweat

After urine and feces, sweat is the most copious bodily secretion and a significant factor in detoxification besides homeostasis. Solute concentration in sweat changes in function of sweat rate and environmental humidity, local skin temperature, muscular

TABLE 5.9. *Range of trace metal levels occurring in adult human hair*

Metal	mg kg^{-1}
Arsenic	0.15–0.30
Cadmium	0.40–1.0
Chromium	0.30–0.80
Copper	0.25–0.40
Lead	2.0–20
Mercury	0.50–2.0
Nickel	0.02–0.20
Selenium	0.50–1.0
Zinc	150–250

TABLE 5.10. *Range of metals in exercise-induced human sweat*

Element	Range
Zn (μg/L)	400–1200
Cu (μg/L)	860–1600
Fe (μg/L)	40–550
Ni (μg/L)	40–80
Pb (μg/L)	40–120
Mn (μg/L)	10–30
Na (mEq/L)	10.6–28.6

activity, or pharmacological stimulation (e.g., by pilocarpine). The rate of component elimination follows idiosyncratic and element-specific patterns, such as acclimatization or, in the case of sodium, reabsorption (Cage and Dobson, 1965). Under extreme environmental conditions, extreme losses of essential nutrients, such as Na, K, Mg, Zn, Cu and Fe, can result, due in part to entrainment of epidermal cells with the flow of sweat (Consolazio et al., 1963; Morris, 1987). The average sodium content of sweat is given as 60 mM and that of potassium as 8 mM, but these values fluctuate significantly as a function of sweat rate, hormonal control or diet (Dobson and Sato, 1972). With respect to detoxification, the levels of lead found in human sweat were found to be equal to those in urine; for nickel and cadmium, the levels seen in sweat exceed those in urine (Cohn and Emmett, 1978; Suzuki, 1976).

Nails

Formed by the keratinization of epidermal cells, nails represent a relatively extensive longitudinal record of elements present in the skin. Analysis of nails for metals thus gives information on metal stores chronically present in the body, a useful diagnostic test for occupational safety purposes. Toenails are the preferable material because they are less exposed to exogenous contamination compared to fingernails or scalp hair. Especially levels of Ni, Pb, Cr, Mn, and V present there give a reliable record of occupational exposure (Mountain et al., 1955, Peters et al., 1991, Zaprianov et al., 1989).

Ear Wax

More recently, ear wax (cerumen) has also been investigated as a medium for the biological monitoring of metals (Krishnan and Que Hee, 1992; Wang et al., 1988). Elemental analysis of samples of human cerumen, a combination of secretions from sebaceous and ceruminous glands and of exfoliated epidermal cells, showed the presence of certain metals that also occur in sweat. The major metals found were K, Na, Ca, and Mg at the level of approximately 1 mg/g dry weight, but also a number of heavy metals of potential toxicological significance were identified in cerumen, in smaller but still quantifiable amounts: As, Au, Cd, Cr, Fe, In, Pb, Pt, Sb, Sn, Sr, Ti, and Tl. Because the ear canal is less exposed and subject to environmental contamination than skin, analysis of ear wax appears to be a potentially useful monitoring medium for occupational exposure to those elements.

OWN RESULTS ON STRATUM CORNEUM PERMEATION BY METAL SALTS (UNPUBLISHED DATA)

The limited and inconsistent database available so far for a number of toxic metals motivated investigation of metal salts into the SC, the dermis' premier barrier to xenobiotics. The salts of nickel, cadmium, and lead were chosen for closer scrutiny of the early-stage absorption dynamics when they are brought into contact with the skin in vivo, as it may occur in the work environment. The priority in focusing on these metal compounds was determined by their idiosyncratic properties.

Nickel

Anthropogenic nickel salts and alloys have turned the nontoxic, insoluble, naturally occurring forms of the element into the most pervasive allergen in alloys and its various salts (Christensen, 1990; Smit and Coenraads, 1993). Dermal and systemic toxicity of this essential trace element is cause for concern from the standpoint of public health, as a state of unremitting hypersensitivity is seen in individuals once they become sensitized (Menné, 1983). Also, the metal was chosen for study because of its apparent paradoxical properties: in vitro diffusion measurements of nickel salts show lag times of 50 hours, and the permeant finally reaches the receptor phase with apparent permeation constants ranging between 10^{-6} and 10^{-5} cm/hr (Fullerton et al., 1988; Fullerton and Hoelgaard, 1988). Diffusion rates, clinically observed as irritancy, also showed dependence on the nature of the counter ion (Wahlberg, 1996). Furthermore, there is evidence of accumulation and reservoir formation of the metal in the SC (Fullerton and Hoelgaard, 1988, Lloyd, 1980, Samitz and Katz, 1976). In contrast, with such observations stands the fact that since earliest epidemiological studies, that metal has been the premier contact sensitizer among the general population, an indication that it nevertheless diffuses promptly through the SC to reach the viable epidermis (Gola et al., 1992). Even mere contact with the metal itself, widely used in alloys, may elicit skin reactions in individuals that are sensitized.

Lead

A metal of significant reproductive toxicity, it is also seen to impair the central nervous system. The ability of lead to also complex with proteins appears to prevent skin penetration of its inorganic salts in any significant amount. Earlier skin penetration experiments in vitro conducted on human skin have shown that exposure times longer than 6 hours are necessary to detect the penetrant beyond the uppermost SC (Odintsova, 1975); in vitro, the Kp for lead acetate was determined at 10^{-7} cm/h. (Moore et al., 1980).

Paradoxically, application of lead nitrate or finely powdered lead metal to the forearm of human volunteers resulted in elevated lead levels in the contralateral arm within 2 hours, and these values reached 10 times background after 2 days. Similar, rapid increases in lead concentration were measured in serum and saliva (Lilley et al., 1988).

Cadmium

This metal, with no known biological function, is a human carcinogen and reproductive toxicant, and its environmental concentration is seen to be on a constant increase due to industrial activities. Thus, besides being an occupational hazard, cadmium represents considerable concern for the safety of the general public also.

In vitro experiments so far have shown that, on skin contact with contaminated water or soil, cadmium binds tightly to epidermal keratin and only insignificant amounts appear to penetrate the human skin (Wester et al., 1991).

Results

In order to determine the concentration profiles of the three metals in the SC following application of several of their salts, the standard experimental protocol of tape stripping was implemented and the removed SC analyzed for metal by inductively coupled plasma—atomic emission spectroscopy. Dosing and stripping was performed on the healthy skin of the volar forearm and intrascapular region of 5 human volunteers, avoiding re-application earlier than one month following previous stripping of the site. Salts applied were $NiCl_2 \cdot 6H_2O$, $NiSO_4 \cdot 6H_2O$, $Ni(NO3)_2 \cdot 6H_2O$, $(CH_3CO_2)_2Ni \cdot 4H_2O$, $(CH_3CO_2)_2Pb \cdot 3H_2O$, $Pb(ClO_4)_2 \cdot 3H_2O$, $Pb(NO_3)_2$, and $CdCl_2 \cdot 21/2 H_2O$.

At intervals of 30 min, 3, 12 and 24 hrs post dosing, the area of application was decontaminated by wiping with wet cotton swabs for separate analysis, to remove all un-absorbed test material from the skin prior to stripping. In order to correctly interpret the absorption profiles obtained for the metals in the SC layers through strip analysis, in a prior experiment the amount of SC removed in the process of sequential tape stripping had been analyzed gravimetrically. Beyond the third strip the weight of SC removed up to the twentieth strip remained constant ($r = 0.99$) (Schwindt et al., 1998).

100 μl of the salt solution in methanol were applied open on the volar forearm and the intrascapular area on the back of human volunteers, on a area of 2.83 cm^2, at concentrations ranging from 0.001% to 1% of the metal ion.

The concentration-versus-depth profiles obtained show that all three metals roughly follow the same pattern in their SC diffusion, and they allow the following conclusions:

Accumulation of metals in the SC was observed in function of exposure time, as the concentration gradient between the superficial and deeper layers of the SC increases, commensurate with duration of exposure.

While the concentration gradients of metal retained in the SC vary with counter ion, anatomical site, dose and exposure time, for all variables tested the profiles converge towards non-detectable levels (<20 ppb) beyond the 15th strip, independent of these parameters. Only notable exception is the concentratioin profile of nickel nitrate, which remains constant at 1% of dose from the 3rd through the 20th strip. The nitrate was also found to be the most lipophilic of the 4 nickel salts, determined by solubility in n-octanol at saturation.

The counter ion in metal salts plays a major role in their passive diffusion through the SC, suggestive of ion pairing. A similar dependence on couter ion was observed for the irritancy potential (skin permeability) for zinc (Lansdown, 1973).

Earlier observations confirmed that transcellular diffusion of these metal ions occurs to a minimal degree, and only after considerable lag times. Accumulation in the uppermost layers of the SC was also observed by others using proton-induced x-ray emission (micro-PIXE) analysis (Forslind et al., 1985, Malmqvist et al., 1987).

Mass balance calculations for nickel salts point to the fact that, particularly at higher concentrations (1%), up to 35% of the applied dose remains unaccounted for. This leads to the tentative conclusion that the metal chooses the alternate shunt pathway for diffusion and deposition to a significant degree, as was observed by other investigators using autoradiography or micro-PIXE analysis (Bos et al., 1985; Lloyd, 1980; Odintsova, 1975). In this context it is also important to note that tape stripping of skin does not result in the quantitative removal of hair follicles, whereby significant "disappearance" values of the metals may elude detection (Finlay and Marks, 1982; Finlay et al., 1982). Because ionic transport through the SC is low in general, these results would point to the importance of nonselective shunt diffusion.

REFERENCES

Adams, R. M. (1983): Contact dermatitis due to irritation and allergic sensitization. In: *Occupational Skin Disease*, pp. 1–26. Grune and Stratton, New York.

Adams, R. M. (1983): Metals. In: *Occupational Skin Disease*, pp. 204–237. Grune and Stratton, New York.

Anderson, B. D., Higuchi, W. I., and Raykar, P. V. (1988): Heterogeneity effects on permeability-partition coefficient relationships in human stratum corneum. *Pharm. Res.*, 5:566–573.

Anderson, B. D., Higuchi, W. I., and Raykar, P. V. (1989): Solute structure-permeability relationships in human stratum corneum. *J. Invest. Dermatol.*, 93:280–286.

Bagniefski, T., and Burnette, R. R. (1990): A comparison of pulsed and continuous current iontophoresis. *J. Controlled Release*, 11:113–122.

Baranowska-Dutkiewicz, B. (1981): Absorption of hexavalent chromium by skin in man. *Arch. Toxicol.*, 47:47–50.

Birmingham, D. J. (1979): Cutaneous reactions to chemicals. In: *Dermatology in General Medicine*. Edited by T. B. Fitzpatrick, pp. 995–1007. McGraw-Hill, New York.

Blank, I. H., and McAuliffe, D. J. (1985): Penetration of benzene through human skin. *J. Invest. Dermatol.*, 85:522–526.

Bos, A. J., van der Stap, C. C., Valkovic, V., Vis, R. D., and Verheul, H. (1985): Incorporation routes of elements into human hair: Implications for hair analysis used for monitoring. *Sci. Total. Environ.*, 42:157–169.

Bronaugh, R. L., Stewart, R. F., and Simon, M. (1986): Methods for in vitro percutaneous absorption studies. VII: Use of excised human skin. *J. Pharm. Sci.*, 75:1094–1097.

Briggs, G. G., Bromilow, R. H., Edmonson, R. H., and Johnston, M. (1977): Distribution coefficients and systemic activity. In: *Special Publication*, vol. 29, edited by the Chemical Society, pp. 129–134. Burlington House, London.

Bucks, D. A. W., Hostýnek, J. J., Hinz, R. S., and Guy, R. H. (1993): Uptake of two zwitterionic surfactants into human skin in vivo. *Toxicol. Appl. Pharmacol.*, 120:224–227.

Bunge, A. L., Flynn, G. L., and Guy, R. H. (1994): Predictive model for dermal exposure assessment. In: *Water Contamination and Health*. Edited by R. G. M. Wang, pp. 347–373. Marcel Dekker, Inc., New York.

Burch, R. E., Hahn, H. K. J., and Sullivan, J. F. (1975): Newer aspects of the roles of zinc, manganese and copper in human nutrition. *Clin. Chem.*, 21:501–520.

Burge, K. M., and Winkelmann, R. K. (1970): Mercury pigmentation: An electron microscopic study. *Arch. Dermatol.*, 102:51–61.

Burnette, R. R., and Bagniefski, T. M. (1988): Influence of constant current iontophoresis on the impedance and passive Na+ permeability of excized nude mouse skin. *J. Pharm. Sci.*, 77:492–497.

Burrows, D. (1972): Prognosis in industrial dermatitis. *Br. J. Dermatol.*, 87:145.

Calvery, H. O., Draize, J. H., and Laug, E. P. (1946): The metabolism and permeability of normal skin. *Physiol. Rev.*, 26:495–540.

Cage, G. W., and Dobson, R. L. (1965): Sodium secretion and reabsorption in the human eccrine sweat gland. *J. Clin. Invest.*, 44:1270–1276.

Christensen, O. B. (1978): An exogenous variety of pseudoxanthoma elasticum in old farmers. *Acta Derm. Venereol. (Stockh)*, 58:319–321.

Christensen, O. B. (1990): Nickel dermatitis: An update. *Dermatol. Clin.*, 8:37–40.

Cohn, J. R., and Emmett, E. A. (1978). The excretion of trace metals in human sweat. *Ann. Clin. Lab. Sci.*, 8:270–275.

Consolazio, C. F., Matoush, L. O., Nelson, R. A., Harding, R. S., and Canham, J. E. (1963): Excretion of sodium, potassium, magnesium and iron in human sweat and the relation of each to balance and requirements. *J. Nutr.*, 79:407–415.

Corbett, G. E., Finley, B. L., Paustenbach, D. J., and Kerger, B. D. (1997): Systemic uptake of chromium in human volunteers following dermal contact with hexavalent chromium (22mg/L). *Journal of Exposure Analysis and Environmental Epidemiology*, 7:179–189.

Cox, A. J., and Marich, K. W. (1973): Gold in the dermis following gold therapy for rheumatoid arthritis. *Arch. Dermatol.*, 108:655–657.

Cousins, R. J. (1985): Absorption, transport, and hepatic metabolism of copper and zinc: Special reference to metallothionein and ceruloplasmin. *Physiol. Rev.*, 65:238–309.

Derry, J. E., McLean, W. M., and Freeman, J. B. (1983): A study of the percutaneous absorption from topically applied zinc oxide ointment. *J. Parenter. Enter. Nutr.*, 7:131–135.

Dobson, R. L., and Sato, K. (1972): The secretion of salt and water by the eccrine sweat gland. *Arch. Dermatol.*, 105:366–370.

Dutkiewicz, T. (1977): Experimental studies on arsenic absorption routes in rats. *Environ. Health Perspect.*, 19:173–177.

Dutkiewicz, T., Dutkiewicz, B., and Balcerska, I. (1971): Dynamics of organ and tissue distribution of selenium after intragastric and dermal administration of sodium selenite. *Bromatol. Chem. Toksykol.*, 4:475–481.

Dutkiewicz, T., and Przechera, M. (1966): Estimation of chromium compounds absorption through the skin. *Ann. Acad. Med. Lodzensis*, 8:189–193.

Emilson, A., Lindberg, M., and Forslind, B. (1993): The temperature effect on in vitro penetration of sodium lauryl sulfate and nickel chloride through human skin. *Acta Derm. Venereol. (Stockh)*, 73:203–207.

Everett, M. A. (1979): Metal discolorations. In: *Metal Discolorations*. Edited by D. J. Demis, pp. 4. Harper and Row, Hagerstown.

Fairlie, D. P., and Whitehouse, M. W. (1991): Transdermal delivery of inorganic complexes as metal drugs or nutritional supplements. *Drug. Des. Deliv.*, 8:83–102.

Feldmann, R. J., and Maibach, H. I. (1970): Absorption of some organic compounds through the skin of man. *J. Invest. Dermatol.*, 54:399–404.

Feldmann, R. J., and Maibach, H. I. (1974): Percutaneous penetration of some pesticides and herbicides in man. *Toxicol. Appl. Pharmacol.*, 28:126–132.

Feldmann, R. J., and Maibach, H. I. (1969): Percutaneous penetration of steroids in man. *J. Invest. Dermatol.*, 52:89–94.

Fick, A. E. (1855): On liquid diffusion. *Philosophical Magazine*, 10:30–39.

Finlay, A., and Marks, R. (1982): Determination of corticosteroid concentration profiles in the stratum corneum using skin surface biopsy technique. *Br. J. Dermatol.*, 107:33.

Finlay, A. Y., Marshall, R. J., and Marks, R. (1982): A fluorescence photographic photometric technique to assess stratum corneum turnover rate and barrier function in vivo. *Br. J. Dermatol.*, 107:35–42.

Fitzgerald, J. J., and Brooks, T. (1979): A new cell for in vitro skin permeability studies: Chromium (III)/(VI) human epidermis investigations. *J. Invest. Dermatol.*, 72:198.

Flynn, G. L. (1990): Physicochemical determinants of skin absorption. In: *Principles of Route-to-Route Extrapolation for Risk Assessment.* Edited by T. R. Gerrity and C. J. Henry, pp. 93–127. Elsevier Science Publishing Co., New York.

Forslind, B., Grundin, T. G., Lindberg, M., Roomans, G. M., and Werner, Y. (1985): Recent advances in X-ray microanalysis in dermatology. *Scanning Electron. Microsc.*, 2:687–695.

Freimuth, H. C., and Fisher, R. S. (1958): The effect of pH and the presence of other elements in solution on the absorption of boron. *J. Invest. Dermatol.*, 30:83–84.

Fullerton, A., Andersen, J. R., Hoelgaard, A., and Menné, T. (1986): Permeation of nickel salts through human skin in vitro. *Contact Dermatitis*, 15:173–177.

Fullerton, A., Andersen, J. R., and Hoelgaard, A. (1988): Permeation of nickel through human skin in vitro: Effect of vehicles. *Br. J. Dermatol.*, 118:509–516.

Fullerton, A., and Hoelgaard, A. (1988): Binding of nickel to human epidermis in vitro. *Br. J. Dermatol.*, 119:675–682.

Gammelgaard, B., Fullerton, A., Avnstorp, C., and Menné, T. (1992): Permeation of chromium salts through human skin in vitro. *Contact Dermatitis*, 27:302–310.

Gilboa, R., Al-Tawil, N. G., and Marcusson, J. A. (1988): Metal allergy in cashiers. *Acta Dermato-Venereologica (Stockh)*, 68:317–324.

Goering, P. L., and Klaassen, C. D. (1984): Tolerance to cadmium-induced toxicity depends on presynthesized metallothionein in liver. *J. Toxicol. Environ. Health*, 14:803–812.

Gola, M., Sertoli, A., Angelini, G., Ayala, F., Deledda, S., Goitre, M., Lisi, P., Meneghini, C. L., Pigatto, P., Rafanelli, A., Santucci, B., Schena, D., Valsecchi, R., and Zavaroni, C. (1992): GIRDCA data bank for occupational and environmental contact dermatitis (1984 to 1988). *Am. J. Contact Dermat.*, 3:179–188.

Granstein, R. D., and Sober, A. J. (1981): Drug- and heavy metal-induced hyperpigmentation. *J. Am. Acad. Dermatol.*, 5:1–18.

Gummer, C. L., Hinz, R. S., and Maibach, H. I. (1987): The skin penetration cell: An update. *Int. J. Pharm.*, 40:101–104.

Hadgraft, J., and Rideout, G. (1987): Development of model membranes for percutaneous absorption measurements. I: Isopropyl myristate. *Int. J. Pharm.*, 39:149–156.

Harr, J. R., and Muth, O. H. (1972): Selenium poisoning in domestic animals and its relationship to man. *Clin. Toxicol.*, 5:175–186.

Hostýnek, J. J., Hinz, R. S., Lorence, C. R., Price, M., and Guy, R. H. (1993): Metals and the skin. *Crit. Rev. Toxicol.*, 23:171–235.

Hursh, J. B., Clarkson, T. W., Miles, E. F., and Goldsmith, L. A. (1989): Percutaneous absorption of mercury vapor by man. *Arch. Environ. Health.*, 44:120–127.

Iyengar, V. (1987): Reference values for trace element concentrations in whole blood, serum, hair, liver, milk, and urine specimens from human subjects. In: *International Symposium on Trace Elements in Man and Animals.* Edited by L. S. Hurley, pp. 535–537. Plenum Press, New York.

Kanerva, L., Estlander, T., and Jolanki, R. (1998): Bank clerk's occupational allergic nickel and cobalt contact dermatitis from coins. *Contacty Dermatitis*, 38:217–218.

Klaassen, C. D. (1976): Absorption, distribution, and excretion of zinc pyridinethione in rabbits. *Toxicol. Appl. Pharmacol.*, 35:581–587.

Kasting, G. B., and Bowman, L. A. (1990): DC electrical properties of frozen, excised human skin. *Pharm. Res.*, 7:134–143.

Kasting, G. B., and Bowman, L. A. (1990): Electrical analysis of fresh, excised human skin: A comparison with frozen skin. *Pharm. Res.*, 7:1141–1146.

Kasting, G. B., and Robinson, P. J. (1993): Can we assign an upper limit to skin permeability? *Pharm. Res.*, 10:930–931.

Kasting, G. B., Smith, R. L., and Cooper, E. R. (1987): Effect of lipid solubility and molecular size on percutaneous absorption. In: *Skin Pharmacokinetics*. Edited by B. Shroot and H. Schaefer, pp. 138–153. Karger, Basel.

Krishnan, U., and Que Hee, S. S. (1992): Ear wax: A new biological monitoring medium for metals? *Bull. Environ. Contam. Toxicol.*, 48:481–486.

Lal, G., Sidhu, N. P. S., Singh, I., Mittal, V. K., and Sahota, H. S. (1987): Neutron activation analysis of trace elements in human hair: Effect of dietary and environmental factors. *Nucl. Med. Biol.*, 14:499–501.

Lamar, L. M., and Bliss, B. O. (1966): Localized pigmentation of the skin due to topical mercury. *Arch. Dermatol. Syphilol.*, 93:450–453.

Lansdown, A. B. G. (1973): Production of epidermal damage in mammalian skins by some simple aluminium compounds. *Br. J. Dermatol.*, 89:67–76.

Lee, S. M., and Lee, S. H. (1994): Generalized argyria after habitual use of AgNO3. *J. Dermatol.*, 21:50–53.

Levantine, A., and Almeyda, J. (1973): Drug induced changes in pigmentation. *Br. J. Dermatol.*, 89:105–112.

Lilley, S. G., Florence, T. M., and Stauber, J. L. (1988): The use of sweat to monitor lead absorption through the skin. *Sci. Total Environ.*, 76:267–278.

Lloyd, G. K. (1980): Dermal absorption and conjugation of nickel in relation to the induction of allergic contact dermatitis: Preliminary results. In: *International Conference on Nickel Toxicology*. Edited by S. S. Brown and F. W. Sunderman, pp. 145–148. Academic Press, London.

Loeffler, R. K. (1951): III. Effect of Concentration and Solubility on Tracer Absorption of Sr^{89} From Intact and Abraded Skin and From Muscle in the Rat: Validity of Surface Counter Technique. *US Naval Radiological Defense Laboratory Report AD-290 (B)*.

Luders, G., Fischer, H., and Hensel, U. (1968): Hydrargyrosis cutis mit allgemeinen Vergiftungserscheinungen nach langdauernder Anwendung quecksilberhaltiger Kosmetica. *Hautarzt.*, 19:61–65.

Maibach, H. I., Feldmann, R. J., Milby, T. H., and Serat, W. F. (1971): Regional variation in percutaneous penetration in man. *Arch. Environ. Health*, 23:208–211.

Mali, J. W. H., van Kooten, W. J., van Neer, F. C. J., and Spruit, D. (1964): Quantitative aspects of chromium sensitization. *Acta Derm. Venereol. (Stockh)*, 44:44–48.

Malmqvist, K. G., Forslind, B., Themner, K., Hylten, G., Grundin, T., and Roomans, G. M. (1987): The use of PIXE in experimental studies of the physiology of human skin epidermis. *Biol. Trace. Element. Res.*, 12:297–308.

Marzulli, F. N., and Brown, D. W. C. (1972): Potential systemic hazards of topically applied mercurials. *J. Soc. Cosmet. Chem.*, 23:875–886.

Matsumura, T., Kumakiri, M., Ohkawara, A., Himeno, H., Numata, T., and Adachi, R. (1992): Detection of selenium in generalized and localized argyria: Report of four cases with X-ray microanalysis. *J. Dermatol.*, 19:87–93.

Menné, T. (1983): Reactions to systemic exposure to contact allergens. In: *Dermatotoxicology*. Edited by F. N. Marzulli and H. I. Maibach, pp. 483–499. Hemisphere Publishing Corporation, Washington.

Minato, A., Fukuzawa, H., Hirose, S., and Matsunaga, Y. (1967): Radioisotopic studies on percutaneous absorption. I: Absorption of water-soluble substances from hydrophilic and absorption ointments through mouse skin. *Chem. Pharm. Bull. (Tokyo)*, 15:1470–1477.

Moon, J., Smith, T. J., Tamaro, S., Enarson, D., Fadl, S., Davison, A. J., and Weldon, L. (1986): Trace metals in scalp hair of children and adults in three Alberta Indian villages. *Sci. Total. Environ.*, 54:107–125.

Moore, M. R., Meredith, P. A., Watson, W. S., Sumner, D. J., Taylor, M. K., and Goldberg, A. (1980): The percutaneous absorption of lead-203 in humans from cosmetic preparations containing lead acetate, as assessed by whole-body counting and other techniques. *Food & Cosmet. Toxicol.*, 18:399–405.

Morris, E. R. (1987): Iron II: Iron metabolism. In: *Trace Elements in Human and*, vol. 1. *Animal Nutrition*, Edited by W. Mertz, pp. 91–108. Academic Press, San Diego.

Mountain, J. T., Stockell, F. R. J., and Stokinger, H. E. (1955): Studies in vanadium toxicology III: Fingernail cystine as an early indicator of metabolic changes in vanadium workers. *AMA Arch. Ind. Health*, 12:494–502.

Nørgaard, O. (1954): Investigations with radioactive Ag^{111} into the resorption of silver through human skin. *Acta Derm. Venereol. (Stockh)*, 34:415–419.

Nørgaard, O. (1955): Investigations with radioactive Ni^{57} into the resorption of nickel through the skin in normal and in nickel-hypersensitive persons. *Acta Derm. Venereol. (Stockh)*, 35:111–117.

Odintsova, N. A. (1975): Permeability of the epidermis for lead acetate according to fluorescence and electron-microscopic studies. *Vestn. Dermatol. Venerol.*, 19–24.

Pariser, R. J. (1978): Generalized argyria. *Arch. Dermatol.*, 114:373–377.

Peters, K., Gammelgaard, B., and Menné, T. (1991): Nickel concentrations in fingernails as a measure of occupational exposure to nickel. *Contact Dermatitis*, 25:237–241.

Phipps, J. B., Padmanabhan, R. V., and Lattin, G. A. (1989): Iontophoretic delivery of model inorganic and drug ions. *J. Pharm. Sci.*, 78:365–369.

Pillai, S., Menon, G. K., Bikle, D. D., and Elias, P. M. (1993): Localization and quantitation of calcium pools and calcium binding sites in cultured human keratinocytes. *Journal of Cellular Physiology*, 154:101–112.

Pollycove, M. (1966): Iron metabolism and kinetics. *Semin. Hematol.*, 3:235–298.

Potts, R. O., and Guy, R. H. (1995): A predicitve algorithm for skin permeability: The effects of molecular size and hydrogen bond activity. *Pharm. Res.*, 12:1628–1633.

Potts, R. O., Guy, R. H., and Francoeur, M. L. (1992): Routes of ionic permeability through mammalian skin. *Solid State Ionics*, 53–56:165–169.

Pugh, W. J., and Hadgraft, J. (1994): Ab initio prediction of human skin permeability coefficients. *Int. J. Pharm.*, 103:163–178.

Rasetti, L., Cappellaro, F., and Gaido, P. (1961): Contribution to the study of saturnism by inhibited oils. *Rass. Med. Ind. Igiene. Lavoro.*, 30:71–75.

Requena, C. L., Sanchez, L. M., and Coca, M. S. (1985): Argiria generalizada: Estudio clinico-histologico, histoquimico y ultrastructural. *Actas Derm.-Sif.*, 76:533–540.

Ridout, G. R., Hinz, R. S., Hostýnek, J. J., Reddy, A. K., Wiersema, R. J., Hodson, C. D., Lorence, C. R., and Guy, R. H. (1991): The effects of zwitterionic surfactants on skin barrier function. *Fundam. Appl. Toxicol.*, 16:41–50.

Rivlin, R. S. (1983): Misuse of hair analysis for nutritional assessment. *Am. J. Med.*, 75:489–493.

Roberts, M. S., Anderson, R. A., Swarbrick, J., and Moore, D. E., 1978. The percutaneous absorption of phenolic compounds: The mechanism of diffusion across the stratum corneum. *J. Pharm. Pharmacol.*, 30:486–490.

Sage, B. H., Hoke, R. A., McFarland, A. C., and Kowalczyk, K. (1993): The importance of skin pH in the iontophoresis of peptides. In: *Predictions of Percutaneous Penetration*, vol. 3b. Edited by K. R. Brain, V. J. James, and K. A. Walters, pp. 410–418. STS, Cardiff.

Samitz, M. H., and Katz, S. (1963): Preliminary studies on the reduction and binding of chromium with skin. *Arch. Dermatol.*, 88:816–819.

Samitz, M. H., and Katz, S. (1964): A study of the chemical reactions between chromium and skin. *J. Invest. Dermatol.*, 43:35–43.

Samitz, M. H., and Katz, S. A. (1976): Nickel–epidermal interactions: Diffusion and binding. *Environ. Res.*, 11:34–39.

Scheuplein, R. J., and Blank, I. H. (1971): Permeability of the skin. *Physiol. Rev.*, 51:707–747.

Schnuch, A., Geier, J., Uter, W., Frosch, P. J., Lehmacher, W., Aberer, W., Agathos, M., Arnold, R., Fuchs, T., and Laubstein, B. (1997): National rates and regional differences in sensitization to allergens of the standard series: Population-adjusted frequencies of sensitization (PAFS) in 40,000 patients from a multicenter study (IVDK). *Contact Dermatitis*, 37:200–209.

Schwindt, D. A., Wilhelm, K. P., and Maibach, H. I. (1998): Water diffusion characteristics of human *stratum corneum* at different anatomical sites *in vivo*. *J. Invest. Dermatol.*, 111:385–389.

Skog, E., and Wahlberg, J. E. (1964): A comparative investigation of the percutaneous absorption of metal compounds in the guinea pig by means of the radioactive isotopes: ^{51}Cr, ^{58}Co, ^{65}Zn, ^{110m}Ag, ^{115m}Cd, ^{203}Hg. *J. Invest. Dermatol.*, 43:187–192.

Smit, H. A., and Coenraads, P. J. (1993): Epidemiology of contact dermatitis. In: *Monographs in Allergy*, vol. 31. Edited by M. L. Burr, pp. 29–48. Karger, Basel.

Sneddon, I. B., and Archibald, R. M. (1958): Traumatic calcinosis of the skin. *Br. J. Dermatol.*, 70:211–214.

Stauber, J. L., Florence, T. M., Gulson, B. L., and Dale, L. S. (1994): Percutaneous absorption of inorganic lead compounds. *Sci. Total Environ.*, 145:55–70.

Suzuki, T. (1976): Study on dermal excretion of metallic elements (Na, K, Ca, Mg, Fe, Mn, Zn, Cu, Cd, Pb). In: *Proceedings of the Tenth International Congress of Nutrition*, pp. 568. International Congress of Nutrition, Kyoto.

Wahlberg, J. E. (1965a): Percutaneous absorption of sodium chromate (51Cr), cobaltous (58Co), and mercuric (203Hg) chlorides through excised human and guinea pig skin. *Acta Derm. Venereol. (Stockh)*, 45:415–426.

Wahlberg, J. E. (1965b): Percutaneous absorption of trivalent and hexavalent chromium. *Arch. Dermat.*, 92:315–318.

Wahlberg, J. E. (1965c): Percutaneous toxicity of metal compounds. *Arch. Environ. Health* 11:201–204.

Wahlberg, J. E. (1965d): Some attempts to influence the percutaneous absorption rate of sodium (^{22}Na) and mercuric (^{203}Hg) chlorides in the guinea pig. *Acta Derm. Venereol. (Stockh)*, 45:335–343.

Wahlberg, J. E. (1965e): Percutaneous toxicity of metal compounds. *Arch. Environ. Health*, 11:201–204.

Wahlberg, J. E. (1965f): pH-Changes in mercuric chloride solutions in contact with human and guinea pig skin, in vivo and in vitro. *Acta Derm. Venereol. (Stockh)*, 45:329–334.

Wahlberg, J. E. (1968a): Percutaneous absorption of radioactive strontium chloride Sr 89 (^{89}SrCl$_2$). A comparison with 11 other metal compounds. *Arch. Dermatol.*, 97:336–339.

Wahlberg, J. E. (1968b): Percutaneous absorption from chromium (^{51}Cr) solutions of different pH, 1.4–12.8. *Dermatologica*, 137:17–25.

Wahlberg, J. E. (1970): Percutaneous absorption of trivalent and hexavalent chromium (^{51}Cr) through excised human and guinea pig skin. *Dermatologica*, 141:288–296.

Wahlberg, J. E. (1971): Vehicle role of petrolatum: Absorption studies with metallic test compounds in guinea pigs. *Acta Derm. Venereol. (Stockh)*, 51:129–134.

Wahlberg, J. E. (1996): Nickel: The search for alternative, optimal and non-irritant patch test preparations: Assessment based on laser Doppler flowmetry. *Skin Research and Technology*, 2:136–141.

Walker, W. R., Reeves, R. R., Brosnan, M., and Coleman, G. D. (1977): Perfusion of intact skin by a saline solution of bis(glycinato) copper(II). *Bioinorg. Chem.*, 7:271–276.

Wang, X. Q., Gao, P. Y., Lin, Y. Z., and Chen, C. M. (1988): Studies on hexachlorocyclohexane and DDT contents in human cerumen and their relationships to cancer mortality. *Biomed. Environ. Sci.*, 1:130–151.

Wester, R. C., and Maibach, H. I. (1980): Regional variation in percutaneous absorption. In: *Regional Variation in Percutaneous Absorption*. Edited by R. L. Bronaugh, and H. I. Maibach, pp. 111–120. Marcel Dekker, Inc., New York.

Wester, R. C., Bucks, D. A. W., and Maibach, H. I. (1991): Percutaneous absorption of hazardous substances from soil. In: *Dermatotoxicology*. Edited by F. Marzulli, and H. I. Maibach, pp. 110–122. Hemisphere Publishing Corporation, New York.

Wester, R. C., Hui, X., Hartway, T., Maibach, H. I., Bell, K., Schnell, M. J., Northington, D. J., Strong, P., and Culver, B. D. (1998): In vivo percutaneous absorption of boric acid, borax, and disodium octaborate tetrahydrate in humans compared to in vitro absorption in human skin from infinite and finite doses. *Toxicological Sciences*, 45:42–51.

Wester, R. C., Maibach, H. I., Sedik, L., Melendres, J., DiZio, S., and Wade, M. (1992): In vitro percutaneous absorption of cadmium from water and soil into human skin. *Fundam. Appl. Toxicol.*, 19:1–5.

Wester, R. C., Maibach, H. I., Sedik, L., Melendres, J., and Wade, M. (1993): In vivo and in vitro percutaneous absorption and skin decontamination of arsenic from water and soil. *Fundam. Appl. Toxicol.*, 20:336–340.

Wheeland, R. G., and Roundtree, J. M. (1985): Calcinosis cutis resulting from percutaneous penetration and deposition of calcium. *J. Am. Acad. Dermatol.*, 12:172–175.

Zaprianov, Z., Tsalev, D., Georgieva, R., Kaloianova, F., and Nikolova, V. (1989): New toxicokinetic exposure tests for metals based on atomic absorption analysis of the nails. *Probl. Khig.*, 14:75–97.

Toxicology of Skin
Edited by Howard I. Maibach
Copyright © 2001 Taylor & Francis

6

Irritant Dermatitis: Subthreshold Irritation

Ethel Tur

Department of Dermatology, Tel Aviv Sourasky Medical Center, and Tel Aviv University, Sackler School of Medicine, Tel Aviv, Israel

- · **Introduction**
- · **The Importance of the Assessment of the Threshold of Irritation**
- · **Cumulative Effect of Subthreshold Irritation**
- · **Combination of Irritants**
- · **Important Variables for Producing an Irritant Reaction**
- · **Age, Gender, Hormonal Status, and Regional Variations**
- · **Seasonal Variations**
- · **Disruption of the Barrier Function**
- · **Dermatological and Systemic Diseases**
- · **Mechanisms**
- · **Conclusion**

INTRODUCTION

Skin reactivity to an irritant depends on the inherent irritation potential of the substance, its concentration, duration and frequency of exposure, and its ability to penetrate the tissue, in addition to the reactivity of the individual as well as extrinsic factors. Repeated or combined exposure to subthreshold concentrations of irritants may cause a clinically apparent irritation. This indicates that subthreshold concentrations have a certain affect, although it is visible only after repetition. Elimination of the offending agent and protection from further exposure are important in both diagnosis and management of contact dermatitis. Identification of the relevant offending agent is difficult when dealing with an effect of subthreshold concentrations of irritants. Visual assessment is not adequate for the detection of subthreshold irritation, but several bioengineering tools are now available and are instrumental in this field.

THE IMPORTANCE OF THE ASSESSMENT OF
THE THRESHOLD OF IRRITATION

Short-term application of subthreshold concentrations of sodium lauryl sulphate (SLS) alters transepidermal water loss and laser Doppler readings, indicating that subclinical changes occur (Dominique et al., 1987). Identifying and characterizing subthreshold irritation is important, as under certain environmental or other circumstances these concentrations may cause clinical irritation.

In addition, allergy patch testing is sometimes difficult or impossible to perform with substances that are irritants, and the assessment of the threshold of irritation is crucial in order to identify and distinguish the allergen. This may be illustrated by calcipotriol studies. Calcipotriol is a mild irritant of the noncorrosive type (i.e., with no influence on the skin barrier). Allergy may also occasionally occur with calcipotriol. To assess the dose-irritation relationship and establish a non-irritant patch test for calcipotriol allergy patch testing, 180 healthy volunteers who were never previously exposed to calcipotriol were tested. In addition to clinical readings of patch tests with a calcipotriol dilution series, bioengineering methods were employed. These included chromametry and laser Doppler flowmetry (Tur, 1999; Tur, 1994). This allowed the assessment of both the optimal concentration of allergy patch testing for calcipotriol and the score that reflects allergy (Fullerton et al., 1998).

Similarly, optimal formaldehyde concentrations for allergy testing were assessed in an attempt to avoid irritancy (Trattner et al., 1998). Formaldehyde is widely used, both in consumer products and in the industry, and such an assessment allows the use of subthreshold irritancy concentrations when testing for allergic contact dermatitis.

CUMULATIVE EFFECT OF SUBTHRESHOLD IRRITATION

Subthreshold irritation is important in every day situations. Repeated exposure to subthreshold concentrations of irritants may develop into perceptible irritation. Such chronic exposure to subthreshold irritation is the most frequent occupational and environmental insult. Studies that take this into consideration may give better treatment options. These studies use a series of bioengineering techniques that provide the means to discriminate subclinical changes. For example, protective creams were tested in a repetitive irritation test, and the reactions were quantified by erythema score, transepidermal water loss, and chromametry. A 1-week period of cumulative irritation was found sufficient for the evaluation of protective creams against 4 irritants sodium lauryl sulphate (SLS), NaOH, lactic acid, toluene (Wigger-Alberti et al., 1998). To answer the need for new predictive tests in order to assess the efficacy of barrier creams, shielded corneosurfametry and corneoxenometry were introduced as novel ex vivo bioassays after four irritants (Goffin et al., 1998).

Cumulative irritation tests are also important for the assessment of the irritation potential of topical formulations in the dermatological armamentarium. Thus, cumulative irritation test was applied to compare several tretinoin preparations and

adapalene (Webster, 1998). The classical 21-day cumulative irritation test usually results in irritation in all tretinoin preparations. Therefore, a modified test was utilized
and could distinguish between the irritation potential of tretinoin formulations and
certain nontretinoin acne treatments. A significantly less irritation was found with the
new formulation that incorporates microsponge technology.

COMBINATION OF IRRITANTS

Depending on the nature of the individual substances, the combination of irritants at
subthreshold concentrations may yield the following:

1. A simple summation of the irritation potential. For instance, volatile organic chemicals, at low exposure concentrations, elicited an additive effect of airborne irritancy
   (Alarie et al., 1996). This study, however, used concentrations above the threshold.
2. An augmentation potential of combinations of subthreshold concentrations of irritants. Such an augmentation was detected in the following study (Tur et al., 1995).
   The cutaneous response to repetitive applications of subthreshold concentrations
   of irritants was assessed by objective noninvasive measurements in addition to visual scoring. Ten subjects were patch tested to determine the minimal irritant dose
   (MID) to dilutions of aqueous SLS and lactic acid. Each subject was then patch-
   tested with half the MID of each chemical for a period of 24 hours. At 25 hours,
   additional patches were applied over the same sites, containing five successive
   twofold dilutions of each irritant, starting with half the MID. Each chemical was
   thus applied onto itself and onto the other chemical as well. In addition, combinations of half the MID of each substance and twofold dilutions of the other were also
   applied for two consecutive periods of 24 hours. At 25 hours and 49 hours the cutaneous changes were monitored by using the noninvasive methods of laser Doppler
   flowmetry and reflectance spectrophotometry, in addition to visual scoring. Significant differences between the various patch-testing combinations were detected by
   the instrumentation only. The reaction over sites treated with half the MID of one
   substance increased upon an additional 24 hour period of occlusion with half the
   MID of the other substance ($p < 0.05$), and in several occasions even with a quarter
   of the MID (Figures 6.1 and 6.2). Repeated application of certain combinations of
   the substances resulted in an elevated blood flow as well. These results suggest an
   augmentation of the response. Although no visual alterations could be detected, the
   noninvasive instruments were able to detect cutaneous responses to consecutive
   applications of subthreshold concentrations of various combinations of two chemical irritants. These responses suggest that subthreshold concentrations of irritants
   may alter the skin, making it susceptible to a further subthreshold insult, resulting
   in a detectable reaction. Some possible mechanisms by which the first irritant may
   prepare the skin for reaction with the consecutive insults are impairment of the
   physical properties of the skin, such as impairment of the barrier function, modification of the inflammatory response, or alteration of the metabolic pathways. This
   study exhibits the importance of instrumentation in the assessment of the effect

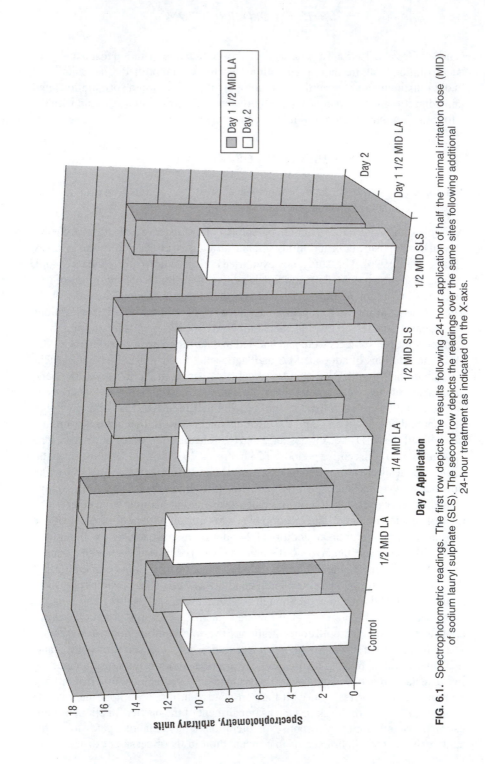

FIG. 6.1. Spectrophotometric readings. The first row depicts the results following 24-hour application of half the minimal irritation dose (MID) of sodium lauryl sulphate (SLS). The second row depicts the readings over the same sites following additional 24-hour treatment as indicated on the X-axis.

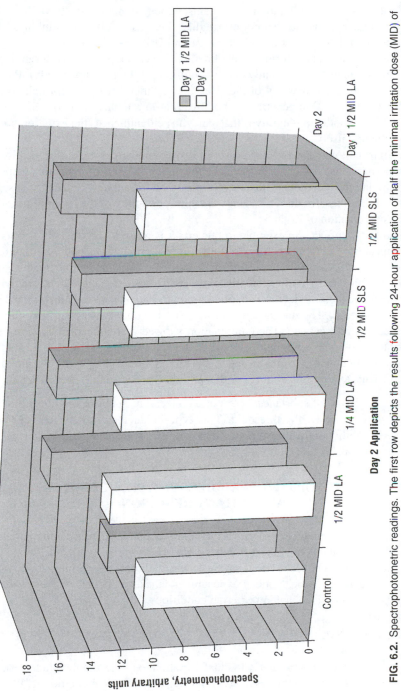

FIG. 6.2. Spectrophotometric readings. The first row depicts the results following 24-hour application of half the minimal irritation dose (MID) of lactic acid (LA). The second row depicts the readings over the same sites following additional 24-hour treatment as indicated on the X-axis.

of subthreshold concentrations of irritants, which are not clinically visible. Other clinical indications are also not useful, like the sensory effect of stinging, which did not correlate with irritancy (Basketter and Griffiths, 1993).

Salicylic acid in the formulation of a dithranol cream promoted dithranol irritation as evaluated both visually and by colorimetry (Goffin et al., 1998). Possible mechanisms are promotion of the release of dithranol from the formulation and enhancement of skin penetration, both leading to a higher skin concentration of dithranol. Another possibility is that salicylic acid influences the irritation process of dithranol itself.

An irritant may enhance the susceptibility to another by impairing the physical properties of the skin or the barrier function, modifying the inflammatory response, or altering the metabolic pathways. Similarly, an irritant may reduce the susceptibility to another irritant. This brings the third possibility of the result of the action of a combination of irritants:

3. A protective effect. An example to this effect is given by mixing surfactants that compete with each other. All current detergent formulations contain mixtures of surfactants. These exhibit lower acute irritation potential than predicted by simple summation of the irritation potential of the individual active ingredients. Individual irritants and their mixture were applied for up to 4 hours in a study involving 31 human volunteers. The substances used were sodium dodecyl sulphate and dimethyl dodecyl amido betaine. The combination showed a significant reduction in the irritation potential. The authors proposed that the critical event leading to skin irritation is binding to skin protein and that in mixed surfactant systems, the individual surfactants exhibit less affinity for this protein (Hall-Manning et al., 1998).

Thus, a mixture of substances, each at a concentration above the threshold, may functionally be below the threshold. This type of interaction between substances draws a possibility of treating or preventing irritation by exposing the skin to the right sequence of substances.

IMPORTANT VARIABLES FOR PRODUCING AN IRRITANT REACTION

Many variables combine to produce an irritant reaction, and insufficiency of any of them may pull the reaction under the threshold. These variables include the potency of the substance, its concentration, physical state, and route of administration, as well as the frequency, duration, mode of exposure, and the area of contact. Other variables depend on the individual and on external conditions.

AGE, GENDER, HORMONAL STATUS, AND REGIONAL VARIATIONS

It is accepted that marked interindividual variation exists in the threshold response to irritants. Differences may result from various physiological conditions, like age, race, gender, hormonal status, and also regional variations.

Age is a factor. Older subjects manifested a delayed and decreased cumulative irritation reaction to consecutive SLS open applications compared to young ones; also, their recovery was prolonged. This was assessed by transepidermal water loss, stratum corneum capacitance, desquamation index, and skin roughness parameters (Schwindt et al., 1998). A recent article reviews the present knowledge of the epidermal permeability barrier as it pertains to irritant dermatitis and discusses how altered barrier function may affect irritant dermatitis in the aged (Ghadially, 1998).

Physiological states, like hormonal changes at menopause, may affect the local conditions, like sweat, sebum, and temperature, and these may affect irritancy. Sweat production upon psychological stimulation decreased significantly after menopause (Ohta et al., 1998). In addition, after menopause, the sebum content of the forehead decreased significantly, but that of the subocular region remained unchanged. Local temperature changes were also not uniform throughout the body. The perimenopausal period was associated with increased skin permeability and vascular responsiveness, both potentially important in lowering the threshold of irritation.

Skin type is another consideration. The threshold for acute irritancy, defined as the lowest concentration of SLS, did not differ with skin type. This was shown on 110 subjects divided into skin type categories according to their sensitivity to sunlight (6 groups) (McFadden et al., 1998). Thus, prediction of acute irritant potential, based on groups of human volunteers, is applicable to all skin types. On the other hand, an earlier study based on determination of the minimal erythema dose (MED) of 44 white subjects found a significant correlation between the MED and the response to seven primary irritants. This relationship was better for water-soluble irritants than for lipid-soluble ones. This study also proved that skin typing based on complexion and sunburn history was less reliable (Frosch and Wissing, 1982).

Regional variations are always very important, as various skin reactions vary with the site over the body. Dithranol irritation varies between individuals and between different parts of the body. The axilla, scrotum, breasts, and the inside of the thighs are the most irritable regions (Lawrence et al., 1984).

Sensory irritation threshold in the eyes to carbon dioxide (CO_2) correlated with skin irritation response to lactic acid applied to the cheek. The irritation to the eyes showed: (i) environmental influence: occupational stress was associated with low threshold; (ii) no gender variations; (iii) no effect of smoking; (iv) an effect of age: subjects under 40 years had lower thresholds; (v) correlation of CO_2 threshold to sensitivity to irritation from airborne pollutants (Kjaergaard et al., 1992).

SEASONAL VARIATIONS

Among the external factors acting in conjunction with irritants, or protecting the skin against irritants, are climatic factors, like temperature, relative humidity, wind, and ultraviolet exposure. Elevation of temperature, as an example, increases the penetration of irritants. Low humidity disrupts the skin barrier, induces epidermal DNA synthesis, and the skin becomes still more sensitive to barrier abrogation, demonstrating both epidermal hyperplasia and alterations in mast cells (Denda et al., 1998). These factors show seasonal variations, like the observed higher reactivity to irritants in winter (Agner and Serup, 1989), and their effect is either direct or indirect.

Repeated natural environmental insults may change the susceptibility to irritation by substances at concentrations that are normally well tolerated. In other words, environmental factors may expose the irritation potential of subthreshold concentrations of irritants, but succeeding such events, adaptation of the skin may occur, with a hardening effect that is expressed by a higher threshold. Other environmental factors, like ultraviolet irradiation (both UVA and UVB), have a protective effect: they increase the resistance of the skin to irritation.

DISRUPTION OF THE BARRIER FUNCTION

Altered barrier function may affect irritant dermatitis, as already mentioned. This is illustrated by the high incidence of irritant contact dermatitis observed in patients who underwent laser skin resurfacing (Levy and Salomon, 1998). In several such patients, delayed irritant contact dermatitis presented, seemingly indicating a cumulative effect.

Individuals who react more readily to irritants exhibited alterations of baseline capacitance values, indicating a tendency to barrier impairment (Seidenari et al., 1998). This supports the view that skin hyperreactivity to water-soluble irritants is induced by a greater amount of irritants absorbed.

DERMATOLOGICAL AND SYSTEMIC DISEASES

Acute irritant reactivity to SLS was similar in atopic and nonatopic patients. Irritancy was assessed by visual grading, chromametry, laser Doppler flowmetry, and transepidermal water loss following application of SLS at a range of concentrations and exposure times (Basketter et al., 1998). This differs from earlier studies that observed significantly lower threshold of irritation by SLS in both atopic dermatitis patients and in patients with allergic rhinitis with no dermatitis (Nassif et al., 1994). As a lower threshold of irritation was observed in atopic individuals without dermatitis, the authors hypothesized that in atopic individuals, abnormal intrinsic hyperreactivity in inflammatory cells, rather than in skin cells, predisposes to a lowered threshold of irritant responsiveness. The discrepancy between the findings of the various studies comparing the irritancy potential of atopic and nonatopic skin may reflect the complexity of the mechanisms involved (Gallacher and Maibach, 1998).

Skin threshold to SLS irritation was reduced in guinea pigs in the presence of allergic contact dermatitis in skin sites distant from the site of the allergic contact dermatitis (Prins et al., 1998). These findings seem to support the theory behind conditioned hyperirritability, which predicts the lowering of the irritation threshold in the presence of preexisting dermatitis. Another study of the irritation thresholds of benzalkonium chloride and trichloroacetic acid found that guinea pigs with even a little reaction to an irritant had lowered irritation thresholds to the same irritant at remote skin sites; however, enlarging the initial area of irritation or causing more severe irritation did not lower the threshold further (Roper and Jones, 1985).

The excited skin syndrome is a state of hyperirritability from clinical or patch test dermatitis, causing the threshold for irritancy to decrease and irritant reactions to increase. As inflammation-modulating phenomena, humoral and cellular, fluctuate during the course of irritant contact dermatitis, the irritability of the skin fluctuates as well. Variable reproducibility is inherent (Mitchell and Maibach, 1997).

The irritant reaction to dithranol was not different between psoriatic patients and healthy controls. There also was no correlation between the irritant dose threshold and minimal erythema dose to ultraviolet radiation, but skin type I patients were more sensitive than other skin types (Kingston and Marks, 1983).

MECHANISMS

The mechanisms underlying the irritant reaction are unknown. The biology of the mechanisms underlying irritant contact dermatitis is under investigation, and hopefully in vitro assays will be developed (Goldberg and Maibach, 1998). Even currently available in vitro methods showed more than 80% predictive value of the in vivo classification of irritants as strong, moderate, or mild irritants and nonirritants (Reinhardt et al., 1985).

The changes induced by subthreshold concentrations of irritants are not of high significance if not accompanied or followed by additional factors. The threshold may only represent a clinical definition, rather than inherent mechanism, so that subthreshold irritation and clinically detectable irritation constitute a continuum of reactions.

Many aspects of the underlying mechanisms are currently studied, including histological and immunohistochemical techniques to determine the expression of various keratins during acute irritation (Le et al., 1998; Willis et al., 1998). Cytokines are intimately linked to the barrier structure and function of the skin. They are key mediators of inflammatory and immunologic reactions and there is evidence that they are modulated in response to changes in the barrier function that leads to irritation (Nickoloff, 1998). A recent study hinted at a downregulation of protein kinase C isoforms as playing a role (Li et al., 1998). Inducible nitric oxide synthase was found in both irritant (SLS induced) and allergic contact dermatitis, with no difference in the distribution of its expression (Ormerod et al., 1997). Both contact and irritant dermatitis are characterized by microvascular hyperpermeability. Vascular endothelial growth factor, an inducer of permeability of endothelial cells, was upregulated by both allergic and irritant contactants, but by different mechanisms. Moreover, this upregulation of vascular endothelial growth factor release by keratinocytes showed an intense dose dependency (Palacio et al., 1997).

CONCLUSION

There are some advances in our knowledge of the changes induced by subthreshold concentrations of irritants and the interactions between such changes, but much more is yet to be studied. As clinical examination does not permit these changes to be identified and characterized, other methods of assessment should be employed, including in vivo and in vitro measurements. Studies would hopefully enable accurate measurements and visualization of as yet imperceptible changes and will elucidate the underlying processes. As our understanding of the mechanisms involved in subthreshold irritation expand, so will our ability to pursue the early events that lead to the visible changes characteristic of these processes. Such information will allow a better design of preventive measures and treatment.

REFERENCES

Agner, T., and Serup, J. (1989): Seasonal variation of skin resistance to irritants. *Br. J. Dermatol.*, 121:323–328.

Alarie, Y., Schaper, M., Nielsen, G. D., and Abraham, M. H. (1996): Estimating the sensory irritating potency of airborne nonreactive volatile organic chemicals and their mixtures. *SAR QSAR Environ. Res.*, 5:151–165.

Arlette, J. P., and Fritzler, M. J. (1984): Reduced skin threshold to irritation in the presence of allergic contact dermatitis in the guinea pig. *Contact Dermatitis*, 11:31–33.

Basketter, D. A., and Griffiths, H. A. (1993): A study of the relationship between susceptibility to skin stinging and skin irritation. *Contact Dermatitis*, 29:185–188.

Basketter, D. A., Miettinen, J., and Lahti, A. (1998): Acute irritant reactivity to sodium lauryl sulfate in atopics and non–atopics. *Contact Dermatitis*, 38:253–257.

Denda, M., Sato, J., Tsuchiya, T., Elias, P. M., and Feingold, K. R. (1998): Low humidity stimulates epidermal DNA synthesis and amplifies the hyperproliferative response to barrier disruption: Implication for seasonal exacerbations of inflammatory dermatoses. *J. Invest. Dermatol.*, 111:873–878.

Dominique, J., Van Neste, D., and de la Cuadra, J. (1987): Non invasive evaluation of functional changes induced by short term application of sodium lauryl sulphate on human skin. *J. Invest. Dermatol.*, 89:320–321.

Frosch, P. J., and Wissing, C. (1982): Cutaneous sensitivity to ultraviolet light and chemical irritants. *Arch. Dermatol. Res.*, 272:269–278.

Fullerton, A., Benfeldt, E., Petersen, J. R., Jensen, S. B., and Serup, J. (1998): The cacipotriol dose-irritation relationship: 48 hour occlusive testing in healthy volunteers using Finn chambers. *Br. J. Dermatol.*, 138:259–265.

Gallacher, G., and Maibach, H. I. (1998): Is atopic dermatitis a predisposing factor for experimental acute irritant contact dermatitis? *Contact Dermatitis*, 38:1–4.

Ghadially, R. (1998): Aging and the epidermal permeability barrier: Implications for contact dermatitis. *Am. J. Contact. Dermat.*, 9:162–169.

Goffin, V., Pierard-Franchimont, C., and Pierard, G. E. (1998): Shielded corneosurfametry and corneoxenometry: Novel bioassays for the assessment of skin barrier products. *Dermatology*, 196:434–437.

Goldberg, A. M., and Maibach, H. I. (1998): Dermal toxicity: Alternative methods for risk assessment. *Environ. Health Perspect.*, 106(Suppl 2):493–496.

Hall-Manning, T. J., Holland, G. H., Rennie, G., Revell, P., Hines, J., Barratt, M. D., and Basketter, D. A. (1998): Skin irritation potential of mixed surfactant systems. *Food Chem. Toxicol.*, 36:233–238.

Kingston, T., and Marks, R. (1983): Irritant reactions to dithranol in normal subjects and psoriatic patients. *Br. J. Dermatol.*, 108:307–313.

Kjaergaard, S., Pedersen, O. F., and Molhave, L. (1992): Sensitivity of the eyes to airborne irritant stimuli: Influence of individual characteristics. *Arch. Environ. Health*, 47:45–50.

Lawrence, C. M., Howel, D., and Shuster, S. (1984): The inflammatory response to anthralin. *Clin. Exp. Dermatol.*, 9:336–341.

Le, T. K., Schalkwijk, J., van de Kerhof, P. C., van Haelst, U., and van der Valk, P. G. (1998): A histological and immunohistochemical study on chronic irritant contact dermatitis. *Am. J. Contact. Dermat.*, 9:23–28.

Levy, P. M., and Salomon, D. (1998): Use of biorbane after laser resurfacing. *Dermatol. Surg.*, 24:729–734.

Li, L. F., Fiedler, V. C., and Kumar, R. (1998): Down-regulation of protein kinase C isoforms in irritant contact dermatitis. *Contact Dermatitis*, 38:319–324.

McFadden, J. P., Wakelin, S. H., and Basketter, D. A. (1998): Acute irritation thresholds in subjects with type I–type VI skin. *Contact Dermatitis*, 38:147–149.

Mitchell, J., and Maibach, H. I. (1997): Managing the excited skin syndrome: Patch testing hyperirritable skin. *Contact Dermatitis*, 37:193–199.

Nassif, A., Chan, S. C., Storrs, F. J., and Hanifin, J. M. (1994): Abnormal skin irritancy in atopic dermatitis and in atopy without dermatitis. *Arch. Dermatol.*, 130:1402–1407.

Nickoloff, B. J. (1998): Immunologic reactions triggered during irritant contact dermatitis. *Am. J. Contact. Dermat.*, 9:107–110.

Ohta, H., Makita, K., Kawashima, T., Kinoshita, S., Takenouchi, M., and Nozawa, S. (1998): Relationship between dermato-physiological changes and hormonal status in pre, peri, and postmenopausal women. *Maturitas*, 30:55–62.

Ormerod, A. D., Dwyer, C. M., Reid, A., Copeland, P., and Thompson, W. D. (1997): Inducible nitric oxide synthase demonstrated in allergic and irritant contact dermatitis. *Acta. Derm. Venereol.*, 77:436–440.

Palacio, S., Schmitt, D., and Viac, J. (1997): Contact allergens and sodium lauryl sulphate upregulate vascular endothelial growth factor in normal keratinocytes. *Br. J. Dermatol.*, 137:540–544.

Prins, M., Swinkels, O. Q. J., Kolkman, E. G. W., Wuis, E. W., Hekster, Y. A., and van der Valk, P. G. M. (1998): Skin irritation by dithranol cream. *Acta Derm. Venereol.*, 78:262–265.

Reinhardt, C. A., Pelli, D. A., and Zbinden, G. (1985): Interpretation of cell toxicity data for the estimation of potential irritation. *Food Chem. Toxicol.*, 23:247–252.

Roper, S. S., and Jones, H. E. (1985): An animal model for altering the irritability threshold of normal skin. *Contact Dermatitis*, 13:91–97.

Schwindt, D. A., Wilhelm, K. P., Miller, D. L., and Maibach, H. I. (1998): Cumulative irritation in older and younger skin: A comparison. *Acta Derm. Venereol.*, 78:279–283.

Seidenari, S., Francomano, M., and Mantovani, L. (1998): Baseline biophysical parameters in subjects with sensitive skin. *Contact Dermatitis*, 38:311–315.

Trattner, A., Johansen, J. D., and Menne, T. (1998): Formaldehyde concentration in diagnostic patch testing: Comparison of 1% with 2%. *Contact Dermatitis*, 38:9–13.

Tur, E., Eshkol, Z., Brenner, S., and Maibach, H. I. (1995): Cumulative effect of subthreshold concentrations of irritants in humans. *Am. J. Contact. Dermatitis*, 6:216–220.

Tur, E. (1999): Blood flow as a technology in percutaneous absorption: The assessment of the cutaneous microcirculation by laser Doppler and photoplethysmographic techniques. In: *Percutaneous Absorption*, 3rd ed. Edited by R. L. Bronaugh and H. I. Maibach, pp. 315–346, Marcel-Dekker, New York.

Tur, E. (1994): Skin pharmacology. In: *Remittance Spectroscopy and Chromametry.* Edited by Berardesca, E., Elsner, P., Maibach, H. I., pp 259–268. CRC Press, Boca Raton, FL.

Webster, G. F., Topical tretinoin in acne therapy. (1998): *J. Am. Acad. Dermatol.*, 39(Suppl):S38–44.

Wigger-Alberti, W., Rougier, A., Richard, A., and Elsner, P. (1998): Efficacy of protective creams in a modified repeated irritation test. *Acta Derm. Venereol.*, 78:270–273.

Willis, C. M., Reiche, L., and Wilkinson, J. D. (1998): Keratin 17 is expressed during the course of SLS-induced irritant contact dermatitis, but unlike keratin 16, the degree of expression is unrelated to the density of dividing keratinocytes. *Contact Dermatitis*, 39:21–27.

Toxicology of Skin
Edited by Howard I. Maibach
Copyright © 2001 Taylor & Francis

7

Cytokines and Irritant Dermatitis Syndrome

Isaak Effendy

Department of Dermatology, University of Marburg, Germany

Howard I. Maibach

School of Medicine, Department of Dermatology, University of California, San Francisco, California, USA

INTRODUCTION

Although nonimmunologic irritant contact dermatitis (ICD) is a major human health problem (Fregert, 1975; Mathias, 1985), the mechanisms underlying this reaction are inadequately understood (Patrick et al., 1987; Bruze and Emmett, 1991). Skin irritancy, previously thought a monomorphous process, is now considered a complex biologic syndrome, with a diverse pathophysiology, natural history, and clinical appearance (Bason et al., 1991; Willis et al., 1992). Numerous factors determine whether a particular substance will cause irritation and inflammation in a given individual. The intensity and clinical features of ICD may also depend on the ability of the irritant to injure the skin surface, penetrate the skin, and reach its target (Olson, 1991). The type of exogenous stimulus may influence the reaction. While certain topically applied chemicals cause irritant dermatitis (Patrick et al., 1987; Wilmer et al., 1994), mechanically induced skin irritancy by tape stripping exhibits no inflammatory cell infiltration during the initial 24 hours (Pinkus, 1951; Nickoloff and Naidu, 1994), although both actions upregulated numerous epidermal cytokines.

Investigations draw attention to molecular events in allergic contact dermatitis (ACD) and ICD; however, most of the latter have just been performed to provide control data for allergic reactions. Mounting evidence suggest that keratinocytes are not only involved in allergic reactions, but also in ICD through the synthesis and release of inflammatory cytokines, chemokines, and growth factors (Nickoloff and Naidu, 1994; Kupper, 1990; McKenzie and Sauder, 1990; Ansel et al., 1990; Schwarz and Luger, 1992). Although there is a distinct pathway between allergic and irritant reactions, it is likely that a connecting network at molecular levels between both types of contact dermatitis exists. This could partly explain why numerous similar epidermal cytokines have been detected in both allergic and irritant responses. The subject of the review is the current state of the cytokines detected in both ICD and ACD.

CYTOKINES

Cytokines, peptides, or (glyco)proteins with molecular weights ranging from 6000 to 60,000 are produced by various cells. Cytokines modulate reactions of the host to foreign antigens or injurious agents by regulating activation, proliferation, and differentiation of immune as well as nonimmune cells. These water-soluble mediators are potent, acting at concentrations of 10^{-10}–10^{-15} mol/L to stimulate target cell functions following specific ligand-receptor interactions (Oppenheim et al., 1991).

Cytokines may exhibit considerable overlap in their biologic effects on target cells. On the other hand, biologically distinct cytokines may have similar effects by initiating the production of a cascade of identical cytokines or of one another. Cytokines regulate each other by competition, interaction, and mutual induction in a series of lymphokine cascades and circuits with positive or negative feedback effects. The effects of cytokines can also be regulated at the level of cell membrane receptors. Hence, agents that influence cytokine receptor expression modulate the activities of these mediators (Oppenheim et al., 1991; Aggarwal and Puri, 1995).

The epidermis is a rich source of cytokines and growth factors (Luger, 1989; McKenzie and Sauder, 1990). Keratinocytes, the major cell mass of human epidermis, not only represents the first target for irritants (DeLeo et al., 1987) but may also act as a "signal tranducer," capable of converting exogenous stimuli into the production of cytokines, adhesion molecules, and chemotactic factors (Barker et al., 1991). Unstimulated keratinocytes express and secrete low levels of cytokines but provide a reservoir of preformed (primary) cytokines such as interleukin 1 (IL 1-α, and IL 1-β) and tumor necrosis factor alpha (TNF-α). In response to exogenous stimuli, however, activated keratinocytes can produce various inflammatory cytokines

TABLE 7.1. *Keratinocyte-derived cytokines*

Cytokine	References
Interleukins	
IL-1α	Kupper et al., 1986
IL-3 (multicolony-stimulating factor)*	Luger et al., 1988
IL-6	Kupper et al., 1989
IL-7	Heufler et al., 1993
IL-8 (chemotactic factor)	Larsen et al., 1989
IL-10	Enk et al., 1992
IL-12 (natural killer stimulatory factor)	Aragane et al., 1994
IL-13	Brown et al., 1989
IL-14	Ambrus et al., 1993
IL-15	Grabstein et al., 1994
IL-17	Yao et al., 1995
IL-18*	Stoll et al., 1997
Tumor necrosis factor	
TNF-α	Kock et al., 1990
Colony stimulating factors (CSF)	
Granulocyte colony-stimulating factor (G-CSF)	Sauder et al., 1988
Macrophage colony-stimulating factor (M-CSF)	Sauder et al., 1988
Granulocyte-macrophage CSF (GM-CSF)	Kupper et al., 1988
Multi-CSF (IL-3)	Luger et al., 1988
Growth factors (GF)	
Transforming growth factor alpha (TGF-α)	Coffey et al., 1987
Transforming growth factor beta (TGF-β)	Akhurst et al., 1988
Basic fibroblast growth factor (bFGF)	Halaban et al., 1988
Platelet-derived growth factor (PDGF)	Ansel et al., 1993
Chemotactic factors	
Interferon-induced protein 10 (IP-10)	Kaplan et al., 1987
Macrophage inflammatory protein 2 (MIP-2)	Tekamp-Olson et al., 1990
Monocyte chemotaxis and activating factor (MCAF)	Barker et al., 1991
Chemotactic factor (IL-8)	Larsen et al.,1989

*Murine keratinocytes.

(Table 7.1), however, the profile of secreted cytokines is highly dependent upon the particular type of T cells; in many diseases it seems that this specific response of T cells to antigenic challenge defines the nature of the immune response (Mosmann and Coffman, 1989). In fact, it was in 1986 when Mosmann et al. began a conceptual revolution in immunology in dividing T helper (Th) cells into two populations with contrasting and cross-regulating cytokine profiles: Th1 and Th2 cytokines. This new paradigm was taken up in every area of immunology and infectious disease and has proved extremely useful. For instance, contact sensitivity (ACD) has generally been regarded as a specific Th1-mediated process. Today, however, there is good evidence that both Th1 and Th2 cytokines, for example, are primarily involved in ACD, suggesting somehow that certain prior distinctions in molecular mechanisms of cell-mediated DTH and ACD need to be revisited (Allen and Maizels, 1997; Werfel et al., 1997; Grabbe and Schwarz, 1998; Agarwal et al., 1999).

UPREGULATED CYTOKINES FOLLOWING CHEMICAL IRRITANT

In Vitro Studies (Table 7.2)

Cytokine Induction in Murine Cell Cultures

Similar to previous studies in mice (Piguet et al., 1991; Piguet, 1992) in vitro findings suggest that TNF-α plays a critical role in ICD as well as in ACD. Lisby and colleagues (1995) reported that the primary irritants dimethylsulfoxide (DMSO) and sodium lauryl sulfate (SLS) upregulated TNF-α mRNA both in Ia$^-$/CD3$^-$ mice epidermal cells and in transformed mice keratinocyte cell lines. Interestingly, nickel, a frequent contact allergen, also upregulated TNF-α mRNA and protein in Ia$^-$/CD3$^-$ epidermal cells; however, while both irritants upregulated TNF-α mRNA via a protein kinase C–dependent increase in promoter activity, nickel salts act through post-transcriptional modulation of the TNF-α mRNA. In addition, the data suggest that some irritants and sensitizers directly induce TNF-α in keratinocytes without intermediate Langerhans' cell (LC)–derived signals.

Investigations on lymph node cells (LNC) using BALB/mice resulted in enhanced IL-6 production by allergen-activated LNC but not by LNC prepared from naïve or vehicle-treated mice (Dearman and Kimber, 1994).

Cytokine Induction in Human Cell Cultures

A 24-hours exposure of cultured human keratinocytes to SLS (a potent irritant), Triton X-100 anionic surfactant (a moderete irritant), or Tween 20 polysorbate (a mild irritant) resulted in an upregulation of IL 1-α mRNA; however, there was no definite rank order stratification of surfactant potency in term of IL 1-α message at 24 hours. Possibly, earlier time points of mRNA assessment might have revealed some stratification (Shivji et al., 1994).

Various irritants, namely SLS, phenol, and croton oil (at noncytotoxic concentrations), induced IL-8 production directly in cultured human epidermal keratinocytes (Wilmer et al., 1994); however, beside these irritants, benzalkonium chloride (an ulcerogenic agent) as well as dinitrofluorobenzene (DNFB), a contact allergen, stimulated the production and intracellular accumulation of IL-1α. Thus, of the cytokine changes detected, increases in intracellular IL-1α and IL-8 secretion were the most remarkable. Phenol and croton oil, but not SLS, stimulated TNF-α production, whereas croton oil was the only agent found to induce GM-CSF production. Interestingly, the cytokines' stimulatory potential of the compounds tested varied. Based on these data, it has been proposed that a given pattern of cytokine production may be chemical-specific.

Others reported IL-8 gene expression was significantly increased in normal human keratinocyte and human keratinocyte cell line HaCaT upon stimulation with contact allergens (DNFB, 3-n-pentadecylcatechol), tolerogen (5-methyl-3-n-pentadecylcatechol), and SLS (Mohamadzadeh et al., 1994). They concluded that the induction and

TABLE 7.2. Upregulated epidermal cell–derived cytokines following chemical irritant, contact allergen, and tolerogen: In vitro studies

Stimulus	Subject	Reaction time	Assessment	Upregulated cytokines	References
SLS (0.0075% w/v), DMSO (20% v/v)	Murine epidermal cells (MEC) and murine keratinocyte cell lines (MKCL)	15 minutes–6 hours	Northern blot, RT-PCR	TNF-α mRNA in MEC and MKCL	Lisby et al., 1995
PMA (10 ng/ml)				TNF-α mRNA and protein in MEC	
NiSO$_4$ (10^{-2} M) [sensitizer]				TNF-α mRNA and protein in MEC	
SLS (5 μg/ml)	Normal human epidermal keratinocytes (NHEK)	3–48 hours	ELISA, RT-PCR	IL-8 mRNA	Wilmer et al., 1994
Phenol (200 μg/ml)				IL-8 mRNA, TNF-α mRNA, and IL-1α (intracell.)	
Croton oil (20 μg/ml)				IL-8 mRNA, TNF-α mRNA, GM-CSF, and IL-1α (intracell.)	
BAC (0.4 μg/ml)				IL-1α (intracell.)	
DNFB (1 μg/ml), oxazolone (2 μg/ml) [sensitizer]				IL-1α (intracell.) (note: IL-8 is not altered)	
SLS, Triton X-100, Tween 20	Normal human epidermal keratinocytes (NHEK)	24 hours	RT-PCR	IL-1α mRNA	Shivji et al., 1994
SLS (10 μg/ml)	NHEK, and KC cell line HaCaT	10 minutes	Northern blot, ELISA	IL-8 mRNA	Mohamadzadeh et al., 1994
5-Me-PDC (10 μg/ml) [tolerogen]				IL-8 mRNA	
DNFB or PDC (10 μg/ml) [sensitizer]				IL-8 mRNA	
SLS (100 μg/ml)	KC cell line HaCaT	4–24 hours	ELISA, RT-PCR	IL-1α (intracell. and mRNA)	Whittle et al., 1995
Croton oil (25 μg/ml)		4 hours		IL-1α (intracell.), and IL-8	
Oxazolone (700 μg/ml) [sensitizer]		4 hours		IL-1α (intracell.)	

5-Me-PDC = 5-methyl-3-n-pentadecylcatechol; BAC = benzalkonium chloride; DMSO = dimethylsulfoxide; DNFB = dinitrofluorobenzene; PDC = 3-n-pentadecylcatechol; PMA = phorbol myristate actate; SLS = sodium lauryl sulphate.

production of IL-8 does not represent a stimulus-specific response but rather a general mediator in tissue injury. Indeed, exposing human keratinocytes to ultraviolet-B (UV-B) has induced IL-8 production (Kondo et al., 1993).

Studies on HaCaT monolayers showed that SLS produced a dose-dependent and time-dependent increase in intracellular IL-1α (Whittle et al., 1995). Typically, 100 μg/ml SLS caused the maximal increase in IL-1α as measured by ELISA (190 pg/ml at 4 hours exposure, and 390 pg/ml at 12 hours exposure). Concentrations of 150 μg/ml and 200 μg/ml SLS caused IL-1α to leak into the media. Exposure to 25 μg/ml croton oil caused a marked increase in intracellular IL-1α levels as well as a release of IL-8 into the media. A contact allergen, oxazolone, has also been shown to increase intracellular IL-1α level.

In Vivo Murine Studies (Table 7.3)

Cytokine Induction in Murine Epidermis

(Enk and Katz, 1992) showed a distinct cascade of epidermal cytokines in ICD caused by SLS when compared with that in an early phase of ACD induced by TNCB, DNFB, or dinitrochlorobenzene (DNCB)-induced ACD in BALB/c mice. TNF-α, interferon (IFN)γ, and granulocyte-macrophage colony-stimulating factor (GM-CSF) mRNAs were upregulated after application of allergens, irritant, and tolerogens; however, class II major histocompatibility I-Aα, IL-1α, IL-1β, inflammatory protein (IP-10), and macrophage inflammatory protein (MIP-2) mRNAs were upregulated only after allergen painting. They concluded that the LC-derived cytokine IL-1β appears specific for the early molecular events in ACD, but not in ICD.

Using a similar experimental approach, (Haas et al., 1992) demonstrated that the contact sensitizers DNFB, oxazolone, urushiol, and 3-n-pentadecylcatechol increased the mRNA levels of TNF-α, and IL-1α but not GM-CSF. The cytokine production induced by tolerogens was comparable with that caused by solvents alone (vehicle controls). They stated that the discrepancies between their data and the findings of (Enk et al., 1992) may be explained by different technical factors. Interestingly, 1% croton oil, a contact irritant, also caused an increase in TNF-α and IL-1α mRNA. This, however, suggests that the cytokine release may not be restricted to diseases in which antigen presentation is crucial.

(Kondo et al., 1994) reported that topical application of 20% SLS onto BALB/c mice skin upregulated not only TNF-α and GM-CSF, as shown earlier, but also IL-1β, IL-6, and IL-10 mRNA at 24 hours after treatment. Intriguingly, an upregulation of IL-1β has also been detected in ICD, as the LC-derived cytokine has been thought to be a specific event in early phase of allergic reaction (Enk and Katz, 1992; Matsue et al., 1992; Enk et al., 1993; Rambukkana et al., 1994). In addition, IL-1α mRNA was elevated in the first 3 hours followed by suppression at least until 24 hours after SLS treatment. The detection of IL-1α and IL-1β mRNA in ICD reported by (Kondo et al., 1994), but not in the investigations by (Enk et al., 1992), could be explained by differences in the sample preparation, namely single-cell suspension

TABLE 7.3. Upregulated epidermal cell–derived cytokines following chemical irritant, contact allergen, and tolerogen: In vivo murine studies

Stimulus	Subject	Reaction time	Assessment	Upregulated cytokines	References
SLS (20% w/v), DNTB (2% w/v) [tolerogen]	BALB/c mice	15 minutes–24 hours	RT-PCR	TNF-α, IFN-γ, and GM-CSF mRNA; TNF-α, IFN-γ, and GM-CSF mRNA	Enk et al., 1992
TNCB (3% w/v), DNFB (0.5% w/v), DNCB (2% w/v), [sensitizer]				TNF-α, IFN-γ, GM-CSF, IL-1α, IL-1β, IP-10, and MIP-2 mRNA	
Croton oil (1%)	BALB/c mice	1–12 hours	RT-PCR	TNF-α ↑ and IL-1α mRNA⇑	Haas et al., 1992
DNTB (2%), 5-Me-PDC (5.9%) [tolerogen]				TNF-α and IL-1α mRNA	
DNFB (0.5%), PDC (5.9%), Urushiol (5.9%), oxazolone (1%) [sensitizer]				TNF-α ⇑ and IL-1α mRNA⇑	
SLS (20% w/v)	BALB/c mice	1–24 hours	RT-PCR	IL-1β, IL-6, and IL-10 mRNA at 24 hours after treatment	Kondo et al., 1994
DNFB (0.5% and 0.2%) [sensitizer]				IL-1β, IL-6, IL-10, and GM-CSF mRNA at 6–24 hours after treatment as well as IL-1α mRNA at 12 hours or 24 hours after treatment	
Acetone pure (repeated applications, acute barrier disruption)	Hairless mice	1–8 hours	Northern blot, Western blot	TNF-α, IL-1α, IL-1β, and GM-CSF mRNA (but not IL-6, nor IFN-γ)	Wood et al., 1992
Croton oil (1%)	BALB/c mice	0–72 hours	RT-PCR and liquid hybrid.	IL-1ra mRNA	Wood et al., 1994
				IL-10 mRNA (at 9 hours after treatment in the epidermis) and IL-2 mRNA (at 1 hour after treatment, dermis)	Asada et al., 1997
TNCB (3% and 1%) [sensitizer]				IFN-γ (at 24 and 72 hours after treatment) and IL-4 mRNA (at 24 hours after treatment): in the epidermis and dermis	
Phorbol ester Oxazolone (1.6%) [sensitizer]	BALB/c mice	0–32 days	RT-PCR	TNF-α; TNF-α, IFN-γ (initial challenge), and IL-4 (at chronic exposure)	Webb et al., 1998

5-Me-PDC = 5-methyl-3-n-pentadecylcatechol; BAC = benzalkonium chloride; DNCB = dinitrochlorobenzene; DNFB = dinitrofluorobenzene; DNTB = dinitrothiocyanobenzene; PDC = 3-n-pentadecylcatechol; SLS = sodium lauryl sulphate; TNCB = trinitrochlorobenzene.

versus epidermal sheet (Kondo et al., 1994). An increase in IL-6 and IL-10 mRNA was also observed in DNFB-induced ACD, suggesting that both cytokines are similarly regulated by allergen as well as irritant. The lack of an increase in TNF-α mRNA in ACD is consistent to certain findings (Oxholm et al., 1991; Boehm et al., 1992) but not to other studies (Barker et al., 1991; Piguet et al., 1991; Griffiths et al., 1991). These discrepancies require further investigation.

Acute disruption of the murine epidermal permeability barrier by repeated topical applications of absolute acetone or tape stripping increased the mRNA levels of TNF-α, IL-1α, IL-1β, IL-1ra, and GM-CSF in the epidermis; however, their kinetics differed (Wood et al., 1992, 1994a, 1994b). In a chronic model of barrier disruption induced by feeding an essential fatty acid-deficient diet (EFAD), the mRNA levels of each of these cytokines was also increased. Moreover, TNF-α protein also increased in the epidermis following both acute and chronic barrier impairment (Wood et al., 1992; Tsai et al., 1994). Except for IL-6, the cascade of cytokines observed in these studies are relatively comparable with that obtained in SLS-induced ICD by Kondo and colleagues (1994). Interestingly, whereas occlusion with an impermeable membrane decreased epidermal cytokine production in normal and EFAD mice, occlusion did not block the upregulation of cytokines in acute model of epidermal barrier disruption (Wood et al., 1994a). The data suggest that cytokine release after acute perturbation may rather be related to skin injury than to barrier status alone. Indeed, further studies exhibited that certain skin manipulations injuring the epidermis stimulate cytokine production leading to cutaneous pathology, independent of barrier repair (Denda et al., 1995).

The function of IL-4 and IL-10 in ICD has been the subject of intriguing investigations: Berg and colleagues (1995) demonstrated that mice with targeted disruptions of the IL-4 (IL-4T) gene, as well as wild-type mice, exhibited equivalent responses to the irritant croton oil. In contrast, the response of the mice with targeted disruptions of the IL-10 (IL-10T) gene was abnormally increased. Similar results were obtained in ACD induced by oxazolone. They concluded that IL-10, but not IL-4, is a natural suppressant of irritant response as well as of contact hypersensitivity. Indeed, IL-10 has widely been accepted as an inhibitor of ACD in mice (Enk et al., 1994; Ferguson et al., 1994; Schwarz et al., 1994; Kondo et al., 1994); however, prior studies indicated that ICD was not regulated by IL-10 (Enk and Katz, 1992; Schwarz et al., 1994). The reason for the discrepancies has not been elucidated. Mounting data imply that IL-4 represents an important downmodulator of contact hypersensitivity or ACD (Asada et al., 1997; Webb et al., 1998; Rowe an Bunker, 1998), contradicting the findings by Berg and colleagues (1995). This difference requires clarification.

In the same studies, Berg and colleagues (1995) have also shown that when IL-10T mice were exposed to a higher dose of the irritant, irreversible tissue damaged occurred. Interestingly, the anti-TNF antibody treatment of IL-10T mice prevented hemorrhage and tissue necrosis but did not significantly reduce edema or influx of inflammatory cells. Probably, the changes were caused by the uncontrolled production of other proinflammatory mediators. Earlier studies by (Piguet et al., 1991, 1992) have demonstrated that primary irritant reactions to trinitrochlorobenzene (TNCB), as well as the elicitation phase of ACD, could be inhibited by administration of antibodies directed to TNF-α or by recombinant soluble TNF receptors. Possibly, these inhibiting effects of anti-TNF antibody on ICD and ACD observed may, at least in part, be explained by the data from Berg and colleagues (1995).

In Vivo Human Studies (Table 7.4)

Cytokine Induction in Human Epidermis

Vejlsgaard and colleagues (1989) reported that ICAM-1 expression is a feature of ACD reactions but does not occur in ICD induced by SLS or croton oil. In contrast,

TABLE 7.4. *Upregulated epidermal cell–derived cytokines following chemical irritant and contact allergen: In vivo human studies*

Stimulus	Subject	Exposure time	Assessment	Upregulated cytokines	References
SLS (5% w/v), BAC (0.5% w/w), croton oil (0.8% w/v), dithranol (0.02% w/w).	Healthy volunteer	48 hours	Immuno-labeling	ICAM-1	Willis et al., 1991
NAA (80% v/v)				(ICAM-1 not altered)	
SLS (2% and 4% w/v)	Healthy volunteer	24 hours	Immuno-labeling	ICAM-1	Lindberg et al., 1991
NAA (20% and 80% v/v)				(ICAM-1 not altered)	
SLS (5%)	Healthy volunteer	48 hours	Immuno-labeling	IL-6 (Note: TNF-α, IL-1α, and IL-1β not altered)	Oxholm et al., 1991
Ni SO$_4$ (5%) [sensitizer]	Patients with nickel allergy			IL-6 (Note: TNF-α, IL-1α, and IL-1β not altered)	
SLS (10%) Formaldehyde (8%)	Healthy volunteer	48 hours	Immuno-labeling	IL-1α, TNF-α, IL-2, and IFN-γ	Hoefakker et al., 1995
Epoxy resin (1%), formaldehyde (1%) [sensitizer]	Patients with epoxy resin or formaldehyde allergy			IL-1α, TNF-α, IL-2, and IFN-γ	
SLS (10%)	Healthy volunteer	36 hours	ELISA	In skin lymph: IL-6 ⇑, TNF-α ⇑, IL-1β, IL-2, IL-2 r, and GM-CSF	Hunziker et al., 1992

NAA = nonanionic acid.

Willis and colleagues (1991) demonstrated an increase in ICAM-1 in the epidermis following application of irritants, including SLS and croton oil, although in a different manner. These findings were confirmed by others (Lindberg et al., 1991), suggesting that exposure to different irritants induces distinct reactions not only in clinical features (Patrick et al., 1987) or histopathology (Willis et al., 1989), but also at the molecular level.

An increase in IL-6, but not TNF-α, IL-1α, or IL-1β, has been detected either after a 48-hour patch test with 5% SLS in healthy volunteers or with 5% nickel sulfate in patients with nickel contact allergy (Oxholm et al., 1991). An unaltered level of TNF-α has also been observed by others (Kondo et al., 1994), but conflicts with previous in-vitro data (Enk and Katz, 1992; Lisby et al., 1995). On the other hand, the lack of IL-1α and IL-1β expression in both ICD and ACD contradict with the data from Kondo and colleagues (1994). Differences in method and technique could perhaps be responsible for these discrepancies.

Another study demonstrated that the production of IL-1α, TNF-α, IL-2, and IFN-γ in the dermis was upregulated in ICD induced by 10% SLS or 8% formaldehyde applied for 48 hours in healthy volunteers (Hoefakker et al., 1995). Remarkably, the enhanced cytokines have also been detected in ACD caused by 1% epoxy resin or 1% formaldehyde in allergic individuals at 72 hours after application. These data elucidate the dynamics of the cytokine production. In the late phase of inflammation, at 72 hours after exogenous stimulation, cytokines may have generally been upregulated regardless of stimulation with contact allergens or irritants.

Cytokine Induction in Human Skin Lymph

IL-6 and TNF-α were markedly increased in human skin lymph derived from the early phase of SLS-induced ICD as assayed by ELISA (Hunziker et al., 1992). In addition, a slight elevation of IL-1β, IL-2, soluble IL-2r, and GM-CSF was also noted, particularly in the late phase of ICD; there was no relevant increase in IL-1α and IL-8. In other studies, however IL-1β did not discriminate allergic from irritant reactions (Brand et al., 1996).

Moreover, the absolute and relative number of LC has been demonstrated to increase eminently in the late phase of SLS-induced ICD in human skin lymph (Brand et al., 1992, 1993). These findings support previous data showing that LC leave the epidermis in ICD (Willis et al., 1986; Gawkrodger et al., 1986; Marks et al., 1987; Mikulowska, 1992). Probably, the increase in TNF-α obtained may account for the phenomenon, as the cytokine has been shown to stimulate LC migration from skin. It has been postulated that SLS may activate the keratinocytes resulting in induction of several cytokines, chemotactic factors, and adhesion molecules. These signals are thought responsible for an increased turnover of epidermal LC as well as for their migration toward the skin lymph (Kimber and Cumberbatch, 1992). Unlike in ACD, LC migration in ICD, in which the antigen-presenting function of LC is not mandatory, seems to be secondarily involved.

UPREGULATED CYTOKINES FOLLOWING MECHANICAL AND PHYSICAL IRRITATION (TABLE 7.5)

Cytokine Induction by a Mechanical Skin Irritation

Acute disruption of the cutaneous permeability by repeated tape stripping in hairless mice increased TNF-α, IL-1α, IL-1β, IL-1ra, and GM-CSF mRNA in the epidermis (Wood et al., 1992, 1994a, 1994b), but not IL-6 or IFN-γ (Wood et al., 1992). As stated above, the enhanced cytokines detected may merely be linked with the epidermis injury and alteration of keratinocytes than with barrier status (Denda et al., 1995).

In man, however, repeated tape stripping of the skin abrogating the epidermal barrier induced an upregulation of mRNA coding for TNF-α, IL-8, IL-10, IFN-γ, TGF-α, TGF-β, ICAM-1 in the epidermis, and IL-2, IL-4, IFN-γ, ICAM-1, TGF-β mRNA in dermal samples as assayed with RT-PCR (Nickoloff and Naidu, 1994).

TABLE 7.5. *Upregulated epidermis cell–derived cytokines following mechanical and physical skin irritation: In vitro and in vivo studies*

Stimulus (mechanical, physical)	Subject	Reaction time	Assessment	Upregulated cytokines	References
Repeated skin stripping	Hairless mice	1–8 hours	Northern blot, Western blot	TNF-α, IL-1α, IL-1β, and GM-CSF mRNA (but not IL-6, nor IFN-γ) IL-1ra mRNA	Wood et al., 1992, 1994a, 1994b
Repeated skin stripping	Healthy volunteer	1–24 hours	RT-PCR	TNF-α, IL-8, IL-10, IFN-γ, TGF-α, TGF-β, and ICAM-1 mRNA (note: decreased IL-1β, IL-5 not altered)	Nickoloff et al., 1994
Repeated skin stripping	Healthy volunteer	Immediate	Immuno-assay	IL-1α and PGE$_2$ (note: decreased TNF-α, unchanged IL-6)	Reilly et al., 1999
UVB (280–320 nm), UVA1 (340–400 nm), UVA2 (320–400 nm)	Murine and human KC	1–24 hours	ELISA, RT-PCR	IL-1α, IL-3, IL-6, IL-8, IL-10, IL-15, GM-CSF, TNF-α, NGF, bFGF, and ICAM-1 ($\Uparrow$ or $\Downarrow$);	Schwarz et al., 1989, 1994; Norris et al., 1990., Aubin et al., 1991; Takashima et al., 1996 (Review)
				IL-1 β	Griswold et al., 1991

In addition, the authors reported that mRNA for IL-2 and IL-4 was not detectable at any time point; IL-5 mRNA was constitutively present at time 0 and did not significantly change during the initial 6 hours. IL-1β mRNA was present at time 0 and at 1 hour, but appeared to decrease after 6 hours. Conversely, studies in mice showed that IL-1β mRNA was significantly increased after tape stripping (Kondo et al., 1994) and SLS-treatment (Wood et al., 1992), respectively.

More recently, in tape-stripped human skin, levels of prostaglandin E$_2$ and IL-1α were significantly increased, TNF-α was decreased, whereas IL-6 and leukotriene B$_4$ in blister fluids remained unchanged (Reilly and Green, 1999). The findings regarding TNF-α and IL-6 are not consistent with the data obtained in mice in prior studies (Wood et al., 1992), probably due to different species and techniques used (Reilly and Green, 1999); however, the authors concluded that PGE$_2$ and IL-α may play key roles in acute responses to mild irritants.

Cytokine Induction by a Physical Skin Irritation

UV radiation represents a well-established modality for the treatment of inflammatory skin diseases as well as triggers for photosensitive dermatoses. In addition, under certain conditions, UV exposure may also stand for an irritant potential, as it can cause acute sunburn and erythema.

UVB irradiation has been reported by many investigators (Ansel et al., 1983; Kupper et al., 1986; Kirnbauer et al., 1991; Kock et al., 1989; Schwarz and Luger, 1989; Aubin et al.,1991; Schwarz et al., 1994) to induce or upregulate numerous cytokines by keratinocytes. These cytokines include IL-1α, IL-3, IL-6, IL-8, IL-10, IL-15, GM-CSF, TNF-α, NGF, and bFGF (see review by Takashima and Bergstresser, 1996). Moreover, UV radiation of cultured human keratinocytes can either suppress or induce expression of ICAM-1 (Norris et al., 1990).

UV-induced suppression of the induction of delayed-type hypersensitivity is mediated primarily through IL-10, while UV-induced immunosuppression of contact hypersensitivity seems to be mediated by TNF-α (Rivas and Ullrich, 1993). IL-10 induced by UVB as well as UVA1 may also be responsible for downregulation of Th-1 cell-derived cytokines in inflammatory skin diseases (Grewe et al., 1995). Studies in mouse models indicated that susceptibility to UV-B–induced immunosuppression is partly governed by the *Tnf* locus (Yoshikawa and Streilein, 1990). TNF-α is involved in LC migration from skin, and induction of this cytokine could lead to alter antigen presentation due to LC depletion. Regulation of IL-12 is also involved in UV-induced immunosuppression (Schmitt et al., 1995; Riemann et al., 1996). Furthermore, IL-15 could play an important role in regulating the local cytokines at UV-exposed sites (Mohamadzadeh et al., 1995).

Recent findings indicate that UVB irradiation induces LC depletion from skin by a dual mechanisms (e.g., by downregulating surface expression of CSF-1 receptor and GM-CSF receptor by LC) and by inducing CSF-1 deficiency in the epidermal microenvironment (Kitayama and Takashima, 1999).

CONCLUSION

Based on these data, upregulated epidermal cytokines can be detected in allergic and irritant reactions. Moreover, even tolerogens (nonsensitizing, nonirritating agents) when applied to the skin are capable of elevating cytokine release in the epidermis. Some cytokines seem restricted to ACD whereas others do not (Enk and Katz, 1992); however, some thought to be ACD-specific mediators could also be detected in the skin upon stimulation with irritants or tolerogens (Haas et al., 1992; Kondo et al., 1994; Brand et al., 1996; Zheng et al., 1995). Although the pathways are distinctly defined, secreted cytokines in ACD are comparable to those in ICD.

Several cytokines are not specific to allergic or to irritant responses. An increase in IL-1α, TNF-α, and GM-CSF have been observed in allergic and irritant as well as in tolerogenic reactions, respectively, as assayed in cell cultures and in murine epidermis (Tables 7.2 and 7.3). Furthermore, the production of TNF-α and GM-CSF were upregulated in human lymph following SLS application (Hunziker et al., 1992). Likewise, IL-6 was detected in allergic and irritant responses (Enk and Kaz, 1992; Lisby et al., 1995), although others have found that IL-6 appeared restricted to allergic reactions (Dearman and Kimber, 1994). Particularly, epidermal IL-8 appears as a nonspecific mediator providing inflammatory cytokine in tissue injury, as shown in in vitro and in vivo investigations (Wilmer et al., 1994; Nickoloff and Naidu, 1994; Mohamadzadeh et al., 1994). Hence, we may suggest that IL-1α, IL-6, IL-8, TNF-α, and GM-CSF may represent nonspecific mediators in cutaneous homeostasis providing inflammatory cytokines, although others have reported a lack of TNF-α in SLS-induced ICD (Oxholm et al., 1991; Boehm et al., 1992) and no elevation in GM-CSF in skin reaction to croton oil (Haas et al., 1992).

The distinct cytokine found to be specific for ACD in mice may be IL-1β (Enk and Katz, 1992; Enk et al., 1993; Mueller et al., 1996); however, an increase of this cytokine in murine epidermis has also been detected in irritant reaction following SLS application (Kondo et al., 1994) as well as after skin stripping (Wood et al., 1992 and 1994), and also after UV-irradiation (Griswold et al., 1991), respectively. Likewise, in human lymph, IL-1β does not distinguish allergic from irritant reactions (Brand et al., 1996). In fact, further studies on the role of IL-1β in ACD (Zheng et al., 1995) showed that IL-1β knock-out mice demonstrated normal contact hypersensitivity (CHS) responses, contradicting the distinction reported by (Enk and Katz, 1992). Furthermore, in vivo removal of epidermal LC (a major source of IL-1β) does not suppress the CHS reactions (Grabbe et al., 1995). Hence, the specific implication of the cytokine IL-1β for contact sensitivity (ACD) remains controversial.

Similarly, IFN-γ, a Th1 cytokine believed to have a key role in delayed-type hypersensitivity or ACD (Fong and Mosmann, 1989), could nevertheless be found in ACD and in irritant responses or ICD (Wilmer et al., 1994; Nickoloff and Naidu, 1994; Enk and Katz, 1992; Haas et al., 1992; Hoefakker et al., 1995).

Various data on cytokines detected in ACD and ICD mirror the complexity at molecular levels involved in skin responses to contact allergens and irritants. There seems no specific cytokines yet that clearly distinguish allergic from irritant reactions.

Even reviewing the possible role and specification of each set of cytokines obtained would herein not necessarily make a conclusion easier when taking into consideration that some former relevant concepts on cytokines have now still been changing: the distinction of CD4+ "helper cells" and CD8+ "suppressor cells" (Kalish and Askenase, 1999), the paradigm of Th1 and Th2 pattern of cytokine expression (Koga et al., 1996; Allen and Maizels, 1997; Werfel et al., 1997; Sallusto et al., 1998; Agarwal et al.,1999), and the corresponding role of certain cytokines in contact sensitivity (e.g., IL-4, IL-10; Gautam et al., 1992; Berg et al., 1995; Salerno et al., 1995; Asada et al., 1997; Webb et al., 1998; Rowe and Bunker, 1998).

It has been proposed that CD4+ and CD8+ T cells have similar capabilities. Both subsets proliferate and can, in turn, produce similar patterns of Th1 and Th2 cytokines. The critical distinction between CD4+ versus CD8+ T cells pertains to recognition of antigens presented by different MHC molecules. The mode of antigen presentation is, in turn, dependent upon the molecular mechanisms of the antigen processing by degradation pathways inside of antigen-presenting cells (Kalish and Askenase, 1999; Bellinghausen et al., 1999). Indeed, there are mounting data indicating that some prior concepts and paradigms in molecular mechanisms of cell-mediated delayed-type hypersensitivity and ACD require revisit (Allen and Maizels, 1997; Werfel et al., 1997; Grabbe and Schwarz, 1998; Agarwal et al., 1999). It is likely that in ACD and other immunological systems, rather the balanced immune response and not the induction of a particular Th cell pathway will become a relevant concept. Nevertheless, cytokines appear in general a floating-and-renewing field of science causing any distinction or paradigm in this context rather nonpermanent.

Further research on cytokines in ICD remains intriguing. For instance, the biological activity of certain cytokines (e.g., IL-4, IL-18) may specifically be implicated in the pathogenesis of contact sensitivity (Gautam et al., 1992; Koga et al., 1996; Stoll et al., 1997; Asada et al., 1997; Rowe and Bunker, 1998, Xu et al., 1999). Stimulating studies should hence focus on the question of whether or not such cytokines will

TABLE 7.6. *Components of future experiments in ACD and ICD*

Subject	Topics that need clarification
Contact allergens	Differences in Potency and physical chemistry (e.g., DNCB vs. benzocain) Dose-response-relation (not linear?) Time course (complex?) Effect on anatomical site (high and less sensitive area?) Secreted cytokines and their receptors (specific?)
Irritants	Differences in Potency, biologic behavior, and physical chemistry (e.g., SLS, BAC vs. NAA) Dose-response-relation (not linear?) Time course (not correlated to each substance?) Solvent effect (anti-irritant?) Effect on anatomical site (high and less sensitive sites) Secreted cytokines and their receptors (not specific?)

be involved in irritant skin responses. On the other hand, to draw a complete figure, long-term investigations on the function and expression pattern of all cytokines, as well as their receptors involved in ACD and ICD, are useful, providing another wide research field of the future (Table 7.6).

Although much information is now available on cytokines in contact dermatitis, this topic needs clarification. Obviously, much additional work is needed to understand the "overlapping" and "flexible" molecular network which leads, for example, to upregulate numerous comparable cytokines in ACD and ICD. This review provides a work in progress only, as new data are continuously obtained, contributing to solve the cytokine puzzle in this field.

REFERENCES

Agarwal, S., Viola, J. P., and Rao, A. (1999): Chromatin-based regulatory mechanisms governing cytokine gene transcription. *Journal of Allergy and Clinical Immunology*, 103:990–9.

Aggarwal, B. B., and Puri, K. J. (1995): Common und uncommon features of cytokines and cytokine receptors: an overview. In: *Human Cytokines: Their Role in Disease and Therapy*. Edited by Aggarwal, B. B., and Puri, K. J., pp. 1–24. Blackwell, Oxford, UK.

Akhurst, R. J., Fee, F., and Balmain, A. (1988): Localized production of TGF-beta mRNA in tumour promoter-stimulated mouse epidermis. *Nature*, 331:363–365.

Allen, J. E., and Maizels, R. M. (1997): Th1-Th2: Reliable paradigm or dangerous dogma? *Immunology. Today*, 18:387–92.

Ambrus, J. L., Jr, Pippin, J., Joseph, Xu, C., Blumenthal, D., Tamayo, A., Claypool, K., McCourt, D., Srikiatchatochorn, A., Ford, R. (1993): Identification of a cDNA for a human high molecular-weight B-cell growth factor (HMW-BCGF). *Proceedings of the National Academy of Sciences of the United States of America*, 90:6330.

Ansel, J. C., Luger T. A., and Green, I. (1983): The effect of in vitro and in vivo UV irradiation on the production of ETAF activity by human and murine keratinocytes. *Journal of Investigative Dermatology*, 81:519–523.

Ansel, J. C., Perry, P., Brown, J., Damm, D., Phan, T., Hart, C., Luger, T., Hefeneider, S. (1990): Cytokine modulation of keratinocyte cytokines. *Journal of Investigative Dermatology*, 94:101s–107s.

Ansel, J. C., Tiesman, J. P., Olerud, J. E., Krueger, J. G., Krane, J. F., Tara, D. C., Shipley, G. D., Gilbertson, D., Usui, M. L., Hart, C. E. (1993): Human keratinocytes are a major source of cutaneous platelet-derived growth factor. *J. Clin. Invest.*, 92:671–678.

Aragane, Y., Riemann, H., Bhardwaj, B., Schwarz, A., Sawada, Y., Yamada, H., Luger, T. A., Kubin, M., Trinchieri, G., Schwarz, T. (1994): IL-12 is expressed and released by human keratinocytes and epidermoid carcinoma cell lines. *Journal of Immunology*, 153:5366–72.

Armitage, R. J., Macduff, B. M., Eisenman, J., Paxton, R., Grabstein, K. H. (1995): IL-15 has stimulatory activity for the induction of B cell proliferation and differentiation. *Journal of Immunology*, 154:483–90.

Asada, H., Linton, J., and Katz, S. I. (1997): Cytokine gene expression during the elicitation phase of contact sensitivity: Regulation by endogenous IL-4. *Journal of Investigative Dermatology*, 108:406–11.

Aubin, F, Kripke, M. L., and Ullrich, S. E. (1991): Activation of keratinocytes with psoralen plus UVA radiation induces the release of soluble factors that suppress delayed and contact hypersensitivity. *J. Invest. Dermatol.*, 97:995–1000.

Barker, J. N., Jones, M. L., Swenson. C. L., Sarma, V., Mitra, R. S., Ward, P. A., Johnson, K. J., Fantone, J. C., Dixit, V. M., and Nickoloff, B. J. (1991): Monocyte chemotaxis and activating factor production by keratinocytes in response to IFN-gamma. *Journal of Immunology*, 146:1192–1197.

Barker, J. N., Mitra, R. S., Griffiths, C. E., Dixit, V. M., Nickoloff, B. J. (1991): Keratinocytes as initiators of inflammation. *Lancet*, 337:211–214.

Bason, M., Lammintausta, K., and Maibach, H. I. (1991): Irritant dermatitis (irritation). In: *Dermatotoxicology, 4th ed.* Edited by Marzulli, F. N., and Maibach, H. I., pp. 223–254. Hemisphere, New York, NY.

Bason, M. M., Gordon, V., and Maibach, H. I. (1991): Skin irritation: In vitro assays. *International Journal of Dermatology*, 30:623–626.

Bellinghausen, I., Brand, U., Enk, A. H., Knop, J., Saloga, J. (1999): Signals involved in the early
TH1/TH2 polarization of an immune response depending on the type of antigen. *Journal of Allergy
and Clinical Immunology*, 103:298–306.
Berg, D. J., Leach, M. W., Kühn, R., Rajewsky, K., Muller, W., Davidson, N. J., Rennick, D. (1995):
Interleukin 10 but not interleukin 4 is a natural suppressant of cutaneous inflammatory responses.
Journal of Experimental Medicine, 182:99–108.
Boehm, K. D., Yun, J. K., Strohl, K. P., et al. (1992): Human epidermal cytokine transcript level changes
in situ following urushiol application [abstract]. *Journal of Investigative Dermatology*, 98:571.
Brand, C. U., Hunziker, T., and Braathen, L. R. (1992): Studies on human skin lymph containing
Langerhans cells from sodium lauryl sulphate contact dermatitis. *Journal of Investigative Dermatology*,
99:109s–110s.
Brand, C. U., Hunziker, T., Schaffner, T., Limat, A., Gerber, H. A., Braathen, L. R. (1995): Activated immunocomponent cells in human skin lymph derived from irritant contact dermatitis: an
immunomorphological study. *British Journal of Dermatology*, 132:39–45.
Brand, C. U., Hunziker, T., Yawalkar, N., Braathen, L. R. (1996): IL-1β protein in human skin lymph
does not discriminate allergic from irritant contact dermatitis. *Contact Dermatitis*, 35:152–156.
Brown, K. D., Zurawski, S. M., Mosmann, T. R., Zurawski, G. (1989): A family of small inducible proteins
secreted by leukocytes are members of a new superfamily that inclues leukocyte and fibroblast-derived
inflammatory agents, growth factors, and indicators of various activation processes. *Journal of
Immunology*, 142:679–87.
Bruze, M., and Emmett, E. A. (1991): Exposures to irritants. In: *Irritant Contact Dermatitis*. Edited by
Jackson, E. M., Goldner, R., pp. 81–99. Marcel Dekker, New York, NY.
Coffey, R. J., Jr., Derynck, R., Wilcox, J. N., Bringman, T. S., Goustin, A. S. (1987): Production and
auto-induction of transforming growth factor-alpha in human keratinocytes. *Nature*, 328:817–820.
Dearman, R. J., and Kimber, I. (1994): Cytokine production and the local lymph node assay. In: *In vitro
skin toxicology: Irritation, Phototoxicity, Sensitization*. pp. 367–372. Edited by Rougier, A., Goldberg,
A. M., Maibach, H. I., Liebert, New York, NY.
DeLeo, V. A., Harbor, L. C., Kong, B. M., DeSalva, S. J. (1987): Surfactant-induced alteration of
arachidonic acid metabolism of mammalian cells in culture. *Proc. Soc. Exp. Biol. Med.*, 184:
477–482.
Denda, M., Emami, S., Wood, L. C., et al. (1995): Epidermal injury vs. barrier disruption as initiators of
epidermal proliferation and inflammation [abstract]. *Journal of Investigative Dermatology*, 104:562.
Dinarello, C. A. (1999): IL-18: A Th1-inducing, proinflammatory cytokine and new member of the IL-1
family. *Journal of Allergy and Clinical Immunology*, 103:11–24.
Enk, A. H., Angeloni, V. L., Udey, M. C., Katz, S. I. (1993): An essential role for Langerhans cell-derived
IL-1 beta in the initiation of primary immune responses in skin. *Journal of Immunology*, 150:
3698–3704.
Enk, A. H., and Katz, S. I. (1992): Early molecular events in the induction phase of contact sensitivity.
Proceedings of the National Academy of Sciences of the United States of America, 89:1398–1402.
Enk, A. H., and Katz, S. I. (1992): Identification and induction of keratinocyte-derived IL-10. *Journal of
Immunology*, 149:92–95.
Enk, A. H., Saloga, J., Becker, D., Baypmadzadeh, M., Knop, J. (1994): Induction of hapten-specific
tolerance by interleukin 10 in vivo. *Journal of Experimental Medicine*, 179:1397–1402.
Ferguson, T. A., Dube, P., and Griffith, T. S. (1994): Regulation of contact hypersensitivity by interleukin
10. *Journal of Experimental Medicine*, 179:1597–1604.
Fong, T. A., and Mosmann, T. R. (1989): The role of IFN-gamma in delayed-type hypersensitivity
mediated by Th1 clones. *Journal of Immunology*, 143:2887–93.
Ford, R., Tamayo, A., Martin, B., Niu, K., Claypool, K., Cabanillas, F., and Ambrus, J., Jr. (1995):
Identification of B-cell growth factors (interleukin-14.; high molecular weight-B-cell growth factors)
in effusion fluids from patients with aggressive B-cell lymphomas. *Blood*, 86:283–93.
Fregert, S. (1975): Occupational dermatitis in a 10-year material. *Contact Dermatitis*, 1:96–107.
Gautam, S. C., Chikkala, N. F., and Hamilton, T. A. (1992): Anti-inflammatory action of IL-4: Negative
regulation of contact sensitivity to trinitrochlorobenzene. *Journal of Immunology*, 148:1411–5.
Gawkrodger, D. J., McVittie, E., Carr, M. M., Ross, J. A., Hunter, J. A. (1986): Phenotypic characterization
of the early cellular responses in allergic and irritant contact dermatitis. *Clinical and Experimental
Immunology*, 66:590–598.
Grabbe, S., and Schwarz, T. (1998): Immunoregulatory mechanisms involved in elicitation of allergic
contact hypersensitivity. *Immunology Today*, 19:37–44.

Grabbe, S., Steinbrink, K., Steinert, M., Luger, T. A., Schwarz, T. (1995): Removal of the majority of epidermal Langerhans cells by topical or systemic steroid application enhances the effector phase of murine contact hypersensitivity. *Journal of Immunology*, 155:4207–17.

Grabstein, K. H., Eisenman, J., Shanebeck, K., Rauch, C., Srinivasan, S., Fung, V., Beers, C., Richardson, J., Schoenborn, M. A., Ahdieh, M., et al. (1994): Cloning of a T cell growth factor that interacts with the beta chain of the interleukin-2 receptor. *Science*, 264:965–8.

Grewe, M., Gyufko, K., and Krutmann, J. (1995): Interleukin-10 production by cultured human keratinocytes: Regulation by ultraviolet B and ultraviolet A1 radiation. *Journal of Investigative Dermatology*, 104:3–6.

Griffiths, C. E., Barker, J. N., Kunkel, S., Nickoloff, B. J. (1991) Modulation of leucocyte adhesion molecules, a T-cell chemotaxin (IL-8) and a regulatory cytokine (TNF-alpha) in allergic contact dermatitis (rhus dermatitis). *Br. J. Dermatol.*, 124:519–526.

Griswold, D. E., Connor, J. R., Dalton, et al. (1991): Activation of the IL-1 gene in UV- irradiated mouse skin: Association with inflammatory sequelae and pharmacologic intervention. *Journal of Investigative Dermatology*, 97:1019–23.

Haas, J., Lipkow, T., Mohamadzadeh, M., Kolde, G., Knop, J. (1992): Induction of inflammatory cytokines in murine keratinocytes upon in vivo stimulation with contact sensitizers and tolerizing analogues. *Experimental Dermatology*, 1:76–83.

Halaban, R., Langdon, R., Birchall, N., Cuono, C., Baird, A., Scott, G., Moellmann, G., McGuire, J. (1988): Basic fibroblast growth factor from human keratinocytes is a natural mitogen for melanocytes. *Journal of Cell Biology*, 107:1611–1619.

Heufler, C., Topar, G., Grasseger, A., Stanzl, U., Koch, F., Romani, N., Namen, A. E., Schuler, G. (1993): Interleukin 7 is produced by murine and human keratinocytes. *Journal of Experimental Medicine*, 178:1109–1104.

Hoefakker, S., Caubo, M., Van't Erve, E. H. M., Roggeveen, M. J., Boersma, W. J., van Joost, T., Notten, W. R., Claassen, E. (1995): In vivo cytokines profiles in allergic and irritant contact dermatitis. *Contact Dermatitis*, 32:258–266.

Hunziker, T., Brand, C. U., Kapp, A., Waelti, E. R., Braathen, L. R. (1992): Increased levels of inflammatory cytokines in human skin lymph derived from sodium lauryl sulphate-induced contact dermatitis. *Br. J. Dermatol.*, 127:254–257.

Kalish, R. S., and Askenase, P. W. (1999): Molecular mechanisms of CD8+ T cell-mediated delayed hypersensitivity: Implications for allergies, asthma, and autoimmunity. *Journal of Allergy and Clinical Immunology*, 103:192–9.

Kaplan, G., Luster, A. D., Hancock, G., Cohn, Z. A. (1987): The expression of a gamma interferon-induced protein (IP-10) in delayed immune responses in human skin. *Journal of Experimental Medicine*, 166:1098–1108.

Kimber, I., and Cumberbatch, M. (1992): Stimulation of Langerhans cell migration by tumor necrosis factor alpha (TNF-alpha). *Journal of Investigative Dermatology*, 99:498–503.

Kirnbauer, R., Köck, A., Nouner, P., Forster, E., Krutmann, J., Urbanski, A., Schauer, E., Ansel, J. C., Schwarz, T., Luger, T. A. (1991): Regulation of epidermal cell interleukin- 6 production by UV light and corticosteroids. *Journal of Investigative Dermatology*, 96:484–489.

Kitajima, T., and Takashima, A. (1999): Langerhans cell responses to ultraviolet B radiation. *Journal of Dermatological Sciences*, 19:153–160.

Köck, A., Schwarz, T., Kirnbauer, R., Urbanski, A., Perry, P., Ansel, J. C., Luger, T. A. (1990): Human keratinocytes are a source for tumor necrosis factor alpha: Evidence for synthesis and release upon stimulation with endotoxin or ultraviolet light. *Journal of Experimental Medicine*, 172:1609–1614.

Köck, A., Urbanski, A., and Luger, T. A. (1989): mRNA expression and release of tumor necrosis factor by human epidermal cells [abstract]. *Journal of Investigative Dermatology*, 92:462.

Koga, T., Fujimura, T., Imayama, S., Katsuoka, K., Toshitani, S., Hori, Y. (1996): The expression of Th1 and Th2 type cytokines in a lesion of allergic contact dermatitis. *Contact Dermatitis*, 35:105–6.

Kondo, S., Kono, T., Sauder, D. N., McKenzie, R. C. (1993): IL-8 gene expression and production in human keratinocytes and their modulation by UVB. *Journal of Investigative Dermatology*, 101:690–694.

Kondo, S., McKenzie, R. C., and Sauder, D. N. (1994): Interleukin-10 inhibits the elicitation phase of allergic contact hypersensitivity. *Journal of Investigative Dermatology*, 103:811–814.

Kondo, S., Pastore, S., Fujisawa, H., Shivji, G. M., McKenzie, R. C., Dinarello, C. A., Sauder, D. N. (1995): Interleukin-1 receptor antagonist suppresses contact hypersensitivity. *Journal of Investigative Dermatology*, 105:334–338.

Kondo, S., Pastore, S., Shivji, G. M., McKenzie, R. C., Sauder, D. N. (1994): Characterization of epidermal cytokine profiles in sensitization and elicitation phases of allergic contact dermatitis as well as irritant contact dermatitis in mouse skin. *Lymphokine, Cytokine, Res.*, 13:367–375.

Kupper, T. S., Ballard, D. W., Chua, A. O., McGuire, J. S., Flood, P. M., Horowitz, M. C., Langdon, R., Lightfoot, L., Gubler, U. (1986): Human keratinocytes contain mRNA indistinguishable from monocyte Interleukin 1 alpha and beta mRNA: Keratinocyte epidermal cell-derived thymocyte-activating factor is identical to interleukin 1. *J. Exp. Med.*, 164:2095–2100.

Kupper, T. S., Lee, F., Coleman, D., Chodakewitz, J., Flood, P., Horowitz, M. (1988): Keratinocyte derived T-cell growth factor (KTGF) is identical to granulocyte macrophage colony stimulating factor (GM-CSF). *Journal of Investigative Dermatology*, 91:185–8.

Kupper, T. S., and McGuire, J. (1986): Hydrocortisone reduces both constitutive and UV elicited release of epidermal thymocyte activating factor (ETAF) by cultured keratinocytes. *J. Invest. Dermatol.*, 87:570–573.

Kupper, T. S., Min, K., Sehgal, P., Mizutani, H., Birchall, N., Ray, A., May, L. (1989): Production of IL-6 by keratinocytes: Implications for epidermal inflammation and immunity. *Ann. N. Y. Acad. Sci.*, 557:454–465.

Kupper, T. S. (1990): Role of epidermal cytokines. In: *Immunophysiology: The Role of Cells and Cytokines in Immunity and Inflammation.* Edited by Oppenheim, J. J., Shevach, E. M., pp. 285–305. Oxford University Press, New York, NY.

Larsen, C. G., Anderson, A. O., Oppenheim, J. J., Matsushima, K. (1989): Production of interleukin 8 by human dermal fibroblasts and keratinocytes in response to interleukin-1 or tumour necrosis factor. *Immunology*, 68:31–36.

Lindberg, M., Farm, G., and Scheynius, A. (1991): Differential effects of sodium lauryl sulphate and non-anoic acid on the expression of CD1a and ICAM-1 in human epidermis. *Acta Dermatology and Venereology (Stockholm)*, 71:384–388.

Lisby, S., Muller, K. M., Jongkeneel, C. V., Saurat, J. H., Hauser, C. (1995): Nickel and skin irritants up-regulate tumor necrosis factor-alpha mRNA in keratinocytes by different but potentially synergistic mechanisms. *Int. Immunol.*, 7:343–352.

Luger, T. A., Köck, A., Kirnbauer, R., Schwarz, T., Ansel, J. C. (1988): Keratinocyte-derived interleukin 3. *Ann. N. Y. Acad. Sci.*, 548:253–261.

Luger, T. A. (1989): Epidermal cytokines. *Acta. Derm. Venereol. Stockh.*, 151:61–76.

Marks, J. G., Jr; Zaino, R. J., Bressler, M. F., Williams, J. V. (1987): Changes in lymphocyte and Langerhans call populations in allergic and irritant contact dermatitis. *International Journal of Dermatology*, 26:354–357.

Mathias, C. G. (1985): The cost of occupational skin disease (editorial). *Arch. Dermatol.*, 121:332–334.

Matsue. H., Cruz, P. D., Jr; Bergstrasser, P. R., Takashima, A. (1992): Langerhans calls are the major source of mRNA for IL-1 beta and MIP-1 alpha among unstimulated mouse epidermal cells. *Journal of Investigative Dermatology*, 99:537–541.

McKenzie, R. C., and Sauder, D. N. (1990): The role of keratinocyte cytokines in inflammation and immunity. *J. Invest. Dermatol.*, 95:105s–107s.

Mikulowska, A. (1992): Reactive changes in human epidermis following simple occlusion with water. *Contact Dermatitis*, 26:224–227.

Mohamadzadeh, M., Müller, M., Hültsch, T., Enk, A., Saloga, J., Knop, J. (1994): Enhanced expression of IL-8 in normal human keratinocytes and human keratinocyte cell line HaCaT in vitro after stimulation with contact sensitizers, tolerogens and irritants. *Exp. Dermatol.*, 3:298–303.

Mohamadzadeh, M., Takashima, A., Dougherty, Knop, J., Bergstrasser, P. R., Cruz, P. D., Jr. (1995): Ultraviolet B radiation up-regulates the expression of IL-15 in human skin. *Journal of Immunology*, 155:4492–6.

Mosmann, T. R., Cherwinski, H., Bond, Giedlin, M. A., Coffman, R. L. (1986): Two types of murine helper T cell clone. I: Definition according to profiles of lymphokine activities and secreted proteins. *Journal of Immunology*, 136:2348–57.

Mosmann, T. R., and Coffman, R. L. (1989): Th1 and Th2 cells: Different patterns of lymphokine secretion lead to different functional properties. *Annual Review of Immunology*, 7:145–73.

Muller, G., Knop, J., and Enk, A. H. (1996): Is cytokine expression responsible for differences between allergens and irritants? *American Journal of Contact Dermatitis*, 3:177–184.

Nickoloff, B. J., and Naidu, Y. (1994): Perturbation of epidermal barrier function correlates with initiation of cytokine cascade in human skin. *J. Am. Acad. Dermatol.*, 30:535–546.

Norris, D. A., Lyons, M. B., Middleton, M. H., Yohn, J. J., Kashihara-Sawami, M. (1990): Ultraviolet radiation can either suppress or induce expression of intercellular adhesion molecule 1 (ICAM-1) on the surface of cultured human keratinocytes. *Journal of Investigative Dermatology*, 95:132–138.

Olson, C. T. (1991): Evaluation of the dermal Irritancy of chemicals. In: *Dermal and Ocular Toxicology: Fundamentals and Methods*. Edited by Hobson, D. W., pp. 135–149. CRC, Boca Raton, FL.

Oppenheim, J. J., Ruscotti, F. W., Faftynek, C., and Cytokines. (1991): In: *Basic and Clinical Immunology*. Edited by Stites, D. P., Terr, A. l., pp. 78–100. Prentice-Hall, Norwalk, NJ.

Oxholm, A., Oxholm, P., Avnstorp, C., Bendtzen, K. (1991): Keratinocyte-expression of interleukin-6 but not of tumour necrosis factor-alpha is increased in the allergic and the irritant patch test reaction. *Acta. Derm. Venereol. Stockh.*, 71:93–98.

Patrick, E., Burkhalter, A., and Maibach, H. I. (1987): Recent investigations of mechanisms of chemically induced skin irritation in laboratory mice. *J. Invest. Dermatol.*, 88:24s–31s.

Piguet, P. F., Grau, G. E., Hauser, C., Vassalli, P. (1991): Tumor necrosis factor is a critical mediator in hapten induced irritant and contact hypersensitivity reactions. *J. Exp. Med.*, 173:673–679.

Piguet, P. F. (1992): Keratinocyte-derived tumor necrosis factor and the physiopathology of the skin. *Springer. Semin. Immunopathol.*, 13:345–354.

Pinkus, H. (1951): Examination of epidermis by the strip method. II: Biometric data on regeneration of the human epidermis. *J. Invest. Dermatol.*, 1951: 431–447.

Rambukkana, A., Pistor, F. H. M., Kapsenberg, M. L., et al. (1994): Role of cytokines in allergen-induced effect on human Langerhans cells (abstract). *Exp. Dermatol.*, 3:140.

Reilly, D. M., and Green, M. R. (1999): Eicosanoid and cytokine levels in acute skin irritation in response to tape stripping and capsaicin. *Acta. Dermatology and Venereology (Stockholm)*, 79:187–190.

Riemann, H., Schwarz, A., Grabbe, S., Aragane, Y., Luger, T. A., Wysocka, M., Kubin, M., Trinchieri, G., Schwarz, T. (1996): Neutralization of IL-12 in vivo prevents induction of contact hypersensitivity and induces hapten-specific tolerance. *Journal of Immunology*, 156:1799–803.

Rivas, J. M., and Ullrich, S. E. (1993): Essential role of keratinocyte-derived IL-10 in the UV- induced suppression of delayed type hypersensitivity but not contact hypersensitivity (abstract). *J. Invest. Dermatol.*, 100:522.

Rowe, A., and Bunker, C. B. (1998): Interleukin-4 and the interleukin-4 receptor in allergic contact dermatitis. *Contact Dermatitis*, 38:36–9.

Salerno, A., Dieli, F., Sireci, G., Bellavia, A., Asherson, G. L. (1995): Interleukin-4 is a critical cytokine in contact sensitivity. *Immunology*, 84:404–9.

Sallusto, F., Lanzavecchia, A., and Mackay, C. R. (1998): Chemokines and chemokine receptors in T-cell priming and Th1/Th2-mediated responses. *Immunology Today*,19:568–74.

Sauder, D. N., Wong, D., McKenzie, R., et al. (1988): The pluripotent keratinocyte: Molecular characterization of epidermal cytokines (abstract). *J. Invest. Dermatol.*, 90:605.

Schmitt, D. A., Owen-Schaub, L., and Ullrich, S. E. (1995): Effect of IL-12 on immune suppression and suppressor cell induction by ultraviolet radiation. *Journal of Immunology*, 154:5114–20.

Schwarz, A., Grabbe, S., Riemann, H., Aragane, Y., Simon, M., Manon, S., Andrade, S., Luger, T. A., Zlotnik, A., Schwarz, T. (1994): In vivo effects of interleukin-10 on contact hypersensitivity and delayed-type hypersensitivity reactions. *J. Invest. Dermatol.*, 103:211–216.

Schwarz, T., and Luger, T.A. (1989): Effect of UV irradiation on epidermal cell cytokine production. *J. Photochem. Photobiol.*, B4:1–13.

Schwarz, T., and Luger, T. A. (1992): Pharmacology of cytokines in the skin. In: *Pharmacology of the Skin*. Edited by Mukhtar, H., pp. 283–313. CRC, Boca Raton, FL.

Schwarz, T., Urbanski, A., and Luger, T. A. (1994): Ultraviolet light and epidermal cell derived cytokines. In: *Epidermal Growth Factors and Cytokines*. Edited by Luger, T. A., Schwarz, T., pp. 303–324. Marcel Dekker, New York, NY.

Shivji, G. M., Gupta, A. K., and Sauder, D. N. (1994): Role of cytokines in irritant contact dermatitis. In: *In Vitro Skin Toxicology: Irritation, Phototoxicity and Sensitization*. Edited by Rougier, A., Goldberg, A. M., Maibach, H. I., pp. 13–22. Liebert, New York, NY.

Stoll, S., Müller, G., Kurimoto, M., Saloga, J., Tanimoto, T., Yamauchi, H., Okamura, H., Knop, J., Enk, A. H. (1997): Production of IL-18 (IFN-gamma-inducing factor) messenger RNA and functional protein by murine keratinocytes. *Journal of Immunology*, 159:298–302.

Takashima, A., and Bergstresser, P. R. (1996): Impact of UVB radiation on the epidermal cytokine network. *Photochem. Photobiol.*, 63: 397–400.

Tekamp-Olson, P., Gallegos, C., Bauer, D., McClain, J., Sherry, B., Fabre, M., van Deventer, S., Cerami, A. (1990): Cloning and characterization of cDNAs for muhne macrophage inflammatory protein 2 and its human homologues. *J. Exp. Med.*, 172:911–919.

Tsai, J. C., Feingold, K. R., Crumrine, D., Wood, L. C., Grunfeld, C., Elias, P. M. (1994): Permeability barrier disruption afters the localization and expression of TNF alpha-protein in the epidermis. *Arch. Dermatol. Res.*, 286:242–248.

Vejlsgaard, G. L., Ralfkiaer, E., Avnstorp, C., Czajkowski, M., Marlin, S. D., Rothlein, R. (1989): Kinetics and characterization of intercellular adhesion molecules 1 (ICAM-1) expression on keratinocytes in various inflammatory skin lesions and malignant cutaneous Lymphomas. *J. Am. Acad. Dermatol.*, 20:782–790.

Webb, E. F., Tzimas, M. N., Newsholme, S. J., Griswold, D. E. (1998): Intralesional cytokines in chronic oxazolone-induced contact sensitivity suggest roles for tumor necrosis factor alpha and interleukin-4. *Journal of Investigative Dermatology*, 111:86–92.

Werfel, T., Hentschel, M., Renz, H., Kapp, A. (1997): Analysis of the phenotype and cytokine pattern of blood- and skin-derived nickel specific T cells in allergic contact dermatitis. *International Archives of Allergy and Immunology*, 113:384–6.

Whittle, E., Carter, J., Wolfreys, A., et al. (1995): HaCaT-derived cytokine response to noxious agents (abstract). *J. Invest. Dermatol.*, 104:562.

Willis, C. M., Stephens, C. J., and Wilkinson, J. D. (1989): Epidermal damage induced by irritants in man: A light and electron microscopic study. *J. Invest. Dermatol.*, 93:695–699.

Willis, C. M., Stephens, C. J., and Wilkinson, J. D. (1991): Selective expression of immune associated surface antigens by keratinocytes in irritant contact dermatitis. *J. Invest. Dermatol.*, 96:505–511.

Willis, C. M., Stephens, C. J., and Wilkinson, J. D. (1992): Differential effects of structurally unrelated chemical irritants on the density of proliferating keratinocytes in 48 h patch test reactions. *J. Invest. Dermatol.*, 99:449–453.

Willis, C. M., Young, E., Brandon, D. R., Wilkinson, J. D. (1986): Immunopathological and ultrastructural findings in human allergic and irritant contact dermatitis. *Br. J. Dermatol.*, 115:305–316.

Wilmer, J. L., Burleson, F. G., Kayama, F., Kanno, J., Luster, M. I. (1994): Cytokine induction in human epidermal keratinocytes exposed to contact irritants and its relation to chemical-induced inflammation in mouse skin. *J. Invest. Dermatol.*, 102:915–922.

Wood, L. C., Elias, P. M., Sequelra-Martin, S. M., Grunfeld, C., Feingold, K. R. (1994): Occlusion lowers cytokine mRNA levels in essential fatty acid-deficient and normal mouse epidermis, but not after acute barrier disruption. *J. Invest. Dermatol.*, 103:834–838.

Wood, L. C., Feingold, K. R., Sequeira-Martin, S. M., Elias, P. M., Grunfeld, C. (1994): Barrier function co-ordinately regulates epidermal IL-1 and IL-1 receptor antagonist mRNA levels. *Exp. Dermatol.* 3:56–60.

Wood, L. C., Jackson, S. M., Elias, P. M., Grunfeld, C., Feingold, K. R. (1992): Cutaneous border perturbation stimulates cytokine production in the epidermis of mice. *J. Clin. Invest.*, 90:482–487.

Xu, B. H., Aoyama, K., Yu, S., Kitani, A., Okamura, H., Kurimoto, M., Matsuyama, T., Matsushita, T. (1999): Expression of Interleukin-18 in murine contact hypersensitivity. *Journal of Interferon and Cytokine Research*, 18: 653–659.

Yao, Z., Painter, S. L., Fanslow, W. C., Ulrich, D., Macduff, B. M., Spriggs, M. K., Armitage, R. J. (1995): Human IL-17: A novel cytokine derived from T cells. *Journal of Immunology*, 155:5483–6.

Yoshikawa, T., and Streilein, J. W. (1990): Genetic basis of the effects of ultraviolet light B on cutaneous immunity: Evidence that polymorphism at the Tnfa and Lps loci governs susceptibility. *Immunogenetics*, 32:398–405.

Zheng, H., Fletcher, D., Kozak, W., Jiang, M., Hofmann, K. J., Conn, C. A., Soszynski, D., Grabiec, C., Trumbauer, M. E., Shaw, A., et al. (1995): Resistance to fever induction and impaired acute-phase response in interleukin-1-β-deficient mice. *Immunity*, 3:9–19.

Zurawski, G., and de Vries, J. E. (1994): Interleukin 13, an interleukin 4-like cytokine that acts on monocytes and B cells, but not on T cells. *Immunology Today*, 15:19–26.

8

Assay to Quantify Subjective Irritation Caused by the Pyrethroid Insecticide Alpha-Cypermethrin

Rudolf A. Herbst

*Department of Dermatology, University of California, San Francisco, California, USA
and Department of Dermatology and Allergology, Hannover Medical University,
Hannover, Germany*

Robert B. Strimling

Department of Dermatology, University of California, San Francisco, California, USA

Howard I. Maibach

*Department of Dermatology and Allergology, Hannover Medical University,
Hannover, Germany*

INTRODUCTION

Synthetic pyrethroid insecticides, widely used in agriculture, cause subjective (sensory) irritation (Flannigan and Tucker, 1985; Knox and Tucker, 1982; Knox et al., 1984; Tucker and Flannigan, 1983). This irritation begins about 1 hour after contact,

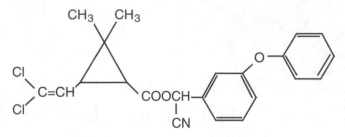

FIG. 8.1. Chemical structure of cypermethrin.

peaks over 3–6 hours, and can last for as long as 24 hours. These compounds penetrate the skin and interfere with axonal function (Lammintausta et al., 1988). Use of Vitamin E oil, if applied directly after contact to the compound, ameliorates the adverse effect of the pyrethroid fenvalerate (Le Quesne et al., 1980). The aim of this study was to establish a subjective irritation assay for a synthetic pyrethroid, alpha-cypermethrin (AC) (see Figure 8.1 for chemical structure), in determining practical test vehicles, concentrations, anatomic sites, and conditions. All subjects had a previous episode of a subjective irritant reaction to a cosmetic. We divided this test subject population into stingers and nonstingers utilizing the lactic acid (LA) assay. This assay has been shown to be one screening test for discovering individuals with sensitive skin who are more likely to experience subjective irritation as well as objective irritation: "stingers" (Frosch and Kligman, 1977; Lammintausta et al., 1988).

MATERIAL AND METHODS

Subjects

All participants denied acute and/or active chronic skin conditions in the area of testing within the period of the tests. Additionally, all were free of active skin disease at the site of compound application by physical examination by the physician tester; informed consent was obtained. The study was approved by the Institutional Review Board of the University of California, San Francisco.

Test subjects were women who had positive histories of experiencing immediate (i.e., onset of symptoms within 1–2 minutes of application), sustained (i.e., at least of 5 minutes' duration) stinging, burning, and/or itching to topically applied cosmetics; 45 participants qualified. Of those 45 screened participants who were tested with the LA test, 11 reacted positively ("stingers") (i.e., experienced stinging, burning, and/or itching of the above described variety and of moderate or greater intensity). Eight stingers and 8 nonstingers were randomly selected from the respective groups and tested further with the test compound, AC, utilizing different vehicles. Both subgroups were further equally divided to be able to also assess the effect of profuse

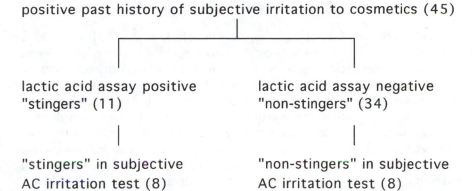

FIG. 8.2. Structure of the test subject population.

facial sweating on the subjective response to the test compound. Figure 8.2 shows the structure of the test subject population.

SUBJECTIVE IRRITATION TESTS

LA Test

The LA test, originally described by Frosch and Kligman (1997) and minimally modified by Lammintausta et al. (1998), was employed. All women were instructed to not wear make-up or any other facial topical product on the day of testing and to have gently cleansed their face with a mild soap that morning. Those who wore a facial cosmetic or other product mistakenly were given a mild soap to gently cleanse their face. Profuse facial sweating was induced with a commercial facial steamer ("The steam facial," Katz Manufacturing Co. Inc., New York) for 15 minutes. Five percent aqueous LA was applied liberally with a soft-tipped cotton applicator on the nasolabial and cheek skin of one side. Water was applied in a similar manner to the corresponding opposite anatomic site. The subject was observed for 15 minutes after application of test compound and control. Immediately after placement of the compounds, the participants gently massaged in a circular motion the corresponding right and left sites of application with their corresponding right and left fingers, so as not to mix the compounds, as instructed for approximately 1–2 minutes. At 5, 10, 15 minutes post-application, participants were asked as to whether or not they were experiencing any discomfort, namely stinging, burning, and/or itching. Additionally, participants identified the site or sites and the quality of discomfort as instructed. Patients graded the discomfort as being either mild, moderate, or marked. Subjects experiencing at least a moderate level of subjective irritation at 5, 10, or 15 minutes, when questioned, were considered stingers.

Alpha-Cypermethrin Irritation Test

Applications of 2 mg/cm^2 of test compound or vehicle control in 1 of 2 vehicles (pesticide-free lanolin or acetone) were delivered via a tuberculin syringe for lanolin or a micro-pipette system for acetone to a 1 cm^2 area in 1 of 4 anatomic sites: left or right lower earlobe or left or right nasolabial region. The corresponding control (vehicle) was delivered to the opposite anatomic site. The test compound in the 2nd vehicle was delivered to the other anatomic site on the same side of the body as the first test compound application site. Similarly, its corresponding control (vehicle) was delivered to the opposite corresponding anatomic site. Application products and sites were randomized, keeping with the above. Grading of the irritant reaction at 15, 30, and 60 minutes and at 1, 2, and 6 hours post-application was on a scale of 0–4, with 0 = no sensation, 1 = mildly felt sensation, 2 = definite sensation, no discomfort, 3 = moderate sensation, little or no discomfort, 4 = intense sensation with discomfort. Test materials were washed with soap and water at the 6th hour. Additionally, those participants utilizing facial steaming underwent 10 minutes of steam treatment prior to application of the compounds, and then an additional 30 minutes of steam therapy post-application. First, AC 5 mg/g in both vehicles (lanolin and acetone) was tested. Half of the participants of either group received the facial steaming protocol. Then, stingers were tested with AC at 2.5 mg/g in acetone and at 25 mg/g in lanolin, and nonstingers with AC at 10 mg/g in acetone and at 25 mg/g in lanolin. This time the other half of the participants in each of those groups received the facial steaming protocol. Finally, AC at 1 mg/g in acetone and at 12.5 mg/g in lanolin was tested in both groups. No participant underwent facial steaming at the latter concentrations. A positive response to the AC irritation test was defined as any responder with a 2 or greater response at the 2-hour and/or 6-hour observation period.

RESULTS

1. Screening: 45 subjects qualified for LA testing by a history of prior subjective irritation to cosmetics. Ages ranged from 18–66 years old (mean: 41.5 years old). By race, 23/45 (51.1%) were Hispanic, 17/45 (37.8%) were white, and 5/45 (11.1%) were black.
2. LA test: 11/45 (24.4%) of the screened participants described a subjective irritant reaction to LA testing. Ages of those stingers ranged from 23–64 years old (mean: 38.5 years old). By race, 5/11 (45.5%) were Hispanic, 5/11 (45.5%) white, 1/11 (9.0%) black. Thus, 21.7% of the Hispanics, 29.4% of the whites, and 20% of the blacks tested in the LA test were found to be stingers.
3. AC irritation test: Eight subjects of either group (stingers and nonstingers) underwent the AC irritation test. Average age was 42.7 years old (42.1 years old for stingers, 43.3 years old for nonstingers). By race, 43.5% were Hispanic (50% for

TABLE 8.1. *AC subjective irritation test results of all tests*

Total no. of tests	192	
	With compound	Without compound
No. of tests	96	96
No. of positive tests	35 = true-positives	2 = false-positives
Percentage of positive tests	36.46	2.08
Positive tests classified		
Face sauna/without*	11 + 24	0 + 2
Stinger/nonstinger	22 + 13	0 + 2
Test site: nose/ear	23 + 12	2 + 0
Test vehicle: acetone/lanolin	23 + 12	2 + 0

*Of the total number of 192 tests, 128 were done with face sauna, 64 without.

stingers/37.5% for nonstingers), 37.5% were white (37.5%/37.5%), 18.8% were black (12.5%/25%).

Table 8.1 shows the results of all tests performed, specifying true positive reactions from false-positive reactions as well as the distribution of those within the different test-subpopulations (stingers/nonstingers), test conditions (with/without face steam protocol), anatomical test sites (nosolabial/ear), and test vehicle (lanolin/acetone). Table 8.2 specifies the positive AC irritation test results as to concentrations of AC used in the two vehicles (lanolin/acetone), as well as classifying them according to test subpopulation, test condition, and anatomical test site. Figures 8.3 and 8.4 show that the AC irritation test responses correlated to the test concentrations used within

TABLE 8.2. *AC subjective irritation test results with compound*

a-Cypermethrin in lanolin	5.0 mg/g	12.5 mg/g	25 mg/g	
Total no. of tests	16	16	16	
No. of positive tests/no. of total tests	0\16	4\16	8\16	
Percentage of positive tests	0	25	50	
Positive tests classified				
Stingers + nonstingers	0 + 0	2 + 2	5 + 3	
With face sauna + without	0 + 0	nd + 4	3 + 5	
Positive test sites: nose + ear	0 + 0	2 + 2	7 + 1	
a-Cypermethrin in acetone	1.0 mg/g	2.5 mg/g	5.0 mg/g	10.0 mg/g
Total no. of tests	16	8	16	8
No. of positive tests/no. of total tests	4/16	5/8	8/16	6/8
Percentage of positive tests	25	62.5	50	75
Positive tests classified				
Stingers + non-stingers	4 + 0	5 + nd	6 + 2	nd + 6
With face sauna + without	nd + 4	2 + 3	3 + 5	3 + 3
Positive test sites: nose + ear	2 + 2	3 + 2	5 + 3	4 + 2

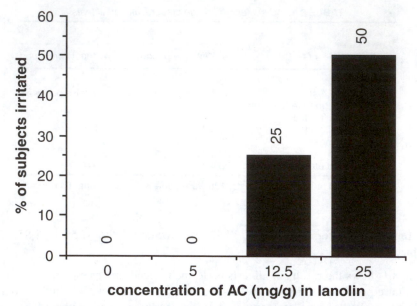

FIG. 8.3. AC in lanolin—response of the subjects to AC in lanolin.

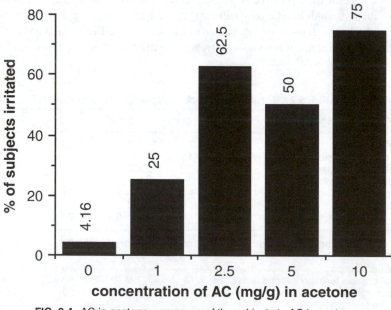

FIG. 8.4. AC in acetone—response of the subjects to AC in acetone.

TABLE 8.3. *Test subjects and positive reactions**

Subject	Concentration of AC in Acetone or Lanolin						
	1.0 A	2.5 A	5.0 A	10 A	5.0 L	12.5 L	25 L
1			N*	nd			N
2	E	N	E*	nd			E
3	N	E	N*	nd			N
4		N		nd			
5			N	nd			N*
6	E	N*	E	nd		N	
7	N	E*	N	nd		E	N*
8				nd			
9		nd	N/N**	E		E	N
10		nd		N			
11		nd	N**				N
12		nd	E	N		N	
13		nd					
14		nd		N*			
15		nd		E*			N*
16		nd		N*			

*Positive irritation reactions to AC of each subject (subjects 1–8 are stingers; 9–16 non-stingers).

1.0 A = test concentration was 1.0 mg/g AC in acetone; 5.0 L = test concentration was 5.0 mg/g AC in lanolin; Blank = negative reaction; E = positive irritation reaction at earlobe; N = positive irritation reaction at nasolabial region; N* and E* = positive with facial steam protocol; N** = positive irritation reaction to control at nasolabial region; nd = not done.

each of the two vehicles. Table 8.3 specifies the positive irritant reaction of each subject with respect to the test modalities.

DISCUSSION

Little is known about the mechanisms of subjective skin irritation; however, Lammintausta et al. demonstrated the importance of previously experienced subjective discomfort to cosmetics, the basis of our screening question, and its correlation to a positive LA assay (Lammintausta et al., 1988). The LA stinging assay is one screening test for identifying individuals with sensitive skin who are more likely to experience subjective as well as objective irritation (Frosch and Kligman, 1977; Lammintausta et al., 1988). Approximately 10%–15% of unscreened individuals experience significant, sustained facial stinging, burning, and/or itching to application of an aqueous solution of 5% LA under sweating conditions. Without sweating conditions, similar responses are obtained utilizing 10% LA (Frosch and Kligman, 1977). We tested 45 persons with the LA assay that all had previously experienced burning, stinging, and/or itching reactions to certain cosmetics and found 11 responders (24.4%). Our screening thus identified a larger population of stingers in our LA-tested participants compared to the numbers in previously performed studies utilizing unscreened subjects, 24.4% vs. 10%–15% (Frosch and Kligman, 1977; Lammintausta et al., 1988).

Purely sensory reactions exist, in particular to the synthetic pyrethroid insecticides. This diverse group of substances, when contacting skin, cause a variety of paresthesias, notably tingling, and numbness, that begin at about 1 hour after contact, peak over the next 3–6 hours, and last approximately 24 hours (Flannigan and Tucker, 1985; Knox et al., 1984). Mechanistically, these compounds penetrate the skin and interfere with axonal function (Le Quesne et al., 1980). In contrast, subjective irritation experienced with LA begins at 5 minutes, is more transient, and is qualitatively different. We performed a subjective irritation test with the synthetic pyrethroid AC on subjects previously divided into stingers and nonstingers by the LA assay, as described above. The purpose was to assess the most practical test vehicle as well as to define optimal test concentrations. Additionally, we employed a facial steam protocol to ascertain if this enhances AC subjective irritation as in the case of LA.

The results (Tables 8.2–8.4; Figures 8.2 and 8.3) permit several conclusions. The specificity of our approach seemed to be reasonable because the number of false-positives (i.e., subjects stating a 2+ reaction in an area where vehicle alone had been applied) was only about 2%. With lanolin as the vehicle, no false-positive reactions occurred.

Utilization of the facial steam protocol did not influence the response (note that the number of tests utilizing the facial steam protocol was half of the number of tests without). Stingers were almost twice as frequently irritated by AC than nonstingers.

The nasolabial area had twice the response of the earlobe. This correlates well with the observation, utilizing various bioengineering techniques that demonstrates the high irritability of this area (Shriner and Maibach, 1996).

With lanolin as the vehicle, AC-subjective irritation responses are likely to be elucidated at concentrations between 5.0–12.5 mg/g. A concentration of 5.0 mg/g or lower is likely to be negative, at least in small test populations. With acetone as the vehicle, subjective irritation was seen at as low a concentration as 1.0 mg/g.

Finally, with both vehicles, an increase of subjective irritation reactions correlated with increasing concentrations of AC in an almost dose-dependent manner. In addition, the LA assay correlated with the AC irritation test in that the likelihood of a subjective irritant reaction to AC was twice as high in LA-positive subjects (stingers) than in LA-negative subjects (nonstingers), although it remains unknown how similar the mechanisms of both phenomena are. In conclusion, it is believed that a practical approach to determine subjective irritation caused by AC could be determined as described above. We do not overgeneralize these observations until experience with other delayed-onset subjective irritants is developed.

REFERENCES

Flannigan, S. A., and Tucker, S. B. (1985): Variation in cutaneous sensation between synthetic pyrethroid insecticides. *Contact Dermatitis*, 13:140–147.

Frosch, P. J., and Kligman, A. M. (1977): A method for appraising the stinging capacity of topically applied substances. *J. Soc. Cosmet. Chem.*, 28:197–209.

Knox, J. M., and Tucker, S. B. (1982): A new cutaneous sensation caused by synthetic pyrethroids. *Clinical Research*, 30:915A.

Knox, J. M., Tucker, S. B., and Flannigan S. A. (1984): Paresthesia from cutaneous exposure to a synthetic pyrethroid insecticide. *Arch. Dermatol.*, 120:744–746.

Lammintausta, K., Maibach, H. I., and Wilson, D. (1988): Mechanisms of subjective (sensory) irritation: Propensity to non-immunologic contact urticaria and objective irritation in stingers. *Dermatosen*, 36: 45–49.

Le Quesne, P. M., Maxwell, I. C., and Butterworth, S. T. G. (1980): Transient facial sensory symptoms following exposure to synthetic pyrethroids: A clinical and electrophysiological assessment. *Neurotoxicology*, 2:1–11.

Shriner, D. L., Maibach, H. I. (1996): Regional variation of nonimmunologic contact urticaria: Functional map of the human face. *Skin Pharmacol.*, 9:312–32.

Tucker, S. B., and Flannigan, S. A. (1983): Cutaneous effects from occupational exposure to fenvalerate. *Arch. Toxicol.*, 54:195–202.

Tucker, S. B., Flannigan, S. A., and Ross, C. E. (1984): Inhibition of cutaneous paresthesia resulting from synthetic pyrethroid exposure. *Int. J. Dermatol.*, 23:686–689.

Toxicology of Skin
Edited by Howard I. Maibach
Copyright © 2001 Taylor & Francis

9

Prediction of Skin Corrosivity Using Quantitative Structure-Activity Relationships

Martin D. Barratt

Marlin Consultancy, Carlton, Bedford, United Kingdom

INTRODUCTION

The principles of Quantitative Structure-Activity Relationships (QSAR) are based on the premise that the properties of a chemical are implicit in its molecular structure. As a consequence, if a mechanistic hypothesis can be proposed that links a group of related chemicals with a particular toxic endpoint, the hypothesis can then be used to define relevant parameters to establish a structure activity relationship. The resulting model is then tested and the hypothesis and parameters refined until an adequate model is obtained.

Within the context of this paper, a "corrosive" substance is one which causes irreversible destruction of skin tissue when applied to rabbit skin for a period of up to 4 hours, as defined in the EC Annex V method (EEC, 1984). This corrosivity classification corresponds to United Nations "packing group" I (UN, 1977).

Skin irritants and corrosives comprise a whole range of different chemistries; therefore a variety of different mechanisms may be postulated for their irritant and/or corrosive properties. Some inorganic acids and bases are highly reactive and are irritant or corrosive by direct action with the external surface of the skin; however, they do not penetrate readily into intact skin by virtue of their high polarity. Cationic and anionic surfactants also have low tissue penetration due to their ionic head groups and relatively large chemical size; irritation or corrosivity for these chemicals can also be expected to result initially from interaction with the surface tissues of the skin. Organic chemicals generally have higher skin permeability than inorganic chemicals and surfactants because of their greater lipophilicity; for these chemicals, skin permeability has a greater influence on their potential to be skin irritants or corrosives. Within the category of organic chemicals, some acids, bases, phenols, and electrophilic chemicals are classified as corrosives.

Most neutral organic chemicals appear to be noncorrosive. Neutral organic chemicals within the scope of these QSAR models are defined as uncharged, carbon-based chemicals that do not possess the potential either to react covalently or to ionise under the conditions prevalent in biological systems. Common chemical classes covered by the definition are hydrocarbons, alcohols, ethers, esters, ketones, amides, unreactive halogenated compounds, unreactive aromatic compounds, and aprotic polar chemicals.

It is well established that for a chemical to be biologically active, it must first be transported from its site of administration to its site of action, and then it must bind to or react with its receptor or target (Dearden, 1990) (i.e., biological activity can be expressed as a function of two terms: partition and reactivity). The mechanistic hypothesis underlying the QSAR models for the skin corrosivity of organic chemicals can therefore be summarized in these terms as follows: The chemical must first penetrate the skin; having done this, the corrosivity of the chemical will also depend on its cytotoxicity.

The partition component of skin corrosivity is the ability of the chemicals to penetrate the skin. The parameters used to model skin permeability (Barratt, 1995a) are log [octanol/water partition coefficient] (logP), molecular volume, and melting point. The reactivity component, which models the cytotoxic potential of the chemicals, varies for the different chemical classes. The pK_a (-log[ionisation constant]) is used to model the cytotoxic potential of organic acids, bases, and phenols (Barratt, 1995b) whereas the dipole moment is used for electrophilic organic chemicals (Barratt, 1996a) (c.f. eye irritation potential of neutral organic chemicals) (Barratt, 1996c).

METHODS AND DATA

Chemical structures were constructed using Sybyl chemical modelling software (Tripos Associates, Milton Keynes, UK). After energy minimization and calculation of logP values using the CHEMICALC system (Suzuki and Kudo, 1990), the chemical structures were imported into the TSAR spreadsheet (Oxford Molecular Ltd, Oxford, UK), where molecular volumes were calculated.

pK_a values were obtained from the Handbook of Chemistry and Physics (Weast, 1970) and the Handbook of Biochemistry and Molecular Biology (Fasman, 1976), except where indicated. Melting points were obtained from suppliers' catalogues. All chemicals with melting points of 37°C or lower were assigned a default value of 37°C on the basis that they would be in the liquid state when applied to rabbit skin (see Suzuki, 1991).

Data sets were analyzed by principal components analysis (TSAR); data were standardized by mean and standard deviation. In principal components analysis, the original variables are transformed into a new orthogonal set of linear combinants called *principal components*. The largest vector of the total variance of the data set is described by the first principal component; the next largest vector, by the second component, and so on. This technique allows multicomponent data sets to be reduced to 2- or 3-dimensional plots without significant loss of information. A more detailed account of the application of principal components analysis can be found elsewhere (Niemi, 1990).

SKIN CORROSIVITY OF ORGANIC ACIDS, BASES, AND PHENOLS

A plot of the first two principal components of logP, molecular volume, melting point, and pK_a for a set of organic acids is shown in Figure 9.1. Organic acids classified as corrosive on the basis of their effects on rabbit skin are clearly separated from those not classified as corrosive. The counterbalancing effects on the corrosivity potential of the different variables can be seen clearly in this plot. Acids with lower logP values, larger molecular volumes, or higher melting points—features associated with lower skin permeability—are less likely to be found in the "corrosive" area of the plot unless they are particularly acidic and therefore more cytotoxic (e.g., oxalic acid (**19** in Figure 9.1), which melts at 106°C but has a very low pK_a).

The effect of logP is more complicated. Acids with low logP values can be expected to have low skin permeability. Skin permeability increases almost linearly with logP up to a point, after which it begins to level off. This effect can be seen in human skin in vitro permeability data for the fatty acids in the data set published by (Flynn, 1990). As logP is increased further, there is evidence that skin permeability actually starts to decrease (Hinz et al., 1991). The consequence of this latter effect, together with the increase in melting point with increasing alkyl chain length, is that fatty acids with alkyl chains of 8 carbons and below (e.g., hexanoic (**9**) and octanoic (**10**)) are corrosive to rabbit skin, whereas those with 10 carbons or more (e.g., decanoic (**11**) and dodecanoic (**12**)), are not.

A plot of the first two principal components of logP, molecular volume, melting point, and pK_a for a set of organic bases is shown in Figure 9.2. In this data set, the effect of pK_a can be seen quite clearly in that while most of the aliphatic amines are classified as corrosive, the aromatic heterocyclic bases are all noncorrosive as a consequence of their lower pK_a values. An interesting feature of this data set is a group of four chemicals—furfurylamine (**36**), N-ethylmorpholine (**37**), N-methylmorpholine (**38**), and triethanolamine (**39**)—labeled as "equivocal" because they are classified as irritant by some suppliers and corrosive by others. Located on the principal components plot on the boundary between corrosive and noncorrosive chemicals, these chemicals provide an interesting example of the effect of biological variability. For much of

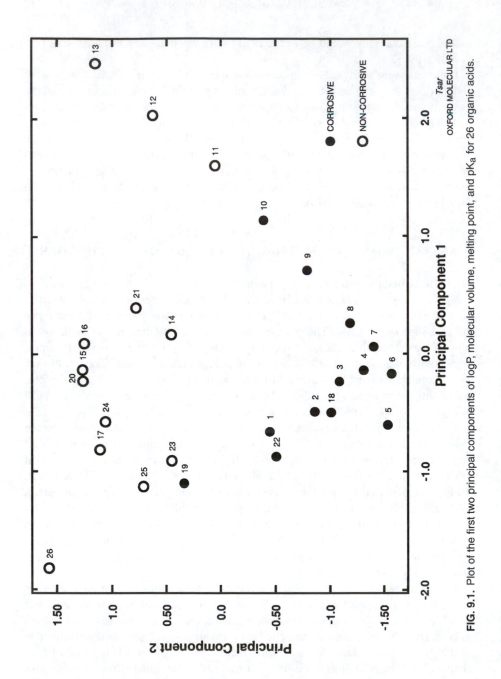

FIG. 9.1. Plot of the first two principal components of logP, molecular volume, melting point, and pK_a for 26 organic acids.

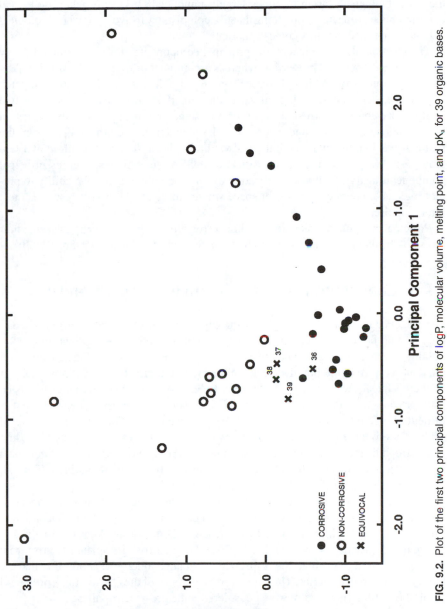

FIG. 9.2. Plot of the first two principal components of logP, molecular volume, melting point, and pK$_a$ for 39 organic bases.

the time, biological variability goes largely unnoticed! It is only when biological experiments are repeated (something that toxicologists try to avoid) or when biological data are correlated that it becomes apparent.

The location of a chemical on the principal components plot is determined solely by its physico-chemical properties, whereas assignment of toxicity classification depends on results from a biological assay, in this case the rabbit 4-hour covered patch test. In these examples, biological variability manifests itself as the results of two well-conducted skin corrosivity tests on the same chemical leading to an irritant classification in one case and a corrosive classification in the other. Away from the boundary region, the inherent biological variability is less likely to result in two separate tests leading to different classifications. QSAR techniques, such as principal components analysis, allow visualization of this phenomenon and the ability to predict regions of chemical parameter space where ambiguity in in vivo results is most likely to arise.

Principal components plots for the skin corrosivity of phenols, using the same four physico-chemical parameters, are to be found elsewhere (Barratt, 1995b, 1996b).

SKIN CORROSIVITY OF ELECTROPHILIC ORGANIC CHEMICALS

For modeling the skin corrosivity of electrophilic organic chemicals (Barratt, 1996a), only two of the three skin permeability (partition) parameters—logP and molecular volume—were used; melting point was omitted because almost all of the chemicals in the data set were in the liquid state at ambient temperature. The total dipole moment of the chemicals was used as the "reactivity" component; this parameter had been used previously to model reactivity in a QSAR study of the eye irritation potential of neutral organic chemicals (Barratt, 1995c). The energy of the lowest unoccupied molecular orbital (E_{LUMO})—a parameter widely used in QSAR studies as a measure of electrophilic reactivity—was also examined as a reactivity parameter for both data sets; however, E_{LUMO} was found not to be a statistically significant variable ($F < 4.0$) in discriminant analysis for skin corrosivity with this data set.

As a general rule, nonelectrophilic neutral organic chemicals do not appear to be classified as skin corrosives; however, two exceptions are propargyl alcohol and 2-butyn-1, 4-diol, which are listed as corrosive chemicals in Annex 1 of the EU Dangerous Substances Directive (EEC, 1993). Their corrosivity appears to arise from their oxidation in the skin to the respective conjugated aldehydes, both of which are highly reactive electrophiles. Therefore, for the purpose of the QSAR study, propargyl alcohol and 2-butyn-1, 4-diol were considered to be pro-electrophiles.

A three-dimensional plot of logP, molecular volume, and dipole moment for 40 organic electrophiles and 2 pro-electrophiles is shown in Figure 9.3. In this plot the corrosive and noncorrosive chemicals in the data set are clearly separated by these three parameters. Initially, it may appear surprising that such a separation can be achieved without using a reactivity parameter such as E_{LUMO}, which truly represents electrophilic reactivity; however, it should be remembered that all of the chemicals

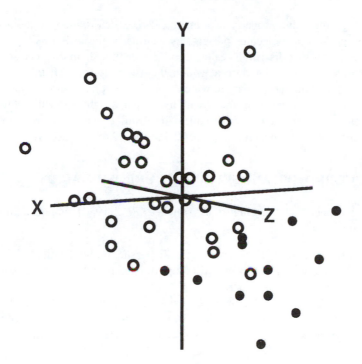

X Axis:clogP (CHEMICALC)
Y Axis:Molecular Volume
Z Axis:Total Dipole
FILLED CIRCLES = CORROSIVE
OPEN CIRCLES = NON-CORROSIVE

FIG. 9.3. Three-dimensional plot of logP, molecular volume, and dipole moment for 40 electrophiles and 2 pro-electrophiles.

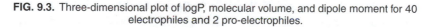

in the data set have been selected because they possess electrophilic reactivity either directly or after metabolism. A large part of their electrophilic reactivity is therefore common to the whole data set and thus will not be a significant discriminating factor between corrosives and noncorrosives.

An interesting feature of this analysis is that the chemicals classified as corrosive tend to be those with lower logP values and lower molecular volumes. This finding appears to conflict with skin permeability QSAR models (Flynn, 1990; Potts and Guy, 1992; Barratt, 1995), which all demonstrate an increase in skin permeability with increasing logP. The origin of this apparent dichotomy lies in the fact that skin corrosivity is a colligative property of a chemical (i.e., determined by the dose in terms of the number of molecules present) as well as being an intrinsic property. The dominant factor that discriminates the corrosive chemicals in this data set from the noncorrosives is the molar dose at which they were tested (c.f. the Annex V method for skin irritation/corrosivity in which a fixed mass or volume of chemical is applied). Molar dose correlates inversely with molecular weight, and within this data set, molar dose is represented by two variables—molecular volume and logP—which show correlations with molecular weight of 72.7% and 37.4%, respectively.

CORROSIVE AND SEVERELY CORROSIVE ACIDS

Corrosive substances as defined in the EU Annex V (EEC, 1984) method are divided into two groups:

Substances that cause irreversible damage to skin tissue, on application of 500 mg (or 500 1), to rabbit skin for a period of up to 4 hours, which are classified as "Corrosive" with R34 (causes burns).

Substances that cause irreversible damage to skin tissue when applied to rabbit skin for a period of up to 3 minutes, which are classified as "corrosive" with R35 (causes severe burns). This corrosivity classification corresponds to United Nations "packing group" III (UN, 1977).

In the final section of this paper, a QSAR study (Barratt et al., 1996) is described that relates the severity of skin corrosivity (designated by the EC risk phrases R34 and R35) of a data set of 27 acids to their log(octanol/water partition coefficient), molecular volume, melting point, pK_a, and an in vitro cytotoxicity parameter, a pEC50 value measured by neutral red uptake (NRU) into Swiss mouse embryo 3T3 cells (Borenfreund and Puerner, 1985).

The four physico-chemical variables, logP, molecular volume, melting point, and pK_a (Barratt, 1995b, 1996b), used in the previous studies were used initially in principal components analysis. Plots of the principal components from this data set failed to resolve the acids classified as R34 clearly from those classified as R35. Chloroacetic, 2-chloropropionic, bromoacetic, and iodoacetic acids (all classified as R35) clustered with acids classified as R34. A putative explanation for the "misclassification" of these halogenated acids was that several of them contain electrophilic groups that confer on them some additional cytotoxicity(i.e., additional to that expressed by the acidity parameter pK_a). To test this hypothesis, in vitro cytotoxicity parameters, expressed as pEC50 values, were determined on the neutralized salts of the 27 acids, using the NRU assay.

The principal components analysis of the data set was then repeated using the original four physico-chemical variables together with the pEC50 data. A three-dimensional plot of the first three principal components (Figure 9.4) shows that 2-chloropropionic acid (**27** in Figure 9.4) is now the only acid in the data set that is misclassified by this analysis.

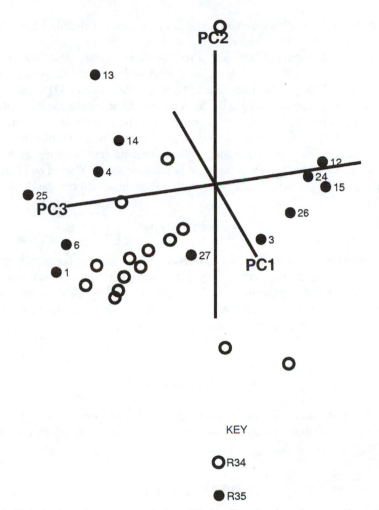

FIG. 9.4. Plot of principal components 1, 2, and 3 of logP, molecular volume, melting point, pK_a, and pEC50 for 27 acids.

Examination of the principal components plots in Figure 9.4 suggests that the acids that are classified as R35 are clustered into three distinct groups. For the largest group, which consists of dichloroacetic (**3**), sulphuric (**12**), trichloroacetic (**15**), nitric (**24**), and trifluoroacetic (**26**) acids, the key factor resulting in their classification as R35 rather than R34 is their low pK_a values (i.e., all are strongly acidic). The group of three acids comprising chloroacetic (**14**), bromoacetic (**4**), and iodoacetic (**13**) acids do indeed appear to be classified as R35 as a result of their pEC50 values, supporting the initial hypothesis underlying the work. For a third group comprising formic (**6**), acetic (**1**), and hydrofluoric (**25**) acids, a major factor contributing to their classification as R35 is their low molecular volume/weight. This appears to be another instance where the application of a fixed mass or volume of chemical has a bearing on its classification for toxic hazard.

The R35 classification of the one outlier, 2-chloropropionic acid, cannot be explained in terms of its pK_a, which is higher than that of any of the acids that appear to be classified as R35 by virtue of their acidity. Its pEC50 value (1.47) suggests that it is not especially cytotoxic to 3T3 cells. Indeed the difference between its cytotoxicity and that of chloroacetic acid (3.05) is entirely consistent with the lower reactivity of secondary alkyl halides with nucleophiles, compared with that of primary alkyl halides. One possible explanation for 2-chloropropionic acid behaving as an outlier in this analysis may be that it was historically classified as R35 based on an assumed similarity to the halogenated acetic acids rather than on animal test data!

CONCLUSION

Mechanistically based QSAR have been developed for the skin corrosivity of acids, bases, phenols, and organic electrophiles and for different degrees of skin corrosivity of acids. In addition to providing useful insight into the chemical mechanisms underlying the skin corrosivity of several classes of chemicals, the models provide a useful method for the prediction of the skin corrosivity potentials of new or untested chemicals.

ACKNOWLEDGMENTS

The author wishes to thank numerous colleagues at the Unilever SEAC Toxicology Unit for their collaboration over a number of years in the development of these concepts.

REFERENCES

Barratt, M. D. (1995a): Quantitative structure activity relationships for skin permeability. *Toxicology In Vitro*, 9:27–37.

Barratt, M. D. (1995b): Quantitative structure activity relationships for skin corrosivity of organic acids, bases and phenols. *Toxicology Letters*, 75:169–176.

Barratt, M. D. (1995c): A quantitative structure activity relationship for the eye irritation potential of neutral organic chemicals. *Toxicology Letters*, 80:69–74.

Barratt, M. D. (1996a): Quantitative structure activity relationships for the skin irritation and corrosivity of neutral and electrophilic organic chemicals. *Toxicology in Vitro*, 10:247–256.

Barratt, M. D. (1996b): QSARs for skin corrosivity of organic acids, bases and phenols: Principal components and neural network analysis of extended datasets. *Toxicology in Vitro*, 10:85–94.

Barratt, M. D., Dixit, M. B., and Jones, P. A. (1996): The use of in vitro cytotoxicity measurement for the prediction of the skin corrosivity potential of acids: A QSAR method. *Toxicology in Vitro*, 10:283–290.

Borenfreund, E., and Puerner, J. A. (1985): Toxicity determined in vitro by morphological determinations and neutral red absorption, *Toxicology Letters*, 24:119–124.

Dearden, J. C. (1990): Physico-chemical descriptors. In: *Practical Applications of Quantitative Structure-Activity Relationships (QSAR) in Environmental Chemistry and Toxicology*. Edited by W. Karcher and J. Devillers, pp.25–59. Kluwer Academic Publishers, Dordrecht, The Netherlands.

EEC 1984, 84/449/EEC (1984): Commission Directive of 25 April 1984 adapting to technical progress for the sixth time Council Directive 67/548/EEC on the approximation of laws, regulations and administrative provisions relating to the classification, packaging and labelling of dangerous substances, *Official Journal of the European Communities*, L251:106–108.

EEC (1993): Annexes I, II, III, IV to Commission Directive 93/21/EEC of 27th April 1993 adapting to technical progress for the 18th time Council Directive 67/548/EEC on the approximation of the laws, regulations and administrative provisions relating to the classification, packaging and labelling of dangerous substances. *Official Journal of the European Communities*, L110A: 1–86.

Fasman, G. R. (1976): *Handbook of Biochemistry and Molecular Biology*, 3rd ed. The Chemical Rubber Co., Cleveland.

Flynn, G. L. (1990): Physicochemical determinants of skin absorption. In: *Principles of Route-to-Route Extrapolation for Risk Assessment*. Edited by T. R. Gerrity and C. J. Henry, pp. 93–127. Elsevier Science Publishing Co., Inc. New York.

Hinz, R. S., Lorence, C. R., Hodson, C. D., Hansch, C., Hall, L. L., and Guy, R. H. (1991): Percutaneous penetration of para-substituted phenols in vitro. *Fundamental and Applied Toxicology*, 17:575–583.

Niemi, G. J. (1990): Multivariate analysis and QSAR: Application of principal components analysis. In: *Practical Applications of Quantitative Structure-Activity Relationships (QSAR) in Environmental Chemistry and Toxicology*. Edited by W. Karcher and J. Devillers, pp. 153–169. Kluwer Academic Publishers, Dordrecht, The Netherlands.

Potts, R. O., and Guy, R. (1992): Predicting skin permeability. *Pharmaceutical Research*, 9:663–669.

Suzuki, T. (1991): Development of an automated estimation system for both partition coefficient and aqueous solubility. *Journal of Computer-Aided Molecular Design*, 5:140–150.

Suzuki, T., and Kudo, Y. (1990): Automatic logP estimation based on combined additive modelling methods, *Journal of Computer-Aided Molecular Design*, 4:155–198.

UN (1977): United Nations Economic and Social Council: Joint meeting of the RID safety committee and the group of experts on the transportation of dangerous goods, *Trans/GE 15/R* 274:2.

Weast, R. C. (1970): *Handbook of Chemistry and Physics,* 51st ed. The Chemical Rubber Co., Cleveland.

SECTION II
Percutaneous Absorption

Toxicology of Skin
Edited by Howard I. Maibach
Copyright © 2001 Taylor & Francis

10

Contamination and Percutaneous Absorption of Chemicals from Water and Soil

Ronald C. Wester and Howard I. Maibach

Department of Dermatology, University of California, San Francisco, California, USA

INTRODUCTION

Contamination of soil and water (ground and surface water) and the transfer of hazardous chemicals are major concerns. When the large surface area of skin is exposed to contaminated soil and water (work, play, swim, daily bath), skin absorption may be significant. (Brown et al., 1984) suggested that skin absorption of contaminants in water has been underestimated and that ingestion may not constitute the sole or even the primary route of exposure. Soil has become an environmental depository for potentially hazardous chemicals. Exposure through work in pesticide-sprayed areas on chemical dump sites seems obvious; however, there may be hidden dangers in weekend gardening or in the child's play area, where the soil has become laden with lead or other hazardous chemicals.

EXPOSURE TO SOIL AND WATER

There are two theories as to how much soil is involved when skin comes in contact during a work or recreational endeavor. Risk analysis favors a constant amount for mathematical assessment. The first is the adherence theory in that soil particles less than 150 μm in diameter have been shown to have greater adherence to skin than larger size fractions, and that a monolayer will be exceeded with a soil loading of 2–5 mg/cm^2. Soil washed from skin will generally show this; however, the concern is with exposure to the hazardous chemicals within all the soil that comes in contact with skin. The adherence theory would suppose that chemical transfer only occurs from the monolayer. The child sitting in the dirt making mud pies will most certainly form a soil monolayer on the hands with the first mud pie; however, soil in successive mud pies certainly does come in contact with the skin. In fact, all activities, from heavy work to gardening to child's play and sports, involve an active interaction with soil; call it the *turnover theory*. The same analogy holds for water, where a monolayer is 2 μg/cm^2. Volume beyond the 2 μg/cm^2 will run off the body. Bathing and swimming certainly involves a turnover in volume exposure, and the initial water monolayer certainly is replaced constantly. Perhaps an open mind needs to be kept when attempting risk analysis; an underestimate is the worst scenario.

The other variable, limited to soil, is the textural triangle in that soil's major components shift between percentage of sand, percentage of clay, and percentage of silt. The organic content of soil is also considered important. Hazardous chemical in soil must partition to skin in order to become bioavailable. Composition of the textural triangle can affect the soil: skin partition.

Water

Tables 10.1 and 10.2 give concentrations of chemical contaminants in drinking water. These lists are extensive but not inclusive, and because the problem of drinking water contamination is just becoming appreciated, we must assume that the lists and concentrations of chemicals are also just a beginning.

Methodology to calculate an acceptable level of a chemical in drinking water has been developed (National Academy of Sciences, 1977), but the underlying assumption

TABLE 10.1. *Drinking water contamination levels*

Compound	Ranges detected in ground water (g/l)	Ranges detected in surface water (g/l)
Trichloroethylene	Trace—35,000	Trace—3.2
Tetrachloroethylene	Trace—3,000	Trace—21
Carbon tetrachloride	Trace—379	Trace—30
1,1,1-trichloroethane	Trace—401,300	Trace—3.3
1,2-Dichloroethane	Trace—400	Trace—4.8
Vinyl chloride	Trace—380	Trace—9.8
Methylene chloride	Trace—3,600	Trace—13
1,1-Dichloroethylene		
cis-1,2-dichloroethylene	Trace—860	Trace—2.2
trans-1,2-Dichlorethylene		
Ethylbenzene	Trace—2,000	
Xylene	Trace—300	
Toluene	Trace—6,400	

TABLE 10.2. *Concentration of drinking water contaminants*

Compound	NAS[a] 10⁻⁶ (g/l)	CAG[b] 10⁻⁶ (g/l)
Acrylonitrile	0.77	0.034
Arsenic		0.004
Benzene		3.0
Benzo[a]pyrene		
Bis(2-chloroethyl)ether	0.83	
Carbon tetrachloride	9.09	0.086
Chlordane	0.056	0.012
Chloroform	0.59	0.48
DDT	0.083	
1,2-Dichlorethane	1.4	1.46
1,1-Dichlorethylene		.28
Dieldrin	0.004	
Ethylenedibromide	0.11	0.0022
ETU	0.46	
Heptachlor	0.024	2.4
Hexachlorobutadiene		1.4
Hexachlorobenzene	0.034	
N-Nitrosodimethylamine		0.0052
Kepone	0.023	
Lindane	0.108	
PCB	0.32	
PCNB	7.14	
TCDD		5.0×10^{-6}
Tetrachloroethylene	0.71	0.82
Trichloroethylene	9.09	5.8
Vinyl chloride	2.13	106.0

CAG = Cancer Assessment Group; NAS = National Academy of Science.
From Office of Technological Assessment, 1981.

of this methodology is that ingestion constitutes the chief route of exposure to the contaminant. Such an assumption disregards other routes of exposure, such as skin absorption during daily bathing or swimming. Indeed, this assumption is so overlooked that it is not uncommon to see bottled water being brought to a home where the well water has been shown to be contaminated. Certainly, this bottled water will not be connected to the shower or swimming pool.

(Brown et al., 1984) showed that the skin absorption rates for solvents are high. They concluded that skin absorption of contaminants in drinking water has been underestimated and that ingestion may not constitute the sole, or even the primary, route of exposure. Table 10.3 gives their estimates of the relative contribution for skin absorption versus ingestion. The dermal route usually equals or exceeds the oral.

(Wester and Maibach, 1985) estimated the body burden of environmental contaminants from short-term exposure to skin absorption while bathing or swimming. They

TABLE 10.3. *Estimated dose and contribution per exposure for skin absorption versus ingestion*

Compound	Concentration (mg/L)	Dose (mg/kg)					
		Case 1*		Case 2†		Case 3‡	
		Dermal	Oral	Dermal	Oral	Dermal	Oral
Toluene	0.005	67	33	44	56	91	9
	0.10	63	37	46	54	89	11
	0.5	59	41	45	55	89	11
Ethylbenzene	0.005	75	25	44	56	91	9
	0.10	63	37	46	54	89	11
	0.5	68	31	45	55	89	11
Styrene	0.005	67	33	29	71	83	17
	0.10	50	50	35	65	84	16
	0.5	59	41	29	71	83	17

*7-kg adult bathing 15 minutes, 80% immersed (skin absorption); 2 l water consumed per day (ingestion).
†10.5-kg infant bathed 15 minutes, 75% immersed (skin absorption); 1 l water consumed per day (ingestion).
‡21.9-kg child swimming 1 hour, 90% immersed (skin absorption); 1 l water consumed per day (ingestion).
From Office of Technological Assessment, 1981.

concluded that, for chemicals for which there was published data, the body burden was low for short-term exposure at water concentrations detected in surface water; however, for higher contaminant concentrations, such as those detected in ground water, skin exposure has a potential for being hazardous to human health.

(Wester et al., 1987) further investigated the interactions of chemical contaminants in water and their skin absorption and potential systemic availability. Table 10.4 shows that when chemicals in water solution come in contact with skin (powdered human stratum corneum), the chemicals partition/bind to human skin, depending on their physiochemical interactions with skin. This partitioning was also shown to be linear for a 10-fold concentration range of nitroaniline.

Table 10.5 shows that each of these chemicals is able to distribute into the inner layers of skin after binding and that a significant portion of the chemical is absorbed. Other experiments showed that the absorption of nitroaniline was linear over a 10-fold

TABLE 10.4. *Partition of contaminants from water solution to powdered human stratum corneum following 30-minute exposure*

Chemical contaminant	Dose partitioned to skin (%)
Benzene (21.7 μg/ml)	16.6 ± 1.4
54% PCB (1.6 μg/ml)	95.7 ± 0.6
Nitroaniline (4.9 μg/ml)	2.5 ± 1.1

TABLE 10.5. *In vitro percutaneous absorption and skin distribution of contaminants in water solution for 30-minute exposure*

Parameter	Chemical contaminant (% dose)[*]		
	Benzene	54% PCB	Nitroaniline
Percutaneous absorption (systemic)	0.045 ± 0.037	0.03 ± 0.00	5.2 ± 1.6
Surface bound/stratum corneum	0.036 ± 0.005	6.8 ± 1.0	0.2 ± 0.17
Epidermis and dermis	0.065 ± 0.057	5.5 ± 0.7	0.61 ± 0.25
Total (skin/systemic)	0.15	12.3	6.1
Skin wash/residual	2.51 ± 0.94	71.2 ± 0.2	0.47 ± 0.03
Apparatus wash	0.006 ± 0.005	0.25 ± 0.2	0.47 ± 0.03
Total (accountability)	2.67	83.3	99.0

[*]Percentage of applied dose ($n = 4$ for each parameter): Benzene, 21.7 μg/ml; 54% PCB, 1.6 μg/ml; Nitroaniline, 4.9 μg/ml.

concentration range. Also, multiple doses of benzene were at least additive and, perhaps, exceeded single-dose predictions; this is due to benzene evaporation into the air during the absorption period. There is a partition of benzene between the air and skin; this is illustrated in Table 10.6, which gives the human skin absorption of 0.1% benzene in water and in toluene. From water vehicle, a total of 5.7% dose benzene was absorbed, whereas only 0.16% dose was absorbed from toluene vehicle. This is because the benzene and toluene rapidly evaporate off the skin, leaving little benzene absorbed. Contrarily, water is less volatile than toluene, so there is more time for benzene absorption.

The in vivo percutaneous absorption of nitroaniline in the Rhesus monkey, an animal model relevant for skin absorption in humans (Wester and Maibach, 1983) was $4.1 \pm 2.3\%$ of applied dose. This compared well with the $2.5 \pm 1.6\%$ binding to powdered human stratum corneum and the $5.2 \pm 1.6\%$ in vitro absorption (Table 10.5).

Table 10.7 gives a hypothetical percutaneous absorption of a chemical contamination from water while bathing or swimming for a single 30-minute period. It appears

TABLE 10.6. *In vitro percutaneous absorption of benzene in human skin*

Parameter	Percentage of dose absorbed	
	Water vehicle	Toluene vehicle
Receptor Fluid (RF)	5.0 ± 1.9	0.11 ± 0.08
Skin		
Epidermis	0.4 ± 0.3	0.04 ± 0.02
Dermis	0.3 ± 0.2	0.01 ± 0.01
RF + skin (absorbed)	5.7	0.16
Skin Surface Wash	1.8 ± 0.9	0.04 ± 0.02
Total 2.67	7.5 ± 1.0	0.19 ± 0.07

1% benzene in water or toluene, dosed at 5μl/cm^2.

TABLE 10.7. *Hypothetical percutaneous absorption of chemical contaminant from water while bathing or swimming for 30 minutes*

Parameters	
Concentration of chemical in water	1.0 μg/ml
Volume of water on skin surface in study	1.5 ml
Skin surface area in study	5.7 cm^2
Total skin surface area of human adult	17,000 cm^2
Percentage of dose absorbed, 30-minute exposure	5%

Calculation of body burden per single bath or swim

$$\frac{1.0\,\mu g}{ml} \times \frac{1.5\,ml}{5.7\,cm^2} \times \frac{17,000\,cm^2}{1} \times 0.05 = 223.7\,\mu g$$

The amount will obviously vary owing to concentration of chemical contaminant in water, percutaneous absorption of contaminant, and frequency of bathing and swimming

that it is possible for milligram amounts to be absorbed for single exposure, and thus would be at least cumulative in daily life.

A more detailed overview of the complexities of percutaneous flux and, more specifically, implications of penetration from bathing and swimming are outlined in (Wester and Maibach, 1984). The estimates provided are tentative; few chemical moieties have been examined. Specialized skin sites (ear canal, genitalia, face, etc.) offer substantial opportunity for penetration enhancement. Workers (e.g., housewives, cooks, mechanics, and the like) may wash their hands and arms many times daily. Damaged skin and occluded skin (diapers), as well as skin in the infant and aged, offer other possibilities for enhanced flux.

Finite and Infinite Doses

Table 10.8 compares the in vitro and in vivo absorption from water for boric acid, borax, and disodium octaborate tetrahydrate (DOT). The Kp values from the infinite dose are 1000-fold higher than from the in vivo study, whereas the finite dose Kp of boric acid was only 10-fold higher. Clearly, use of a similar dose volume was critical to the comparison. The in vitro system used a phosphate-buffered saline perfusion as well as a water solution dose. The in vivo dose was applied and allowed to dry, and absorption took place in whatever water content the skin of each volunteer contained. The preponderance of water in the infinite dose may have influenced the skin membrane during 24-hour continual exposure, and this may have increased the permeability of the borate chemicals. This would make the in vivo data and the in vitro finite dose data more relevant for the usual exposure conditions. The only exception might be where a subject bathed in a boric acid, borax, or DOT solution for an extended period of time. The in vitro infinite dosing conditions may better predict the bathing situation. Note that there was a 1000-fold difference between in vivo absorption and in vitro absorption where the in vitro dose was infinite. General risk assessment from in vitro infinite dosing may be greatly overestimated.

TABLE 10.8. *In vivo and In vitro percutaneous absorption of boron administered as boric acid, borax, and disodium octaborate tetrahydrate (DOT) in human skin*

Dose	In vitro		In vivo	
	Flux (μg/cm^2/hr)	Kp (cm/hr)	Flux (μg/cm^2/hr)	Kp (cm/hr)
Boric acid				
5% at 2 μl/cm^2	0.07	1.4×10^{-6}	0.009	1.9×10^{-7}
5% at 1 ml/cm^2	14.6	2.9×10^{-4}		
Borax				
5% at 2 μl/cm^2			0.009	1.8×10^{-7}
5% at 1 ml/cm^2	8.5	1.7×10^{-4}		
DOT				
5% at 2 μl/cm^2			0.010	1.0×10^{-7}
5% at 1 ml/cm^2	7.9	0.8×10^{-4}		

2 μl/cm^2 is a finite water dose on in vivo skin. More will just run off the body.

1 ml/cm^2 is an infinite dose only capable of being used with an in vitro system.

Soil

The study design should be such that percutaneous absorption from soil is relative to actual exposure conditions. The gardener, or the child playing in dirt, will have some stationary contact with the soil (sitting on it; dust which settles on the skin). There also will be dirt manipulated with the hands where skin contacts an everchanging layer of dirt. The exposed worker will certainly have airborne dust settle on skin; manipulated dirt will depend on the job. In all cases, the dirt covering skin is exposed to air (dirt in clothing is noted as an exception). Laboratory studies, to be relevant, should have the soil on skin open to air exchange. This is simple with in vitro studies where the soil is contained. With in vivo studies the soil needs to be contained, and the best method is to use a nonocclusive cover that allows free passage of air and moisture; a Gore-Tex (Flagstaff, AZ) film accomplishes this (Wester et al., 1990a). Use of an occlusive cover such as glass (Turkall et al., 1994) will cause changes in the microenvironment under the cover, which will enhance skin absorption. The in vitro diffusion cell is a truly stationary system where chemical passage from soil to skin is physically undisturbed and will function according to the chemical kinetics of the components. An in vivo study may have some animal or human movement which could shift/rotate soil confined under the nonocclusive patch if space under the patch permits this (the patch can be sufficiently concave to allow soil movement in the open space).

It is important to know the relevant characteristics of the test soil. This should include percentage of sand, clay, silt, and organic content. Soil can be passed through mesh sieves to be uniform in size. Mixing of the test chemical added in solvent to the soil should be done open to air to allow dissipation of the solvent. The "dust" fraction of soil can be avoided by the sieving method. This is for safety purposes so that

TABLE 10.9. *In vitro and in vivo percutaneous absorption of organic chemicals*

Compound	Vehicle	Percentage of Dose		In vivo
		In vitro		
		Skin	Receptor fluid	
DDT	Acetone	18.1 ± 13.4	0.08 ± 0.02	18.9 ± 9.4
	Soil	1.0 ± 0.7	0.04 ± 0.01	3.3 ± 0.5
Benzo[a]pyrene	Acetone	23.7 ± 9.7	0.09 ± 0.06	51.0 ± 22.0
	Soil	1.4 ± 0.9	0.01 ± 0.06	13.2 ± 3.4
Chlordane	Acetone	10.8 ± 8.2	0.07 ± 0.06	6.0 ± 2.8
	Soil	0.3 ± 0.3	0.04 ± 0.05	4.2 ± 1.8
Pentachlorophenol	Acetone	3.7 ± 1.7	0.6 ± 0.09	29.2 ± 5.8
	Soil	0.11 ± 0.04	0.01 ± 0.00	24.4 ± 6.4

airborne particles containing radioactivity do not contaminate laboratory personnel. An uncontrolled airborne fraction may also affect study results.

Table 10.9 gives the in vitro (human skin) and in vivo (Rhesus monkey) percutaneous absorption of organic chemicals from soil and a comparative vehicle (water or solvent, depending on vehicle). The soil is the same source for all chemicals. For each chemical the concentration of mass (μg) per unit skin area (cm^2) is the same for each vehicle. Receptor fluid (human plasma) accumulation of DDT was negligible in the in vitro study due to solubility restriction. Chemicals with higher logPs are lipophilic and therefore are not soluble in biological fluid receptor fluid (plasma, buffered saline) (Wester et al., 1990a, 1990b, 1992a). Human skin content was 18.1% dose from acetone vehicle. in vivo absorption in the Rhesus monkey was 18.9% dose from acetone vehicle. These values are comparable to the published 10% dose absorbed in vivo in man from acetone vehicle. Percutaneous absorption from soil was predicted to be 1.0% dose in human skin in vitro and a comparative 3.3% dose in vivo in Rhesus monkey.

In vivo percutaneous absorption of benzo[a]pyrene is high: 51.0% reported here for Rhesus monkey and 48.3% (Bronaugh and Stewart, 1986) and 35.3% (Yang et al., 1989) for the rat. Benzo[a]pyrene absorption from soil was approximately one fourth that of solvent vehicle (Wester et al., 1990a; Shu et al., 1988).

For chlordane and pentachlorophenol, the in vivo percutaneous absorption in Rhesus monkey from soil was equal to or slightly less than that obtained from solvent vehicle. Validation to man in vivo is available for 2,4-D, where the percutaneous absorption is the same for Rhesus monkey and man (Wester et al., 1996). In vitro percutaneous absorption is variable, probably due to solubility problems relative to high lipophilicity.

In vivo studies have an advantage over in vitro studies in that in vivo pharmacokinetic data can be obtained and these data applied to better understand the potential toxicokinetics of a chemical (Table 10. 10). The percutaneous absorption of PCP from acetone vehicle was $29.2 \pm 5.8\%$ of total dose applied for a 24-hour exposure period. Compared to other compounds, the absorption of PCP would be considered high. In vivo absorption from soil was 3.3% for DDT, 13.2% for benzo[a]pyrene, and 4.2% for

TABLE 10.10. *In vivo percutaneous absorption of pentachlorophenol in Rhesus monkey*

	Percent dose*		
	Topical		
	Soil	Acetone	Intravenous
Percent dose absorbed	24.4 ± 6.4	29.2 ± 5.8	—
Surface recovery[†]	38.0 ± 13.4	59.6 ± 4.1	—
Half-life (days)	4.5	4.5	4.5

*Means ± SD for four animals.
[†]Includes chamber, residue, and surface washes.

chlordane. Additionally, the ^{14}C excretion for PCP in urine was slow; a half-life of 4.5 days for both intravenous and topical application. If biological exposure is considered in terms of dose X time, then PCP biological exposure can be considered high. The percutaneous absorption of PCP from soil vehicle was also high (24.4 ± 6.4%) and not statistically different. The study of (Reigner et al., 1991) and this study show PCP to have good bioavailability, both topical and oral, and PCP also exhibits an extensive half-life. This suggests that PCP has the potential for extensive biological interactions.

Table 10.11 gives the in vitro and in vivo percutaneous absorption of PCBs (Wester et al., 1993a). As with the other organic chemicals with high log P, receptor fluid accumulation in vitro was essentially nil. Skin accumulation in vitro did exhibit some PCB accumulation. in vivo PCB percutaneous absorption for both Aroclor 1242 and 1254 (Electric Power Research Institute, Palo Alto, CA) was (a) high, ranging from 14%–21%, and (b) generally independent of formulation vehicle. PCBs thus have a strong affinity for skin and are relatively easily absorbed into and through skin.

Selected salts of arsenic, cadmium, and mercury are soluble in water, and thus are amenable to in vitro percutaneous absorption with human skin (Table 10.12) (Wester et al., 1992b, 1993b). Arsenic absorption in vitro was 2.0% (1.0% plus 0.9% receptor

TABLE 10.11. *In vitro and in vivo percutaneous absorption of PCBs*

		Percent dose		
		In vitro		In vivo
Compound	Vehicle	Skin	Receptor fluid	
PCBs (1242)	Acetone	—	—	21.4 ± 8.5
	TCB	—	—	18.0 ± 8.3
	Mineral oil	6.4 ± 6.3	0.3 ± 0.6	20.8 ± 8.3
	Soil	1.6 ± 1.1	0.04 ± 0.05	14.1 ± 1.0
PCBs (1254)	Acetone	—	—	14.6 ± 3.6
	TCB	—	—	20.8 ± 8.3
	Mineral oil	10.0 ± 16.5	0.1 ± 0.07	20.4 ± 8.5
	Soil	2.8 ± 2.8	0.04 ± 0.05	13.8 ± 2.7

TABLE 10.12. *In vitro and in vivo percutaneous absorption of metals*

		Percent dose		
		In vitro		
Compound	Vehicle	Skin	Receptor fluid	In vivo
Arsenic	Water	1.0±1.0	0.9±1.1	2.0±1.2
	Soil	0.3±0.2	0.4±0.5	3.2±1.9
Cadmium	Water	6.7±4.8	0.4±0.2	—
	Soil	0.09±0.03	0.03±0.02	—
Mercury	Water	28.5±6.3	0.07±0.01	—
	Soil	7.9±2.2	0.06±0.01	—

fluid) and the same in vivo in Rhesus monkey. Absorption from soil was equal to (in vivo) or approximately one third (in vitro). Cadmium and mercury both accumulate in human skin and are slowly absorbed into the body. (Note that in vivo studies with cadmium and mercury are difficult to perform; cadmium accumulates in the body and mercury is not excreted via urine.) Note the high skin content with cadmium and mercury.

Soil Load

Percutaneous absorption of 2,4-D for 24-hour exposure was $8.6 \pm 2.1\%$ dose for a dose load of 4.2 μg/cm^2 in acetone vehicle. The same 2,4-D dose was then loaded into 1 mg soil or 40 mg soil per cm^2 skin surface area. Percutaneous absorption for 24 hours from the 1 mg soil load was $9.8 \pm 4.0\%$ dose (Table 10.13) (Wester et al., 1996).

Percutaneous absorption from the 40 mg soil load for 24 hours was $15.9 \pm 4.7\%$ dose ($P = 0.178$ nonsignificant for paired t test). Thus, the in vivo percutaneous absorption of 2,4-D was not affected by soil load (1 mg vs 40 mg soil; chemical dose constant). Additionally, the percutaneous absorption from the two soil loads was the same as from acetone vehicle.

TABLE 10.13. *Effect of soil load on 2,4-D percutaneous absorption*

Compound absorbed[‡]	Soil Load[†] Percent dose (mg/cm)	Percent dose absorbed
In vivo, Rhesus monkey[*]	1	9.8±4.0
	40	15.9±4.7
In vitro, human skin[*]	5	1.8±1.7
	10	1.7±1.3
	40	1.4±1.2

[*]Mean ± SD ($n=4$).
[†]Concentration of 2,4-D chemical per cm^2 skin area was kept constant, whereas soil load per cm^2 skin area was varied.
[‡]In vivo percutaneous absorption measured by urinary ^{14}C accumulation; in vitro absorption determined by ^{14}C skin content.

2,4-D at a constant chemical dose 2 μg/cm^2 was applied to human skin in vitro in soil loads of 5, 10, and 40 mg. Dose accumulation in buffered saline receptor fluid was low (0.02 ± 0.02%), presumably due to relative low solubility of 2,4-D in water. The 2,4-D human skin content was analyzed after the 24-hour exposure period, and the results show no difference relative to the three soil loads. The in vitro skin content would predict a 2,4-D percentage dose absorption from soil of approximately 2%, which is approximately one fifth of that in monkey in vivo.

Skin Contact Time

Table 10.14 provides the effect of skin deposition time on 2,4-D percutaneous absorption. In acetone vehicles, the percutaneous absorption of 2,4-D over 8 hours was 3.2 ± 1.0% dose. This is approximately one third of the absorption seen for 24 hours (8.6 ± 2.1%). Thus, with acetone vehicle (where the dose is in immediate contact with the skin), the percutaneous absorption was linear over time. With soil as the vehicle, the absorption was only 0.05 ± 0.04 for 1 mg soil/cm^2 and 0.03 ± 0.02% for 40 mg/cm^2 at 8 hours. The 16-hour absorption for 1 mg/cm^2 soil was 2.2 ± 1.2 percent. This suggests that percutaneous absorption of 2,4-D from soil was not linear over-time. There may be a "lag" time where chemical must partition from soil into skin.

Soil and Risk Assessment Assumptions

In the study with 2.4-D and the other compounds mentioned previously, an experimental soil load of 40 mg/cm^2 skin surface area was used. Studies on dermal soil adherence show that less than 1 mg to perhaps 5 mg of soil will adhere to skin (EPA, 1992). It then becomes the practice in estimating potential health hazard to assume linearity and divide the results of 40 mg by a soil adhesion factor, thus reducing estimated body burden by 1/40 or 5/40, etc. (EPA, 1992). The data generated here for 2,4-D, where soil load (range 1–40 mg) had no effect on percutaneous absorption, suggest that this mathematical practice can severely underestimate absorption.

TABLE 10.14. *Effect of skin contact time on in vivo percutaneous absorption of 2,4–D*

Skin contact time	Acetone vehicle	Percent dose absorbed* Soil vehicle	
		1 mg/cm^2	40 mg/cm^2
8 hour	3.2 ± 1.0	0.05 ± 0.04	0.03 ± 0.02
16 hour	—	2.2 ± 1.2	—
24 hour	8.6—2.1		15.9 ± 4.7

*Mean ± SD ($n = 4$).

2,4-D IN VIVO PERCUTANEOUS ABSORPTION FROM SOIL AND ACETONE VEHICLES

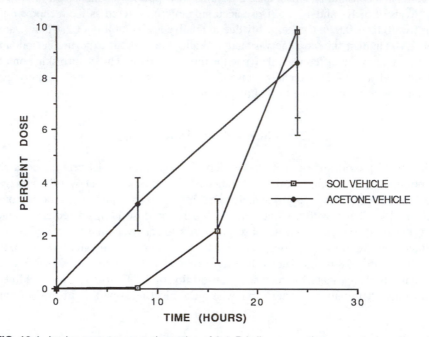

FIG. 10.1. In vivo percutaneous absorption of 2,4–D is linear over time, except where there is an initial lag time from soil vehicle.

A second assumption, which transferred to the laboratory, is that chemical and soil are "static." The gardener planting summer flowers will certainly have multiple contacts with soil. Chemicals are also mobile within soil. On the short term, as seen in the data, chemicals move from soil to skin. (Calderbank, 1989) reports that with time, adsorbed residues in soil will become more stable. Therefore, some dynamics in the system need to be understood.

A third assumption of linearity is also placed on time—that 2,4–D absorption in acetone vehicle was $3.2 \pm 1.0\%$ for 8 hours and 8.6 ± 2.1 for 24 hours suggests that absorption is linear over time and that the compound can be removed with washing (Figure 10.1). With acetone vehicle, the total dose is deposited on the skin surface and percutaneous absorption is dependent upon skin barrier diffusion. With soil vehicle, a lag time is needed for the 2,4–D to "mobilize" within soil and become available for absorption. Sufficient time is needed for 2,4–D to become available, then skin barrier function becomes the rate-limiting step in absorption. Should it turn out that a substantial lag time does exist for transfer from soil to skin, this would favor risk assessment for the worker; however, the data to date show that substantial amounts of hazardous chemicals can be absorbed from soil (Figure 10.2).

IN VIVO PERCUTANEOUS ABSORPTION FROM SOLVENT AND SOIL

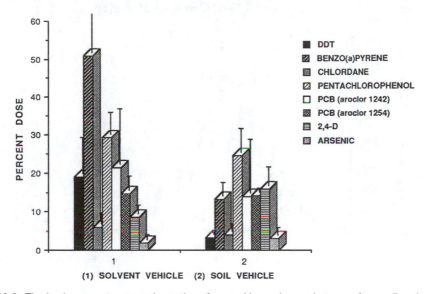

FIG. 10.2. The in vivo percutaneous absorption of several hazardous substances from soil and solvent (either acetone or water). Overall, soil reduced absorption to about 60%, compared to solvent. There is caution, however, because the absorption of some compounds is the same for soil and solvent.

NEW METHODOLOGY

The future for dermal barrier assessment may be here; a system to provide real-time measurement for dermal exposure. Breath analysis is used to obtain real-time measurements of volatile organics in expired air following exposure in animals and humans. Exhaled breath data are analyzed using physiologically based pharmacokinetic (PBPK) models to determine the dermal bioavailability of organic solvents under realistic exposure conditions. The end product of this research will be a tested framework for the rapid screening of real and potential exposures while simultaneously developing PBPK models to comprehensively evaluate and compare exposures to organic chemicals. In the new system, human volunteers and animals breathe fresh air via a new breath-inlet system that allows for continuous real-time analysis of undiluted exhaled air. The air supply system is self-contained and separated from the exposure solvent-laden environment. The system uses a Teledyne (Mountain view, CA) 3DQ Discovery ion trap mass spectrometer (MS/MS) equipped with an atmospheric sampling glow discharge ionization source (ASGDI). The MS/MS system provides an appraisal of individual chemical components in the breath stream in the single-digit parts-per-billion (ppb) detectable range for each of the compounds proposed for study while maintaining linearity of response over a wide dynamic range. Figure 10.3 shows human volunteer exposure to 0.1% methyl chloroform (MC) (1,1,1-trichlorethane) by

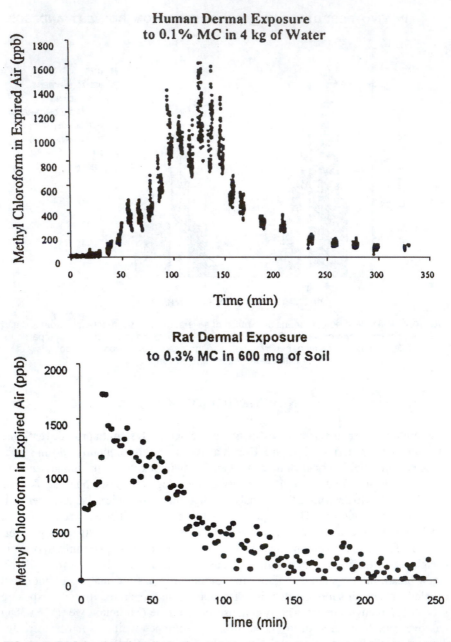

FIG. 10.3. Real-time breath analysis of methylchlorofrom in human volunteer (hand in bucket of water containing 0.1% MC) and rat (with soil containing 0.3% MC in a nonocclusive patch adhered to skin).

a single hand immersion into a bucket of water containing the MC. Breath analyses are done every 4 seconds. Hand immersion was stopped after 2 hours of exposure. Figure 10.3 also shows rat dermal exposure to 0.3% MC in 600 mg soil contained in a nonocclusive patch applied to the skin. Envisioned for the future is a portable real-time system where people suspected of exposure can be immediately tested (Wester and Maibach, 1998).

DISCUSSION

The evolution of skin resulted in a tissue that protects precious body fluids and constituents from excessive uptake of water and contaminants in the external environment. The outermost surface of the skin that emerged for humans is the stratum corneum, which restricts but does not prevent penetration of water and other molecules. This is a complex lipid-protein structure that is exposed to contaminants during bathing, swimming, and exposure to the environment. Industrial growth has resulted in the production of organic chemical and toxic metals whose disposal results in contamination of water supplies. As one settles into a tub or pool, the skin with a surface area of approximately 18,000 cm^2 acts as a lipid sink (stratum corneum) for the lipid-soluble contaminants. Skin also serves as transfer membrane for water and whatever contaminants may be dissolved in it. Note that (a) water transfers through skin and can carry chemicals, and (b) the outer layer of skin is lipid in nature. Thus, highly lipophilic chemicals such as DDT, PCBs, and chlordane residing in soil will quickly transfer to skin. Percutaneous absorption can be linear, orderly, and predictive (a measured flux from water); however, evidence exists that chemicals may transfer to skin with short-term exposure.

REFERENCES

Bronaugh, R. L., and Stewart, R. F. (1986): Methods for *in vitro* percutaneous absorption studies. VI: Preparation of the barrier layer. *J. Pharm. Sci.*, 75:487–491.

Brown, H. S., Bishop, D. R., and Rowan, C. A. (1984): The role of skin absorption as a route of exposure for volatile organic compounds (VOCs) in drinking water. *Am. J. Public. Health.*, 74:479–484.

Calderbank, A. (1989): The occurrence and significance of bound pesticide residues in soil. *Rev Environ. Contam. Toxicol.*, 108:71–103.

EPA (1992): Dermal Exposure Assessment: Principles and Applications, EPA/600/8–91/011B, Office of Research and Development, Washington, DC, Section 5:48–49.

National Academy of Sciences (1977): *Drinking Water and Health.* NAS/NRC, Washington.

Office of Technological Assessment (1981): Factors associated with cancer. In: *Assessment of Technologies for Determining Cancer Risks from the Environment,* Chap. 3. Office of Technological Assessment, Washington.

Reigner, B. G., Gungon, R. A., Hoag, M. K., and Tozer, T. N. (1991): Pentachlorophenol toxicokinetics after intravenous and oral administration to rat. *Zenobiotica*, 21:1547–1558.

Shu, H., Teitelbaum, P., Webb, A. S., Marple, L., Brunck, B., Del Rossi, D., Murray, F. J., and Paustenbach, D. (1988): Bioavailability of soil-bound TCDD: Dermal bioavailability in the rat. *Fundam. Appl. Toxicol.*, 10:335–343.

Turkall, R. M., Skowronski, G. A., Kadry, A. M., and Abdel-Rahman, M. S. (1994): A comparative study of the kinetics and bioavailability of pure and soil-absorbed naphthalene in dermally exposed male rates. *Arch. Environ. Contam. Toxicol.*, 26:504–509.

Wester, R. C., and Maibach, H. I. (1985): Body burden of environmental contaminants from acute exposure to percutaneous absorption while bathing or swimming. *Toxicologist*, 5:177.

Wester, R. C., and Maibach, H. I. (1983): Cutaneous pharmacokinetics: 10 steps to percutaneous absorption. *Drug. Metab. Rev.*, 14:169–205

Wester, R. C., and Maibach, H. I. (1984): Assessment of dermal absorption of contaminants in drinking water. EPA contract No. 68002–3168.

Wester, R. C., Mobayen, M., and Maibach, H. I. (1987): *In vivo* and *in vitro* absorption and binding to powdered stratum corneum as methods to evaluate skin absorption of environmental chemical contaminants from ground and surface water. *J. Toxicol. Environ. Health.*, 21:367–374.

Wester, R. C., Maibach, H. I., Bucks, D. A. W., Sedik, L., Melendres, J., Liao, C., and Di Zio, S. (1990a): Percutaneous absorption of [^{14}C] DDT and [^{14}C] benzo[a]pyrene from soil. *Fundam. Appl. Toxicol.*, 15:510–516.

Wester, R. C., Maibach, H. I., Sedik, L., Melendres, J., Wade, M., and DiZio, S. (1990b): *In vitro* percutaneous absorption of pentachlorophenol from soil. *Fundam. Appl. Toxicol.*, 19:68–71.

Wester, R. C., Maibach, H. I., Sedik, L., Melendres, J., Liao, C., and DiZio, S. (1992a) Percutaneous absorption of [^{14}C] chlordane from soil. *J. Toxicol. Environ. Health.*, 35:269–277.

Wester, R. C., Maibach, H. I., Sedik, L., Melendres, J., DiZio, S., and Wade, M. (1992b): *In vitro* percutaneous absorption of cadmium from water and soil into human skin. *Fundam. Appl. Toxicol.*, 19:1–5.

Wester, R. C., Maibach, H. I., Sedik, L., Melendres, J., and Wade, M. (1993a): Percutaneous absorption of PCBs from soil: *In vivo* Rhesus monkey, *in vitro* human skin, and binding to powdered human stratum corneum. *J. Toxicol. Environ. Health.*, 39:375–382.

Wester, R. C., Maibach, H. I., Sedik, L., Melendres, J., and Wade, M. (1993b): *In vivo* and *in vitro* percutaneous absorption and skin decontamination of arsenic from water and soil. *Fundam. Appl. Toxicol.*, 20:336–340.

Wester, R. C., Melendres, J., Logan, F., Hui, X., and Maibach, H. I. (1996): Percutaneous absorption of 2,4-dichlorophenoxyacetic acid from soil with respect to soil load and skin contact time: *In vivo* absorption in Rhesus monkey and *in vitro* absorption in human skin. *J. Toxicol. Environ. Health.*, 47:335–344.

Wester, R. C., and Maibach, H. I. (1998): *US DOE environmental Management Science Program Workshop*, July, 48.

Yang, J. J., Roy, T. A., Kruger, A. J., Neil, W., and Mackerer, C. R. (1989): *In vitro* and *in vivo* percutaneous absorption of benzo[a]pyrene from petroleum crude-fortified soil in the rat. *Bull. Environ. Contam. Toxicol.*, 43:207–214.

11

Determining an Occupational Dermal Exposure Biomarker for Atrazine through Measurement of Metabolites in Human Urine by HPLC-Accelerator Mass Spectrometry

Bruce A. Buchholz, John S. Vogel

Center for Accelerator Mass Spectrometry, Lawrence Livermore National Laboratory, Livermore, California, USA

S. Douglas Gilman, Shirley J. Gee, Bruce D. Hammock

Department of Entomology, University of California, Davis, California, USA

Xiaoying Hui, Ronald C. Wester, Howard I. Maibach

Department of Dermatology, University of California, San Francisco, California, USA

INTRODUCTION

Atrazine is among the most widely used herbicides throughout the world. It is effective at controlling broadleaf grasses, and approximately 70 million pounds is applied annually in the United States (Gianessi and Anderson, 1995). The chemistry and transport of atrazine and its by-products in soil and in ground and surface water has been studied in detail during four decades of commercial use (Koskinen and Clay, 1997; Ma and Selim, 1996). Although often found in agricultural areas (Ribaudo and Bouzaher, 1994; Kolpin et al., 1996; Solomon et al., 1996; Environmental Protection Agency, 1994; Clark et al., 1999), atrazine is generally regarded as effective and safe to the general population when used as directed. As with all pesticides, occupational exposures to agricultural and chemical production workers can be considerably higher than exposures to the general population. A survey of farmers found that pesticide exposures occur commonly by inhalation and dermal absorption, but use of personal protective equipment was not routine (Perry and Layde, 1998). Dermal absorption studies with rats (Ikonen et al., 1988) and in vitro (Ademola et al., 1993; Lang et al., 1996) showed that atrazine is absorbed and metabolized by this exposure route, and little, if any, parent compound was found in rat urine (Ikonen et al., 1988) or in human urine in a limited number of exposure studies (Lucas et al., 1993; Catenacci et al., 1993; Jaeger et al., 1998; Driskell et al., 1997). A practical assay to assess occupational dermal exposure to atrazine thus requires knowledge of the pharmacokinetics and metabolic profile of atrazine in humans.

The ideal field exposure assessment method should rely on samples obtained non-invasively (e.g., urine or saliva) and be sensitive to environmental doses. It should identify a specific metabolite or class of metabolites dominant within the time frame of the target exposure and sampling window. An assay which identifies a class of compounds is often more desirable than one which identifies a specific one. Variations in dose and metabolism among individuals is less problematic when a class of compounds can be detected.

An in vivo human clinical study using ^{14}C labeled atrazine was designed to measure its percutaneous absorption and clearance and determine the metabolites excreted in urine following a dermal exposure to atrazine. Doses were selected based on data from in vitro and rodent dermal absorption experiments (Ikonen et al., 1988; Ademola et al., 1993; Lang et al., 1996; Lucas et al., 1993). The doses and specific activities were designed for detection by liquid scintillation counting (LSC). The human subjects absorbed the labeled atrazine less than anticipated, so the radioactivity of the collected urine was lower than expected. Although the neat urine could usually be measured by LSC for the early time points, the activity was insufficient in all but one subject to measure any metabolites after HPLC separation (Gilman et al., 1998). The ^{14}C levels in the urine samples of later time points were often at or below the detection limits for LSC for practical sample sizes and analysis times (3 replicates of 2 mL urine counted 15 minutes each).

We used these archived urine samples in a demonstration of high performance liquid chromatography-accelerator mass spectrometry (HPLC-AMS): metabolite separation and identification by HPLC followed by ^{14}C quantitation with AMS. AMS can accurately measure less than 10^6 atoms (1 amol) of ^{14}C in a sample containing $\approx$1 mg total carbon and analyze hundreds of samples per day (Vogel et al., 1995; Vogel, 1998). A carbon carrier with a low ^{14}C level (petrochemical) can be added to samples with small amounts of carbon (Gilman et al., 1998; Buchholz et al., 1999b), such as HPLC fractions, to bring them to the preferred sample size. Samples with carrier carbon containing 10% the contemporary ^{14}C level routinely achieve limits of quantitation near 10 amol, sufficient to quantify the ^{14}C content of HPLC eluents. Urine collected at three post-exposure time points and a predose control were analyzed from 3 of the 10 subjects who completed the clinical study. The time points were selected to include urine samples whose ^{14}C concentration was approximately 50% of the peak value during application of the dose; the peak value; and less than 50% of the peak value several days after the removal of the dermal dose. The archived urine samples were selected from subjects receiving low and high doses.

EXPERIMENTAL SECTION

Chemicals

All chemicals were checked for ^{14}C content by AMS prior to use to ensure no inadvertently labeled or contaminated solvents were used. The solvents for the HPLC mobile phases were glacial acetic acid, acetonitrile, and 18.2 megaohm de-ionized water for the aqueous phase. Tributyrin (glycerol tributyrate) was dissolved in methanol for use as the carbon carrier for AMS sample combustion and graphitization. The atrazine tracer was uniformly labeled with ^{14}C on the ring structure to retain the tag during metabolism (Figure 11.1). The radiolabeled atrazine was mixed with the end-use formulation to make dosing solutions with specific activities of 8.34 and 2.70 mCi $\cdot$ mmol^{-1}.

The atrazine was an end-use formulation (AATREX-4L) prepared by Novartis Crop Protection (Greensboro, NC). The following HPLC standards were prepared and provided by Novartis: 2-chloro-4-ethylamino-6-[(1-methyl-ethyl)amino]-1,3,5-triazine (atrazine), *N*-acetyl-*S*-{4-ethylamino-6-[(1-methyl-ethyl)amino]-1,3,5-triazin-2-yl}-L-cysteine (atrazine mercapturate), 2-amino-4-chloro-6-[(1-methyl-ethyl)amino]-1,3,5-triazine (deethylatrazine), *N*-acetyl-*S*-{4-amino-6-[(1-methyl-ethyl)amino]-

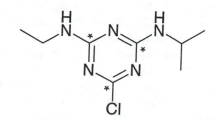

FIG. 11.1. Uniformly ring ^{14}C-labeled (∗) atrazine.

1,3,5-triazin-2-yl}-L-cysteine (deethylatrazine mercapturate), 2-amino-4-chloro-6-ethylamino-1,3,5-triazine (deisopropylatrazine), N-acetyl-S-{4-ethylamino-6-amino-1,3,5-triazin-2-yl}-L-cysteine (deisopropylatrazine mercapturate), 2-chloro-4,6-diamino-1,3,5-triazine (didealkylatrazine), N-acetyl-S-{4,6-diamino-1,3,5-triazin-2-yl}-L-cysteine (didealkylatrazine mercapturate), 2-ethylamino-4-hydroxy-6-[(1-methyl-ethyl)amino]-1,3,5-triazine (hydroxy atrazine), 2,4,6-triamino-1,3,5-triazine (melamine), 2,4-diamino-6-hydroxy-1,3,5-triazine (ammeline), and 2-amino-4,6-dihydroxy-1,3,5-triazine (ammelide).

Clinical

All clinical work was performed at the University of California, San Francisco, after approval by the UCSF Committee on Human Research. Twelve normal, healthy males aged 43 to 74 served as volunteers, with 10 completing the trial. The atrazine was applied for 24 hours to a 25 cm^2 area of skin on the left ventral forearm of each volunteer at low (subjects 1–4, 0.167 mg, 6.45 μCi) and high (subjects 5–10, 1.98 mg, 24.7 μCi) doses. The protective nonocclusive cover was removed at 24 hours, excess atrazine was washed from the skin with soap and water, and the rinse was analyzed by LSC. A series of 10 cellophane tape strippings were conducted on the skin at the site of application on day 7 following the dosing to remove any residual atrazine remaining in the epidermis. These tapes and the nonocclusive cover were also measured for ^{14}C by LSC. Urine samples were collected a day before the dose and 7 days following it (24-hour predose, 0–4 hours, 4–8 hours, 8–12 hours, 12–24 hours, 24–48 hours, 48–72 hours, 72–96 hours, 96–120 hours, 120–144 hours, 144–168 hours). Total urine volume for each sample was measured before separating into aliquots and storing at $-20°$C. Samples were thawed at room temperature before analysis and immediately returned to $-20°$C after use. Fecal samples were collected throughout the trial, homogenized, and measured for ^{14}C by LSC.

HPLC Sample Preparation

Predose and three post-dose samples from three subjects (2, 5, 6) were selected for HPLC separation and AMS quantitation. For each time point, approximately 3 ml of urine was centrifuged at 10,000 g to remove solids, and a 500 μL aliquot of the supernatant was measured by LSC to determine the specific activity. The amount of ^{14}C was limited to less than 0.17 Bq (4.5 pCi) for each HPLC injection, restricting injected volumes of supernatant to 25–100 μL. Supernatant was stored in new 1.5 ml Eppendorf tubes at $-35°$C and thawed as needed.

We reserve an HPLC system solely for separations prior to AMS quantitation. The HPLC system includes pumps, an analytical octadecylsilane chromatography column and guard column, and a diode array UV detector with a dedicated PC for spectra collection and analysis. The mobile phase used a reverse phase gradient of 0.1% acetic acid in water and 0.1% acetic acid in acetonitrile with a 1 mL/min flow rate.

Fractions were collected each minute from 4–42 minutes and were labeled with the end of the collection period (fraction 8 contained eluent from 7–8 minutes). Peaks were tentatively identified based on cochromatography. Urine samples were mixed with authentic standards and the UV profile of HPLC elution compared with the histogram of ^{14}C measured by AMS.

Accelerator Mass Spectrometry Sample Preparation and Measurement

Half of each collected fraction was analyzed for ^{14}C content using AMS. HPLC eluent was transferred with carbon carrier solution (50 μL of 40.0 mg mL^{-1} tributyrin in methanol to yield 1.19 mg carrier C) to a small quartz vial nested inside two borosilicate tubes and dried in a vacuum centrifuge, leaving the eluted metabolite dispersed in tributyrin. A small glass fiber filter (e.g., Whatman GF/A 21 mm) was placed into the top of each set of nested tubes during centrifugation to minimize intersample contamination by aerosols. Tributyrin is used for the carbon carrier because it has a relatively low ^{14}C content, high carbon content (60% C by mass), low nitrogen content, low vapor pressure, high solubility in alcohol, and retains target compounds well. Three tributyrin carrier blanks were prepared with each set of samples. AMS measures an isotope ratio so the addition of a carrier is an isotope dilution. The precision of the ^{14}C determination depends on the precision of the added carbon mass.

The inner quartz vials were transferred to quartz combustion tubes, which were evacuated and sealed. The samples were combusted to CO_2 and reduced to carbon using Vogel's method (Vogel, 1992). Graphite samples were packed into aluminum sample holders, and carbon isotope ratios were measured on the LLNL spectrometer. Typical AMS measurement times were 3 min/sample, with a counting precision of 1.4%–2.0% and a standard deviation among 3–7 measurements of 1%–3%. The ^{14}C/^{13}C ratios of the unknowns were normalized to measurements of four identically prepared standards of known isotope concentration (Australian National University sucrose).

RESULTS AND DISCUSSION

All 10 subjects absorbed atrazine similarly, based on the ^{14}C recovered from the skin and nonocclusive cover after 24 hours, but the excretion varied considerably. Recovery of the ^{14}C was 101 $\pm$ 3% for the low-dose group and 92 $\pm$ 3% high-dose group. The unaccounted ^{14}C in the high-dose group may have been retained for more than 7 days, may have been excreted in an insoluble form that was inefficiently measured by LSC, or may have adhered to container surfaces. Urine was the dominant excretion route for dermally absorbed atrazine. The ^{14}C concentrations in urine ranged over 1.8–4300 fmol/ml (0.25–600 DPM/ml) among the 10 subjects. The lower levels could only be measured by AMS (Gilman et al., 1998).

The absorption and clearance of atrazine were measured by LSC and AMS (Figure 11.2). The two analysis methods agreed well on ^{14}C concentrations. The

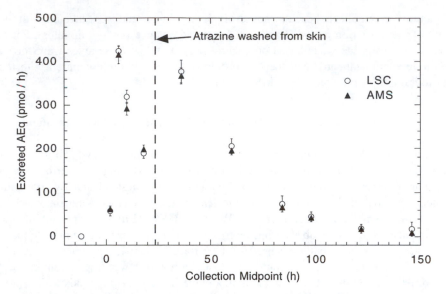

FIG. 11.2. Urinary excretion rates of molar atrazine equivalents (AEq) for subject 6 as measured by LSC and AMS. The dose was applied at 0 hours and removed at 24 hours. Low urine output may be responsible for the dip in the excretion rate before removal of the dose.

excretion profile of subject 6 depicted in Figure 11.2 had an unusual absorption phase, probably due to very low urinary output. The excretion phase was similar among the subjects. The excretion half-life of atrazine was measured to be 26 ± 3 hours by both methods in 9 of 10 subjects (Gilman et al., 1998). AMS achieved superior concentration (2.2 vs. 27 fmol/ml) and mass (5.5 vs. 54000 amol) detection limits compared to LSC when determining the clearance kinetics of unseparated urine (Gilman et al., 1998). Precision was similar for samples containing higher levels of [14]C. The lower concentration detection limit for AMS kept uncertainties near 1%–2% for all samples, whereas LSC uncertainties ballooned to 10%–20% near the detection limit (Gilman et al., 1998).

The synthetic HPLC standards eluted from the column between 3.8 and 45 minutes after injection. The most highly substituted compounds, melamine, ammeline, and ammelide, were very poorly soluble in aqueous and alcohol solutions and unlikely to appear in urine at high concentrations. No excess ^{14}C was observed at the retention times corresponding to melamine, ammeline, and ammelide. Figure 11.3 shows the UV spectra of standards of probable metabolites that were readily soluble in aqueous solution (Buchholz et al., 1999b). The peaks were obtained from multiple injections and are not scaled. Varying combinations of standards were injected and retention volumes monitored. This procedure ensured consistency of both relative and absolute retention volumes of standards. The absorbance at 254 nm and solubility of the standards varied greatly, but elution time could be easily determined. The dealkylated atrazine metabolites and their mercapturates eluted between 17 and 29 minutes.

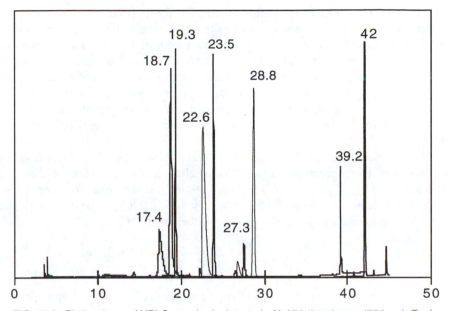

FIG. 11.3. Elution times of HPLC standards detected with UV absorbance (254 nm). Peak heights are not normalized. The UV absorbance spectra were used to identify peaks but not quantify levels. The standards shown in order of retention time were deethylatrazine (17.4), didealkylatrazine (18.7), didealkylatrazine mercapturate (19.3), hydroxy atrazine (22.6), deisopropylatrazine mercapturate (23.5), deethylatrazine mercapturate (27.3), deisopropylatrazine (28.8), atrazine mercapturate (39.1), and atrazine (42.1). The UV absorbance peaks of the standards were narrow and stable. Elution times were within 0.1–0.2 minutes and the entire spectrum shifted.

Although hydroxy atrazine also elutes in this interval, it is produced by hydrolysis with mineral acids and is not expected in urine after dermal exposure. Metabolites were tentatively identified based on cochromatography of the radioactive fraction from urine with an authentic synthetic standard. Cochromatography involved co-injection of the synthetic standard and urine with monitoring by both AMS and UV absorbance. The best separated standard peaks of likely major metabolites were deisopropylatrazine mercapturate (23.5 minutes) and atrazine mercapturate (39.2 minutes). The congestion of metabolite peaks in the central range and the coarse fraction sampling for AMS present difficulties in deconvolution of individual compounds. The concentration and purity of the metabolites were too low to verify the structure by classical mass spectroscopy methods.

The predose samples of each of the three subjects were analyzed by AMS for ^{14}C content neat and after separation by HPLC. Each urine supernatant contained slightly higher ^{14}C concentrations (118 ± 5, 124 ± 6, and 117 ± 2 amol ^{14}C/mg C) than the contemporary level at the time of the clinical exposure (112 amol ^{14}C/mg C). This elevation may be caused by the subjects' diets, the rate of carbon turnover in their bodies (Libby et al., 1964; Stenhouse and Baxter, 1977), or slight contamination at the clinic, but it did not adversely affect the AMS analysis. Tributyrin carrier provides the majority of the carbon contained in each HPLC-AMS sample (1.19 mg C with 11.7 amol ^{14}C). The ^{14}C background and the mean detection limit for the fractions depended on the mass consistency and the reproducible measurement of the tributyrin carrier. The contribution of carrier ^{14}C was subtracted and the predose samples of the three subjects were pooled to determine the average predose fraction contained 4.2 ± 5.0 amol ^{14}C above the known carrier. A reliable limit of quantitation 2 standard deviations above the blank is 14 amol ^{14}C per fraction above carrier (Buchholz et al., 1999b).

The measured ratio of ^{14}C to total C for each sample (R_{sample}) is described in equation (1). The concentration of ^{14}C labeled atrazine and metabolites in the urine is represented by $^{14}C_{tracer}/C_{tracer}$. The contributions from the urine and carrier tributyrin to the measured ratio are $^{14}C_{issue}/C_{issue}$ and $^{14}C_{carrier}/C_{carrier}$, respectively. The possibility of contamination to the sample is indicated as $^{14}C_{unknown}/C_{unknown}$:

$$R_{sample} = \frac{^{14}C_{tracer} + {}^{14}C_{issue} + {}^{14}C_{carrier} + {}^{14}C_{unknown}}{C_{tracer} + C_{issue} + C_{carrier} + C_{unknown}}. \tag{1}$$

Some components of equation (11.1) can be minimized by experimental design. In the case of HPLC eluent samples, $C_{carrier}$ is much larger than C_{tracer} and C_{tisue}. Furthermore, $^{14}C_{tracer}$ is much greater than $^{14}C_{issue}$ and greater than $^{14}C_{carrier}$ for samples with a signal. If contamination is assumed negligible, equation (11.1) then reduces to equation (11.2):

$$R_{sample} = \frac{^{14}C_{tracer} + {}^{14}C_{carrier}}{C_{carrier}}. \tag{2}$$

In the case of tissue samples, carrier components of equation (11.1) disappear and C_{tissue} dominates the total carbon inventory. Both $^{14}C_{tracer}$ and $^{14}C_{issue}$ are significant in tissue samples, so equation (11.1) then reduces to equation (11.3):

$$R_{sample} = \frac{^{14}C_{tracer} + {}^{14}C_{tissue}}{C_{tissue}}. \tag{3}$$

Unfolding the contribution of tracer to the measured ratio requires knowledge of the ^{14}C and total carbon content of control tissue.

Carbon analysis of individual urine samples was not necessary for the HPLC separation, as most of the carbon in the sample was from the tributyrin carrier. Urine typically contains 1% carbon by mass and has a density near $1 \text{ g} \cdot \text{ml}^{-1}$. The 25, 50, or 100 μL of urine injected on the HPLC column contained 0.25–1.0 mg carbon spread over 40–50 fractions. The distribution of carbon among the fractions is not known, but urea or inorganic carbon should reside as polar salts and pass through the column quickly. The carbon contributed by the sample (C_{tissue}) in any fraction is estimated to be less than 0.05 mg. This corresponds to only 4% of the total carbon after the addition of 1.19 mg carrier.

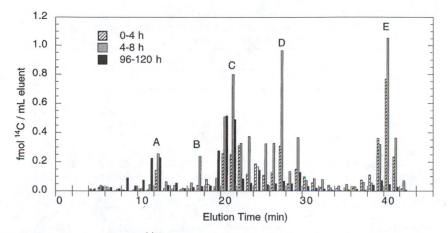

FIG. 11.4. Concentrations of ^{14}C in HPLC eluents of three time points for subject 6 (1.98 mg dose) at 0–4 hours, 4–8 hours, and 96–120 hours. Possible biomarker peaks are tentatively identified as (A) unknown, (B) deethylatrazine, (C) didealkylatrazine and didealkylatrazine mercapturate, (D) deethylatrazine mercapturate and deisopropylatrazine, and (E) atrazine mercapturate. Error bars depict 1 σ uncertainty in quantifying the ^{14}C signal.

The HPLC-AMS histograms of ^{14}C content vs. time (Figure 11.4) show the shift in atrazine metabolites from nonpolar to polar over the 7-day sampling period. Metabolite shifts were observed in all subjects, as the alkyl groups were cleaved to produce amine groups. The histogram peaks are much wider than the corresponding UV absorbance peaks. Coarse fraction sampling widens peaks that fall on boundaries between fractions. Every ^{14}C peak was wider than our 1-minute collection resolution. AMS has linear response over four orders of magnitude and measures the real width of the ^{14}C peaks, giving a true picture of the elution wavefront moving through the column at these low levels. Conventional UV absorbance is less sensitive than AMS, burying tails of the elution peaks in the background and making the peaks appear thinner. Peak broadening destroys resolution in the 20–30 minute elution range in our case, but it does not necessarily cause problems with less complicated mixtures of compounds. In addition, the ^{14}C elution spectrum could contain metabolites for which we did not have HPLC standards.

Table 11.1 reports the quantity of atrazine equivalents per milliliter of urine in identifiable peaks (Buchholz et al., 1999b). Uncertainties are expressed as the square root of the sum of the squares of uncertainties of individual fractions. Atrazine metabolites are quantified in molar atrazine equivalents (AEq) in Table 11.1 because changes in side chains have no effect on the ring-sited ^{14}C label. The two best separated metabolite peaks were an unknown at 10–13 minutes and atrazine mercapturate at 39–41 minutes. Deethylatrazine (17–18 minutes) can be identified on most plots, and a likely mixture of didealkylatrazine and didealkylatrazine mercapturate at 19–21 minutes also appear to be usable as markers of exposure. Other peaks seen in some histogram spectra were 23–24 minutes (deisopropylatrazine mercapturate), 26–28

TABLE 11.1. *Quantity of atrazine metabolites in distinct peaks expressed in fmol AEq/(ml urine). These peaks are possible biomarkers for dermal exposure to atrazine*

Sample	Elution time (min)				
	$10–13^a$	$17–18^b$	$19–21^c$	$26–28^d$	$39–41^e$
Subject 2					
4–8 hours	36 ± 1	33 ± 1	126 ± 1	130 ± 2	157 ± 2
Day 2	500 ± 4	163 ± 4	626 ± 4	345 ± 3	236 ± 2
Day 7	37 ± 1	4 ± 1	27 ± 1	11 ± 1	9 ± 1
Subject 5					
0–4 hours	1140 ± 10	1090 ± 10	4120 ± 40	2760 ± 20	2030 ± 10
4–8 hours	5110 ± 40	2850 ± 30	13000 ± 100	8190 ± 60	4690 ± 30
Day 5	318 ± 3	29 ± 1	483 ± 4	42 ± 1	45 ± 2
Subject 6					
0–4 hours	21 ± 1	14 ± 1	95 ± 1	51 ± 1	171 ± 1
4–8 hours	175 ± 3	130 ± 2	797 ± 6	660 ± 5	832 ± 6
Day 5	65 ± 1	8 ± 1	148 ± 1	19 ± 1	22 ± 1

[a] Unknown
[b] Deethylatrazine
[c] Didealkylatrazine and didealkylatrazine mercapturate
[d] Deethylatrazine mercapturate and deisopropyl atrazine
[e] Atrazine mercapturate

minutes (deethylatrazine mercapturate and possibly some deisopropylatrazine), and 30–31 minutes (possibly deisopropylatrazine or an unknown). No parent compound was observed in any sample.

The AEq/(ml urine) listed in Table 11.1 were calculated with equation (11.4), where the isotope ratios R_{sample}, $R_{carrier}$, and R_{urine} were expressed in mol ^{14}C/mg C, the masses $m_{carrier}$ and m_{urine} were expressed in mg C, V_{urine} is the volume of urine injected on the HPLC column in milliliters, and F_{label} is the fraction of atrazine labeled in the dosing solution:

$$\frac{AEq}{mL\,urine} = \frac{(R_{sample}(m_{carrier} + m_{urine}) - R_{carrier}m_{carrier} - R_{urine}m_{urine})}{V_{urine} \times F_{label}}. \tag{4}$$

The mass of urine was assumed to be distributed evenly over the 40 HPLC fractions. Although not strictly correct, this assumption has little effect due to $m_{carrier}$ overwhelming the carbon inventory.

The first (10–13 minutes) and last (39–41 minutes) peaks are sufficiently well separated that the tails in the AMS spectra could be included in the area under the curve. All other peaks overlapped, so the magnitudes of the peaks between 17 and 31 minutes are less certain. The small peaks are most vulnerable to tails of large peaks. The 19–21 minutes peak is the largest in all but one sample (subject 2, day 7) and is least likely to be affected by tails of neighboring peaks. In all cases the peaks were well above the background. The 14 amol per fraction limit of quantitation determined from 100 μL injections of predose urine corresponds to 140 amol ^{14}C $\cdot$ (mL urine)$^{-1}$.

CONCLUSIONS

The sensitivity of AMS salvaged the original aim of the clinical experiment: to determine the major metabolites of atrazine in human urine after a dermal exposure. The low levels of ^{14}C in the urine precluded the use of a flow through radiation detector in conjunction with an HPLC. Fraction collection with liquid scintillation counting was also not possible because the volume of urine injected on the HPLC column is limited to a few hundred microliters in practice. Traditional UV absorbance as a monitoring system for HPLC was useful for determining the retention times of standards, but was not sufficiently sensitive nor selective to detect fmol quantities of atrazine metabolites in fractions of crude urine. It did not detect all the ^{14}C peaks and identified numerous peaks that did not contain any ^{14}C tracer. AMS is the only detection method capable of quantifying the tracer atrazine and its by-products at these levels.

Atrazine mercapturate was the only metabolite that could be identified with certainty. Its dominance at the earliest time points and the large peak including didealkylatrazine and its mercapturate suggests that mercapturate metabolites of atrazine are suitable biomarkers for dermal exposure. The dealkylated atrazine metabolites may also be suitable as a class biomarker. The unknown metabolite that eluted at 10–13 minutes is well separated and easy to quantify, but it is only a major component several days after exposure when the concentrations of all metabolites are relatively low, possibly too low to be detected with routine assays (e.g., ELISA). Because this peak grows in proportion to the other peaks for the later sampling time points and occurs at the polar end of the elution gradient, it might be a metabolic end product.

HPLC-AMS is a sensitive new technique for metabolite determination in small samples of easily obtained fluids, such as urine. AMS can quantitate individual metabolites from human exposures to environmental levels of isotope-labeled compounds. This experiment was originally designed for LSC detection and used an exposure level higher than required for AMS quantitation. The effective radiation doses to these subjects was generally in the 1–30 μSv range, similar to that delivered in a dental x-ray. The activity of the applied dermal doses could have been reduced to 500 nCi, even with the poor absorption observed. An oral or IV dose containing 100 nCi would still allow use of 100 μL neat samples on the HPLC (Clifford et al., 1998; Buchholz et al., 1999a). Identification and quantitation of major metabolites are necessary to determine the most appropriate biomarker of exposure for field assays. Ultimately, these field assays can be used to protect farm worker health. HPLC-AMS assists in finding the most useful species for assay development. Furthermore, AMS could be used to calibrate immunoassays for improved precision.

The current HPLC-AMS sample preparation and graphitization method is cumbersome; however, batch processing of numerous samples is no more time consuming than waiting for LSC measurement. In this study, approximately 500 HPLC fractions were converted to graphite and measured on the LLNL AMS spectrometer with about 24 hours of machine time. A small single isotope spectrometer and an ion source that could accept gas injection from a pyrolized LC eluent or GC output is needed to make HPLC-AMS a common laboratory technique. Work on these improvements is underway at LLNL and elsewhere (Hughey et al., 1998).

These studies indicate that it is possible to carry out human metabolism studies of agricultural chemicals at levels that are commensurate with actual exposures. Such studies can be done with levels of ^{14}C that are far below those currently used and pose no radiation hazard to the subjects.

ACKNOWLEDGMENT

This work was supported in part by the NIEHS Superfund Basic Research Program (P42 ES04699), UC Davis NIEHS Center for Environmental Health Sciences (1 P30 ES05707), U.S. EPA Center for Ecological Health Research (CR819658), NIEHS ES05697, NIEHS RO1 ES02710, NIEHS R01 ES09690-01, and funds from the University of California and the Department of Energy, LLNL (UC Presidents Campus Laboratory Collaboration 95-103 and Intra-University Agreement B29 1419, LLNL-UCD). Work completed at LLNL was supported under DOE contract W-7405-ENG-48.

S. Douglas Gilman is currently at Department of Chemistry, University of Tennessee, Knoxville, Tennesee, USA.

REFERENCES

Ademola, J. I., Sedik, L. E., Wester, R. C., Maibach, H. I. (1993): In vitro percutaneous absorption and metabolism in man of 2-chloro–ethylamino-6-isopropylamine-s-triazine (atrazine). *Arch. Toxicol.*, 67:85–91.

Buchholz, B. A., Arjomand, A., Dueker, S. R., Schneider, P. D., Clifford, A. J., Vogel, J. S. (1999a): Intrinsic erythrocyte labeling and attomole pharmacokinetic tracing of 14C-labeled folic acid with accelerator mass spectrometry. *Anal. Biochem.*, 269:348–352.

Buchholz, B. A., Fultz, E., Haack, K. W., Vogel, J. S., Gilman, S. D., Gee, S. J., Hammock, B. D., Hui, X., Wester, R. C., Maibach, H. I. (1999b): HPLC-AMS measurement of atrazine metabolites in human urine after dermal exposure. *Anal. Chem.*, 71:3519–3525.

Catenacci, G., Barberi, F., Bersani, M., Ferioli, A., Cottica, D., Maroni, M. (1993): Biological monitoring of human exposure to atrazine. *Toxicol. Lett.*, 69:217–222.

Clark, G. M., Goolsby, D. A., Battaglin, W. A. (1999): Seasonal and annual load of herbicides from the Mississippi River basin to the Gulf of Mexico. *Environ. Sci. Tech.*, 33:981–986.

Clifford, A. J., Arjomand, A., Dueker, S. R., Schneider, P. D., Buchholz, B. A., Vogel, J. S. (1998): The dynamics of folic acid metabolism in an adult given a small tracer dose of ^{14}C-folic acid. *Adv. Exp. Med. Biol.*, 445:239–251.

Driskell, W. J., Head, S. L., Hill, R. H. Jr. (1997): LC-MS/MS analysis of the human urinary metabolites of atrazine, alachlor, and metolachlor. In: *Proceedings of the 45th Conference on Mass Spectrometry and Allied Topics*, p. 913, June 1–5, 1997, Palm Springs, CA.

Environmental Protection Agency. 1994. *Federal Register*, 59:60412–43.

Gianessi, L. P., Anderson, J. E. *Pesticide Use in U.S. Crop Protection*, National Center for Food and Agricultural Policy: Washington, DC, 1995.

Gilman, S. D., Gee, S. J., Hammock, B. D., Vogel, J. S., Haack, K., Buchholz, B. A., Freeman, S. P H. T., Wester, R. C., Hui, X., Maibach, H. I. (1998): Analytical performance of accelerator mass spectrometry and liquid scintillation counting for detection of ^{14}C-labeled atrazine metabolites in human urine. *Anal. Chem.*, 70:3463–3469.

Hughey, B. J., Klinkowstein, R. E., Shefer, R. E., Fried, N. A., Skipper, P. L., Wishnok, J. S., Jiao, K., Tannenbaum, S. R. (1998): Development of a GC-AMS system for biomedical research. In: *Synthesis and Applications of Isotopically Labeled Compounds 1997*. Edited by J. R. Heys and D. G. Melillo, pp. 107–109. John Wiley and Sons Ltd., Chichester.

Ikonen, R., Kangas, J., Savolainen, H. (1988): Urinary atrazine metabolites as indicators for rat and human exposure to atrazine. *Toxicol. Lett.*, 44:109–112.

Jaeger, L. L., Jones, A. D., Hammock, B. D. (1998): Development of an enzyme-linked immunosorbent assay for atrazine mercapturic acid in human urine. *Chem. Res. Toxicol.*, 11:342–352.

Kolpin, D. W., Thurman, E. M., Goolsby, D. A. (1996): Occurrence of selected pesticides and their metabolites in near-surface aquifers of the midwestern United States. *Environ. Sci. Technol.*, 30:335–40.

Koskinen, W. C., Clay, S. A. (1997): Factors affecting atrazine fate in north central U.S. soils. *Rev. Environ. Contam. Toxicol.*, 151:117–165.

Lang, D., Criegee, D., Grothusen, A., Saalfrank, R. W., Bocker, B. H. (1996): In vitro metabolism of atrazine, terbutylazine, ametryne, and terbutryne in rats, pigs, and humans. *Drug. Metab. Dispos.*, 24:859–865.

Libby, W. F., Berger, R., Mead, J. F., Alexander, G. V., Ross, J. F. (1964): Replacement rates for human tissue from atmospheric radiocarbon. *Science*, 146:1170–1172.

Lucas, A. D., Jones, A. D., Goodrow, M. H., Saiz, S. G., Blewett, C., Seiber, J. N., Hammock, B. D. (1993): Determination of atrazine metabolites in human urine: Development of a biomarker of exposure. *Chem. Res. Toxicol.*, 6:107–116.

Ma, L., Selim, H. M. (1996): Atrazine retention and transport in soils. *Rev. Environ. Contam. Toxicol.*, 145:129–173.

Perry, M. J., Layde, P. M. (1998): Sources, routes and frequency of pesticide exposure among farmers. *J. Occup. Environ. Med.*, 40:697–701.

Ribaudo, M. O., Bouzaher, A. *Atrazine: Environmental Characteristics and Economics of Management*. Agricultural Economic Report 699, United States Department of Agriculture, Washington, DC, 1994.

Solomon, K. R., Baker, D. B., Richards, R. P., Dixon, K. R., Klaine, S. J., La Point, T. W., Kendall, R. J., Weisskopf, C. P., Giddings, J. M., Giesy, J. P., Hall, L. W., Jr., Williams, W. M. (1996): Ecological risk assessment of atrazine in North American surface waters. *Environ. Toxicol. Chem.*, 15:31–76.

Stenhouse, M. J., Baxter, M. S. (1977): Bomb ^{14}C as a biological tracer. *Nature*, 267:828–832.

Vogel, J. S. (1992): Rapid production of graphite without contamination for biological AMS. *Radiocarbon*, 34:344–350.

Vogel, J. S., Turteltaub, K. W., Finkel, R., Nelson, D. E. (1995): Accelerator mass spectrometry: Isotope quantification at attomole sensitivity. *Anal. Chem.*, 67:353–9A.

Vogel, J. S. (1998): Accelerator mass spectrometry as a bioanalytical tool. In: *Synthesis and Applications of Isotopically Labeled Compounds (1997)*. Edited by J. R. Heys and D. G. Melillo, pp. 11–20. John Wiley and Sons Ltd., Chichester.

Toxicology of Skin
Edited by Howard I. Maibach
Copyright © 2001 Taylor & Francis

12

Partitioning of Chemicals from Water into Powdered Human Stratum Corneum (Callus)

Xiaoying Hui, Ronald C. Wester, Philip S. Magee, Howard I. Maibach

Department of Dermatology, University of California, San Francisco, California, USA

INTRODUCTION

Chemical delivery/absorption into and through skin is important for both dermato-pharmacology and dermatotoxicology. A major endeavor is underway to enable predictions of percutaneous absorption from molecular properties. This is especially important to the Environmental Protection Agency (EPA), where chemicals produced by industry outdistance the ability to experimentally document absorption properties (EPA, 1992). Early correlations of percutaneous absorption were restricted to octanol/water partition coefficients (PC o/w) as a desorptor. These correlations were not completely successful in reducing the experimental variance, and other parameters, such as molecular weight, have been added to the predictive equations (EPA, 1992). The substitution of powdered human stratum corneum (PHSC) prepared from callus (sole) for the intact membranous stratum corneum is expected to bring equation predictions closer to actual absorption values. To validate the PHSC system, the first step is to correlate PHSC/water partitioning with octanol/water partitioning.

The human stratum corneum (SC) includes the horny pads of the palms and soles (callus) and the membranous stratum corneum covering the remainder of the body (Barry, 1983). It represents a rate-limiting barrier in the transport of most chemicals across the skin (Blank, 1965). Chemicals must first partition into the SC before entering deeper layers of the SC, epidermis, and dermis to reach the vascular system. This partitioning process occurs much faster than complete diffusion through the whole SC, and the process quickly reaches equilibrium (Scheuplein and Bronaugh, 1985). In addition to binding within the SC, a chemical can also be retained within the SC as a reservoir (Zatz, 1993). Thus, understanding the process of chemical partitioning into the SC becomes important in developing an insight into the barrier properties and transport mechanisms. Many experiments have been conducted to predict chemical partitioning into the SC in vitro; however, most were based on quantitative structure activity relationships (QSAR's) of related chemicals to determine the partitioning process, and few studies focus on structurally unrelated chemicals (Guy and Hadgraft, 1991). Because the range of molecular structure and physico-chemical properties is very broad, any predictive model must address a broad scope of partitioning behavior.

This study assesses the relationship of a limited number of chemicals with a broad scope of physico-chemical properties in the partitioning mechanism between the PHSC and water. Uniqueness and experimental accuracy is added by using the PHSC (Wester et al., 1987). The experimental approach is designed to determine how the PC PHSC/w is affected by (1) chemical concentration, (2) incubation time, (3) chemical lipophilicity (or hydrophilicity), and other factors. These parameters are used to develop an in vitro model and may aid in predicting chemical dermal exposure and in developing dermally applied drugs.

MATERIALS AND METHODS

Materials

[Ring UL-^{14}C] atrazine (purity >98%, specific activity 11.5 mCi/mmol), [ring UL-^{14}C] 2,4-dichlorophenoxyacetic acid (2,4-D, >98%, 20.2 mCi/mmol), [8-^{14}C] dopamine hydrochloride (>98%, 15.6 mCi/mmol), and [methoxy-^{14}C] malathion (95%, 8.8 mCi/mmol) were obtained from Sigma Chemical Company (St. Louis, MO). [Dimethylamine-^{14}C] aminopyrine (98%, 107 mCi/mmol), [^{14}C] glyphosate (>98%, 51 mCi/mmol), [8-^{14}C] theophylline (>98%, 51 mCi/mmol), and [U-^{14}C] polychlorinated biphenyls (PCB, aroclor 1254, 54% chlorine, 32 mCi/mmol) were obtained from Amersham Corporation (Arlington Heights, IL) [U-^{14}C] glycine (98%, 63 mCi/mmol), and [^{14}C] urea (98%, 56 mCi/mmol) were obtained from ICN Biomedical Inc (Irvine, CA). [4-^{14}C] Hydrocortisone (97.6%, 52 mCi/mmol) and [4-^{14}C] estradiol (99%, 54.1 mCi/mmol) were obtained from Du Pont Company (Wilmington, DE). [Ring ^{14}-C(U)] alachlor (21 mCi/mmol) was a gift from Monsanto Company

(St. Louis, MO). [7-^{14}C] Acitretin (>98%, 58.2 mCi/mmol) was a gift from Roche Dermatologics (Nutley, NJ). The pH values of these chemical solutions in distilled deionized water were determined by a Corning pH/ion Analyzer 250 (Corning Science Products, Corning, NY).

Preparation of Powdered Human Stratum Corneum (Callus)

The procedure was based on the method of (Wester et al., 1987). Adult foot callus was ground in the Micro-Mill Grinder (Bel-Art Products, Pequannock, NJ) in the presence of liquid nitrogen to form a powder. Particles of the PHSC that passed through a 50-mesh sieve but not an 80-mesh were used. Particle sizes within this range were shown to favor the experimental conditions. The sample was stored in the freezer until using.

Depleting Lipid Content of the PHSC

PHSC was mixed with n-hexane for 30 minutes, followed by chloroform and methanol mixture (2:1, v/v) overnight for extraction of intercellular lipids (Raykar et al., 1988; Kurihara-Bergstrom et al., 1990). The delipidized PHSC was stored in the freezer. The lipid content of the PHSC was determined by the change in weight before and after solvent extraction (Raykar et al., 1988).

Prewetting of the PHSC with Water or Ethanol

To determine effect of hydrated or perturbed PHSC on compound partitioning, powdered stratum corneum was wetted by 10 μl water or ethanol for 30 minutes prior to incubation.

Incubation Procedure

The experiment was performed by modifying the method of (Wester et al., 1987). A given dosage of radiolabeled chemical in 4.0 ml vehicle (distilled deionized water) was mixed with 10.0 mg of the powdered callus in a glass vial and incubated in a water bath with moderate shaking at 35°C for a given time period. The mixture then was separated by centrifugation at $1,500 \times g$ for 15 minutes and the supernatant carefully removed. The PHSC pellet was resuspended three times in the same volume of deionized distilled water to remove any excess material clinging to the surface. Experiments indicated that further washing (four or five times) did not significantly change the PC value. The radioactivity of the chemical bound to the PHSC and that remaining in the vehicles was determined by scintillation counting. Five samples were used for each test.

Scintillation Counting

The scintillation cocktail was Universal-ES (ICN, Costa Mesa, CA). Background controls, and test samples were counted in a Packard model 4640 counter (Packard Instruments). The data was transferred to a computer program (Appleworks/Apple IIE computer; Apple Computer Co., Mountain View, CA) that subtracted background control samples and generated a spreadsheet for statistical analysis. The counting process and a computer program have been verified to be accurate by a quality assurance officer at the University of California at San Francisco.

Powdered Human Stratum Corneum/Water Partition Coefficient

The value is determined by the following equation:

$$PC_{PHSC/w} = C_{PHSC}/C_{water},\qquad(1)$$

where C_{PHSC} is the amount (μg) of the chemical absorbed in 10 mg of the PHSC and C_{water} is the amount (μg) remaining in 4000 mg of the water after removing the PHSC pellet. Because the micrograms of the chemical in the PHSC or water is proportional to the counts (dpms) determined by the scintillation counter and partitioning of chemical into the PHSC exactly equaled the decrease of the chemical concentration in the vehicle (water), an alternative calculation method was also used:

$$PC_{PHSC/w} = \frac{(dpm_{PHSC}/mg_{PHSC})}{(dpm_i - dpm_{PHSC})/(mg_{water})},\qquad(2)$$

where dpm_i is the initial chemical concentration in the vehicle (water). These two calculation methods were shown to give similar results, and the reported experiments are calculated by the second method.

Statistical Analysis

Statistical analysis (students' t-test, linear and multiple regressions) were performed in version 6.1 of MINITAB (Minitab Inc, State College, PA) on an IBM PC-compatible computer. When P value was smaller than 0.05, it was considered as statistical significance.

RESULTS

Table 12.1 shows the effect of varying initial chemical concentrations on the PC PHSC/w of 12 test compounds. Under the fixed experimental conditions (2 hours incubation time and 35°C incubation temperature), the concentration required to reach a peak value of the partition coefficient varied from chemical to chemical. After reaching the maximum, increasing the chemical concentration in the vehicle did not elevate

TABLE 12.1. *Effect of initial aqueous phase chemical concentration on powdered human callus/water partition coefficient*

Chemical	Concentration* (%, w/v)	Partition (mean)	Coefficient[†] (S.D.)
Dopamine	0.23	5.42	0.22
	0.46	6.04	0.28
	0.92	5.74	0.28
Glycine	0.05	0.36	0.01
	0.10	0.40	0.02
Urea	0.03	0.26	0.02
	0.06	0.15	0.02
	0.12	0.17	0.02
Glyphosate	0.02	0.79	0.04
	0.04	0.68	0.04
	0.08	0.70	0.01
Theophylline	0.18	0.37	0.02
	0.36	0.43	0.03
	0.54	0.42	0.02
Aminopyrine	0.07	0.44	0.09
	0.14	0.46	0.03
Hydrocortisone	0.09	0.37	0.01
	0.18	0.34	0.01
	0.36	0.29	0.02
Malathion	0.47	0.50	0.09
	0.94	0.40	0.03
	1.88	0.53	0.04
Atrazine	0.09	0.53	0.06
	0.14	0.59	0.07
	0.19	0.58	0.03
2,4-D	0.27	7.52	0.81
	0.54	7.53	1.01
	0.82	8.39	1.67
Alachlor	0.32	1.11	0.05
	0.64	1.08	0.04
	1.28	1.96	0.15
PCB	0.04	1237.61	145.52
	0.08	1325.44	167.03
	0.16	1442.72	181.40

*Concentration expressed as gram of chemical in 100 ml water; 2 hours' incubation time at 35°C.
[†] Stratum corneum/water partition coefficient represented the mean of each test (n = 5) +/− S.D.

the PC value; instead, it slightly decreased or was held at approximately the same level. The PC value of each chemical correlated to its lipophilicity or hydrophilicity as described by log PC o/w. For example, the lipophilicity of malathion (log PC o/w = 2.36) is approximately the same as that of atrazine (log PC o/w = 2.75). Therefore, their PC PHSC/w values are close to each other; however, the concentration of malathion to achieve a peak value was approximately 10-fold higher than that of atrazine.

Figure 12.1 shows that the maximum amount of chemical partitioning into the SC is limited when the experimental condition is fixed (30 minutes preincubation followed

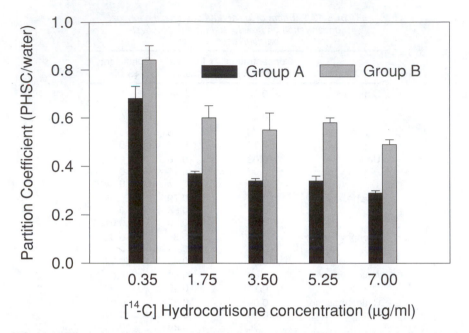

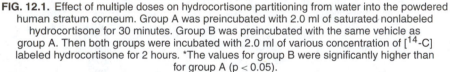

FIG. 12.1. Effect of multiple doses on hydrocortisone partitioning from water into the powdered human stratum corneum. Group A was preincubated with 2.0 ml of saturated nonlabeled hydrocortisone for 30 minutes. Group B was preincubated with the same vehicle as group A. Then both groups were incubated with 2.0 ml of various concentration of [^{14}-C] labeled hydrocortisone for 2 hours. *The values for group B were significantly higher than for group A ($p < 0.05$).

by 2 hours incubation time). Preliminary experiments demonstrated no statistical differences of the PC values of [^{14}C]-labeled hydrocortisone between 30 minutes' and 2 hours' incubation. Group A was preincubated with saturated unlabeled hydrocortisone while group B was incubated with the same vehicle without hydrocortisone. Both groups were then challenged by different concentrations of [^{14}C]-labeled hydrocortisone. The peak value of group A was reduced by approximately 40% when compared with group B.

Figures 12.2 and 12.3 show the changes in PC PHSC/w of 12 test compounds as a function of varying incubation time. Theophylline and atrazine reached their maximum values as early as 6 hours' incubation. For most chemicals, however, increasing incubation time resulted in elevation of PC PHSC/w. Higher lipophilic compounds, such as 2,4-D, alachlor, and PCB (log PC o/w = 2.8–6.4), and very hydrophilic compounds, such as dopamine (log PC o/w = −3.4), reach their maximum partitioning into the SC around 12 hours of incubation. For those compounds of relatively lower lipophilicities or hydrophilicities compounds, PC PHSC/w continually increased with incubation time up to 24 hours.

Figure 12.4 describes a smooth, partially curvilinear relationship between the log PC PHSC/w and the log PC o/w of these chemicals. In this study, lipophilicities

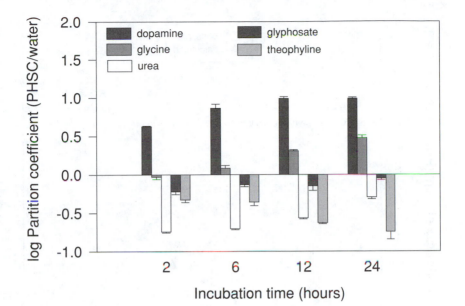

FIG. 12.2. Logarithm of stratum corneum/water partition coefficients of hydrophilic compounds (log PC o/w < 0) determined as a function of incubation time. Each bar expressed as the mean +/− S.D., n = 5.

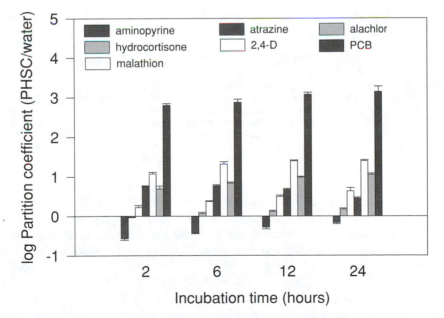

FIG. 12.3. Logarithm of stratum corneum/water partition coefficients of lipophilic compounds (log PC o/w > 0) determined as a function of incubation time. Each bar expressed as the mean +/− S.D., n = 5.

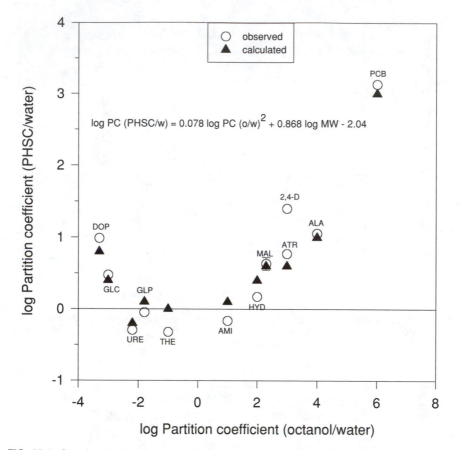

FIG. 12.4. Correlation of the logarithm of stratum corneum/water partition coefficients (log PC sc/w) and logarithm of octanol/water partition coefficients of the 12 test chemicals. Open symbols expressed as observed values, and each represented the mean of a test chemical +/− S.D. (n = 5). Close symbols expressed as calculated values by the equation 6. 2,4-D = 2,4-dichlorophenoxyacetic acid; ALA = alachlor; AMI = aminopyrine; ATR = atrazine; DOP = dopamin; GLC = glycine; GLP = glyphosate; HYD = hydrocortisone; MAL = malathion; PCB = polychlorinated biphenyls; THE = theophylline; URE = urea.

and hydrophilicities of compounds were defined as log PC o/w larger or smaller than 0, respectively. For lipophilic chemicals, such as aminopyrine, hydrocortisone, malathion, atrazine, 2,4-D, alachlor, and PCB, the logarithms of PHSC/water partition coefficients are proportional to the logarithms of the octanol/water partition coefficients:

$$\log PC\ PHSC/w = 0.59 \log PC\ o/w - 0.72,$$

Student T values: 9.93, (3)

$$n = 7, \quad r^2 = 0.95, \quad S = 0.26, \quad F = 98.61.$$

For hydrophilic chemicals, such as theophylline, glyphosate, urea, glycine, and dopamine, the log PC PHSC/w values are approximately and reversely proportional to log PC o/w:

$$\log \text{PC PHSC/w} = -0.60 \log \text{PC o/w} - 0.27,$$

Student T value: -4.86, (4)

$$n = 5, \quad r^2 = 0.88, \quad s = 0.26, \quad F = 23.61;$$

however, the overall relationship of the PC PHSC/w of these chemicals and their PC o/w is nonlinear. This nonlinear relationship is adequately described by the following equation:

$$\log \text{PC PHSC/w} = 0.078 \log \text{PC o/w}^2 + 0.868 \log \text{MW} - 2.04$$

Student T values: 8.29 2.04 (5)

$$n = 12, \quad r^2 = 0.90, \quad s = 0.33, \quad F = 42.59$$

The logarithm of molecular weight (MW) gave a stronger correlation in this regression than MW (T = 1.55). In Figure 12.4, the calculated log PC PHSC/w (Y estimate) are compared to the corresponding observed values for these chemicals. As shown, the calculated values are acceptably close to the observed values. The correspondence with minimal scatter suggests that this equation would be useful in predicting in vitro partitioning in the PHSC for important environmental chemicals.

Table 12.2 shows that average lipid content of the dry PHSC samples derived from various sources was 2.29 ± 0.25 weight percent after extraction. Water uptake capacities of untreated PHSC, delipidized PHSC (as the protein fraction), and the lipid content are also shown in Table 12.3. The determination was made by measuring the amount $[^3H]$-water (μg equivalent) per mg PHSC after equilibration. Results showed that no statistical differences (P > 0.05) were observed for untreated

TABLE 12.2. *Lipid content and water uptake of powdered human stratum corneum*

Stratum corneum source	Lipid content (%w/w dry PHSC)	Water uptake (μg/mg dry PHSC)			
		Untreated PHSC	Delipidized PHSC		
			Lipid*	Protein[†]	Total
1	2.38	495.85	26.44	452.40	478.84
2	2.21	452.49	39.26	364.96	404.22
3	2.39	585.62	23.09	498.40	521.49
4	2.69	554.27	40.05	492.31	532.36
5	2.08	490.04	49.86	363.30	413.16
6	2.01	381.61	14.82	324.18	339.00
Mean	2.29	493.31	32.26	415.92	448.18
SD	0.25	72.66	12.97	74.50	75.47

*Lipid part extracted from the PHSC.
[†] Rest part of the PHSC after lipid extraction.

TABLE 12.3. *Effects of water and ethanol pretreatment on the water-binding capacities of untreated and delipidized PHSC*

Sample pretreatment	Water absorbed by the PHSC (μg equivalent/mg PHSC)	
	Untreated PHSC	Delipidized PHSC
Dry sample	468.00 ± 71.80	463.02 ± 25.17
Water prewetted	314.40 ± 65.89*	308.67 ± 48.06[†]
Ethanol prewetted	222.14 ± 17.87*	222.39 ± 24.01[†]

*Statistical significance between prewetted and dry untreated PHSC ($p < 0.05$).

[†] Statistical significance between prewetted and dry delipidized PHSC ($p < 0.05$).

PHSC, delipidized PHSC, and the combination of delipidized PHSC and the lipid content.

Table 12.3 shows the effects of water or ethanol prewetting on dry PHSC for water uptake. For both untreated and delipidized PHSC pre-wetted with water or ethanol, their water uptake capacities were significantly reduced when compared with corresponding dry PHSC groups ($P < 0.05$).

Figure 12.5 plots the uptake of [^{3}H]-water versus ethanol concentration for untreated and delipidized PHSC. Both groups exhibited slightly increasing water uptake capacities in low ethanol concentrations ($< 40\%$) and then decreasing as ethanol concentration continuously increased (40%–80%). Now statistical differences between untreated and delipidized PHSC for uptake water were observed ($P > 0.05$).

Table 12.4 shows partitioning of eight compounds with different lipophilicities (or hydrophilicities) as described by their log PC (octanol/w) values into untreated and delipidized PHSC. Depleting the lipid content from the PHSC did not equally change partition coefficients of lipophilic compounds. Partition coefficients of acitretin, atrazine, and hydrocortisone reduced significantly in the delipidized group when compared with the untreated group ($P < 0.05$); however, alachlor significantly increased partitioning into delipidized PHSC ($P < 0.01$). Estradiol did not show a difference between untreated and delipidized PHSC. Glyphosate, a hydrophilic compound, also did not show a difference between untreated and delipidized PHSC, but the other hydrophilic compound, theophylline showed lower partitioning into delipidized PHSC than into untreated PHSC ($P < 0.05$).

Figure 12.6 plots partitioning of four compounds with different lipophilicities into untreated and delipidized PHSC as a function of ethanol concentration. Although alachlor, atrazine, and glyphosate showed different lipophilicities, their partitioning patterns were similar. Their partition coefficients decreased as ethanol concentration increased. The partitioning pattern of malathion, however, is similar to that of water, as described above (see Figure 12.5). When ethanol concentration was lower than 40%, lipophilic compounds atrazine and malathion partitioning into untreated PHSC were significantly higher that those in delipidized PHSC ($P < 0.05$). Glyphosate, a high

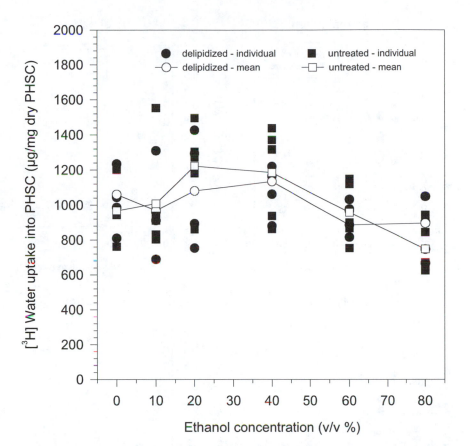

FIG. 12.5. Comparisons of water uptake capacities of untreated and delipidized PHSC as a function of ethanol concentration. The sample size of each group = 6.

TABLE 12.4. *Comparisons of partition coefficients between untreated and delipidized PHSC*

Chemical	log PC (o/w)	log PC (PHSC/w)		p Value
		Untreated	Delipidized	
Acitretin	6.40	2.69 ± 0.04	2.33 ± 0.10	0.01
Alachlor	3.52	0.84 ± 0.02	0.99 ± 0.03	0.01
Atrazine	2.75	0.47 ± 0.05	0.35 ± 0.04	0.05
Estradiol	2.70	1.23 ± 0.02	1.29 ± 0.10	N
Glyphosate	−1.70	−0.03 ± 0.04	−0.03 ± 0.03	N
Hydrocortisone	1.61	−0.16 ± 0.04	−0.56 ± 0.05	0.01
Malathion	2.36	0.59 ± 0.05	0.36 ± 0.06	0.01
Theophylline	−0.76	−0.86 ± 0.01	−0.09 ± 0.02	0.05

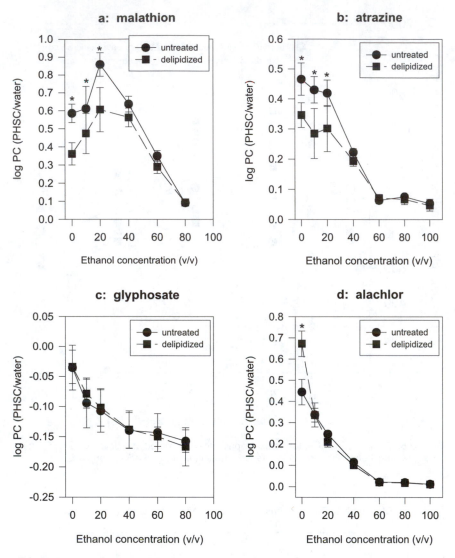

FIG. 12.6. Logarithm of PHSC/water partition coefficients of four compounds with different lipophilicities as a function of ethanol concentration. Each symbol represents the mean ± S.D., n = 5.

hydrophilic compound, showed no difference in partition coefficients between untreated and delipidized PHSC ($P > 0.05$). Alachlor, as shown in Table 12.4 had significant difference between untreated and delipidized PHSC when water was only solvent ($P < 0.05$), and no statistical difference was observed when ethanol was employed ($P > 0.05$).

Figure 12.7 shows effects of water or ethanol prewetting on dry PHSC for lipophilic compounds hydrocortisone, acitretin, and alachlor partitioning. Only the ethanol

a: hydroquinone

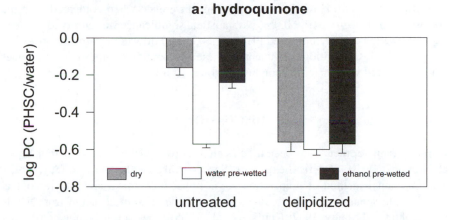

b: acitretin

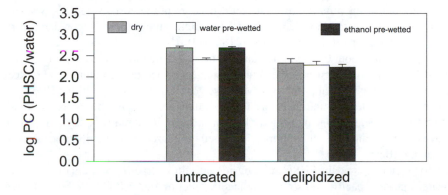

c: alachlor

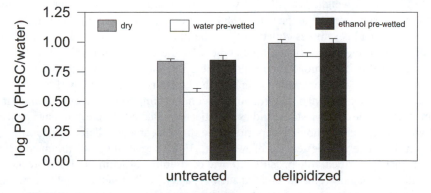

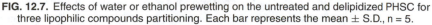

FIG. 12.7. Effects of water or ethanol prewetting on the untreated and delipidized PHSC for three lipophilic compounds partitioning. Each bar represents the mean ± S.D., n = 5.

prewetted untreated PHSC showed significant difference when compared with the corresponding dry group (P < 0.05). No statistical significance was observed between water prewetted and dry untreated PHSC (P > 0.05). For water or ethanol prewetted delipidized PHSC, partition coefficients of these compounds did not show differences when compared with the corresponding dry ones (P > 0.05).

DISCUSSION

Human stratum corneum has been used as an in vitro model to explore percutaneous absorption and risk of dermal exposure (Surber et al., 1990; Potts and Guy, 1992). The traditional method of preparation is via physico-chemical and enzymological processes to separate the membranous layers of the stratum corneum from whole skin (Juhlin and Shelly, 1977; Knufson et al., 1985); however, it is time consuming and in some cases difficult to control the size and thickness of a sheet of stratum corneum.

In this study, the human callus (stratum corneum of the horny sole pads) was employed as an in vitro model. It is easier to obtain and prepare in powdered form (PHSC), and the particle sizes between 180–300 μM were selected. Because a corneocyte is only about 0.5 μM thick and about 30–40 μM long, the PHSC is considered to contain intact corneocytes plus intercellular medium structures and to keep its physical-biochemical properties. Moreover, the exterior surface of the PHSC has been greatly extended and therefore favors solutes penetrating the PHSC from any direction. A disadvantage in using the human callus is that it may display some differences in water and chemical permeation when compared with membranous stratum corneum (Barry, 1982).

To best understand the PHSC made by human callus, stratum corneum lipid content, water uptake capacity, and solvent (ethanol) effect on the PHSC have been studied. Stratum corneum lipid plays an important role in the determination of skin functions; however, average lipid content of the stratum corneum varies regionally from 2.0, 4.3, 6.5., to 7.2 weight percent of dry stratum corneum for plantar, leg, abdomen, and face, respectively (Lampe et al., 1983). To test such variations and/or individual differences, callus samples collected from the plantar region of six different human subjects were measured for the lipid content by the change in weight of the PHSC before and after extraction. The result (Table 12.2) shows that average lipid content of the PHSC is consistent with that in human plantar as determined by (Lampe et al., 1983). The water content of the stratum corneum is of importance in maintaining stratum corneum flexibility. Many studies have been conducted to determine mechanism of water uptake and/or holding capacity of the stratum corneum. Three possible mechanisms have been suggested. (Imokawa et al., 1986) suggested that stratum corneum lipids played a critical role because removal of stratum corneum lipids following applications of acetone/ether decreased such capacity. (Friberg et al., 1992), however, considered that protein may also play an important role in holding water in the stratum corneum. They found that the additional water taken up, after reaggregation of equilibrated lipids and proteins, was equally partitioned between the protein and the natural lipid fraction of human stratum corneum. (Middleton, 1968) considered that the water-soluble substances were responsible for water holding and for most of the extensibility of corneum. (Middleton, 1968) found that powdered stratum corneum, but not the intact corneum extracted by water, could reduce its water uptake capacity. He suggested this was because the powdering procedure ruptured the walls of the corneum cells and allowed water to extract the water-soluble substances without a prior solvent extraction. Data from this study show that the PHSC can take water up to 49% of weight of the dry untreated PHSC (Table 12.2), which is consistent with literature reports. (Middleton, 1968) found that the amount of water bound to intact, small pieces and powdered Guinea-pig footpad stratum corneum was 40%, 40%, and 43% dry corneum weight. (Leveque and Rasseneur, 1988) measured that the human stratum corneum was able to absorb water up to 50% of its dry weight. Our results suggest that the protein domain of the PHSC plays an important role in the take up of water, as depleting the lipid content of the PHSC did not change water uptake capacity (Table 12.2). Although the lipid of the PHSC takes water up to 140% its own weight, only 2.29% of the PHSC by weight is composed of lipids (Table 12.2). Data of this study also show that the role of the water-soluble substances in holding water, at least under in vitro conditions, may not be important. For example, in this study the PHSC was immersed in water solution to reach equilibration, not as described by (Middleton, 1968) set in a humidity-controlled environment to reach equilibration. Therefore, it is possible to lose the water-soluble substances during the equilibration and later the washing process; however, the data from this study, as discussed above, are the same as the literature reports. An interesting observation from this study is that the PHSC prewetted by water or ethanol can significantly reduce water uptake capacity of both untreated and delipidized PHSC ($P < 0.05$). This is similar to the results of Middleton's study (1968), when he pretreated intact and powdered stratum corneum with water or ether extraction. Our suggestion is that, for a unit of PHSC, its water uptake (or holding) capacity depends on the availability of inter- and intracellular spaces. When the PHSC was hydrated, its available space was limited and the uptake capacity decreased. Effect of ethanol on the PHSC may result in alterations of the stratum corneum's lipoidal and keratinoidal constituents (Knutson et al., 1990; Hatanaka et al., 1993).

The effect of varying initial chemical concentration in the vehicle shows a parallel increase in the partition coefficient (PC PHSC/w) until reaching the maximum and then remains approximately the same or slightly decreased (Table 12.1 and Figure 12.1). This is consistent with results of (Surber et al., 1990a, 1990b) on whole stratum corneum. Chemical partitioning from the vehicle into the SC includes processes in which molecular binding occurs at certain sites of the SC as well as simple partitioning. Equilibration of partitioning is largely dependent upon the saturation of the chemical binding sites of the SC (Surber et al., 1990a; Rieger, 1993). The results from Table 12.1 and Figure 12.1 also indicate that under a given experimental condition, the maximum degree of partitioning was compound specific. As the SC contains protein, lipids, and various lower molecular weight substances with widely differing properties, the many available binding sites will display different selective affinities with each chemical. Thus, the degree of maximum binding or of equilibration varies naturally with molecular structure (Rieger, 1993).

PC PHSC/w values of chemicals are sensitive to their lipophilicity or hydrophilicity (Figure 12.4). Thus, two types were classified: those for lipophilic compounds, where PHSC/w partitioning is proportional to their lipo-solubilities (as determined by log PC o/w, Equation 3), and those for hydrophilic compounds where partitioning appears to relate to water solubility or reverse lipophilicities (Table 12.2 and Equation 4). This result demonstrated that the solubility limit of a compound in the SC was important in determining the degree of partitioning, as suggested by (Potts and Guy, 1993). On the basis of the solubility limit of a chemical, the uptake process of water-soluble or lipid-soluble substances was controlled by the protein domain or the lipid domain, respectively, or a combination of two (Raykar et al., 1988). The protein and lipid domains in the SC are histologically revealed as a mosaic of cornified cells containing cross-linked keratin filaments and intercellular lipid-containing regions (Elias, 1981). Because the lipophilicity of the lipid domain in the SC is much higher than that of water, a lipophilic compound would partition into the SC in preference to water. Thus, when water is employed as the vehicle, the PC PHSC/w will increase with increasing lipophilicity of solute (Scheuplein and Bronaugh, 1983). Conversely, the protein domain of the SC is significantly more polar than octanol and governs the uptake process of hydrophilic chemicals (Rayker et al., 1988). For very lipophilic compounds, low solubility in water rather than increased solubility in the SC can be an important factor (Scheuplein and Bronaugh, 1983). Moreover, in addition to partitioning into these two domains, some amount of chemicals may be taken into the SC as the result of water hydration. This is the so-called "sponge domain" by (Raykar et al., 1988). They assume that this water, having the properties of bulk water, carries an amount of solute into the SC equal to the amount of solute in the same volume of bathing solution. Therefore, for hydrophilic compounds and some lower lipophilic compounds, the partitioning process may include both the protein domain and sponge domain.

It is of interest to compare chemical partition coefficients (stratum corneum/w) between untreated and delipidized stratum corneum groups with the chemical lipophilicities. (Raykar et al., 1988) found that the partition coefficient (stratum corneum/w) of a lipophilic compound, hydrocortisone 21-ester, was reduced when the delipidized stratum corneum was employed. (Surber et al., 1990), however, reported that partition coefficients (stratum corenum/w) of hydrocortisone and four other lipophilic compounds increased when the delipidized stratum corneum was used. Results of the study (Table 12.4) show that lipophilic compounds could decrease partitioning into the delipidized PHSC (such as atrazine, acitretin, and malathion), or increase (alachlor), or keep the same as the untreated PHSC (estradiol). Hydrophilic compounds, theophylline and glyphosate, showed similar behaviors when the delipidized PHSC was employed. A possible explanation of such changes as described by (Surber et al., 1990) is that the structure of the delipidized stratum corneum was perturbed or damaged after treatment with organic solvent; therefore solute partitioning into such membrane was different from that of intact membrane.

Ethanol is a widely used skin penetration enhancer. (Hatanaka et al., 1993) found that stratum corneum permeability coefficients of lipophilic but not hydrophilic compounds markedly changed when ethanol-water solvent system was employed as the vehicle. Such change was dependent not on the lipophilicity of the compound, but on the ethanol fraction in the solvent. Data from this study showed that when ethanol-water solvent system was employed as the vehicle, PC (PHSC/w) of lipophilic compounds atrazine and alachlor and a hydrophilic compound, glyphosate, were reduced (Figure 12.6). PC (PHSC/w) of a lipophilic compound, malathion (Figure 12.6), and water uptake (Figure 12.5) were increased at a low concentration of ethanol ($< 40\%$) and then decreased at high concentration ($> 40\%$). Effect on the stratum corneum uptake or permeability of a compound by ethanol at low concentration could result from ethanol physically perturbing the lipoidal barrier region in the stratum corneum (Hatanaka et al., 1993). At high concentration, ethanol could cause dehydration or shrinkage of the protein domain and partial extraction of the stratum corneum lipid (Berner et al., 1989; Hatanaka et al., 1993). Data of the study demonstrated that absolute ethanol (100%) prewetted PHSC significantly reduced its uptake capacity for hydroquinone, acitretin, and alachlor (Figure 12.7) when compared with that of dry or water prewetted PHSC ($P < 0.05$).

The data of Figures 12.2 and 12.3, which displays the PC PHSC/w as a function of incubation time, suggests that the time required to reach PC equilibrium varies chemically. This is consistent with the reports of (Surber, 1990a, 1990b). Two chemicals containing a similar nitrogen ring-structure, theophylline and atrazine, both reach equilibration in around 6 hours of incubation. Most higher lipophilic compounds, such as 2,4-D, atrazine, and PCB (log PC o/w $= 2.75 - 6.40$), and higher hydrophilic compounds, such as dopamine (log PC o/w $= -3.40$), reached equilibration in around 12 hours. For the other compounds with relatively lower lipophilicities and hydrophilicities, partition coefficients continued to increase up to 24 hours, which suggested that the time to reach equilibration may be beyond 24 hours; however, in comparing the ratio of the PC value at 12 hours' incubation to 2 or 6 hours' and the PC value at 24 hours–12 hours, the latter was much smaller than the former (Figures 12.2 and 12.3). This suggests that the equilibrium may not be far from 24-hour incubation times. Moreover, for those compounds with log PC o/w in the range of -2 to $+1.5$, the log PC PHSC/w were approximately the same (Table 12.1 and Figure 12.4). These results suggested that for higher lipophilic and hydrophilic compounds, the uptake process might be controlled by the lipid and protein domain, respectively, while for intermediate lipophilic and hydrophilic compounds, partitioning into the SC may partially depend on the amount of water of hydration in addition to participation of both domains. Another possibility relates to the swelling of whole SC from prolonged hydration (Scheuplein and Morgan, 1967; Robbins and Fermee, 1983). To the extent that progressive hydration occurs within the SC, the number of sites accessible for solute sorption is increased (Rieger, 1993), which may result in increasing PC PHSC/w for some chemicals. Theophylline (Figure 12.2) and atrazine (Figure 12.3) reached their equilibration in around 6 hours. These compounds contain a similar nitrogen ring-structure with hydrogen-bonding capacity, which may facilitate binding to the protein domain. Full hydration may disrupt this type of binding and lead to some reversal of adsorption.

In this study, powdered stratum corneum was immersed in the vehicle (water) containing a known concentration of a solute for certain times. Ideally, the model should mimic the physiological pH that exists across human skin in vivo. Experimentally, however, distilled deionized water was employed as the vehicle for chemical

TABLE 12.5. *MW, log PC o/w, water solubility, pKa, pH values of the 12 test chemicals*

Chemical	MW*	Log PC[†] (o/w)	pKa*	pH (in H_2O)	Ionization (in H_2O)
Dopamine	153	−3.40	2[‡]	4.0	I
Glycine	75	−3.20	2.34, 9.6	4.5	PI
Urea	60	−2.11	0.18[‡]	2.1	PI or I
Glyphosate	169	−1.70	2.27, 5.58, 10.25[§]	7.3	N or PI?
Theophylline	180	−0.76	8.77	7.6	I (mono-ion)
Aminopyrine	231	0.84	5	4.8	N
Hydrocortisone	362	1.61	7[†]	8.1	N
Malathion	330	2.36	7.5[†]	5.2	N
Atrazine	216	2.75	8.15[†]	6.2	N
2,4-D	221	2.81	4[†]	5.9	P
Alachlor	270	3.52	?	6.6	N
PCB	326	6.40	?	6.5	N

*Molecular weight and water solubility of chemical edited from *The Merck Index*, 11th edition, 1989.
[†]Logarithm partition coefficient cited in Hansch and Leo, 1979.
[‡]Data from Sigma Technical Service.
[§]Data from Monsanto Inc.
+ = full dissolved in water; − = almost insoluble in water; ? = unknown; I = ionized in water; PI = partially ionized in water; N = nonionized in water.

partitioning into the SC, and natural pH's were allowed to prevail. Table 12.5 shows the logarithm of octanol/water partition coefficients, the pKa values, and the pH values (in distilled deionized water) of the 12 test chemicals. Seven of the chemicals are neutral or nearly neutral in water, while three are partially ionized and two are fully ionized. There are reports suggesting that the pH value of the vehicle can influence the PC sc/w of some chemicals (Pardo et al., 1992; Downing et al., 1993).

In conclusion, the powdered human stratum corneum (callus) offered an experimentally easy in vitro model for determination of chemical partitioning from water into the SC. Due to the heterogeneous nature of the SC, the number and affinity of the SC binding sites may vary from chemical to chemical, depending on molecular structure. For the most lipophilic compounds, the PC PHSC/w were governed by the lipid domain, whereas PCs of the more hydrophilic compounds are determined by the protein domain and possibly by the sponge domain. These relationships can be expressed by the log PC PHSC/w of these chemicals as a function of the corresponding square of log PC o/w and log MW (Equation 5), which is useful in predicting various chemical partitioning into the SC in vitro. Further studies are needed to determine the roles of the lipid domain and/or the sponge domain in more detail.

REFERENCES

Berner, B., Juang, R. H., and Mazzenga, G. C. (1989): Ethanol and water sorption into stratum corneum and model systems. *J. Pharmaceutical Sci.*, 78:472–476.
Blank, J. H. (1965): Cutaneous barriers. *J. Invest. Dermatol.*, 45:249–256.

Downing, D. T., Abraham, W., Wegner, B. K., Willman, K. W., and Marchall, J. L. (1993): Partition of sodium dodecyl sulfate into stratum corneum lipid liposomes. *Arch. Dermatol. Res.*, 285:151–157.

Elias, P. M. (1981): Lipids and epidermal permeability barrier. *Arch. Dermatol. Res.*, 270:95–117.

E. P. A. (1992): Interim report: Dermal exposure assessment: Principles and applications.

Friberg, S. E., Kayali, I., Suhery, T., Rhein, L. D., and Simion, F. A. (1992): Water uptake into stratum corneum: Partition between lipids and proteins. *J. Dispersion Science and Technology*. 13: 337–347.

Guy, R. H., and Hadgraft, J. (1991): Principles of skin permeability relevant to chemical exposure. In: *Dermal and Ocular Toxicology*. Edited by D. W. Hobson, pp. 221–246. CPC Press Inc., Boston.

Hansch, C., and Leo, A. (1979): *Substituent Constants for Correlation Analysis in Chemistry and Biology*. John Wiley and Sons, New York.

Hatanaka, T., Shimoyama, M., Sugibayashi, K., and Morimoto, Y. (1993): Effect of vehicle on the skin permeability of drugs: Polyethylene glycol 400-water and ethanol-water binary solvents. *J. Controlled Release*, 23:247–260.

Jmokawa, G., Akasaki, S., Hattori, M., and Yoshizuka, N. (1986): Selective recovery of deranged water-holding properties by stratum corneum lipids. *J. Investigative Dermatology*, 87:758–761.

Juhlin, L., and Shelly, W. B. (1977): New staining techniques for the Langerhans cell. Acta Dermatovener (Stockholm) 57:289–296.

Knutson, K., Potts, R. O., Guzek, D. B., Golden, G. M., Mckie, J. E., Lambert, W. J., and Higuchi, W. I. (1985): Macro and molecular physical-chemical considerations in understanding drug transport in the stratum corneum. *J. Contr. Rel.*, 2:67–87.

Knutson, K., Krill, S. L., and Zhang, J. (1990): Solvent-mediated alterations of the stratum corneum. *J. Controlled Release*, 11:93–103.

Kurihara-Bergstrom, T., Knutson, K., Denoble, L. J., and Goates, C. Y. (1990): Percutaneous absorption enhancement of an ionic molecule by ethanol-water systems in human skin. *Pharmaceutical Research*, 7: 762–766.

Lampe, M. A., Burlingame, A. L., Whitney, J., Williams, M. L., Brown, B. E., Roitmen, E., and Elias, P. M. (1983): Human stratum corneum lipids: Characterization and regional variations. *J. Lipid. Res.*, 24:120–130.

Leveque, J. L., and Rasseneur, L. (1988): Mechanical properties of stratum corneum: Influence of water and lipids. In: *The Physical Nature of the Skin*. Edited by R. M. Marks, S. P. Barton, and C. Edwards. MTP Press Limited, Norwell, MA, USA. Chapter 17.

Middleton, J. D. (1968): The mechanism of water binding in stratum corneum. *Brit. J. Derm.*, 80:437–450.

Pardo, A., Shiri, Y., and Cohen, S. (1992): Kinetics of transdermal penetration of an organic ion pair: Physostigmine salicylate. *J. Pharmaceut. Sci.*, 81:990–995.

Potts, R. O., and Guy, R. H. (1992): Predicting skin permeability. *Pharmaceut. Res.*, 9:663–669.

Raykar, P. V., Fung, M. C., and Anderson, B. D. (1988): The role of protein and lipid domains in the uptake of solutes of human stratum corneum. *Pharmaceutical. Res.*, 5:140–150.

Rieger, M. (1993): Factors affecting sorption of topically applied substances. In: *Skin Permeation Fundamentals and Application*. Edited by J. L. Zatz, pp. 33–72. Allured Publishing Co., Wheaten.

Robbins, C. R., and Fermee, K. M. (1983): Some observations on the swelling of human epidermal membrane. *J. Soc. Cosm. Chem.*, 34:21–34.

Scheuplein, R. J., and Morgan, L. J. (1967): Bound water in keratin membranes measured by a microbalance technique. *Nature*, 214:456.

Scheuplein, R. J., and Bronaugh, R. L. (1983): Percutaneous absorption. In: *Biochemistry and Physiology of the Skin*. vol. 1. Edited by Goldsmith L. A., pp. 1255–1294, Oxford University Press, Oxford.

Surber, C., Wilhelm, K. P., Hori, M., Maibach, H. I., and Guy, R. H. (1990): Optimization of topical therapy: Partitioning of drugs into stratum corneum. *Pharmceut Res.*, 7:1320–1324.

Surber, C., Wilhelm, K. P., Maibach, H. I., Hall, L., and Guy, R. H. (1990): Partitioning of chemicals into human stratum corneum: Implications for risk assessment following dermal exposure. *Fundamental and Applied Toxicology*, 15:99–107.

Wester, R. C., Mobayen, M., and Maibach, H. I. (1987): In vivo and in vitro absorption and binding to powdered stratum corneum as methods to evaluate skin absorption of environmental chemical contaminants from ground and surface water. *J. Toxicol. and Environ. Health*, 21:367–374.

Zatz, J. L. (1993): Scratching the surface: Rationale approaches to skin permeation. In: *Skin Permeation Fundamentals and Application*. Edited by J. L. Zatz, pp. 11–32. Allured Publishing Co., Wheaten.

Toxicology of Skin
Edited by Howard I. Maibach
Copyright © 2001 Taylor & Francis

13

Dermal Exposure of Hands to Pesticides

Graham A. Matthews

Imperial College at Silwood Park, Ascot, Berkshire, United Kingdom

Over the last 50 years, the use of chemical pesticides has increased not only in agriculture and forestry, but also for control of pests in homes and gardens. During application, hands are the part of the body most at risk to exposure to pesticide when sprays are applied. This can occur when opening containers of the concentrated active ingredient as well as while spraying. Traditionally, most pesticides were formulated as emulsifiable concentrates in which a high concentration of active ingredient, dissolved in a solvent with emulsifier, readily forms an emulsion on mixing with water; however, concern about the environmental effects of the solvents in emulsions has led to greater use of particulate suspensions, with the active ingredient preferably formulated as a dry dispersible granule or a suspension concentrate. Irrespective of the formulation, the trend has been to improve packaging and introduce closed transfer systems to avoid exposure of the user to the concentrated active ingredient. In tropical countries, small-scale farmers remain more exposed due to cheaper containers, from which the pesticide is often measured into a small hand-held cup. Small sachets of pesticide for a knapsack load, water-soluble sachets, and a new design of container with built-in dispenser are all techniques for reducing exposure to the hands. In practice, users of hand-operated sprayers generally are more exposed than those with tractor equipment, but the hands of both groups can contact deposits on the surface of equipment, especially if nozzles are blocked. Users can be exposed particularly to leakages, especially at the trigger valve of poor quality equipment (Chester, 1993a).

The increased risk of exposure by users of small-scale equipment has been recognized by most registration authorities, so they limit the number of pesticides available for home and garden use.

Many studies of operator exposure have used absorbent pads attached to different parts of the body or disposable overalls, cut into sections to measure the quantity of pesticide deposited during field applications (Chester, 1993b, 1995). Exposure is often on the lower part of the body, especially when operators walk into treated foliage (Thornhill et al., 1996). This exposure is significantly reduced by using appropriate personal protective equipment (PPE), but few spray operators like wearing

TABLE 13.1. *Data on hand immersion in water plus surfactant (mean of 5 replicates)*

Person	Area of hand (cm^2)	Amount of liquid retained (by weight)		Amount retained in weighing towel (ml)
		(ml)	(ml/cm^2)	
1	316	1.19	0.00377	0.898
2	330	1.70	0.00515	1.21
3	440	1.53	0.00348	0.97
4	451	2.16	0.00479	1.51
5	448	2.36	0.00527	1.84

it, especially in hot climates. In particular, often hands are not protected by rubber gloves, so users are advised to wash off immediately any deposits, especially if contaminated by the concentrate when preparing sprays. When assessing operator exposure, artificial surfaces, such as pads, provide practical information relevant to the use of PPE, but they do not reflect the actual conditions on an operator's skin. An absorbent pad can accumulate liquid arriving over a period, whereas the skin can retain only a thin film of liquid. To assess the amount of liquid that can be retained on the surface of a hand, a small sample of hands were immersed in a beaker of water containing a surfactant (washing-up liquid). The beaker was placed on the pan of a sensitive balance to enable the amount of liquid on the hand to be calculated by the difference in weight before immersion and afterwards, holding the hand vertically to allow excess liquid to drip back into the beaker. Immediately after this the wet hand was dried on a weighed paper towel, which was reweighed to determine the weight gain. The area of the hand was determined approximately by measuring the length, breadth, and thickness of individual fingers and palm.

The hands of two females were smaller than male hands (Table 13.1), but overall retention averaged 0.0045 ml/cm^2. Using the towel gave lower values, as not all the moisture on the hand was removed. A similar result was achieved when one person immersed his hand in a suspension of a pesticide diluted according to the instruction on the label (Table 13.2). By wearing a vinyl glove, the amount of liquid retained was far less than on the unprotected skin (Table 13.2). One test indicated the amount of liquid liable to drip off from the surface of a hand by removing the hand away from the beaker immediately after immersion.

In a second series of tests, a known volume of water in a watering can was applied through a "rose" to sprinkle water plus surfactant as large droplets over a hand held

TABLE 13.2. *Amount of liquid retained on a hand ml/cm^2*

Using a pesticide	0.00554
Wearing a vinyl glove	0.0032
No allowance for liquid dripping from hand	0.0137

horizontally above a funnel to collect the liquid not retained on it. The hand was turned over to wet both upper and lower surfaces of the hand. The mean amount of liquid retained on the hand (448 cm^2) was 5.5 ml (i.e., 0.0123 ml/cm^2), but the data were more variable (range, 3–10 ml). With this technique, the amount of liquid was reduced when the hand was moved deliberately to "shake-off" surplus liquid. Depending on the concentration of the pesticide in the spray, the amount of active ingredient deposited on the hand can then be calculated.

These measurements indicate the maximum amount of water plus surfactant, simulating normal use of a diluted pesticide, that can be retained on the unprotected skin of a hand. Once the surface of the skin is wetted, any extra liquid will either drip off or be shaken off by movement of the hand. Further retention can occur when the surface dries. Apart from the size of the hand, the amount retained on its surface will be affected by the texture of the skin, its hairiness, and differences in the shape of the fingers. Touching clean surfaces may remove dislodgeable deposits, whereas contact with deposits on spray equipment may increase dermal exposure.

Similar tests to quantify the amount of liquid deposited on the surface of hands were reported by (Cinalli et al., 1992). In their tests using three different oils, the amount retained on the hands after immersion was measured by weighing a dry cloth before and after each test. On average, the amount retained was 10.33 mg/cm^2 (mineral oil), 6.02 mg/cm^2 (cooking oil), and 5.94 mg/cm^2 (bath oil). More variable results were obtained with aqueous liquids due to evaporation from the cloth, which may partially explain lower values in the towel technique compared with direct weighing of the liquid. (Roff, 1997) also referred to a single-hand immersion test with one female subject carried out by (Velsar Inc., 1984), in which 5 ml was retained on a hand. From this test, the U.S. Environmental Protection Agency estimated that up to 6 ml could be retained on a man's larger hand. An increase in viscosity is expected to increase retention due to a thicker film on the surface. In measuring dermal exposure during brushing a fluid on a wooden fence, (Roff, 1997) removed an insecticide tracer by washing the hands with soap and water. In this example, the amount of permethrin varied in relation to the amount of splashes deposited on the hands from 0.03 to 4.7 mg. This technique was compared with a fluorescence technique that allows visualization of deposits over all parts of the person.

Subsequent movement of the deposit through the skin depends on the formulation of the pesticide and other factors, such as whether the deposit remains wet or dries, and is described elsewhere.

REFERENCES

Chester, G. (1993): Operator exposure to pesticides. In: *Application Technology for Crop Protection.* Edited by Matthews, G. A., Hislop, E. C., pp. 123–144. CAB International, Wallingford.

Chester, G. (1993): Evalution of agricultural worker exposure to, and absorption of, pesticides. *Ann. Occup. Hyg.,* 37: 509–523.

Chester, G. (1995): Revised guidance document for the conduct of field studies of exposure to pesticides in use. In: *Methods of Pesticide Exposure Assessment.* Edited by Curry, P. B. et al., pp. 179–215. Plenum Press, New York.

Cinalli, C., Carter, C., Clark, A., and Dixon, D. (1992): A laboratory method to determine the retention of liquids on the surface of hands. Exposure Evaluation Division, Washington DC., U.S. Environmental Protection Agency, Contract No. 68-02-4254.

Roff, M. W. (1997): Dermal exposure of amateur or non-occuptional users to wood-preservative fluids applied by brushing outdoors. *Ann. Occup. Hyg.,* 41: 297–311.

Thornhill, E. W., Matthews, G. A., and Clayton, J. S. (1996): Potential operator exposure to insecticides: A comparison between knapsack and CDA spinning disc sprayers. Proc. Brighton Crop Protection Conference—Pests and Diseases, British Crop Protection Council, pp. 1175–1180.

Velsar Inc. (1984): Exposure assessment for retention of chemical liquids on hands. Exposure Evaluation Division, Washington DC., U.S. Environmental Protection Agency, Contract No. 68-01-6271.

Toxicology of Skin
Edited by Howard I. Maibach
Copyright © 2001 Taylor & Francis

14

Iontophoresis

Angela N. Anigbogu, Howard I. Maibach

Department of Dermatology, University of California, Davis, California, USA

INTRODUCTION AND HISTORICAL PERSPECTIVES

Drug delivery through the skin for systemic effects, though limited, is a well-established branch of pharmaceutics. The excellent barrier properties of the stratum corneum, the outermost layer of the skin, limits the number of drug candidates for passive transdermal delivery to usually small, potent, and lipophilic compounds. Physical and chemical techniques have been used to improve the permeability of the skin to applied substances. Dermal iontophoresis is one of such physical techniques.

Iontophoresis may be defined as the facilitated transport of ions of soluble salts across membranes under the influence of an applied electric field. It was first introduced by Pivati to treat arthritis in the 1740s (Licht, 1983). The technique temporarily lost its importance, partly due to safety considerations. Munch earlier demonstrated the systemic application of this technique in 1879, when strychnine delivered under the positive electrode in rabbit killed the animal within 15 minutes of current passage. (Leduc, 1900) described some of the earliest experiments outlining the usefulness of iontophoresis in systemic drug delivery. He placed a solution of strychnine sulphate (positively charged strychnine ion) in the positive electrode (anode) of an iontophoresis set up on one rabbit with the negative electrode filled with water, and a solution of potassium cyanide (negatively charged cyanide ion) in the negative electrode (cathode) of a set up on another rabbit with the positive electrode filled with water. The animals were connected, and when a constant current of 40–50 mA was applied, both animals died due to strychnine and cyanide poisoning, respectively. In a subsequent experiment reversing the polarity of the delivery electrodes (i.e., strychnine in the cathode and cyanide in the anode), neither animal died, demonstrating that in the first case, the electric current delivered the lethal ions.

Since the early years, there has been a resurgence of interest in iontophoresis. (Gibson and Cooke, 1959) used iontophoretic delivery of pilocarpine to induce sweating, and the procedure is now used for the diagnosis of cystic fibrosis. Iontophoresis has been used for the treatment of palmoplantar hyperhydrosis. In addition to this and other local applications of the technique, the present focus on iontophoresis is for systemic drug delivery. With interest in controlled drug delivery surfacing in the last two decades, and the inability to deliver a great number of drugs, especially proteins and peptides passively, iontophoresis appears to be particularly attractive and holds great commercial promise for noninvasive rate-controlled transdermal drug delivery.

THEORY

Biological tissues including skin consist of membrane barriers made up of lipids and proteins. Transport through these membranes is better suited to un-ionized than ionized compounds. Many potential drug candidates are ionized at skin pH (4–5) and therefore cannot be transported across membranes passively. As stated previously, the stratum corneum provides an excellent barrier to transport across the skin. In addition, passive diffusion depends on a concentration gradient across the membrane. Membrane transport of drugs can be facilitated by the application of an external energy source (active transport).

Iontophoresis by utilizing electric current provides an excellent source of this external energy. It operates on the general principles of electricity (i.e., opposite charges attract and like charges repel). Thus, if the drug of interest is cationic, for delivery across the skin, it is placed in the anode reservoir. When a voltage is applied, the positively charged drug is repelled from the anode through the skin and into the systemic circulation. Conversely, an anionic drug is placed in the cathode reservoir. The transport of neutral and uncharged molecules can also be facilitated by iontophoresis by the process of electroosmosis (Gangarosa et al., 1980). Figure 14.1 is an illustration of an iontophoretic set-up.

In this section, the underlying principles of iontophoretic transport will be described briefly. The Nernst–Planck flux equation, as applied in iontophoresis, provides that the flux of an ion across a membrane under the influence of an applied charge is due to a combination of iontophoretic (electrical potential difference), diffusive (increased skin permeability induced by the applied field), and electroosmotic (current-induced

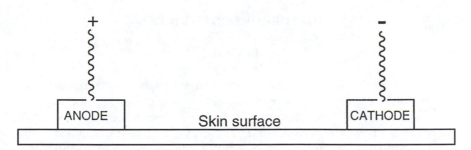

Positively charged
ions are delivered
by the anode

Negatively charged
ions are delivered
by the cathode

FIG. 14.1. A schematic illustration of transdermal iontophoresis.

water transport) components (Schultz, 1980):

$$Jion = Je + Jp + Jc, \tag{1}$$

where Je is the flux due to electrical potential difference and is given by

$$Je = \frac{ZiDiF}{RT}Ci\frac{\partial E}{\partial x}; \tag{2}$$

Jp is the flux due to passive delivery and is given by

$$Jp = KsDs\frac{\partial C}{\partial x}; \tag{3}$$

Jc is the flux due to electroosmosis and is given by

$$Jc = kCsI, \tag{4}$$

where Zi is the valence of the ionic species, Di is the diffusivity of the ionic species, i is the skin, F is the Faraday constant, T is the absolute temperature, and R is the gas constant, $\partial E, \partial x$ is the electrical potential gradient across the skin, Ci is the donor concentration of the ionic species, Ks is the partition coefficient between donor solution and stratum corneum, Ds is the diffusivity across the skin, $\partial C, \partial x$ is the concentration gradient across the skin, Cs is the concentration in the skin, I is the current density, k is the proportionality constant (Chien et al., 1990).

In iontophoretic drug delivery, the major contribution to the overall flux of a compound would be that due to electrical potential gradient. The contribution to the flux due to electroosmosis is likely to be small, and (Roberts et al., 1990) have suggested that only about 5% of the overall flux is due to convective solvent flow.

IONTOPHORESIS DEVICES

In Vitro

As the technique is still in development, there are relatively few descriptions in the literature of different apparatuses used in in vitro iontophoresis experiments. Examples include (Molitor, 1943; Burnette and Marrero, 1986; Bellantone et al., 1986; Masada et al., 1989; Green et al., 1991; Thysman et al., 1991; Chang and Banga, 1998). Usually these involve modifications of the two-compartment in vitro passive diffusion set-up. Two electrodes connected to a power supply are used, and in some instances one is inserted in each compartment separated by the mounted skin and voltage or current measurements are made between the electrodes. In other instances, using vertical flow through diffusion cells, a horizontally mounted piece of skin separates the positive and negative electrode chambers with the epidermal side of the skin from the receptor phase bathing the dermal side.

A four-electrode potentiostat system designed to maintain a constant voltage drop across a membrane in a two-chamber diffusion cell has been described by (Masada et al., 1989). As with passive diffusion studies, the whole assembly is kept at 37°C with the aid of a constant-temperature water bath.

In Vivo

Devices used in iontophoresis are designed for rate-controlled delivery of therapeutic agents. Essentially, they consist of a power source to provide current, anode, and cathode reservoirs. Generally they are operated at a constant voltage, allowing the current to be varied for patient comfort and compliance. As with in vitro apparatuses, various devices have been described for use in iontophoresis in vivo (Molitor and Fernandez, 1939; Barner, 1961; Rapperport et al., 1965). (Rattenbury and Worthy, 1996) described systems used in the United Kingdom. Hidrex (Gessellschaft für Medizin and Technik, Wuppertal, Germany) has been described by (Hölzle and Alberti, 1987).

Available in the United States is a portable battery-operated power supply unit called a Phoresor (Dermion Drug Delivery Research, Salt Lake City, Utah), which is suitable for home use. Platinum electrodes or patches consisting of zinc/zinc chloride or silver/silver chloride electrodes are used. The choice of electrode material depends on several factors including good conductivity, malleability, and the ability to maintain a stable pH. Silver/silver chloride electrodes, also referred to as reversible electrodes, are the most commonly used, as they satisfy these requirements. (Phipps et al., 1989) described a custom-made, battery-operated device with two hydrogel electrodes for in vivo delivery of pyridostigmine. These devices deliver direct steady current, which have been postulated to be responsible for skin irritation arising from iontophoresis. To minimize this, others advocate devices delivering pulsed current (Sanderson et al., 1987).

The U.S. Food and Drug Administration has categorized iontophoretic devices into those for specialized uses (class II) and others (class III) (Tyle, 1986). These include Drionic (General Medical Company, Los Angeles, CA), Macroduct (Wescor Inc., Logan, UT), Iontophor-PM (Life-Tech Inc., Houston, TX), Model IPS-25 (Farrall Instruments Inc., Grand Island, NE), Electro-Medicator (Medtherm Corporation, Huntrille, AL), Dagan (Dagan Corporation, MN), Desensitron II (Parkell, Farmingdale, NY).

Animal Models

The ultimate goal of any research done in the field of iontophoresis is the application in humans for drug delivery. For obvious reasons, animals, not human subjects, are the first choice for experimental purposes. There is no consensus as to which of the animal models used in passive uptake studies is suitable for iontophoresis. Hairless mouse has been the most commonly used model (e.g., Bellantone et al., 1986). Other models that have been investigated include hairless guinea pig (Walberg, 1970), dog (McEvan-Jenkinson et al., 1974), furry rat (Siddiqui et al., 1987), pig (Monteiro-Riviere, 1991), hairless rat (Thysman and Preat, 1993), Rabbit (Lau et al., 1994; Anigbogu et al., 2000). There is therefore the need to establish which model closely resembles human skin for both penetration and toxicological studies.

PATHWAYS OF ION TRANSPORT

The predominant pathway for ion transport through the skin remains controversial. Appendages: sweat ducts and hair follicles are thought to be the major pathway for iontophoretic transport through the skin (Grimnes, 1984; Burnette, 1989). This is obviously so in the use of pilocarpine for the diagnosis of cystic fibrosis. (Abramson and Gorin, 1940) showed that charged dyes delivered iontophoretically produced a dot-like pattern on human skin, and the dots were identified as sweat glands. (Papa and Kligman, 1966) observed a direct link between methylene blue staining of the skin and the location of sweat ducts. (Monteiro-Riviere, 1991) demonstrated the appendageal pathway for the iontophoretic delivery of mercuric chloride. Laser scanning confocal microscopy has been used to elucidate the pathway for the iontophoretic transport of Fe^{2+} and Fe^{3+} ions (Cullander, 1992) as being the sweat glands, hair follicles, and sebaceous glands. Based on these and other studies, the sweat ducts and glands, however, appear to be more important than hair follicles in the transport of ions through the shunts.

It is not correct to assume that all charge transport takes place through the appendages. (Walberg, 1968) demonstrated that Na^+ and Hg^{2+} could penetrate through guinea pig skin in areas devoid of sweat glands and hair follicles. (Millard and Barry, 1988) compared the iontophoretic delivery of water and glutamic acid through full-thickness human skin and shed snakeskin, which is largely devoid of sweat glands and hair follicles. Iontophoresis was shown to increase the delivery of both materials through snakeskin. (Sharata and Burnette, 1989) showed that mercuric and nickel ions can diffuse passively between the keratinocytes. (Jadoul et al., 1996) concluded from results of Fourier transform infrared spectroscopy (FTIR) and small-angle x-ray scattering (SAXS) studies on isolated rat and human cadaver skin following prolonged in vitro iontophoresis that iontophoresis transport is related to lipid bilayer stacking disorganization.

FACTORS AFFECTING IONTOPHORETIC DRUG ADMINISTRATION

Several factors come into play when considering iontophoresis for drug delivery, including the physicochemical properties of the drug in question: the charge, molecular size and concentration; formulation parameters: choice of vehicle, pH range in which drug is ionic, presence of competing or parasitic ions, viscosity or mobility. Others include physiologic considerations, such as appropriate skin site for application; instrumentation (e.g., type of current source, pulsed or constant), current density. This list is by no means exhaustive but includes some of the more critical factors, which will be considered briefly in this section.

Transdermal iontophoresis achieves the transport of drug molecules into and through the skin under the influence of an applied electric field. This means that the drug candidate be charged to allow for delivery in therapeutically relevant levels through the skin. The optimum pH for delivery of a drug by iontophoresis is that at which it exists predominantly in the ionic form. This has been demonstrated by (Siddiqui et al., 1985, 1989). The pH of peptides, proteins, and other amphoteric substances characterized by their isoelectric point is of particular significance (i.e., a pH above which the molecule is anionic and below which it is cationic). For instance, the skin permeability of insulin has been shown to be greater at a pH below its isoelectric point (Siddiqui et al., 1987). Furthermore, the pH gradient encountered in the skin is an important factor in iontophoretic transport. The pH of the skin ranges from 4–6 on the outside to about 7.3 in the viable tissues. If at any time the drug encounters an environment in which it becomes uncharged, its transport becomes impeded. Thus, for a molecule to be delivered efficiently by iontophoresis, it must remain charged during its transport into and through the skin. For proteins and peptides, iontophoretic transport may be limited to those with isoelectric points below 4 or above 7.3.

The molecular size of the compound of interest is crucial in predicting the efficiency of its iontophoretic delivery (Srinivasan et al., 1989,1990; Yoshida and Roberts, 1993). The greater the molecular size, the lower the permeability coefficient. Nevertheless, high-molecular-weight proteins and peptide drugs with a molecular weight of 3000–5000 daltons have been delivered effectively by iontophoresis. The concentration of the drug in the formulation also affects the flux achieved by iontophoresis. A relationship between concentration of drug in the donor solution and flux has been established for gonadotropin-releasing hormone (GnRH) and sodium benzoate with flux increasing linearly with increasing concentration (Bellantone et al., 1986).

The fraction of current carried by each type of ion in solution is called the transference or transport number. When a migrating ion carries 100% of the current through the membrane, its rate of transport is maximal and its transport number is unity. To control the pH of the donor solution, buffers are often employed. The buffers, however, introduce extraneous ions, which may be of different type but are of the same charge as the drug ion. These are called co-ions and usually are more mobile than the drug ion. The co-ions reduce the fraction of current carried by the drug ion, thus resulting in a diminished transdermal flux of the drug. Some workers also employ antioxidants and antimicrobials, which themselves contain co-ions. In addition to these, co-ions can also be introduced from reactions occurring at the electrodes if, for example, platinum is the conducting material. Hydrolysis of water occurs, resulting in the generation of hydronium ion at the anode and hydroxyl ion at the cathode. Reducing the amount of competing ions in the drug donor solution will increase the transport efficiency of the drug ions, but as there are also endogenous ions in the skin (e.g., sodium, potassium, chloride, bicarbonate, lactate, etc.) which carry an appreciable fraction of the ionic current (Phipps and Gyory, 1992), the transport number of any drug will always be less than unity.

A linear relationship has been established between the iontophoretic fluxes of a number of compounds and the current applied. Examples include lithium (Phipps et al., 1989), thyrotropin-releasing hormone (Burnette and Marrero, 1986), mannitol (Burnette and Ongpipattanakul, 1987), GnRH (Miller et al., 1990), and verapamil (Wearley and Chien, 1989). Tissue distribution of phosphorus following iontophoretic delivery was shown to be proportional to current density (O'Malley and Oester, 1955). The limiting factor (especially in humans), however, is safety, comfort, and acceptability. The upper limit of current tolerable to humans is thought to be 0.5 mA/cm^2 (Abramson and Gorin, 1941; Ledger, 1992). Increasing the surface area of the electrodes allows for increasing current, therefore improving the delivery of some drugs. This is not a linear relationship, however, and may not apply to all drugs (Phipps et al., 1989).

Iontophoretic deliveries of lithium and pyridostigmine have been found to be comparable in pig, rabbit, and human skin (Phipps et al., 1989). (Burnette and Ongpipattanakul, 1987) found the iontophoretic fluxes of sodium chloride and mannitol through thigh skin from male and female cadavers to be comparable. Successive iontophoretic delivery of iodine through the same knee in a human volunteer resulted in a constant uptake (Puttemans et al., 1982). Iontophoresis therefore decreases the intra- and intersubject variability as well as influence of site usually observed with passive diffusion. Further studies need to be done, however, to establish the degree to which factors such as race, age, skin thickness, hydration, and status of the skin (healthy or diseased) affect iontophoretic drug delivery.

Whether pulsed or continuous direct current should be used is one of the controversies that exist in the field of iontophoresis. Continuous direct current causes skin polarization with time, and this reduces the efficiency of delivery. This can be avoided by using pulsed direct current (i.e., direct current delivered periodically). During the "off period," the skin becomes depolarized, returning to near its original state. Ion transport using pulsed current is affected by the frequency, however. If the frequency is high, the efficiency of pulsed delivery is reduced.

ADVANTAGES OF IONTOPHORESIS

Considering the complexity of iontophoresis compared to traditional dosage forms, such as tablets, liquids, injections, ointments, and even passive transdermal patches, it must have advantages to enjoy a resurgence of interest.

Transdermal iontophoresis shares many of the advantages of passive transdermal drug delivery. An important consideration is patient compliance. The dosage regimens of many pharmacologic agents available for delivery through other routes pose a challenge to patients (e.g., the need to take with or without food, dosing frequency etc.). In addition, the injectable route is particularly uncomfortable to many patients. Many drugs which are available for systemic therapy cannot be delivered through many of the existing traditional dosage forms, as they are subject to extensive hepatic first-pass metabolism and variable gut absorption. Many drugs, including new biotech drugs (proteins, peptides, oligonucleosides, etc.) (Meyer, 1988; Merino et al., 1997) and local anesthetics, such as lidocaine (Gangarosa, 1981), which would have to be injected to derive maximum benefit, have been delivered efficiently using iontophoresis.

Because the rate of drug delivery is generally proportional to the applied current, the rate of input can therefore be preprogrammed on an individual basis (Banga and Chien, 1988). The controllability of the device would eliminate the peaks and troughs in blood levels seen with oral dosing and injections. Patients can titrate their intake of drugs as required.

PROBLEMS ASSOCIATED WITH IONTOPHORESIS

Only a fraction of the charge introduced in iontophoresis is delivered, suggesting that iontophoresis is not necessarily as efficient as theoretically proposed (Sage and Riviere, 1992). Of more serious consideration, however, are the unwanted skin effects of iontophoresis arising from the system itself and/or drug formulation. Typically, side effects of iontophoresis with low voltage electrodes properly used are minimal but, nevertheless, must be considered. These include itching, erythema, edema, small punctate lesions, and sometimes burns. A slight feeling of warmth and tingling is generally associated with iontophoresis (Kellog et al., 1989; Zeltzer et al., 1991; Ledger, 1992; Maloney et al., 1992). Erythema is also commonly reported and is thought to arise from skin polarization associated with continuous direct current. To minimize this, pulsed direct current has been advocated. Shocks occur when high current density is directed at the skin. To minimize this, the current should be increased slowly from zero to the maximum desired current level acceptable to the patient. Similarly, at the end of the procedure, current should be returned from the maximum to zero in a stepwise manner. The effect of current on nerve fibers is thought to be responsible for the itching, tingling, and erythema.

The histological and functional changes that occur in animal skin following iontophoresis have been studied. Under similar delivery conditions (i.e., drug concentration, current density, duration), as are used in humans, (Moteiro-Riviere, 1990) studied structural changes in porcine skin following iontophoresis of lidocaine. Light microscopy revealed epidermal changes. He, however, noted that similar changes were not observed following iontophoresis of other compounds, suggesting the effects were largely due to the lidocaine rather than the electric current. (Cho and Kitamura, 1988) iontophoresing lidocaine through the tympanic membrane of the guinea pig, observed a loss of adhesion of the epidermis to underlying connective tissue and retraction of noncornified epidermal cells. (Jadoul et al., 1996) used FTIR and small-angle x-ray scattering (SAXS) to study isolated rat skin and human skin from cadaver following prolonged iontophoresis. While FTIR revealed transient increases in the hydration of the outer layers of the stratum corneum but no increase in lipid fluidity, SAXS showed that iontophoresis induced a disorganization of the lipid layers; this was also reversible within days of the procedure. Using wide-angle x-ray scattering (WAXS), the authors did not find evidence of modification of the intralamellar crystalline packing of lipids nor of keratin.

The answer to what should be the upper limit of current tolerable to humans is not very straightforward, as what may be just discernible to one patient may be uncomfortable to another. Generally, however, 0.5 mA/cm^2 is cited (Abramson and Gorin, 1941; Banga and Chien, 1988; Ledger, 1992). (Molitor and Fernandez, 1939) found that the greater the surface area of the electrode, the larger the tolerable current, but the relationship is curvilinear. Small punctate lesions are associated with electric current traveling through a path of least resistance in the skin. Common sense dictates that iontophoresis not be used on skin showing signs of damage. Pain and burns arising from iontophoresis are linked to electrochemical reactions which occur at the electrodes and involve the electrolysis of water to generate hydronium and hydroxyl ions resulting in pH changes (Sanderson et al., 1989). Much earlier, however, (Molitor and Fernandez, 1939), using continuous flow electrodes which did not generate hydroxyl and hydronium ions, and therefore did not produce any pH changes, showed that burns could not solely be related to pH changes.

Erythema is the most common side effect associated with iontophoresis and could be due to nonspecific skin irritation, such as that which occur with the delivery of an irritant drug. Erythema may be due to a direct effect of electric current on blood vessels and/or current-induced release of histamine, prostaglandins, or other neurotransmitters leading to local vasodilatation of the affected area. It has also been suggested that electric current can stimulate specific classes of nociceptors, the C-fibers causing them to release the potent vasodilators, substance P, and calcitonin gene-related peptide (CGRP) (Brain and Edwardson, 1989; Dalsgaard et al., 1989). Whatever the cause of the erythema, it usually is transient and not associated with any permanent changes in the skin.

Delayed-type contact sensitivity to components of the iontophoresis system, electrodes, electrode gels, etc. (Schwartz and Clendenning, 1988), or to the drug being delivered (Holdiness, 1989), have been reported.

Another consideration in choosing iontophoresis for drug delivery is cost. Iontophoresis requires a power source to supply electrical energy. Even though the power requirement for a unit delivery may be small, repeated applications would require a considerable investment in battery supply. Better batteries than those currently used need to be developed.

Iontophoresis is contraindicated in patients with high susceptibility to applied currents and in patients with known hypersensisitivity to the drug in question. Iontophoresis should be avoided in patients with electrically sensitive implants, such as pacemakers.

To improve acceptability by both prescribers and patients, more studies need to be done in the field of iontophoresis to minimize unwanted side effects and improve safety.

APPLICATIONS OF IONTOPHORESIS IN DERMATOLOGY

In the past, iontophoresis was found useful in local delivery of pharmacologic agents. Iontophoresis has been used for the treatment of various dermatologic conditions, including lupus vulgaris using zinc. Before the advent of antibiotics, infections were treated by the iontophoresis of metals (e.g., the treatment of streptococcal infections with copper sulphate). Other conditions that have benefited from the use of iontophoresis include lichen planus, scleroderma, plantar warts, hyperhydrosis, and infected burn wounds, achieving local anesthesia. Bursitis and other musculoskeletal conditions have been treated with iontophoresed corticoids (Harris, 1982). A nice summary of dermatologic applications of iontophoresis has been made by (Sloan and Soltani, 1986). Of greater interest in this era is the use of iontophoresis for controlled systemic drug delivery and for targeting deep tissue penetration. Recently, "reverse iontophoresis" involving the extraction of material from the body for the purposes of clinical chemistry has been described (Guy, 1995). Although glucose is not charged, iontophoresis can markedly increase its passage across the skin by electroosmosis (Merino et al., 1997), and this has been applied for the noninvasive monitoring of diabetics' blood sugar levels (Tamada et al., 1995; Svedman and Svedman, 1997). In addition to drug delivery, with the availability of sensitive assay methods, iontophoresis is being touted as a diagnostic tool.

CONCLUSIONS

Iontophoresis is an attractive alternative to existing dosage forms in delivering drugs for both local as well as systemic indications. The fact that it could allow for a programmable rate-controlled delivery of drugs is particularly relevant. Iontophoresis appears to be promising for the delivery of proteins and peptides, which make up a big chunk of new biotech drugs. Like any new technology, more work is needed to clearly define the parameters that would maximize the safety, acceptability, and efficiency of iontophoresis as a dosage form.

REFERENCES

Abramson, H. A., and Gorin, M. H. (1940): Skin reactions IX: The electrophoretic demonstration of the patent pores of the living human skin: Its relation to the charge of the skin. *J. Phys. Chem.*, 44:1094–1102.

Abramson, H. A., and Gorin, M. H. (1941): Skin reactions. X: Preseasonal treatment of hay fever by electrophoresis of ragweed pollen extracts into the skin: Preliminary report. *J. Allergy*, 12:169–175.

Anigbogu, A., Patil, S., Singh, P., Liu, P., Dihn, S., Maibach, H. (2000): An in vivo investigation of the rabbit skin responses to transdermal iontophoresis. *Int. J. Pharm.* 200:195–206.

Banga, A., and Chien, Y. W. (1988): Iontophoretic delivery of drugs: Fundamentals, developments and biomedical applications. *J. Cont. Rel.*, 7:1–24.

Bellantone, N. H., Rim, S., Francouer, M. L., and Rasadi, B. (1986): Enhanced percutaneous absorption via iontophoresis I: Evaluation of an in vitro system and transport of model compounds. *Int. J. Pharm.*, 30:63–72.

Brain, S. D., and Edwardson, J. A. (1989): Neuropeptides and the skin. In: *Pharmacology of the Skin I.* Edited by M. W., Greaves and S., Shuster, pp. 409–422. Springer Verlag, Berlin.

Burnette, R., and Marrero, D. (1986): Comparison between the iontophoretic and passive transport of thyrotropin releasing hormone across excised nude mouse skin. *J. Pharm. Sci.*, 75:738–743.

Burnette, R. R. (1989): Iontophoresis. In: *Transdermal Drug Delivery.* Edited by J. Hadgraft and R. H. Guy, pp. 247–291. Marcel Dekker, New York.

Burnette, R. R., and Ongpipattanakul, B. (1987): Characterization of the permselective properties of excised human skin during iontophoresis. *J. Pharm. Sci.*, 77:132–137.

Chang, S., and Banga, A. K. (1998): Transdermal iontophoretic delivery of hydrocortisone from cyclodextrin solutions. *J. Pharm. Pharmacol.*, 50:635–640.

Chien, Y. W., Lelawongs, P., Siddiqui, O., Sun, Y., and Shi, W. M. (1990): Facilitated transdermal delivery of therapeutic peptides and proteins by iontophoretic delivery of devices. *J. Cont. Rel.*, 13:263–278.

Cho, Y. B., and Kitamura, K. (1988): Short-term effects of iontophoresis on the structure of the guinea pig tympanic membrane. *Acta. Otolaryngol. (Stockh.)*, 106:161–170.

Cullander, C. (1992): What are the pathways of iontophoretic current flow through mammalian skin? *Adv. Drug Deliv. Rev.*, 9:119–135.

Dalsgaard, C. J., Jernbeck, J., Stains, W., Kjartansson, J., Haegerstrand, A., Hökfelt, T., Brodin, E., Cuello, A. C., and Brown, J. C. (1989): Calcitonin gene-related peptide-like immunoreactivity in nerve fibers in human skin. *Histochemistry*, 91:35–38.

Gangarosa, L. P. (1981): Defining a practical solution for iontophoretic local anesthesia of skin. *Methods Fund. Exp. Clin. Pharmacol.*, 3:83–94.

Gangarosa, L. P., Park, N. H., Wiggins, C. A., and Hill, J. M. (1980): Increased penetration of nonelectrolytes into hairless mouse skin during iontophoretic water transport (iontohydrokinesis). *J. Pharmacol. Exp. Ther.*, 212:377–381.

Gibson, L. E., and Cooke, R. E. (1959): A test for the concentration of electrolytes in sweat in cystic fibrosis of the pancreas utilizing pilocarpine by iontophoresis. *Pediatrics*, 23:545–549.

Green, P. G., Hinz, R. S., Cullander, C., Yamane, G, and Guy, R. H. (1991): Iontophoretic delivery of amino acids and amino acid derivatives across the skin in vitro. *Pharm. Res.*, 8:1113–1120.

Grimnes, S. (1984): Pathways of ionic flow through the human skin in vivo. *Acta Derm. Venereol. (Stockh.)*, 64:93–98.

Guy, R.H. (1995): A sweeter life for diabetics? *Nature Med.*, 1:1132–1133.

Harris, P. R. (1982): Iontophoresis: Clinical research in musculoskeletal inflammatory conditions. *J. Orth. Sport Phy. Ther.*, 4:109–112.

Holdiness, M. R. (1989): A review of contact dermatitis associated with transdermal therapeutic systems. *Contact Dermatitis.*, 20:3–9.

Hölzle, E., and Alberti, N. (1987): Long-term efficacy and side effects of tap water iontophoresis of palmoplantar hyperhydrosis: The usefulness of home therapy. *Dermatologica*, 175:126–135.

Jadoul, A., Doucet, J., Durand, D., and Preat, V. (1996): Modifications induced on stratum corneum structure after in vitro iontophoresis: ATR-FTIR and X-ray scattering studies. *J. Cont. Rel.*, 42:165–173.

Kellog, D. L., Johnson, J. M., and Kosiba, W. A. (1989): Selective abolition of adrenergic vasoconstrictor responses in skin by local iontophoresis by bretylium. *Am. J. Physiol.*, 257: H1599–H1606.

Lau, D. T., Sharkey, J. W., Petryk, L., Mancuso, F. A., Yu, Z., and Tse, F. L. S. (1994): Effect of current magnitude and drug concentration on iontophoretic delivery of octreotide acetate (sandostatin) in the rabbit. *Pharm. Res.*, 11:1742–1746.

Ledger, P. W. (1992): Skin biological issues in electrically enhanced transdermal delivery. *Adv. Drug Deliv. Rev.*, 9:289–307.

Leduc, S. (1900): Introduction of medicinal substances into the depth of tissues by electric current. *Ann. D'electrobiol.*, 3:545–560.

Licht, S. (1983): History of electrotherapy. In: *Therapeutic Electricity and Ultraviolet Radiation.* Edited by G. K. Stillwell, pp. 1–64. Williams & Wilkins, Baltimore.

Maloney, J. M., Bezzant, J. L., Stephen, R. L., and Petelentz, T. J. (1992): Iontophoretic administration of lidocaine anesthesia in office practice: An appraisal. *J. Dermatol. Surg. Oncol.*, 18:937–940.

Masada, T., Higuchi, W. I., Srinivasan, V., Rohr, U., Fox, J., Behl, C. R., and Pons, S. (1989): Examination of iontophoretic transport of drugs across skin: Baseline studies with the four electrode system. *Int. J. Pharm.*, 49:57–62.

McEvan-Jenkinson, D., McLeon, J. A., Walton, G. S. (1974): The potential use of iontophoresis in the treatment of skin disorders. *Vet. Rec.*, 94:8–11.

Merino, V., Kalia, Y. N., and Guy, R. H. (1997): Transdermal therapy and diagnosis by iontophoresis. *Trends Biotechnol.*, 15:288–290.

Meyer, R. B. (1988): Successful transdermal administration of therapeutic doses of a polypeptide to normal human volunteers. *Clin. Pharmacol. Ther.*, 44:607–612.

Millard, J., and Barry, B. W. (1988): The iontophoresis of water and glutamic acid across full thickness human skin and shed snake skin. *J. Pharm. Pharmacol.*, 40(suppl.):41.

Miller, L. L., Kolaskie, C. J., Smith, G. A., and Riviere, J. (1990): Transdermal iontophoresis of gonadotropin releasing hormone (LHRH) and two analogues. *J. Pharm. Sci.*, 79:490–493.

Molitor, H. (1943): Pharmacologic aspects of drug administration by ion-transfer. *The Merck Report.* 52: 22–29.

Molitor, H., and Fernandez, L. (1939): Studies on iontophoresis. I: Experimental studies on the causes and prevention of burns. *Am. J. Med. Sci.*, 198:778–785.

Monteiro-Riviere, N. A. (1991): Identification of the pathway of transdermal iontophoretic drug delivery: Ultrastructural studies using mercuric chloride in vivo in pigs. *Pharm. Res.*, 8:S41.

Moteiro-Riviere, N. A. (1990): Altered epidermal morphology secondary to lidocaine iontophoresis: In vivo and in vitro studies in porcine skin. *Fundam. Appl. Toxicol.*, 15:174–175.

O'Malley, E. P., and Oester, Y. T. (1955): Influence of some physical chemical factors on iontophoresis using radio isotopes. *Arch. Phys. Med. Rehabil.*, 36:310–316.

Papa, C. M., and Kligman, A. M. (1966): Mechanism of eccrine anhydrosis. *J. Invest. Dermatol.*, 47:1–9.

Phipps, J. B., and Gyory, J. R. (1992): Transdermal ion migration. *Adv. Drug Deliv. Rev.*, 9:137–176.

Phipps, J. B., Padmanabhan, R. V., and Lattin, G. A. (1989): Iontophoretic delivery of model inorganic and drug ions. *J. Pharm. Sci.*, 78:365–369.

Phipps, J. B., Padmanabhan, R. V., Lattin, G. A. (1989): Iontophoretic delivery of model inorganics and drug ions. *J. Pharm. Sci.*, 78:365–369.

Puttemans, F. J. M., Massart, D. L., Gilles, F., Lievens, P. C., and Jonckeer, M. H. (1982): Iontophoresis: Mechanism of action studied by potentiometry and x-ray fluorescence. *Arch. Phys. Med. Med. Rehabil.*, 63:176–180.

Rattenbury, J. M., and Worthy, E. (1996): Is the sweat test safe? Some instances of burns received during pilocarpine iontophoresis. *Ann. Clin. Biochem.*, 33:456–458.

Roberts, M. S., Singh, J., Yoshida, N. H., and Curries, K. I. (1990): Iontophoretic transport of selected solutes through human epidermis. In: *Prediction of Percutaneous Absorption.* Edited by R. C. Scott, R. H. Guy, and J. Hadgraft, pp. 230–241. IBC Technical Services, London.

Sage, B. H., and Riviere, J. E. (1992): Model systems in iontophoresis-transport efficacy. *Adv. Drug. Deli. Rev.*, 9:265–287.

Sanderson, J. E., Caldwell, R. W., Hsiao, J., Dison, R., and Tuttle, R. R. (1987): Noninvasive delivery of a novel ionotropic catecholamine: Iontophoretic versus intravenous infusion in dogs. *J. Pharm. Sci.*, 76:215–218.

Sanderson, J. E., De Riel, S., and Dixon, R. (1989): Iontophoretic delivery of nonpeptide drugs: Formulation optimization for maximum skin permeability. *J. Pharm. Sci.*, 78:361–364.

Schultz, S. G. (1980): *Basic Principles of Membrane Transport*, pp. 21–30. Cambridge University Press, New York.

Schwartz, B. K., and Clendenning, W. E. (1988): Allergic contact dermatitis from hydroxypropyl cellulose in a transdermal estradiol patch. *Contact Dermatitis*, 18:106–107.

Sharata, H., and Burnette, R. R. (1989): Percutaneous absorption of electron-dense ions across normal and chemically perturbed skin. *J. Pharm. Sci.*, 77:27–32.

Siddiqui, O., Roberts, M. S., and Polack, A. E. (1989): Iontophoretic transport of weak electrolytes through excised human stratum corneum. *J. Pharm. Pharmacol.*, 41:430–432.

Siddiqui, O., Roberts, M. S., and Polack, A. E. (1985): The effect of iontophoresis and vehicle pH on the in vitro permeation of lignocaine through human stratum corneum. *J. Pharm. Pharmacol.*, 37:732–735.

Siddiqui, O., Sun, Y., Liu, J. C., and Chien, Y. W. (1987): Facilitated transdermal transport of insulin. *J. Pharm. Sci.*, 76:341–345.

Sloan, J. B., and Soltani, K. (1986): Iontophoresis in dermatology. *J. Am. Acad. Dermatol.*, 15:671–684.

Srinivasan, V., Higuchi, W. I., Sims, S. M., Ghanem, A. H., Behl, C. R., and Pons, S. (1989): Transdermal iontophoretic drug delivery: Mechanistic analysis and application to polypeptide delivery. *Int. J. Pharm.*, 78:370–375.

Srinivasan, V., Su, M-H., Higuchi, W. I., Sims, S. M., Ghanem, A. H., and Behl, C. R. (1990): Iontophoresis of polypeptides: Effects of ethanol pretreatment of human skin. *Int. J. Pharm.*, 79:588–591.

Svedman, P., and Svedman, C. (1997): Skin mini-erosion sampling technique: Feasibility study with regard to serial glucose measurement. *Pharm. Res.*, 15:883–888.

Tamada, J., Bohannon, N. J. V., and Potts, R.O. (1995): Measurement of glucose in diabetic subjects using noninvasive transdermal extraction. *Nature Med.*, 1:1198–1201.

Thysman, S., and Preat, V. (1993): In vivo iontophoresis of fentanyl and sufentanyl in rats: Pharmacokinetics and acute antinociceptive effects. *Anesth. Analg.*, 77:61–66.

Thysman, S., Preat, V., and Roland, M. (1991): Factors affecting iontophoretic mobility of metoprolol. *J. Pharm. Sci.*, 81:670–675.

Tyle, P. (1986): Iontophoretic devices for drug delivery. *Pharm. Res.*, 3:318–326.

Walberg, J. E. (1968): Transepidermal or transfollicular absorption. *Acta Derm. Venereol. (Stockh.)*, 48:336–344.

Walberg, J. E. (1970): Skin clearance of iontophoretically administered chromium (51Cr) and sodium (22Na) ions in the guinea pig. *Acta Derm. Venereol. (Stockh.)*, 50:255–262.

Wearley, L. L., and Chien, Y. W. (1989): Iontophoretic transdermal permeation of verapamil II: Factors affecting the reversibility of skin permeability. *J. Control. Rel.*, 9:231–281.

Yoshida, N. H., and Roberts, M. S. (1993): Solute molecular size and transdermal iontophoresis across excised human skin. *J. Cont. Rel.*, 25:177–195.

Zeltzer, L., Regalado, M., Nitchter, L. S., Barton, D., Jennings, S., and Pitt, L. (1991): Iontophoresis versus subcutaneous injection: A comparison of the two methods of local anesthesia in children. *Pain*, 44:73–84.

Toxicology of Skin
Edited by Howard I. Maibach
Copyright © 2001 Taylor & Francis

15

Penetration and Barrier Creams

Hongbo Zhai, Howard I. Maibach

Department of Dermatology, University of California, San Francisco, California, USA

INTRODUCTION

Defining the penetration of chemicals into and through the skin is important in dermatotoxicology and dermatopharmacology, as local and systemic toxicity depend on a chemical penetrating the skin. Dermatopharmacokinetics (DPK) deals with penetration or percutaneous absorption, as well as the distribution, metabolism, and excretion of the applied chemical (Wester and Maibach, 1991). DPK models describe drug concentrations as a function of dose and time. DPK studies may provide kinetic details and mechanistic insights that have the potential to supplement traditional clinical studies. (Wester and Maibach, 1983) described percutaneous absorption in terms of 10 steps with examples of bioavailability. According to the functions of topical products, most may be divided into two major types: to treat disease and/or to alleviate symptomatology, and to prevent disease. The latter includes barrier creams (BC) designed to prevent or reduce the penetration and absorption of hazardous materials into skin, preventing skin lesions and/or other toxic effects from dermal exposure (Orchard, 1984; Frosch et al., 1993a, 1993d; Lachapelle, 1996; Zhai and Maibach, 1996a, 1996b). DPK models may be used in the determination of bioequivalence of topical dermatological products (Shah et al., 1991, 1993; Gupta et al, 1993), in particular, to quantify the penetration (percutaneous absorption) of chemicals; therefore, they could be utilized to determine BC efficacy.

Most of the preceding discussion of DPK has focused on bioavailability (rate and amount of absorption) from a percutaneous route of administration (Wester and Maibach, 1983, 1991, 1993, 1997; Jamoulle and Schaefer, 1993). This chapter highlights the penetration of chemicals in the evaluation of the BC efficacy; however, the BC effect may depend on the DPK of the chemical and other factors (Packham, 1994; Wigger-Alberti et al., 1997).

WHY USE PK MODEL?

One advantage of DPK models over traditional in vivo methods is the ability to accurately describe nonlinear biochemical and physical processes (McDougal, 1991, 1997). Describing skin penetration based on blood concentrations or excretion rates as "percent absorbed" assumes all processes have a simple linear relationship with the exposure concentration. When nonlinear processes occur in the absorption, distribution, metabolism, or elimination of a chemical, describing absorption as percent absorbed may be misleading. Dermal absorption may not be linear when there is binding or metabolism in the skin or when skin blood flow is a limiting factor. Many biochemical processes in the body are nonlinear. A quantitative description of saturable kinetics in the model may allow it to be predictive of blood or tissue concentrations from various doses. A complete mathematical description of DPK takes mass balance throughout the animal into account and makes it possible to estimate fluxes (amount/time) and permeability constants (distance/time). These expressions of the penetration process are required to accurately predict penetration in other situations (i.e., different exposure area, time, or concentrations) when nonlinear processes are present. Another advantage is to estimate the metabolic parameters in the skin after application to the skin at several concentrations. DPK models can often be used to form predictions that will help in designing experimental doses and sampling times. It also may allow the use of fewer animals because it may not be necessary to sacrifice animals at various time points to obtain tissue concentrations (McDougal, 1991, 1997). Additionally, the movement of the dose of chemicals into the first organ system (skin) and the systemic circulation can be determined by physical dermatotechniques (skin washing, skin tape stripping) and bioengineering techniques, such as transepidermal water loss (TEWL). Singly, each can provide a segment of DPK information. Multiple integrated segments can truly define DPKs (Wester and Maibach, 1997).

PK MODELS

Numerous DPK models have been developed to detect the penetration of chemicals. Most methods may utilize radioactivity for analytic purposes (Wester and Maibach, 1993, 1997; Sznitowska, 1996; Bronaugh, 1997). Assay usually is done following topical application by directly measuring from a collection vessel (in vitro) (Wester and Maibach, 1993; Sznitowska, 1996; Bronaugh, 1997) or indirectly in biologic fluids or tissues (in vivo) (Shah et al., 1991; Wester and Maibach, 1993, 1997). Other methods determine dye penetration (Marks et al., 1989; Treffel et al., 1994;

Zhai and Maibach, 1996a), observe histologic change (Lupulescu and Birmingham, 1976; Mahmoud and Lachapelle, 1985; Lachapelle et al., 1990), and quantify irritant response by bioengineering techniques such as TEWL (Frosch et al., 1993a, 1993b, 1993c, 1993d; Frosch and Kurte, 1994; Treffel and Gabard, 1996).

In Vitro Models and Data

Most in vitro technique entails placing excised skin in a diffusion chamber, applying compound to its surface, and then assaying for compound in the collection vessel on the other side. Excised human and animal skin are used, and the skin may be entirely intact or separated into epidermis or dermis. Artificial membranes have also been used (Wester and Maibach, 1993; Sznitowska, 1996; Bronaugh, 1997). A major advantage of in vitro systems is that they allow for absorption measurements through skin of chemicals too toxic to test in man. Absorption rates and skin metabolism can be measured in the in vitro system because sampling is performed directly beneath the barrier layer. Skin metabolism can be studied in viable skin without interference from systemic metabolic processes. Finally, absorption measurements are obtained more easily from diffusion cells than from analysis of biological specimens from clinical studies. Other advantages of the in vitro technique are that the method is easy to use and results are obtained rapidly. A disadvantage is that the collection bath is saline, compatible with hydrophilic compounds but not with hydrophobic compounds. In vivo, the penetrating compound does not pass completely through the dermis but may be removed by capillaries in the dermis.

(Langford, 1978) conducted in vitro studies to determine the efficacy of the formulated fluorochemical (FC)-resin complex included solvent penetration through treated filter paper, solvent repellency on treated pigskin, and penetration of radio-tagged sodium lauryl sulfate through treated hairless mouse skin. FC-resin complex provided the penetration resistance against a range of solvents.

(Reiner et al., 1982) examined the protective effect of ointments on guinea pig skin in vitro. The permeation values of *Toxic Agent 4* through unprotected and protected skin within 10 hours as a function of time was determined radiologically and enzymatically. Permeation of Toxic Agent 4 was markedly reduced by polyethylene glycol ointment base and ointments containing active substance.

(Loden, 1986) evaluated the effect of four barrier creams on the absorption of (^{3}H)-water, (^{14}C) benzene, and (^{14}C) formaldehyde into excised human skin. The control and the barrier-cream—treated skins were exposed to the test substance for 0.5 hours and the amount absorbed determined. The results indicated that the experimental cream *water barrier* reduced the absorption of water and benzene but not formaldehyde. Kerodex 71 cream slightly reduced the absorption of benzene and formaldehyde. Petrogard and *Solvent Barrier* did not affect the absorption of the substances. One advantage of the method is the use of human skin. The effects of the barrier cream on the skin and the test substance mimic the in vivo situation. Another advantage is that the method is not only qualitative (i.e., distinguishes between the creams), but also quantitative (i.e., a certain degree of alteration in absorption is obtained).

(Fullerton and Menne, 1995) tested that the protective effect of ethylenediaminete-traacetate (EDTA) barrier gels against nickel contact allergy using in vitro and in vivo methods. In an vitro study, about 30 mg of barrier gel, was applied on the epidermal side of the skin and a nickel disc applied. After 24 hours' application, the disc was removed and the epidermis separated from the dermis. Nickel content in epidermis and dermis was quantified by adsorption differential pulse voltammetry (ADPV). The amount of nickel in the epidermal skin layer on barrier gel–treated skins was significantly reduced compared to the untreated control.

(Zhai et al., 1998) utilized an in vitro diffusion system to measure the protective effective of quaternium-18 bentonite (Q18B) gels to prevent 1% concentration of $[^{35}S]$ sodium lauryl sulfate (SLS) penetration by human cadaver skin. The accumulated amount of $[^{35}S]$-SLS in receptor cell fluid were counted to evaluate the efficacy of the Q-18B gels over 24 hours. These test gels significantly decreased SLS absorption when compared to the unprotected skin control samples. The percentage of protection effect of three test gels against SLS percutaneous absorption was 88%, 81%, and 65%, respectively.

(Sadler and Marriott, 1946) introduced some facile tests to evaluate BC efficiency. One method used the fluorescence of a dyestuff and eosin as an indicator to measure penetration and the rates of penetration of water through BC; this is rapid and simple but provides only a qualitative estimate.

(Treffel et al., 1994) measured in vitro on human skin the effectiveness of BC against three dyes (eosin, methylviolet, oil red O) with varying n-octanol/water partition coefficients (0.19, 29.8, and 165, respectively). BC efficacy was assayed by measurements of the dyes in the epidermis of protected skin samples after 30 minutes' application. The efficacy of BC against the three dyes showed in some data contrary to manufacturer's information. There was no correlation between the galenic parameters of the assayed products and the protection level, indicating that neither the water content nor the consistency of the formulations influenced protection effectiveness.

In Vivo Models and Data

Percutaneous absorption in vivo usually is determined by the indirect method of measuring radioactivity in excreta following topical application. Plasma levels of compounds after topical application are often below assay detection level. The chemical, often labeled, is applied to the skin and the total amount of radioactivity excreted in urine or urine plus feces determined (Shah et al., 1991; Wester and Maibach, 1993, 1997). The amount of radioactivity retained in the body or excreted by some route not assayed (CO_2) is corrected for by determining the amount of radioactivity excreted following parental administration. (Feldmann and Maibach, 1969, 1970, 1974) developed a protocol for measurement of percutaneous absorption in man and animal. This method involves quantifying absorption on the basis of the percentage of radioactivity excreted in urine following application of a known amount of the labeled compound to the skin.

(Lupulescu and Birmingham, 1976) observed the ultrastructural and relief changes of human epidermis following exposure to a protective gel and acetone and kerosene on humans. Unprotected skin produced cell damage and a disorganized pattern in the upper layers of epidermis. Protective agent prior to solvent exposure substantially reduced the ultrastructural and relief changes of epidermis cells.

(Mahmoud and Lachapelle, 1985; Lachapelle et al., 1990) utilized a guinea pig model to evaluate the protective value of BC and/or gels by laser Doppler flowmetry and histological assessment. The histopathological damage after 10 minutes of contact to toluene was mainly confined to the epidermis whereas the dermis was almost normal. Dermal blood flow changes were relatively high on the control site compared to the gel pretreated sites. In addition, the blood concentration of n-hexane of the control group and the gel-pretreated group was determined. It was possible to correlate results obtained by invasive (blood levels) and noninvasive techniques.

(Frosch et al., 1993a, 1993b, 1993c, 1993d; Frosch and Kurte, 1994) developed the repetitive irritation test (RIT) in the guinea pig and in humans to evaluate the efficacy of BC using a series bioengineering technique. The cream pretreated and untreated test skin (guinea pig or humans) were exposed daily to the irritants for 2 weeks. The resulting irritation was scored on a visual scale and assessed by biophysical techniques' parameters. Some test creams suppressed irritation with all test parameters; some failed to show such an effect and even increased scores.

(Zhai and Maibach, 1996a) utilized an in vivo human model to measure the effectiveness of BC against dye indicator solutions: methylene blue in water and oil red O in ethanol, representative of model hydrophilic and lipophilic compounds. Solutions of 5% methylene blue and 5% oil red O were applied to untreated and BC pretreated skin, with the aid of aluminum occlusive chambers, for 0 hours and 4 hours. At the end of the application time, the materials were removed and consecutive skin surface biopsies (SSB) obtained. The amount of dye penetrating into each strip was determined by colorimetry. Two creams exhibited effectiveness, but one cream enhanced cumulative amount of dye.

Brief data of BC efficacy from recent experiments are summarized in Table 15.1.

CONCLUSION

The accuracy of measurements of the efficacy of barrier creams depends on the use of proper methodology. The choice of methods for measurement of the protective effect of barrier creams usually is guided by two main criteria: efficacy and safety. In vitro methods are simple, rapid, and safe, and are recommended as a screening procedure for BC candidates. With radiolabeled methods we may determine the accurate protective and penetration results even with low levels of chemicals due to the sensitivity of radiolabeled counting. Animal's experiment may be used to generate kinetic data because of a similarity between man and some animals (pigs, monkeys, etc.) in percutaneous

TABLE 15.1. *Brief data of BC efficacy from recent experiments*

DPK models		Irritants or allergens	Barrier creams	Evaluations by	Efficacy	Reference
In vitro	In vivo animals or humans					
Human skin		Dyes (eosin, methylviolet, oil red O)	16 barrier creams	Amount of dyes in the epidermis	Various percentage protection effects	Treffel et al., 1994
Human skin	Nickel-sensitive patients	Nickel disc	Ethylenediami- netetraacetate (EDTA) gels	Nickel content	Significantly reduced the amount of nickel in the epidermis in vitro, and significantly reduced positive reactions in vivo	Fullerton and Menne, 1995
Human skin		[^{35}S]-SLS	3 quaternium-18 bentonite (Q-18B) gels	Amount of [^{35}S]-SLS	Percentage protection effect was 88%, 81%, and 65%, respectively	Zhai et al., 1999
	Guinea pigs	n-Hexane, trichlorethylene, toluene	3 water-miscible creams	Morphological assessment	Limited protective effects	Lachapelle et al., 1990
	Guinea pigs and humans	SLS, sodium hydroxide, toluene, lactic acid	Several barrier creams	Various bioengi- neering techniques	Some of them suppressed irritation, some failed	Frosch et al., 1993a, 1993b, 1993c, 1993d, 1994
	Humans	Dyes (methylene blue and oil red O)	3 barrier creams	Amount of dye penetrating into strips	Two of them exhibited effectiveness, one enhanced cumulative amount of dye	Zhai and Maibach, 1996a

absorption and penetration for some compounds, but no one animal, with its complex anatomy and biology, will simulate the penetration in humans for all compounds. Therefore, the best estimate of human percutaneous absorption is determined by in vivo studies in humans. The histological assessments may define what layers of skin are damaged or protected and may provide insight into BC mechanism. Noninvasive bioengineering techniques may provide accurate, highly reproducible, and objective observations in quantifying the inflammation response to irritants and allergens when BC are to be evaluated, and could assess subtle differences to supplement traditional clinical studies.

To validate these models, well-controlled field trials are required to define the relationship of the model to the occupational setting. Finally, the clinical efficacy of BC should be assessed in the workplace rather than only in experimental circumstances. We believe that sensible, standardized, and, hopefully, widely accepted methods are available and capable of refinement.

REFERENCES

Bronaugh, R. L. (1997): Methods for in vitro percutaneous absorption. In: *Dermatotoxicology Methods: The Laboratory Worker's vade mecum*. Edited by F. N. Marzulli and H. I. Malbach, pp. 1–14. Taylor & Francis, Washington.

Feldmann, R. J., and Maibach, H. I. (1969): Percutaneous penetration of steroids in man. *J. Invest. Derm.*, 52:89–94.

Feldmann, R. J., and Maibach, H. I. (1970): Absorption of some organic compounds through the skin in man. *J. Invest. Derm.*, 54:339–404.

Feldmann, R. J., and Maibach, H. I. (1974): Percutaneous penetration of some pesticides and herbicides in man. *Toxicol. Appl. Pharmacol.*, 28:126–132.

Frosch, P. J., and Kurte, A. (1994): Efficacy of skin barrier creams. IV: The repetitive irritation test (RIT) with a set of 4 standard irritants. *Contact. Dermatitis*, 31:161-168.

Frosch, P. J., Kurte, A., and Pilz, B. (1993a): Biophysical techniques for the evaluation of skin protective creams. In: *Noninvasive Methods for the Quantification of Skin Functions*. Edited by P. J. Frosch and A. M. Kligman, pp. 214–222. Springer-Verlag, Berlin.

Frosch, P. J., Kurte, A., and Pilz, B. (1993b): Efficacy of skin barrier creams. III: The repetitive irritation test (RIT) in humans. *Contact Dermatitis*, 29:113–118.

Frosch, P. J., Schulze-Dirks, A., Hoffmann, M., and Axthelm, I. (1993c): Efficacy of skin barrier creams. II: Ineffectiveness of a popular "skin protector" against various irritants in the repetitive irritation test in the guinea pig. *Contact Dermatitis*, 29:74–77.

Frosch, P. J., Schulze-Dirks, A., Hoffmann, M., Axthelm, I., and Kurte, A. (1993d): Efficacy of skin barrier creams. I: The repetitive irritation test (RIT) in the guinea pig. *Contact. Dermatitis*, 28:94–100.

Fullerton, A., and Menne, T. (1995): In vitro and in vivo evaluation of the effect of barrier gels in nickel contact allergy. *Contact. Dermatitis*, 32:100–106.

Gupta, S. K., Bashaw, E. D., and Hwang, S. S. (1993): Pharmacokinetic and pharmacodynamic modeling of transdermal products. In: *Topical Drug Bioavailability, Bioequivalence, and Penetration*. Edited by V. P. Shah and H. I. Maibach, pp. 311–332. Plenum Press, New York.

Jamoulle, J. C., and Schaefer, H. (1993): Cutaneous bioavailability, bioequivalence, and percutaneous absorption: In vivo methods, problems, and pitfalls. In: *Topical Drug Bioavailability, Bioequivalence, and Penetration*. Edited by V. P. Shah and H. I. Maibach, pp. 129–153. Plenum Press, New York.

Lachapelle, J. M. (1996): Efficacy of protective creams and/or gels. In: *Prevention of Contact Dermatitis, Curr Probl Dermatol*. Edited by P. Elsner, J. M. Lachapelle, J. M. Wahlberg, and H. I. Maibach, pp. 182–192. Karger, Basel.

Lachapelle, J. M., Nouaigui, H., and Marot, L. (1990): Experimental study of the effects of a new protective cream against skin irritation provoked by the organic solvents n-hexane, trichlorethylene and toluene. *Dermatosen*, 38:19–23.

Langford, N. P. (1978): Fluorochemical resin complexes for use in solvent repellent hand creams. *Am. Ind. Hyg. Assoc. J.*, 39:33–40.

Loden, M. (1986): The effect of 4 barrier creams on the absorption of water, benzene, and formaldehyde into excised human skin. *Contact Dermatitis*, 14:292–296.

Lupulescu, A. P., and Birmingham, D. J. (1976): Effect of protective agent against lipid-solvent-induced damages: Ultrastructural and scanning electron microscopical study of human epidermis. *Arch. Environ. Health*, 31:29–32.

Mahmoud, G., and Lachapelle, J. M. (1985): Evaluation of the protective value of an antisolvent gel by laser Doppler flowmetry and histology. *Contact Dermatitis*, 13:14–19.

Marks, R., Dykes, P. J., and Hamami, I. (1989): Two novel techniques for the evaluation of barrier creams. *Br. J. Dermatol.*, 120: 655–660.

McDougal, J. N. (1991): Physiologically based pharmacokinetic modeling. In: *Dermatotoxicology*. Edited by F. N. Marzulli and H. I. Maibach, pp. 37–60. Hemisphere Publishing Inc., New York.

McDougal, J. N. (1997): Methods in physiologically based pharmacokinetic modeling. In: *Dermatotoxicology Methods: The Laboratory Worker's vade mecum*. Edited by F. N. Marzulli and H. I. Maibach, pp. 51–68. Taylor & Francis, Washington.

Orchard, S. (1984): Barrier creams. *Dermatologic Clinics*, 2:619–629.

Packham, C. L. (1994): Evaluation of barrier creams: An in vitro technique on human skin (letter). *Acta. Derm. Venereol.*, 74: 405.

Reiner, R., Roβmann, K., Hooidonk, C. V., Ceulen, B. I., and Bock, J. (1982): Ointments for the protection against organophosphate poisoning. *Arzneim.-Forsch./Drug Res.*, 32:630–633.

Sadler, C. G. A., and Marriott, R. H. (1946): The evaluation of barrier creams. *Br. Med. J.*, 23:769–773.

Shah, V. P., Flynn, G. L., Guy, R. H., Maibach, H. I., Schaefer, H., Skelly, J. P., Wester, R. C., and Yacobi, A. (1991): In vivo percutaneous penetration/absorption. *Pharm. Res.*, 8:1071–1075.

Shah, V. P., Hare, D., Dighe, S. V., and Williams, R. L. (1993): Bioequivalence of topical dermatological products. In: *Topical Drug Bioavailability, Bioequivalence, and Penetration*. Edited by V. P. Shah and H. I. Maibach, pp. 393–413. Plenum Press, New York.

Sznitowska, M. (1996): Some aspects of in vitro percutaneous penetration studies using radioisotopes. In: *Dermatologic Research Techniques*. Edited by H. I. Maibach, pp. 207–216. CRC Press, Boca Raton.

Treffel, P., and Gabard, B. (1996): Bioengineering measurements of barrier creams efficacy against toluene and NaOH in an in vivo single irritation test. *Skin Research and Technology*, 2:83–87.

Treffel, P., Gabard, B., and Juch, R. (1994): Evaluation of barrier creams: An in vitro technique on human skin. *Acta. Derm. Venereol.*, 74: 7–11.

Wester, R. C., and Maibach, H. I. (1983): Cutaneous pharmacokinetics: 10 steps to percutaneous absorption. *Drug. Metab. Rev.*, 14:169-205.

Wester, R. C., and Maibach, H. I. (1991): In vivo percutaneous absorption. In: *Dermatotoxicology*. Edited by F. N. Marzulli and H. I. Maibach, pp. 75–96. Hemisphere Publishing Inc., New York.

Wester, R. C., and Maibach, H. I. (1993): Percutaneous absorption. In: *Topical Drug Bioavailability, Bioequivalence, and Penetration*. Edited by V. P. Shah and H. I. Maibach, pp. 3–15. Plenum Press, New York.

Wester, R. C., and Maibach, H. I. (1997): Percutaneous absorption in humans. In: *Dermatotoxicology Methods: The Laboratory Worker's vade mecum*. Edited by F. N. Marzulli and H. I. Maibach, pp. 15–27. Taylor & Francis, Washington.

Wigger-Alberti, W., Maraffio, B., Wernli, M., and Elsner, P. (1997): Self-application of a protective cream. Pitfalls of occupational skin protection. *Arch. Dermatol.*, 133:861–864.

Zhai, H., and Maibach, H. I. (1996a): Effect of barrier creams: Human skin in vivo. *Contact Dermatitis*, 35:92–96.

Zhai, H., and Maibach, H.I. (1996b): Percutaneous penetration (Dermatopharmacokinetics) in evaluating barrier creams. In: *Prevention of Contact Dermatitis, Curr Probl Dermatol*. Edited by P. Elsner, J. M. Lachapelle, J. M. Wahlberg, and H. I. Maibach, pp. 193–205. Karger, Basel.

Zhai, H., Buddrus, D. J., Schulz, A. A., Wester, R. C., Hartway, T., Serranzana, S., and Maibach, H. I. (1999): In vitro percutaneous absorption of sodium lauryl sulfate (SLS) in human skin decreased by Quaternium-18 bentonite gels. *In Vitro Mol. Toxicol.* 12:11–15.

SECTION III
Allergic Contact Dermatitis

16

Fiber

Kathryn L. Hatch

College of Agriculture, The University of Arizona, Tucson, Arizona, USA

Howard I. Maibach

University of California, San Francisco, California, USA

INTRODUCTION

The published account of textiles as a cause of skin disease begins in 1869 with Wilson's diagnosis (Wilson, 1869) that his patient's skin problem was due to dyes in the socks he wore. During the 130 years that have passed since this report, numerous dermatologists (individually and in teams) have described their investigations to determine the textile origin of the problem and determine prevalence of textile-caused skin disease. Fortunately, reviews of this information have been written (Cavelier et al., 1988; Cronin, 1980; Farrell-Beck and Callan-Noble, 1998; Foussereau, 1995; Hatch and Maibach, 1985a; 1985b; 1986; 1994; 1995; 1999; 2000; Hatch et al., 1992; Hatch, 1984a; 1984b; 1988; 1995; Storrs, 1986. This review of textiles as a cause of textile dermatitis (a) provides updated and comprehensive summaries of textiles as irritants and textile chemicals as allergens, and (b) comments on and makes recommendations about investigatory procedures that would improve understanding of textiles as irritants and as contact allergens.

TEXTILES AS IRRITANTS

Irritant contact dermatitis (ICD) is a nonallergic inflammation of the skin caused by contact with primary or cumulative irritants. In the case of textiles, primary irritants may be chemical compounds, fibers that protrude above a fabric's surface, the surface character of the fabric such as its frictional coefficient, or physical, permeability, and sorptive properties of the fabric.

Irritation to the skin may arise from (a) contact with textile chemicals prior to application to the fiber/fabric, (b) contact with chemically treated fabric, (c) transfer of irritant chemicals from the textile substrate to the skin, (d) scratching of the skin as a reaction to uncomfortable feeling fabrics, and (e) repeated sliding or movement of fabric over the skin. Further, inflammation may be a consequence of skin changes introduced when fabric alters the skin environment by its presence in a static or dynamic state.

Contact with Chemical Irritants

Reports of chemically induced irritant dermatitis from wearing of fabrics are few. (Redmond and Schappert, 1987) provide substantive evidence of ICD caused by wearing of garments with high concentrations of residual perchloroethylene (PCE) remaining in the fabric following dry cleaning and storage in plastic bags. In their investigation they collected garments that a semiconductor manufacturer required cleanroom workers to wear and analyzed them for PCE concentration. They state that there is potential for PCE-cleaned garments to affect the general public if PCE remains on the fabric. The recommendation made was that people should not wear dry cleaned garments that have been in sealed plastic bags, as was the case with the semiconductor workers' work garments. (Rietschel and Fowler, 1995) mention the possibility of irritant dermatitis arising from the wearing of fabrics to which starch, glue, vegetable gum, rosin, or shellac had been added. They note that Levi Strauss 501 "shrink to fit" jeans are made from a fabric containing cornstarch; in fact, 10% of the weight of the fabric when new is due to cornstarch; however, they cite no specific cases in which inflammation of the skin was observed when people wore these garments. Further, (Matthies et al., 1990) attempted to induce irritant and/or allergic contact dermatitis (ACD) by placing swatches of laundered fabrics on the skin of seborrheic and sebostatic volunteers for 48 hours and on babies and small children; no reactions were observed, so the conclusion drawn was that "the contribution of detergent residues as a releaser of skin reactions is rather insignificant." In 1993, (Matthies, 1993) patch tested 20 healthy volunteers to determine the influence of pH of laundered cotton fabrics on the skin. He noted that normally the pH of laundered work clothes was between 7 and 8.5. When the pH of the fabric swatches placed on the skin was between 4.6 and 10.1 there were few reactions, but when it was 10.1 and above, the number of reactions rose rapidly. He concluded that pH 10.4 was the alkaline threshold value and expected few problems to workers, as most wash and rinse procedures would render work-garment fabrics with pH values lower than 10.4.

Reports of chemically induced irritation from contact with chemicals during the manufacture of textiles are also few in number. In 1989, Gasperini et al. examined 11 men and 93 women who were garment industry workers with occupational contact dermatitis. They used 30 chemicals, including 8 disperse dyes and 1 acid dye and 2 formaldehyde-based durable-press reins which they prepared according to instructions in the literature and applied using the Rapid Patch test technique. The data for the garment workers showed an ICD incidence of 68.3%, which was higher than this group's ACD incidence (31.7%). The incidence data for the textile workers was 50.8% ICD and 49.2% ACD. In 1996, Soni and Sherertz (1996) studied skin responses to potential irritants and allergens of 28 men and 44 women who worked for 56 different textile manufacturers; 40 had daily work-related exposure to raw textiles (they were dye machine operators, dye mixers, sewing machine operators, or spinners), 19 had exposure to finished textiles (they were engaged in inspecting or packaging garments), and 13 had essentially no textile product exposure (they were housekeepers, painters, and office workers). All 72 patients underwent patch testing with the Hermal (Delmar, NY) standard screening series. Many (70%) were also patch tested with the Chemotechnique AB textile dye and finish series. Twenty-seven patients (38%) were diagnosed as having a predominantly work-related ICD, and 21 patients (29%) were diagnosed as having a predominantly work-related ACD. The worker's hands were the most common site of involvement whether the reaction from the chemical was irritant or allergenic. It is unclear in both the (Gasperini et al., 1989) and the (Soni and Sherertz, 1996) article which chemicals were responsible for the ICD in the textile workers.

Prickliness and Scratching

When wearing garments, fabric moves repeatedly across the skin surface and it is pressed onto the skin surface. People may express dissatisfaction with certain garments when these interactions are judged unpleasant. Words often used to describe the sensations include itchy, prickly, rough, and scratchy. During the 1980s, research sponsored by the Commonwealth Scientific and Industrial Organization (CSIRO) and undertaken by (Garnesworthy et al., 1998) assisted in understanding the neurological basis for sensations of discomfort and in identifying the parameters of fabric which were responsible for one sensation, that one called prickle, which they defined as "feeling many very gentle pin-pricks."

The first phase of their work involved discovering which nerve fibers in the skin were triggered when fabric/fibers contacted them. As a result of nerve-block experiments on 12 human volunteers, they thought it was the small-diameter nerve fibers of both the thermal or pain groups that were responsible for prickle; however, as a result of making recordings of responses of single nerve fibers in rabbits as they applied fabrics of different prickle intensities to an innervated area of their skin, they noted that only the pain group of receptors responded differently. Thus, it was the pain group of nerves that were triggered by fabrics/fibers. Similar results were achieved from single fiber recordings conducted with human subjects, demonstrating that pain nerves are involved in prickle sensation.

The second phase of the work involved discovering parameters that would assist wool fabric manufacturers in producing fabric free of prickle sensation. These experiments showed that a force of about 100 mgf was required at the end of a 40-μm diameter fiber to trigger a pain receptor. Thus, the stimuli on fabrics which trigger pain receptors to cause prickle are protruding fiber ends capable of supporting loads of about 100 mgf or more without buckling against the skin.

The third phase of the work consisted of psychophysical experiments in which fabrics of different prickle sensation were pressed onto and slid over different skin areas of human volunteers. CSIRO researchers found that (a) prickle results only when fabric is pressed onto a skin surface, not when the fabric is slid over the skin, and (b) the skin must have hair follicles for prickle to be felt, so prickle can be felt on the forearm but not on fingers and palms. Further, they discovered that prickle is an elusive sensation and may fluctuate in intensity over a period of seconds. Several seconds usually elapse between the time when fabric capable of eliciting a prickly sensation is pressed on the skin and when prickle is felt. They found a pressure of 4 g/cm^2 to be near the upper limit of typical downward pressure at which prickle is sensed. Other results were that prickle cannot be felt if the skin is uncomfortably cold or the skin contact area is smaller than about 1 cm^2. Moisture on the skin greatly increases the magnitude of the sensation. It was also found that chemicals are released at or near the nerve endings, which dilate surface blood capillaries. Reddening first occurs in the vicinity of activated pain nerve endings and then diffuses over larger areas. Inflammation may appear within 1 hour or more slowly after several hours of skin contact. It usually subsides rapidly after removal of the irritant from the skin unless the skin/fabric contact over several days has produced a severe inflammation. Fabrics that are abrasive or poke the skin may produce this irritant response.

Here one might note first that prickle also tends to irritate the wearer, who then scratches his/her skin as a means to alleviate the uncomfortable feeling, further irritating the skin, and one might secondly note that one of the primary reasons consumers end up purchasing garments that induce prickle is that they do not try on the garment at the retail store. Feeling the fabric with the hands is insufficient to determine fabric prickle, as the palms and fingers do not have hair follicles.

Movement of Fabric Over the Skin

When garments are worn, fabric generally moves over the skin surface. The degree of friction between the fabric and skin has a potential to cause skin damage when the coefficient of friction is high and the action is continuously repeated. A notable case report is that of (Ramam et al., 1998), in which patients living in New Delhi, India presented with a "striking pattern of dermatitis occurring in the summer, localized to the skin directly beneath the undergarments." They called the dermatitis "frictional sweat dermatitis" because they thought that movement of the garments over skin that was wet with sweat was the combination of factors that cause the dermatitis of their New Delhi patients. The typical skin lesion was "a slightly elevated plaque with a glazed, wrinkled surface and sharp margins corresponding to the edges of the undergarment. Erythema was mild and obvious only in early lesions. There usually was a network of superficial fissures running across the plaque, with delineated hyperpigmented brown-black scales. The lesions usually scaled off completely in 1–2 weeks with no sequelae."

Additionally, LaMotte (1977) investigated response of cutaneous receptors when fabric was moved over the skin surface, and (Gwosdow et al., 1986) measured the friction between skin and fabric and assessed degree of unpleasantness of the fabrics. While these studies do not show that irritant dermatitis results from friction between fabric and skin, they provide key elements as the links are pursued.

In LaMotte's study, fabrics and a bare metal plate were slid over a monkey's hand using an electromechanical device that displaced the monkey's skin. The response of the nerve fibers to fabric was always greater than the response to the smooth, uncovered metal plate. The intensity of response of the mechanoreceptive nerve fibers to the fabrics depended on the amount of skin displacement, the weave of the fabric, the density of the fabric, and the rate of movement of the fabric. Precise relationships between nerve response and fabric construction parameters were not obtained.

In the Gwosdow et al. study (Gwosdow et al., 1986), the degree of unpleasantness experienced by eight male volunteers when six different fabrics were slid across their forearms was studied. Force required to pull each fabric over the skin increased as the skin became wetter. As force and skin wetness increased, subjects rated all fabrics as feeling rougher and less pleasant. The researchers concluded that "moisture on the skin surface increases skin friction, which enhances perception of roughness."

Fabric as a Stationary Skin Cover

Placing fabric over the skin surface without any downward or lateral movement of the fabric relative to the skin may cause changes in the skin. One of the primary ones is hydration of the stratum corneum (SC) because (a) the fabric usually acts as a barrier to the rate of loss of transepidermal water evaporated at the skin surface, or (b) the fabric is a source of water which may contribute to stratum corneum hydration. The primary change, an increase in SC hydration, may lead to other events, such as increased susceptibility to microbial growth, abrasive damage, and absorption of chemicals, and altered pH with inflammation of the skin as a final outcome or manifestation.

Two main types of assessment have been used to determine the effect of fabric on moisture in or on the skin. The first type involves the use of instruments that measure the water content of the stratum corneum or measure the rate of evaporative water loss as an indication of SC water content. Hydration due to fabric covering the skin is the difference between measurements taken after covering the skin with fabric and those taken on uncovered skin. The second type involves the use of rating scales by which subjects indicate skin wetness or wet comfort.

Studies in which the effect on the skin of diaper and incontinence materials/products while dry, wet with water, and soiled have been reviewed by (Hatch et al., 1992) and most recently in the introduction to a research study by (Cottenden et al., 1998). In the latter reference, about 30 studies are cited.

In general, researchers have found that wearing diapers, incontinence pads, or patches cut from these products does elevate skin wetness as determined instrumentally. This is not surprising in that these products are designed to hold urine (liquid water). The inner layers of the product are designed to absorb the urine and the outer materials usually are water resistant (waterproof) and not water vapor permeable. The products fit tightly to the skin, allowing little opportunity for ventilation. SC hydration differs, depending on the materials used in the product. In general, products containing superabsorbent polymers keep the skin driest. A notable exception occurred in the study by (Dallas and Wilson, 1992), in which no significant differences in skin wetness were found for disposable incontinence pads containing superabsorbent and not containing superabsorbent and reusable pads. There were several significant differences within the groups, however.

(Cottenden et al., 1998) focused on discovering what pad characteristics (absorption capacity, strike-through time, and rewet property) resulted in the best wet comfort. To do so, they used experimental pads composed of different combinations of materials but the same design (hourglass-shaped). Twenty lightly incontinent women used the range of experimental pads. In the first phase, the women used a random mix of three of the six pad variants, they logged the time at which they put on and took off each pad, and scored it for leakage performance, wet comfort, and absorbance using a three-point scale (good, ok, poor). Used pads were weighed and pad performance studied as a function of urine weight, wear time, and put-on time. In phase two, the women were asked to compare the overall performance and leakage performance of different pairs of pad variants after a week of use. Pads with high absorption capacity and low rewet were rated best; however, they concluded that "differences are smaller than conventional wisdom would have predicted and often fail to reach statistical significance."

(Hatch et al., 1990; 1992; 1997) and (Cameron et al., 1997) have reported their studies in which SC hydration was assessed when everyday garment fabrics were worn by subjects. In an initial study by (Hatch et al., 1987), subjects wore patches of dress-weight woven fabrics composed of triacetate and polyester. The longer the patches were in place, the higher the SC hydration became. SC hydration was higher when a water vapor–impermeable film was placed over the swatches than when it was absent. SC hydration differed depending on the moisture-absorbing properties of the fabrics when the water vapor impermeable film was in place but did not differ when the impermeable film was not used. This initial study was followed by a series of studies in which the same set of three jersey knit fabrics, the first of all cotton fibers, the second of all 1.5-denier polyester fibers, and the third of all 3.5 polyester fibers, were used. In the first phase of the study, subjects exercised while wearing garments made from the three fabrics. At intervals during the exercise protocol, subjects reported contact comfort sensation and researchers instrumentally determined SC water content and rate of evaporative water loss. There were no significant differences in water evaporation or SC water content due to the type of fabric, a result that was not expected due to significant differences in water absorption capability of the fabrics. Further, subjects did not indicate differences between fabrics for wetness sensation. Because the fabrics did not lie on the skin, where the SC assessments were made, the next phase of the study involved placing swatches of the fabrics with different amounts of moisture in them on the volar forearm of sedentary volunteer subjects. These swatches were covered with a water vapor impermeable film. Under these conditions, the trend was that the wetter the fabric applied to the skin, the higher the SC hydration at fabric removal.

In the Cameron et al. study (Cameron et al., 1997), placement of dry fabrics on the skin with a water vapor–impermeable cover did not significantly affect the hydration level of the SC, though all dry fabrics did increase the hydration level slightly. Wet wool and cotton fabrics significantly hydrated the SC when levels were compared to either uncovered skin or skin covered by dry fabrics. Of the seven synthetic-fiber fabrics tested in a wet state, three (acrylic, PTFE, and spun nylon) significantly increased the SC hydration level. These three fabrics and the natural fabrics had comparable wetted moisture content.

TEXTILE CHEMICALS AS CONTACT ALLERGENS

Textile chemicals may be classed as fiber polymers and polymer-forming chemicals, natural-fiber coating materials, manufactured-fiber additives, chemical-finishing compounds and chemical finishes, azoic components, dyes, pigments and binders, degradation products on fabrics, and miscellaneous. Table 16.1 lists all textile chemicals, except dyes, reported to be contact allergens. Disperse dyes reported to be allergic contact allergens are listed in Table 16.2, nondisperse dyes with C.I. names in Table 16.3, and dyes without published C.I. names in Table 16.4. The basis for inclusion of a chemical was at least one case report or at least one positive patch test in a prevalence study. In the case of dyes, the report or study needed to appear in the time interval 1950 to 1999, and in the case of all other chemicals, the time period was 1980 to 1999 to make the latter list reflect chemistries likely to be in existence today.

TABLE 16.1. *Textile chemicals other than colorants reported to be contact allergens (1985 to 1999)*

Category: Chemical	Exposure	Reference
Fiber polymers: epsilon-aminocaprioic acid (EACA), a polymer used to make a type of nylon fiber	A 43-year-old housewife who wore nylon body stockings during a 2-year period	Tanaka et al., 1993
Natural-fiber coating materials: sericin, a compound which covers silk fiber, usually removed in finishing (degumming)	A 44-year-old woman wearing a silk garment	Inque et al., 1997
Manufactured-fiber additives: Tinuvin P in spandex fiber 1-(2-hydroxy-5-methylphenyl-benzothiozole) a UV absorber that slows the rate of decomposition of spandex fiber	(1) One patient who wore a t-shirt that had a polyurethane elastomer tape attached to it. (2) A 54-year-old female sleeping in underwear containing spandex	Kaniwa et al., 1991 Arisu et al., 1992
Chemical finishing compounds and finishes (a) Durable press Not identified	13 suspected cases revealed in a review of clinic records	Sherertz, 1992
Not identified	19 cases of ACD from wearing of durable press treated fabrics	Fowler et al., 1992
Not identified	A 19-year-old female who handled finished fabrics during work was patch-test positive to quaternium 15, formaldehdye, urea-formaldehyde, melamine-formaldehyde, and ethylene urea melamine-formaldehyde	Bracamonte et al., 1995
(b) Flame retardants not identified	Not known as French manuscript not read	Guinnepain, 1990
Flamentin ASN	A 37-year-old painter who wore a white cotton hat finished with this flame retardant to protect his head	Moreau et al., 1994
Diammonium hydrogen phosphate (DAP)	12 surgical personnel who wore surgical caps containing excessive residual amounts of the offending chemical	Belsito, 1990
(c) Biocides Isothiazolinone; 4,5-dichloro-2-n-octyl-4-isothiazolin-3-one, a biocide marketed as Kathon 930	2 males and 6 females who worked in a finishing mill	Kojima and Momma, 1989
Isothiazolinone, a biocide marketed as Acticide SPX	20 of 100 workers in a flax spinning mill who handled fiber wet with the biocide	Ramam et al., 1998

TABLE 16.1. (*Continued*)

Category: Chemical	Exposure	Reference
(d) Other 1,6 diisocyanatohexane (HDI), a component of a finish used to slow the rate of pill formation	19 operatives at 2 clothing manufacturing firms were patch-test positive	Wilkinson et al., 1991
Tinofix S, a cationic condensation product of dicyandiamide, formaldehyde, ammonium chloride, and ethylenediamine	A 55-year-old male wearing the trousers to which 8 chemical finishes had been applied causing urticaria (hives)	DeGroot and Gerkens, 1989
Dicyclohexylmethane-4-4'-diisocyanate, HMDI	A 47-year-old engaged in wool processing	Thompson and Belsito, 1997
Degradation products: phosgene (2,5-dichlorophenyl) hyrazone, a product formed from bleaching fabric containing Pigment Yellow 16 (C.I. 20040)	12 young males developed ACD from wearing yellow cotton sweaters	Kojima and Momma, 1989; Kojima et al., 1990
Formaldehyde	As reported above.	Fowler et al., 1992; Sherertz, 1992
Azoic components: 2-hydroxy-3-naphthoic acid analide marketed as Naphthol AS, a coupling agent	(1) A 60- and 63-year-old female wearing a cotton flannel nightshirt	Hayakawa et al., 1985
	(2) A 60-year-old male laying his head on a cotton pillowcase	Osawa et al., 1997
	(3) 5 patients sensitized from wearing of a special kind of shirt	Roed-Petersen et al., 1990
2,4,5-trichloroaniline, an azoic diazo component	A chemical worker who had direct exposure to the compound	Sano et el., 1994
4-benzamide-2-5-diethoxyaniline, an azoic diazo component	A chemical worker who had direct exposure to the compound	Sano et al., 1994
Miscellaneous: Isophoromediamine (IPDA), 3-amino-methyl-3,5,5-trimethylcyclohexylamine, a hardener component in a lamination glue	A non atopic 39-year-old male worker engaged in lamination of a film to fabric	Tarvainen et al., 1998
Ethylene glycol dimethacrylate (EGDMA) and BIS-GMA, epoxy resin, a UV-curable emulsion for screen printing ink. Trade named Saatigral.	A 40-year-old male engaged in prepared stencils for printing athletic shirts	Goossens et al., 1998

Readers interested in listings of textile contact allergens reported at earlier times are referred to reviews referenced in the chapter introduction.

Table 16.1 provides a record for each chemical category, including the name of the specific allergen, a description of the exposure, and the original reference. Tables 16.2–16.4, the dye tables, provide a detailed record for each dye listed. That record includes its chemistry, how many case reports were located, the total number of

TABLE 16.2. *Disperse dyes reported to be contact allergens (1950–1999)*

C.I. Name and number/ Chemical class	Case report data[*]: number of reports/ total positive patch tests reported	Prevalence study data: number of positive patch tests/population size
Disperse Red 1 11110 Monoazo	20/106	13/721 (Balato et al., 1990) 29/2,752 (Seidenari et al., 1991) 2/3,336 (Dooms-Goossens, 1992) 1/10,191 (Goncalo et al., 1992) 1/72 (1/5)[§] (Soni and Sherertz, 1996) 67/6,203 (Seidenari et al., 1997) 5/1,012 (5/31)[§] (Lodi et al., 1998)
Disperse Red 11 62015 Anthraquinone	2/22	-----------
Disperse Red 15 60710 Anthraquinone	1/15	-----------
Disperse Red 17 11210 Monozao	13/24	3/145 (Balato et al., 1990) 20/2,752 (Seidenari et al., 1991) 5/159 (Dooms-Goossens, 1992)
Disperse Red 19 11130 Monoazo	1/1	-----------
Disperse Red 137 ---[¶] Monoazo	1/1	-----------
Disperse Red 153 --- Monozao (heterocyclic)	1/1	-----------
Disperse Orange 1 11080 Monoazo	6/23	3/721 (Balato et al., 1990) 20/2,752 (Seidenari et al., 1991) 5/159 (Dooms-Goossens, 1993)
Disperse Orange 3 11005 Monoazo	17/77	11/721 (Balato et al., 1990) 1/569 (Manzini et al., 1991) 28/2,752 (Seidenari et al., 1991) 1/462 (Sherertz, 1992) 5/159 (Dooms-Goossens, 1992) 1/10,191 (Goncalo et al., 1992) 2/72 (2/5)[§] (Soni and Sherertz, 1996) 107/6,203 (Seidenari et al., 1997)
Disperse Orange 13 --- Disazo	3/6	2/159 (Dooms-Goossens, 1992)
Disperse Orange 76[†] --- Monoazo	5/14	3/145 (Balato et al., 1990) 11/2,752 (Seidenari et al., 1991)
Disperse Yellow 1 10345 Nitrodiphenylamine	1/15	-----------
Disperse Yellow 3 11855 Monoazo	23/140	1/159 (Dooms-Goossens, 1992) 1/10,191 (Goncalo et al., 1992) 1/72 (1/5) (Soni and Sherertz, 1996) 42/6,203 (Seidenari et al., 1997) 8/1012 (8/31)[§] (Lodi et al., 1998)
Disperse Yellow 4 12770 Monoazo (heterocyclic)	1/1	-----------
Disperse Yellow 9 10375 Nitrodiphenylamine	7/33	11/2,752 (Seidenari et al., 1991) 2/159 (Dooms-Goossens, 1992)

TABLE 16.2. (*Continued*)

C.I. Name and number/ Chemical class	Case report data[*]: number of reports/ total positive patch tests reported	Prevalence study data: number of positive patch tests/population size
Disperse Yellow 39 --- Indigoid	3/22	-----------
Disperse Yellow 49 --- Indigoid	1/15	-----------
Disperse Yellow 54 --- Quinophthalone	1/2	3/2,752 (Seidenari et al., 1991)
Disperse Yellow 64 47023 Quinophthalone	1/1	-----------
Disperse Blue 1 64500 Anthraquinone	3/19	-----------
Disperse Blue 3 61505 Anthraquinone	6/40	4/2,752 (Seidenari et al., 1991)
Disperse Blue 7 62500 Anthraquinone	3/23	-----------
Disperse Blue 26 63305 Anthraquinone	1/1	-----------
Disperse Blue 35 ---[†] Anthraquinone	13/50	2/145 (Balato et al., 1990) 5/2,752 (Seidenari et al., 1991) 6/159 (Dooms-Goossens, 1992)
Disperse Blue 85 ---[‡] Monoazo	5/6	2/159 (Dooms-Goossens, 1992)
Disperse Blue 102 --- Monoazo (heterocyclic)	1/1	-----------
Disperse Blue 106 --- Monoazo (heterocyclic)	16/61	16/159 (Dooms-Goossens, 1992)
Disperse Blue 124 --- Monoazo (heterocyclic)	13/42	23/721 (Balato et al., 1990) 36/2,752 (Seidenari et al., 1991) 6/159 (Dooms-Goossens, 1992) 2/462 (Sherertz, 1992) 1/72 (1/5)[§] (Soni and Sherertz, 1996) 104/6,203 (Seidenari et al., 1997) 22/1012 (22/31)[§] (Lodi et al., 1998)
Disperse Blue 153 --- Anthraquinone	2/2	3/159 (Dooms-Goossens, 1992)
Disperse Brown 1 11152 Monoazo	6/12	2/145 (Balato et al., 1990) 4/159 (Dooms-Goossens, 1992)
Disperse Black 1 11365 Monoazo	8/28	12/2,752 (Seidenari et al., 1991) 1/569 (Manzini et al., 1991)
Disperse Black 2 11255 Monoazo	3/8	-----------

[*] Case report references are provided in (Hatch and Maibach, 1999).

[†] In one case report, the dye was probably a phototoxin, as the skin eruption occurred upon exposure of the skin to light.

[‡] In one case report, this dye caused a pigmented purpuric eruption rather than that of classical ezematous dermatitis.

[§] Incidence for the dye-sensitive patients.

[¶] This dye is identical to Disperse Orange 37.

TABLE 16.3. *Nondisperse dyes with C.I. names reported to be contact allergens (1950 to 1999).*

C.I. name and number/ Tradename[†]/ Chemical class	Case report data[*]: number of reports/ positive patch tests reported	Prevalence study data: number of positive patch tests/population size
ACID		
Acid Red 85 22245 Disazo	1/1	-----------
Acid Red 118 --- Supramine Red GW Monoazo (heterocyclic)	---	1/569 (Manzini et al., 1991) 1/1,814 (Seidenari et al., 1995)
Acid Red 359 --- Neutrichrome Red Neutrichrome Red SGN Monoazo (1:2 chromium complex azo)	---	1/569 (Manzini et al., 1991) 2/1,814 (Seidenari et al., 1995)
Acid Yellow 23 19140 Azo tartrazine Monoazo (pyrazolone)	1/1	-----------
Acid Yellow 36 13065 Metanil yellow Monoazo	2/2	-----------
Acid Yellow 61 18968 Supramine Yellow GW Monoazo (heterocyclic)	---	1/569 (Manzini et al., 1991) 5/1,814 (Seidenari et al., 1995)
Acid Violet 17 42650 Triarylmethane	1/1	-----------
Acid Black 48 65005 Anthraquinone	4/6	3/145 (Balato et al., 1990) 4/2,752 (Seidenari et al., 1991) 1/569 (Manzini et al., 1991)
BASIC		
Basic Red 22 --- Synacril Red 3B Monoazo	1/1	-----------
Basic Red 46 --- Monoazo	2/3	1/569 (Manzini et al., 1991) 1/159 (Dooms-Goossens, 1992) 1/72 (2/5)[‡] (Soni and Sherertz, 1996)
Basic Brown 1 21000 Disazo dye mixture	1/1	2/145 (Balato et al., 1990) 1/10,191 (Goncalo et al., 1992)
Basic Black 1 50431 Azine dye mixture	---	9/2,752 (Seidenari et al., 1991)
DIRECT		
Direct Orange 34 40215, 40220 Arancio Diazol Luce 7 JL, ICI Stilbene	---	8/1,814 (Seidenari et al., 1995)
Direct Black 38 30235 Triazo	1/22	-----------

TABLE 16.3. (*Continued*)

C.I. name and number/ Tradename[†]/ Chemical class	Case report data[*]: number of reports/ positive patch tests reported	Prevalence study data: number of positive patch tests/population size
REACTIVE		
Reactive Red 123 --- Scarlet Drimaren K2G Monozo with unpublished reactive system	1/1	1/1,813 (Manzini et al., 1996)
Reactive Red 238 --- Red Cibacron CR Unpublished	---	1/1,813 (Manzini et al., 1996)
Reactive Red 244 --- Red Cibacron C4G Unpublished	---	1/1,813 (Manzini et al., 1996)
Reactive Orange 107 --- Gold Yellow Remazol RNL Unpublished	1/1	2/1,813 (Manzini et al., 1996)
Reactive Yellow 17 18852 Gold Yellow Remazol G Monoazo with vinylsulfonyl reactive system	1/1	1/1,813 (Manzini et al., 1996)
Reactive Yellow 56 ---Cibacron Pront Yellow 4R Monozo with a chlorotrizinyl reactive system	1/1	----------
Reactive Blue 21 ---Turquoise Blue Remazol G133 Phthalocyanine with vinylsulphonyl reactive group	1/1	1/1,813 (Manzini et al., 1996)
Reactive Blue 74 --- Cibacron Pront Blue 3R Anthraquinone with chlorotrizinyl reactive system	1/1	----------
Reactive Blue 75 --- Cibacron Pront Turquoise G Anthraquinone with chlorotrizinyl reactive system	1/1	----------
Reactive Blue 122 --- Printing Blue Remazol RR Azo (metal complex)	---	3/1,813 (Manzini et al., 1996)
Reactive Blue 225 Azo	1/1	----------
Reactive Blue 238 --- Marine Cibacron CB Unpublished	---	2/1,813 (Manzini et al., 1996)
Reactive Violet 5 18097 Violet Remazol 5R Monoazo with vinylsulfonyl reactive system	1/1	5/1,813 (Manzini et al., 1996)
Reactive Black 5 20505 Black Remazol B Gran Disazo with vinylsulfonyl reactive system	2/3	2/1,813 (Manzini et al., 1996)

(*Continued*)

TABLE 16.3. (*Continued*)

C.I. name and number/ Tradename[†]/ Chemical class	Case report data:[*] number of reports/ positive patch tests reported	Prevalence study data: number of positive patch tests/population size
SOLVENT		
Solvent Orange 8 12175 Monoazo	1/2	-----------
Solvent Yellow 1 11000 Monoazo	1/9	-----------
Solvent Yellow 14 12055 Monoazo	1/2	-----------
VAT		
Vat Green 1 59825-6 Dibenzanthrone	1/5	-----------

[*]Case report references are provided in Hatch and Maibach (Hatch and Maibach, 1999).
[†]Manufacturers of trademarked dyes are as follows: Supramine (Bayer), Neutrichrome and Diazol (ICI), Synacril (Synthron Inc.), Drimaren (Sandoz), Cibacron and Tinofix (Ciba), and Remazol (Hoechst). Neutrichrome is now a product of Crompton and Knowles, offered under the tradename Neutrilan.
[‡]Incidence for dye-sensitive patients.

patient's diagnosed, reference to prevalence studies in which the dye was used, and the results of the patch testing in the prevalence study.

As is readily apparent from studying Table 16.1, few textile chemicals have been reported to be allergic contact allergens. The list contains 17 compounds. Most are finishing compounds and finishes.

Fiber Polymers

Fiber polymers are the chemical units that comprise fibers, the small cylindrical units which comprise all fabrics. In the main, polymers within any fiber are chemically identical but vary in molecular weight (or number of repeat units per polymer).[*] Cotton, flax (linen), rayon, and lyocell are fibers composed of cellulose polymers. Wool, silk, cashmere, and other animal fibers are composed of protein polymers. Polyester fibers are composed of various polyester polymers (PET, PCCT, etc.), nylon (polyamide) of various polyamide polymers (6.6, 6, etc.), and spandex (elastane) of various segmented polyurethane polymers.

(Tanaka et al., 1993) report their diagnoses of a patient having ACD to a fiber polymer. The fabric worn by the patient was composed of nylon fibers made from epsilon-aminocaprioic acid polymers.

---

[*]The exception is the use of polymers of different chemistries to create manufactured fibers known as bicomponents. Wool, a natural fiber, also contains polymers with two distinct chemistries making it the only natural bicomponent fiber.

TABLE 16.4. *Dyes without C.I. names reported to be contact allergens (1950–1999)*

Dye identity provided in case report or incidence study	Case report data[*]: number of reports/ total positive patch tests reported	Prevalence study data: number of positive patch tests/population size
p-amino-acetanilide-p-cresol	1/1	----------
methylamino-4-(2-hydroxy-methylamino) anthraquinone	1/1	----------
Turquoise Reactive	---	1/569 (Manzini et al., 1991)
Remazol DR Brown RR (azo with unpublished reactive group)	1/1	----------
Brilliant Green	---	2/569 (Manzini et al., 1991)
Diazol Orange	---	2/569 (Manzini et al., 1991)
Foron Blue	1/1	----------
4 blue and 1 violet dye with structures similar to Disperse Blue 106 and 125(thiazol-azoyl-paraphenylendiamine chemistry)	1/1	----------
1 orange and 1 blue dye with chemistry similar to Disperse Red 1, 17 and Orange 3, and Brown 1	1/1	----------

[*] Case report references are provided in (Hatch and Maibach, 1999).

Natural-fiber Coating Materials

Natural-fiber coating materials are those compounds that surround or enclose a natural fiber in its natural state. Examples of coating materials are waxes (pectins) on cotton fiber, sericin (commonly called gum) on silk fiber, and lanolin on wool fiber. Natural coating materials may or may not be completely removed in processing of the fiber.

Inque and colleagues (1997) reported their finding that sericin on silk caused one patient to have ACD. This patient wore a fabric made of silk fibers from which the sericin was not completely removed.

Fiber Additives

Fiber additives are chemicals such as flame retardants, UV-radiation absorbers, delustrants, and optical brighteners which are incorporated into manufactured fibers to provide improved fabric performance (enhanced flame resistance, decreased rate of

TABLE 16.5. *Dyes used in patch testing that were found not be contact allergens*

C.I. name and number and/or tradename*	Chemistry	Reference
CLASS NOT PUBLISHED		
Yorkshire Joracril Yellow	Unpublished	Manzini et al., 1991
Yorkshire Red	Unpublished	Manzini et al., 1991
Yorkshire Joracril Blue	Unpublished	Manzini et al., 1991
Sandolan Yellow	Unpublished	Manzini et al., 1991
Sandolan Blue	Unpublished	Manzini et al., 1991
Diazol Black	Unpublished	Manzini et al., 1991
Brilliant Red	Unpublished	Manzini et al., 1991
Marine Foron Blue	Unpublished	Manzini et al., 1991
Supramine Blue	Unpublished	Manzini et al., 1991
Astrazon MS Green	Unpublished	Manzini et al., 1991
Astrazon M Green	Unpublished	Manzini et al., 1991
REACTIVE		
Green 12 --- Green Drimaren X3G	Unpublished	Manzini et al., 1996
Blue 18 --- Turquoise Drimaren X2G	Unpublished	Manzini et al., 1996
Red 198 --- Red Remazol RB 133	Azo	Manzini et al., 1996
Red 180 --- Remazol F3B	Azo	Manzini et al., 1996
Orange 16 --- Brilliant Orange, Remazol 3R	Azo	Manzini et al., 1996
Orange 96 17757 Gold Yellow, Remazol 3R	Azo	Manzini et al., 1996
Blue 220 --- Formazan Brilliant Blue, Remazol BB	Unpublished	Manzini et al., 1996
DIRECT		
Yellow 169 --- Solophenil Yellow AGL	Azo	Seidenari et al., 1995

*Remazol, Cibacron and Solophenil, and Drimaren are tradenames of Hoechst, Ciba, and Sandoz, respectively.

degradation due to ultraviolet radiation exposure) and/or to improve fabric aesthetics (decreased shine and enhanced whiteness or brightness). Fiber additives are embedded between the fiber's polymers and uniformly distributed through the fiber mass because they are included in the fiber spinning solution or melt.

Although fiber additives are commonly used, Tinuvin P, a compound that absorbs ultraviolet radiation, is the only fiber additive reported to be a contact allergen (Arisu et al., 1992, Kaniwa et al., 1991). This compound was added to spandex (elastane) fiber to slow its decomposition by ultraviolet radiation.

Chemical Finishing Compounds and Chemical Finishes

Chemical finishing compounds are those chemicals used in wet processing of textiles, a manufacturing step in which yarns or fabrics are usually submersed into an aqueous bath containing the finish formulation. In durable-press finishing, for example, the bath usually contains formaldehyde and an N-methylol compound. Residual finishing chemicals are those chemicals that are unintentionally left on the fabric at the time it is shipped from the finishing mill. On durable-press fabrics, formaldehyde would be a residual finishing chemical.

A chemical finish is "a chemical material other than colorants and residual processing chemicals added to textiles to impart desired functional and aesthetic properties to the textile product" (American Association of Textile Chemists and Colorists, 1998). The compound on the fabric, the finish, may or may not be the same as those included in the bath. In water-repellent treating of fabrics, the finishing compound and the finish are the same, but in durable-press finishing, they are not the same, as the idea is to have the chemicals in the bath react with each other to form a resin.

Chemical finishes may be topical, essentially forming a thin coating on fibers. Examples of topical finishes are stain repellents and water repellents. Others, such as durable-press and flame-resistant finishes for cottons are not topical, as the chemicals reside largely within the fibers where they are physically entrapped between polymers or covalently bonded to the fiber polymers. Chemical finishes may degrade or decompose during use. Formaldehyde may be released due to the degradation of durable-press resins. Curing may be done to affix the chemical finish to the fabric.

Durable-Press Chemicals and Finishes

Durable-press resins are those resins which make fabrics containing cotton fibers have a smoother (more wrinkle free) appearance during wear and following laundry than would be possible without the resin treatment. The fabric might be composed of only cotton fibers or composed of cotton fibers blended with polyester fibers. Many different chemicals have been reacted with formaldehyde to achieve the wrinkle-free performance, and these have differed in amount of formaldehyde release in end-use. Most DP finishing systems used today are based on a reaction between dimethyldihydroxyethyleneurea (DMDHEU) and formaldehyde or between modified DMDHEU and formaldehyde with magnesium chloride as a catalyst producing formaldehyde-based resins, or they are based on nonformaldehyde alternatives, the most attractive being those based on polycarboxylic acids, creating nonformaldehyde durable-press resins. In all cases, the resin forms crosslinks within cotton fibers, actually covalently bonding adjacent cellulose polymers within a cotton fiber. The applied resin may be cured directly after application to the fabric or it may be cured after the fabric has been made into a garment or other product.

Durable-press fabrics made using formaldehyde in the reaction bath may contain unreacted (free) formaldehyde. Formaldehyde is also formed as these resins degrade.

Usually labels on products do not reveal whether durable-press finishing has been done unless the process was low-formaldehyde or nonformaldehyde.

(Sherertz, 1992), (Fowler et al., 1992), and (Bracamonte et al., 1995) have reported ACD occurring from the wearing of durable-press fabrics. The thirteen cases reported by Sherertz were patients during the years 1988 to 1991. The nineteen cases reported by (Fowler et al., 1992) were patients during the years 1988 to 1990. Patient details are not provided. The one remaining case is that of (Bracamonte et al., 1995). The specific durable-press resin or specific residual chemicals on the fabrics worn by the patients was not identified by any of the dermatologists. The fabrics were not tested for amount of formaldehyde release.

Textile finishing compounds and resins included in the Chemotechnique and/or Hermal textile series have been used to patch test patients to determine incidence in a general eczematous population (Sherertz, 1992; Fowler et al., 1992) and in an eczematous textile worker population (Soni and Sherertz, 1996). Prevalence of textile-formaldehyde resin dermatitis was calculated to be 2.3% of eczematous patients using the Sherertz data and 1.2% of eczematous patients using the Fowler et al. data. Six of the patients tested by (Soni and Sherertz, 1996) patch tested positive to formaldehyde, and five to textile resins (3 of the 5 to dimethylol propylene urea, 2 of 5 to tetramethylol acetylene diurea, one of 5 to urea formaldehyde, and 1 of 5 to p-tert-butylphenol formaldehyde resin).

Flame Retardants

Flame retardants are chemicals added to fabric to change burning behavior, usually to make the fabric more resistant to ignition, to burn slower, and/or to self-extinguish when the flame source is removed. (Rietschel and Fowler, 1995) provide a listing of the major flame-retardant chemicals used to finish fabrics.

In the United States today, flame retardants are rarely used as a means to make consumer apparel fabrics flame resistant (meaning make a fabric capable of passing mandatory or voluntary standards for burning behavior). Children's sleepwear fabrics, all of which must pass mandatory standards, are usually engineered to pass the flame resistance tests by appropriate selection of fiber and careful yarn and fabric constructions, rather than by the addition of flame retardant chemicals to the fabric.

Flame retardants remain a viable means to make fabrics for protective garments worn by firefighters, race car drivers, and others who may be exposed to heat and flame while working. When protective garments are worn, the fabric of the protective garment will have limited contact with the wearer's skin, as several layers of undergarments are worn. Flame-retardant chemicals are also widely used in finishing fabrics for tents, but again these fabrics rarely come into direct contact with human skin other than during tent set up and take-down and during the sewing of the fabric into the tent.

(Guinnepain and Bombaron, 1990), (Moreau et al., 1994), and (Belsito, 1990) report cases of ACD to various flame-retardant chemicals. In all cases, the exposure was to the chemical finish during wearing of the fabric. Interestingly, cotton hats treated with flame retardant were the offending garment. Belsito discovered that the chemical finish was present in excessive concentration. Those hats were recalled by the manufacturer.

Biocides

Biocides are used primarily to prevent degradation of fabric by micro-organisms. (Kawai et al., 1993) and (Podmore, 1998) diagnosed textile mill workers as having ACD to a biocide sold under different tradenames. The biocide came into direct contact with the textile worker's skin, mainly their hands.

Other

Other chemical-finishing compounds reported to be allergens were 1,6 diisocyana-tohexane (Wilkinson et al., 1991), Tinofix S (DeGroot and Gerkens, 1989), and dicyclohexylmethane-4-4'-diisocyanate (Thompson and Belsito, 1997). In all cases, exposure was direct in a textile finishing mill.

Degradation Products

Degradation products are those compounds that result from the decomposition of colorants and chemical finishes on the fabric which subsequently transfer to the skin. Bleaching of fabric is one potential process that can degrade compounds on fabric.

(Kojima and Momma, 1989; Kojima et al., 1990) provides two reports in which the degradation product of Pigment Yellow 16 caused ACD for 12 males who wore yellow sweaters. The pigment was degraded in bleaching. Additionally, most patients seen by (Sherertz, 1992) and (Fowler et al., 1992) with suspected ACD to durable-press treated fabrics patch tested positive to formaldehyde. They did not determine how much formaldehyde was being leased by the fabrics worn by their patients.

Azoic Components

Azoic components are those chemicals included in a dye bath when azoic dyes will be developed in situ within fibers, usually cotton fibers. The bath usually contains a coupling agent and an azoic diazo component. (Hayakawa et al., 1985), (Osawa et al., 1997), and (Roed-Petersen et al., 1990) identified Naphthol AS as the contact allergen causing ACD for seven patients. (Sano et al., 1994) reported that two different azoic diazo compounds caused their patient's skin to react.

Miscellaneous Chemicals

(Tarvainen et al., 1998) and (Goossens et al., 1998) report that a lamination glue and a printing auxiliary compound, respectively, were the cause of ACD. In each case, one textile/apparel worker was diagnosed.

Also to be noted are two studies (Matthies and Krächter, 1993; Rodriquez et al., 1994) in which potential chemicals were studied and reported to not be allergic, nor irritant, contact allergens. The compounds were liquid fabric softeners and detergent residues.

Dyes

The two types of chemicals that are used to create colored fibers/fabrics are pigments and dyes. Pigments, inert materials, are adhered to the base substrate (fibers) by binders without in themselves reacting with the substrate. Pigments may be added to fabrics in dyeing processes but more commonly are added in printing processes. Pigment printing is the application of colored pastes consisting of binders, thickeners and specialized chemicals, and the pigments to fabrics. In contrast, dyes are large organic molecules that have the ability to selectively absorb and reflect light. They usually are added to textiles by submersion of the textile (loose fiber, yarn, fabric, or garment) into an aqueous dyebath in processes known as dyeing. Dyes can also be used to print colored designs onto a fabric's surface when prepared as print pastes. Dyes may migrate into the fiber and/or position themselves on or near the fiber surface. Only reactive dyes form covalent bonds with fiber polymers. Other dyes are physically entrapped to various degrees within the fiber or attached to the fiber surface.

Even though exposure to both pigments and dyes is high, it is dyes which have a record of reports about their allergenicity. Those disperse dyes reported to be allergens are listed in Table 16.2, those nondisperse dyes with C. I. names reported to be contact allergens are listed in Table 16.3, and those nondisperse dyes without C. I. names are listed in Table 16.4. There is no table in which pigment contact allergens are listed because only one has been implicated: Pigment Yellow 54, a pigment that degraded upon bleaching and produced the hapten causing an outbreak of ACD in Japan (Kojima and Momma, 1989; Kojima et al., 1990).

Disperse Dyes

Thirty-two disperse dyes have been reported to be contact allergens in case reports and prevalence studies (Table 16.2). By chemical class, 18 are azo, 8 are anthraquinone, 2 nitrodiphenylamine, 2 indigoid, and 2 quinophthalone. By color, 7 are red dyes, 4 are orange, 8 are yellow, 10 are blue, 1 is brown, and 2 are black. Therefore, 78% are blues (10), yellows (8), and reds (7).

The disperse dyes reported most often to be contact allergens, based on case reports, are Disperse Red 1 (20 times) and Disperse Red 17 (13 times), Disperse Orange 1 (6 times), Disperse Orange 3 (17 times), Disperse Yellow 3 (23 times), Disperse Yellow 9 (7 times), Disperse Blue 3 (6 times), Disperse Blue 35 (13 times), Disperse Blue 106 (16 times), Disperse Blue 124 (13 times), Disperse Brown 1 (6 times), and Disperse Black 1 (8 times). Each of these dyes was also used in one or more prevalence studies in which at least one patient was patch-test positive to the dye (Balato et al., 1990; Dooms-Goossens, 1992; Goncalo et al., 1992; Lodi et al., 1998; Manzini et al., 1991; Seidenari et al., 1991; Seidenari et al., 1997; Sherertz, 1992; Soni and Sherertz, 1996). In fact, prevalence values are available for 53% (17 of 32) of the disperse dyes. Prevalence varies from dye to dye and for each dye as the type of population in the studies differed; however, prevalence has not been reported to be above 4% of any eczematous population.

The disperse dyes whose record consists of only one patient being diagnosed as allergic to the dye are Disperse Reds 19, 137, and 153; Disperse Yellows 4 and 64; and Disperse Blues 26, 102. None of these seven dyes have been used in a prevalence study.

A dye that is not listed in Table 16.1, but might be in the future, is Disperse Blue 14. This dye was used in a quantitative structure-activity relationship study by (Hatch and Magee, 1998) and was predicted to be an allergic contact sensitizer. There are no published reports of its allergenic nature in case reports or prevalence studies.

Nondisperse Dyes with C.I. Names

Thirty-three nondisperse dyes have been reported to be allergic contact allergens in case reports and prevalence studies (Balato et al., 1990; Dooms-Goossens, 1992; Goncalo et al., 1992; Manzini et al., 1991; Manzini et al., 1996; Manzini et al., 1996; Seidenari et al., 1991; Seidenari et al., 1995) (Table 16.3). By application class, 8 are acid, 4 basic, 2 direct, 15 reactive, 3 solvent, and 1 vat. By chemical class, 20 are azo, 3 anthraquinone, 1 azine, 1 dibenzanthrone, 1 phthalocyanine, 1 triarylmethane, 1 stilbene, and 5 unpublished. By color, 9 are red, 6 yellow, 6 blue, 4 black, 3 orange, 2 violet, 1 brown, 1 turquoise, 1 green. Therefore, 63.6% of the nondisperse dyes are red, yellow, or blue.

The nondisperse dyes most often reported to be allergic contact allergens are Acid Black 48 (4 times), Acid Yellow 36 (2 times), Basic Red 46 (2 times), and Reactive Black 5 (2 times). Each of these dyes has been included in a prevalence study with one positive patch test resulting. Ten dyes are included based solely on one case report/one patient. Two dyes, Reactive Reds 238 and 244, are included based on one positive patch test in a prevalence study. The remaining 20 dyes have intermediate records of allergenicity.

Prevalence values are available for 51.5% (17 of 33) of the nondisperse dyes. Prevalence rates provided are less than 3%.

Dyes Without C.I. Names

Information about 14 dyes reported to allergens in case reports and prevalence studies was not sufficient to determine a C.I. name for the dye. In Table 16.4, these dyes are named as they are in the case report or prevalence study.

THE ALLERGEN INVESTIGATIONS

Throughout this chapter we have been careful to say that "the dyes and chemicals have been reported to be contact allergens" rather then saying "the chemicals and dyes listed are allergic contact allergens." The two reasons for this cautious language, limited evidence of allergenicity for many chemicals and dyes and colorant identify and purity problems, are discussed in this section. Also discussed are two recent techniques for predicting the allergenicity of dyes.

Weakness of Some Records

A first reason to speak carefully about number of contact allergens is that the records for many chemicals and dyes listed in Tables 16.1–16.4 consist of only one case report about one patient who was patch-test positive to the chemical or dye. As stated earlier, all that was required for a chemical or dye to be listed was either one case report in the literature in which a dermatologist concluded that the compounds was an allergen or that the compound was used in a prevalence study in which at least one positive patch test was obtained. In Table 16.1, where textile chemicals reported to be allergens are listed, 9 of the 23 (39%) records show one case report about one patient. In Table 16.2, where disperse dyes are listed, 7 of the 32 (21.9%) records consist of one case report about one patient. In Table 16.3, where the nondisperse dyes with C.I. names are listed, 8 of the 34 (24%) records consist of one case report about one patient. Additionally, 2 of the 34 (5.9%) records consist of one positive patch test in a prevalence study. So, in total, 26 of 102 (25.5%) textile chemicals and dyes listed have minimal records to support an argument that they are contact allergens of textile origin.

Colorant Identity and Purity Questions

A second reason to speak carefully about number of contact allergens and specific allergens arises from patch test procedural oversights. In most cases, the investigator failed to check the identity of the dye used in the patch and/or the "purity" of the dye.

What was discovered in exploring journal articles beyond the textile dermatitis case reports and prevalence studies were reports of problems with dye purity (in general and in commercial patch testing products), with dyes sold with identical Colour Index names being the same formulation, and correct labeling of dyes.

In 1984, (Brandle et al., 1984) extracted Disperse Orange 3, Disperse Yellow 3, and para-aminoazobenzene from the petrolatum in which they were mixed by Trolle-Lassen (Norway), a supplier of textile dyes for patch testing. The extracted dyes were then analyzed by thin layer chromatography (TLC). While para-aminoazobenzene and Disperse Yellow 3 were pure (one spot appeared on the chromatogram for each), the Disperse Orange 3 dye was not pure, as two spots appeared.

Further, (Foussereau and Dallara, 1986) in 1986 selected 12 disperse dyes included in the Chemotechnique series of dye allergens (Disperse Reds 1 and 17, Disperse

Oranges 1, 3, and 13, Disperse Yellows 3 and 9, Disperse Blues 3, 35, 85, and 153, and Disperse Brown 1). They extracted each dye from the petrolatum in which it was mixed and then analyzed each extracted dye using TLC. The five dyes that were pure and had only one spot on the chromatogram were Disperse Oranges 1, 3, and 13, Disperse Yellow 3, and Disperse Blue 153. The chromatogram for Disperse Red 17 and Disperse Blue 3 showed these dyes as composed of four components, and Disperse Red 1, Disperse Yellow 9, Disperse Blue 35 and 85, and Disperse Brown 1 as composed of three components.

Additionally, Foussereau and Dallara (1986) obtained dye directly from dye manufacturers and used TLC to analyze the dyes. One sample of Disperse Orange 1, Disperse Blue 35, 85, and 153 were obtained, three samples of Disperse Yellow 3 and Disperse Orange 3 were obtained from different manufacturers, and 2 samples were obtained from different manufacturers for the remaining 5 dyes. Identical chromatogram results were obtained for seven of the dyes: Disperse Red 1 (3 graphs); Disperse Orange 1, Disperse Yellow 3 (4 graphs); Disperse Blue 3 (3 graphs); Disperse Blue 85 (2 graphs); Disperse Blue 153 (2 graphs); and Disperse Brown 1 (3 graphs). Chromatograms differed for Disperse Blue 35 (the 2 graphs from each other), Disperse Orange 3 (one manufacturer's sample from the other three samples), and Disperse Yellow 9 (one manufacturer's sample from the Chemotechnique sample and a second manufacturer's sample).

In 1996, (Bide and McConnell, 1996) reported results of their investigation into variation in dye composition. They stated that "Dye is a product which can vary considerably. Little opportunity arises for significant purification of these products, and indeed, a pure product can be a disadvantage [to the dyehouse]." Sources of variation included batch-to-batch variation and supplier-to-supplier variation due to manufacture of the dye, to addition of compounds that alter shade, and to the addition of diluent that alters dye strength. Thin-layer chromatograms of four disperse dyes obtained from various dye manufacturers included in their article illustrate that dyes are not pure (not comprised of one compound), vary in formulation/composition, and may not be correctly labeled. All five chromatograms for Disperse Blue 3 they showed were different, and each was comprised of multiple spots. The two chromatograms for Disperse Blue 60 were different. One sample was pure but the other was not. The chromatograms for a disperse black mixture dye (Colour Index name not given) were also not identical. In the case of Disperse Violet 1, one sample was not pure and the other sample, while pure, was not Disperse Violet 1 but rather Disperse Violet 27. The sample was mislabeled.

New Predictive Techniques

(Ikarashi et al., 1996) developed and used a sensitive mouse lymph node assay (SLNA) to predict the sensitization capacity of 11 disperse textile dyes (and 9 cosmetic and food dyes as well). This method is particularly appropriate to studying dyes because it is said to be capable of detecting contact allergy to colored compounds that have weak allergenic potential. In this method the sensitization potential is determined by assessing total lymph node activation response induced by application of the chemical rather than by visual assessment of erythema and/or edema induced following challenge to the skin. The difficulty with visual assessment of colored chemicals is that the challenge site is colored by the chemical, thus making it difficult to detect erythema. Further, because many dyes are weak sensitizers, repeated exposure is required to induce an allergic response; a more sensitive test was needed to detect their allergenic nature.

The textile dyes used by (Ikarashi et al., 1996) were Disperse Reds 1, 11, and 17, Disperse Orange 3, Disperse Yellows 3 and 54, and Disperse Blues 1, 3, 7, 35, and a mixture of Blues 106 and 124. All of these dyes have been reported to be contact allergens. A sample of each dye was obtained from various Japanese dye manufacturers and tested, unfortunately without purification nor identity verification. An equivocal result was obtained for Disperse Red 1. The test predicted that all the other dyes were allergic contact allergens except for Disperse Reds 11 and 17 and Disperse Blues 3 and 7.

Another technique that might prove useful for predicting or verifying the allergenicity of dyes is that of quantum mechanical modeling. This technique seeks to identify whether the chemical has the molecular parameters favorable for absorption into the skin and subsequent reaction with cell surface proteins of the epidermal Langerhans' cells to mark the beginning of the ACD immune response. When used by (Magee et al., 1994) to investigate the model's discriminating power for 72 compounds (half of which were known contact allergens and half of which were compounds with no known evidence of being contact allergens), all the compounds were placed into their established categories of being haptens and nonhaptens. In the 1994 study, none of the compounds were textile dyes or other textile chemicals. In 1998, Hatch and Magee used quantum mechanical modeling to study anthraquinone dye contact allergens (Hatch and Magee, 1998).

In the anthraquinone dye study, nine anthraquinone dyes that had been reported to be contact allergens were selected, as well as 10 dyes for which no history of contact allergy was known. The nine contact allergens were Disperse Blues 1, 3, 7, 26, 27, 35, 134, 153; Reds 11 and 15, and Acid Black. The nonallergens were Disperse Blues 14, 56/2, 56/3, 72, and 73, and Disperse Reds 4, 9, 22, 60, and 86. The result of the study was that the method discriminated the contact chemicals from the noncontact chemicals with one notable exception; Disperse Blue 14. Disperse Blue 14, included as a dye without a record of contact sensitization, was predicted to be a sensitizer. Perhaps contact sensitization to Disperse Blue 14 has been missed because this dye is not part of a patch test series.

CONCLUSIONS AND RECOMMENDATIONS

Not nearly enough is known about the irritant and allergic potential of fibers and fabrics and the chemicals they contain. Researchers are just beginning to understand (a) how fabrics cause unpleasant sensations which may lead to skin damage when wearers scratch their skins to relieve the discomfort, and (b) how fabrics can be constructed to ensure that those sensations do not occur. They are just beginning to understand changes in the skin when fabric covering it slows or stops the rate of evaporation of insensible perspiration. These issues are critical to address as more and more people wear garments with great potential to hydrate the skin above its uncovered level. There are babies wearing tight-fitting diapers with cover/outer fabrics that limit the escape of water vapor, elderly persons wearing incontinence products made with nonbreathable or low breathability outer covers, firefighters, chemical clean-up crew members, pesticide applicators and other workers in impermeable protective garments, and health-care workers who must wear impermeable gloves.

Research is needed to confirm which of the chemicals and, especially, dyes reported to be contact allergens are allergens. Such confirmation might best begin with dermatology groups, including those dyes whose record of allergenicity consists of only one positive patch test in a prevalence study. Also necessary is having all dermatologists write up diagnoses of ACD, urticaria, or photoallergy which they believe due to textiles. *Contact Dermatitis, Dermatosen,* and *American Journal of Contact Dermatitis* are appropriate journals in which all dermatologists can record such investigations.

It is imperative that researchers perform at least thin-layer chromatography to establish the identity and purity of the dye of chemical compound to be used. This step is especially important when dye or compound is obtained directly from a manufacturer. A study to establish that dyes with identical Colour Index names in the three commercially available patch test diagnostic trays Chemotechnique, FIRMA, and Trolab are identical in formulation seems appropriate. Additionally, it seems reasonable that the literature contain thin-layer chromatograms of each of the dyes in a standard patch test tray.

Risk of a dye or textile chemical causing a consumer ACD has not been addressed. Some of the dyes reported to be contact allergens, the reactive dyes in particular, probably present little risk to consumers wearing fabrics colored with them because they form covalent bonds within fibers and are therefore not available for transfer to the skin. The case reports for the reactive dyes are about workers who had direct contact with the dye. The same situation applies to durable-press resins, as these resins crosslink polymers within cotton fibers. People wearing garments with properly applied and cured resins probably face little risk of dermatitis from them; however, the risk to textile workers who have direct exposure to the finishing compounds or to the resin prior to curing would be higher.

Related to the issue of risk is the issue that investigators seldom confirmed that the dye(s) or chemicals to which a person was patch test positive were a component in the garment/fabric suspected to be the offending product. Extraction of compounds from the suspected fabric is encouraged to clarify issues of risk.

Researchers are encouraged to refine methods that will predict the allergenic nature of textile chemicals. This effort might begin with further work with the SLNA and quantitative activity-structure relationship (QSAR) modeling, two methods mentioned previously in this chapter. Surely, far more work in cooperation with textile chemical, fiber, and fabric manufacturers is indicated.

REFERENCES

American Association of Textile Chemists and Colorists. (1998): *1997 Technical Manual*. Research Triangle Park, NC:AATCC.

Arisu, K., Hayakawa, R., Ogino, Y., Matsunaga, K., and Kaniwa, M.-A. (1992): Tinuvin P in a spandex tape as a cause of clothing dermatitis. *Contact Dermatitis*, 26:311–316.

Balato, N., Lembo, G., Patruno, C., and Avala, F. (1990): Prevalence of textile dye contact sensitization. *Contact Dermatitis*, 23:111–126.

Belsito, D. V. (1990): Contact dermatitis from diammonium hydrogen phosphate in surgical garb. *Contact Dermatitis*, 23:267–268.

Bide, M. J., and McConnell, B. L. (1996): In-house testing of dyes. *Textile Chem. Colorist*, 28:11–15.

Bracamonte, B. G., Ortiz-Frutos, F. J., and Diez, L. I. (1995): Occupational allergic contact dermatitis due to formaldehyde and textile finish resins. *Contact Dermatitis*, 33:139–140.

Brandle, I., Stampf, J. L., and Foussereau, J. (1984): Thin-layer chromatography study of organic dye allergens. *Contact Dermatitis*, 10:254–255.

Cameron, B. A., Brown, D. M., Dallas, M. J., and Brandt, B. (1997): Effect of natural and synthetic fibers and film and moisture content on stratum corneum hydration in an occlusive system. *Textile Res. J.*, 67:585–592.

Cavelier, C., Foussereau, J., and Tomb, R. (1988): *Allergie de Contact et Colorants*. Paris, Institut National de Recherche ed de Se'curite'.

Cottenden, A. M., Thornburn, P. H., Dean, G. E., Fader, M. J., and Brooks, R. J. (1998): Wet comfort of small disposable incontinence pads. *Textile Res. J.*, 68:479–487.

Cronin, E. (1980): *Contact Dermatitis*. Edinburgh, Churchill Livingstone.

Dallas, M. J., and Wilson, P. A. (1992): Adult incontinence products: Performance evaluation on healthy skin. *INDA J. Nonwovens Res.*, 4:26–32.

DeGroot, A. C., and Gerkens, F. (1989): Contact urticaria from a chemical textile finish. *Contact Dermatitis*, 20:63–64.

Dooms-Goossens, A. (1992): Textile dye dermatitis. *Contact Dermatitis*, 27:321–323.

Farrell-Beck, J., and Callan-Noble, E. (1998): Textiles and apparel in the etiology of skin disease. *Intern. J. Dermatol.*, 37:309–314.

Foussereau J. Clothing. (1995): In: *Textbook of Contact Dermatitis*, 2nd ed. Edited by Rycroft, R. J. G, Menne' T., Frosch P. J. Berlin, Springer-Verlag, pp. 503–514.

Foussereau, J., and Dallara, J. M. (1986): Purity of standardized textile dye allergens: A thin-layer chromatography study. *Contact Dermatitis*, 14:303–306.

Fowler, J. F., Skinner, S., and Belsito, D. V. (1992): Allergic contact dermatitis from formaldehyde resins in permanent press clothing: An underdiagnosed cause of generalized dermatitis. *J. Am. Acad. Dermatol.*, 27:962–968.

Garnesworthy, R. K., Gully, R. L., Kandiah, R. P., Kennis, P., Mayfield, R. J., and Westerman, R. A. (1988): *Understanding the causes of prickle and itch from the skin contact of fabrics (Report NO. G64)*. CSIRO Division of Wool Technology, Belmont, Australia.

Gasperini, M., Farli, M., Lombardi, P., and Sertoli, A. (1989): Contact dermatitis in the textile and garment industry. In: *Current Topics in Contact Dermatitis*. Edited by Frosch P. Berlin, Springler-Verlag. pp. 326–329.

Goncalo, S., Goncalo, M., Azenha, M. A., Barros, A., Bastos, A. S., Brandão, F. M., Faria, A., Marques, M. S. J., Pecegueiro, M., Rodriques, J. B., Salgueiro, E., and Torres, V. (1992): Allergic contact dermatitis in children. *Contact Dermatitis*, 26:112–115.

Goossens, A., Connix, D., Rommens, and Verhamme, B. (1998): Occupational dermatitis in a silk-screen maker. *Contact Dermatitis*, 39:40

Guinnepain, M. T., and Bombaron, M. (1990): Eczéma professionel a un ignifuge. *Lett. GERDA*, 7:74–76.

Gwosdow, A. R., Stevens, J. D., Bergland, L. G., and Stolwijk, J. A. J. (1986): Skin friction and fabric sensations in neutral and warm environments. *Textile Res. J.*, 54:574–580.

Hatch, K. L., and Maibach, H. I. (1985a): Textile fiber dermatitis. *Contact Dermatitis*, 12:1–11.

Hatch, K. L., and Maibach, H. I. (1985b): Textile dye dermatitis. *J. Am. Acad. Dermatol.*, 12:1079–1092.

Hatch, K. L., and Maibach, H. I. (1986): Textile chemical finish dermatitis. *Contact Dermatitis*, 14:1–13.

Hatch, K. L., and Maibach, H. I. (1994): Textile dermatitis: An update (I): Resins, additives, and fibers. *Contact Dermatitis*, 32:1–8.

Hatch, K. L., and Maibach, H. I. (1995): Textile dye dermatitis. *J. Am. Acad. Dermatol.*, 32:631–639.

Hatch, K. L., and Maibach, H. I. (1998) Textile dyes as contact allergens: Part I. *Textile Chem. Colorist*, 30(3):22–29.

Hatch, K.L., and Maibach, H. I. (1999): Dyes as contact allergens: A Comprehensive Record. *Textile Chem. Colorist and Am. Dyestuff Reporter*, 1(2):53–59.

Hatch, K. L, and Maibach, H. I. (2000): Textiles. In: *Occupational Dermatoses Handbook*. (pp. 622–636). Edited by Kanerva, L., Elsner, P., Wahlberg, J. E., Maibach H.I. Basel, Switzerland, S. Karger AG.

Hatch, K. L., Markee, N. L., and Maibach, H. I. (1992): Skin response to fabric: A review of studies and assessment methods. *Clothing Textile Res. J.*, 10:54–63.

Hatch, K. L., Markee, N. L., Maibach, H. I., Barker, R., Radhakrishnaish, P., and Woo, S. (1990): In vivo cutaneous and perceived comfort response to fabric. Part II: Blood flow and water content under garments worn by exercising subects ina hot, humid environment. *Textile Res. J.*, 60:510–519.

Hatch, K. L., Markee, N. L., Prato, H., Zeronian, H., Maibach, H. I., Kuehl, R. O., and Axelson, R. D. (1992): In vivo cutaneous and perceived comfort response to fabric. Part V: Effect of fiber type and fabric moisture content on the hydration state of human stratum corneum. *Textile Res. J.*, 66:638–647.

Hatch, K. L., Prato, H. H., Zeronian, S. H. (1997): In vivo cutaneous and perceived comfort response to fabric, Part VI: The effect of moist fabrics on stratum corneum hydration. *Textile Res. J.*, 67:926–931.

Hatch, K. L., Wilson, D., and Maibach, H. I. (1987): Fabric-caused changes in human skin: In vivo water content and water evaporation. *Textile Res. J.*, 57:583–591.

Hatch, K. L. (1984a): Chemicals and textiles, Part I: Dermatological problems related to fiber content and dyes. *Textile Res. J.*, 54:664–682.

Hatch, K. L. (1984b): Chemicals and textiles, Part II: Dermatological problems related to finishes. *Textile Res. J.*, 54:721–732.

Hatch, K. L., and Magee, P. (1998): A discriminant model for allergic contact dermatitis in anthraquinone disperse dyes. *Quant. Struct-Act. Relat.*, 17:20–26.

Hatch, K. L. (1988): Textile dermatitis from formaldehyde. In: *Formaldehyde Sensitivity and Toxicity*. Edited by Feinman S. E., Boca Raton, FL, CRC Press, pp. 91–103.

Hatch, K. L. (1995): Textile dye contact allergens. *Curr. Prob. Dermatol.*, 22:8–16.

Hayakawa, R., Matsunaga, K., Kojima, S. (1985): Pigmented contact dermatitis due to cotton flannel nightdress. *Nippon Hifuka Gakkai Zasshi*, 95:1441–1446.

Ikarashi, Y., Tsuchiya, T., Nakamura, A. (1996): Application of sensitive mouse lymph node assay for detection of contact sensitization capacity of dyes. *J. Applied Toxicology*, 16:349–354.

Inque, A., Ishido, I., Sihoh, A., Yamada, H. (1997): Textile dermatitis from silk. *Contact Dermatitis*, 37:185.

Kaniwa, M.-A., Isama, K., Kojima, S., Nakamura, A., Arisu, K., and Hayakawa, R. (1991): Chemical approach to contact dermatitis caused by household products. VIII: UV absorber Tinuvin P in polyurethane elastomers for fabric products. *Jap. J. Toxicol. Environ. Health*, 37:218–228.

Kawai, K., Nakagawa, M., Sasaki, Y., and Kawai, K. (1993): Occupational contact dermatitis from kathon 930. *Contact Dermatitis*, 28:117–118.

Kojima, S., and Momma, J. (1989): Phosgene (2,5-dichlorophenyl) hydrzone, a new strong sensitizer. *Contact Dermatitis*, 20:235–236.

Kojima, S., Monmma, J., Kaniwa, M.-A., Ikarashi, Y., Sato, M., Nakaji, Y., Kurokawa, Y., and Nakamura, A. (1990): Phosgene (chlorophenyl) hydrozones, strong sensitizers found in yellow sweaters bleached with sodium hypochlorite, defined as causative allergens for contact dermatitis by an experimental screening method in animals. *Contact Dermatitis*, 23:129–141. (published errata in *Contact Dermatitis*, 23:383.)

LaMotte, R. H. (1977): Psycophysical and neurophysical studies of tactile sensibility. In: *Clothing Comfort*. Edited by Hollies N.R.S., Goldman RF Ann Arbor MI:Ann Arbor Science, pp 83–106.

Lodi, A., Ambonati, M., Coassini, A., Chirarelli, G., Mancini, L. L., and Crosti, C. (1998): Textile dye contact dermatitis in an allergic population. *Contact Dermatitis*, 39:314–315.

Magee, P. S., Hostynek, J. J. and Maibach, H. I. (1994): A classification model for allergic contact dermatitis. *Quant. Struct-Act. Relat.*, 13:22–33.

Manzini, B. M., Seidenari, S., Danese, P., and Motolese, A. (1991): Contact sensitization to newly patch tested non-disperse textile dyes. *Contact Dermatitis*, 25:331–332.

Manzini, B. M., Motolese, A., Conti, A., Ferdani, G., and Seidenari, S. (1996): Sensitization to reactive textile dyes in patients with contact dermatitis. *Contact Dermatitis*, 34:172–175.

Manzini, B. M., Donini, M., Motolese, A., and Seidenari, S. (1996): A study of 5 newly patch-tested reactive textile dyes. *Contact Dermatitis*, 35:313.

Matthies, V. W., and Krächter, H. U. (1993): Hautreaktionen auf Berufskleidung. *Dermatosen*, 41:137–144.

Matthies, V. W., Löhr, A., and Ippen, H. (1990): Bedeutung von rückständen von texilwaschmitteln aus dermatotoxikologischer Sicht. *Dermatosen*, 38:184–185.

Matthies, V. W. (1993): Influence of pH values on the skin compatibility of cotton textiles. *Dermatosen*, 41:97–100.

Moreau, A., Dompmartin, A., Castel, B., Remond, B., Michel, M., and Leroy, D. (1994): Contact dermatitis from a textile flame retardant. *Contact Dermatitis*, 31:86–88.

Osawa, J., Takekawa, K., Onuma, S., Kitamura, K., and Ikezawa, Z. (1997): Pigmented contact dermatitis due to Naphthol AS in a pillowcase. *Contact Dermatitis*, 37:37–38.

Podmore, P. (1998): An epidemic of isothiazoline senstization in a flax spinning mill. *Contact Dermatitis*, 38:165.

Ramam, M., Khaitan, B. K., Singh, M. K., and Gupta, S. D. (1998): Frictional sweat dermatitis. *Contact Dermatitis*, 38:4.9

Redmond, S. F., and Schappert, K. R. (1987): Occupational dermatitis associated with garments. *J. Occup. Med.*, 29:243–244

Rietschel, R. L., and Fowler, Jr. J. E. (1995): Textile and shoe dermatitis. In: *Fisher's Contact Dermatitis*, 4th ed. Edited by Rietschel, R. L., Fowler J. E. Jr Baltimore MD, Williams & Wilkins, pp. 358–413.

Rodriquez, C., Daskaleros, P. A., Sauers, L. J., Innis, J. D., Laurie, R. D., and Tronnier, H. (1994): Effects of fabric softeners on the skin. *Dermatosen*, 42:58–61.

Roed-Petersen, J., Batsberg, W., and Larsen, E. (1990): Contact dermatitis from Naphthol-AS. *Contact Dermatitis*, 22:161–163.

Sano, S., Kume, A., Nishitani, N., and Higaski, N. (1994): Occupational pigmented dermatitis from raw materials of azo dyes, 2,4,5-trichloroaniline and 4-benzamide-2,5-diethoxyaniline. *Environ Dermatol*, 1:195–198.

Seidenari, S., Manzini, B. M., and Danese, P. (1991): Contact sensitization to textile dyes: Description of 100 subjects. *Contact Dermatitis*, 24:253–258.

Seidenari, S., Mantovani, L., Manzini, B. M., and Pignatti, M. (1997): Cross-sensitization between azo dyes and para-amino compound. *Contact Dermatitis*, 36:91–96.

Seidenari, S., Manzini, M., Schiavi, M. E., and Motolese, A. (1995): Prevalence of contact allergy to non-disperse azo dyes for natural fibers: A study in 1814 consecutive patients. *Contact Dermatitis*, 33:118–122.

Sherertz, E. F. (1992): Clothing dermatitis: Practical aspects for the clinician. *Am. J. Contact Dermatitis*, 3:55–64.

Soni, B. P., and Sherertz, E. F. (1996): Contact dermatitis in the textile industry: A review of 72 patients, *Am. J. Contact Dermatitis*, 7:226–230.

Storrs, F. J. (1986): Dermatitis from clothing and shoes. In: *Contact Dermatitis*, 3rd ed. Edited by Fisher, A. A. Philadelphia, Lea and Febiger., pp 283–337.

Tanaka, M., Kobayashi, S., and Miyakawa, S. (1993): Contact dermatitis from nylon 6 in Japan. *Contact Dermatitis*, 28:250.

Tarvainen, K., Jolanki, R., Eckerman, M-LH., and Estlander, T. (1998): Occupational allergic contact dermatitis from isophoronediamine (IPDA) in operative-clothing manufacture. *Contact Dermatitis*, 39:46–47.

Thompson, T., and Belsito, D. V. (1997): Allergic contact dermatitis from a diisocyanate in wool processing. *Contact Dermatitis*, 37:239.

Wilkinson, S. M., Cartwright, P. H., Armitage, J., and English, J. S. C. (1991): Allergic contact dermatitis from 1,6-diisocyanatohexane in an anti-pill finish. *Contact Dermatitis*, 25:94–96.

Wilson, E. (1869): Des stries et des macules atrophiques au fausses cicatrices de la peau. *Annuals of Dermatology and Syphology*, 141.

Toxicology of Skin
Edited by Howard I. Maibach
Copyright © 2001 Taylor & Francis

17

Novel Topical Agents for Prevention and Treatment of Allergic and Irritant Contact Dermatitis*

John J. Wille

Bioderm Technologies Incorporated, Trenton, New Jersey, USA

Agis F. Kydonieus

Samos Pharmaceuticals, LLC, Kendall Park, New Jersey, USA

INTRODUCTION

The risk of skin exposure to environmental and industrial allergens and irritants has greatly increased over the past few decades (Adams, 1983). Traditionally, the approach to contact allergens and irritants has been to avoid and reduce the risk of exposure through use of barrier creams alone or in conjunction with protective clothes,

---

*Portions of the work reviewed here were previously published by CRC Press in the book *Biochemical Modulation of Skin Reactions in Dermal and Transdermal Drug Delivery* and are reproduced here with permission.

such as latex-free gloves (Halker-Sorensen, 1996). Although, this approach has been moderately successful in reducing occupational exposure to known skin allergens and irritants, it has little or no benefit to the wearer of medical devices that must come in direct contact with intact and/or damaged skin to perform their function. This includes long-term users of ostomy products (skin barrier adhesives) and a growing number of users of transdermal drug delivery devices (Maibach, 1985). Further, there are now a considerable number of cosmetic preparations that incorporate agents such as retinoids and alpha-hydroxy acid that cause adverse skin reactions among a significant number of consumers. Finally, there are many otherwise useful therapeutic drugs that cannot be delivered dermally or transdermally because they are irritating or sensitizing to skin at concentrations required to achieve their intended effects (Fischer, 1982).

Among the current armamentarium of anti-inflammatory drugs, there are only a few topical agents that are generally accepted as useful for the prevention and/or treatment of allergic and irritant contact dermatitis. The glucocorticosteroids are by far the most widely used, and the low-potency hydrocortisone preparations are now commercially available in many over-the-counter drug products, whereas higher-potency synthetic analogs (triamcinalone acetinide, betamethasone, clobetasol, prednisone) generally are prescribed for more chronic skin disorders, such as hand eczema and atopic dermatitis (Menne and Maibach, 1994). Treatment of skin with glucocorticosteroids is not considered optimal for contact dermatitis arising from persistent and prolonged contact with the suspected or known causative agent, as it leads to profound thinning of the epidermal layer and disruption of the skin barrier, thus limiting its effective use to transient exposures.

The search for new agents to treat skin exposed to dermatotoxins (allergens, irritants, drugs) has been severely limited by lack of knowledge of how the skin, through its immune system, generates sensitization and elicitation responses and equally by ignorance of the fundamental mechanism of skin irritation. Nevertheless, recent advances have been made on the role of the skin immune system (SIS) (Bos, 1989), the role of epidermal and dermal cells in producing cytokines and responding to them in irritant- and allergen-stimulated contact dermatitis, and by studies on the mechanism of action of ultraviolet light on the SIS (Schaefer and Rougier, 1998). Consequently, new molecular details are now emerging on the cellular processes involved in irritant- and antigen-provoked contact dermatitis that have presented new opportunities to target one or more steps in the cytokine signaling cascade that coordinates cell trafficking through intra- and intercellular signaling during the sensitization and elicitation phases of allergic contact dermatitis (ACD) and, to a lesser extent, the specific molecular processes underlying contact irritation (Wakem and Gaspari, in press).

The aim of this chapter is to provide the reader with a current survey of some of the most promising new agents that inhibit irritant and allergic contact dermatitis and to present in-depth information about mast cell degranulating agents (Wille and Kydonieus, in press) and ion channel blocking agents (Wille and Kydonieus, in press) as two new classes of topical immunomodulators that show promise in the prevention and treatment of contact dermatitis. An authoritative review of the broad subject of biochemical modulations of skin reactions has recently been published (Kydonieus and Wille, in press).

SURVEY OF PATENT LITERATURE

In this section, we survey the current patent literature (1975 to present) pertinent to claimed methods and agents for the treatment of skin irritation and skin sensitization (Kydonieus and Wille, in press). It is hoped that this survey will provide a broad perspective of commercially useful methods and agents that may hold future promise in the treatment of contact dermatitis.

For clarity and simplicity, we have classified the various agents into 11 different groups based on their chemical affiliations or imputed mode of action. This scheme of classification is not meant to imply that some other grouping would not provide a more rational and predictive method of classifying known anti-irritants, countersensitizers, and the like; nor is it conceived as being embrassive of suspected or putative molecular targets underlying the mechanisms of skin irritation and sensitization, as these are yet largely unknown or speculative (Rowe and Bunker, 1998; Yawalkar et al., 1998).

Plant Extracts

Natural products, specifically plant extracts, have been used in cosmetic formulations, as well as in other topical applications, to avoid irritation, including the sensations of stinging, itching, and burning and the clinical signs of redness and peeling. Rosmarrinic acid is an anti-irritant for alpha hydroxy carboxylate compounds, such as lactic acid, glycolic acid, and hydroxycaprilic acid (Castro, 1995). The overall stinging response to a 10% lactic acid formulation was reduced by more than threefold by the addition of 5% rosmarrinic acid (as a sage extract).

Hydroalcoholic extracts of Cola nitida have also been used to prevent or treat both irritation and sensitization. Cola nitida extracts are most effective when applied half an hour before or half an hour after the application of topical cosmetic and pharmaceutical formulations or physical irritants (e.g., skin waxing) (Smith et al., 1989). A 10% Cola nitida extract formulation gave excellent results against two known irritants: para-amino benzoic acid and balsam of Peru.

The oil extracted from the yerba plant (*Eriodictyon californicum*) has also been claimed to minimize or completely eliminate the irritation and sensitization that accompanies topical, transdermal or transmucosal drug delivery. Formulations containing eriodictyon fluid extract (fluid extracted from dried yerba santa leaves) have been used to minimize irritation and sensitization of topically applied substances, such as dihydroergotamine mesylate, acetaminophen, oxymetazoline, diphenhydramine, nystatin, clindamycin, and para amino benzoic acid (Parnell, 1991).

Antipruritic plant extracts have been codelivered with nicotine (Govil and Kohlman, 1990). Nicotine transdermal patches are known to cause minor irritation and itching. Among other antipruritics, oil of chamomile, chamazulene, and bisabolol are claimed in the above-mentioned patent. Oil of chamomile is extracted from the dried flower heads of Anthemis nobilis. Chamazulene (7-ethyl-1, 4-dimethylylazulene) is an anti-inflammatory compound found in chamomile, wormwood, and yarrow.

Many studies in the scientific literature have also investigated the anti-irritant and countersensitizer effects of plant extracts. For example, pretreatment of the skin with a Ginkgo biloba extract mitigated contact dermatitis to the allergen's nickel sulfate, balsam of Peru, and methyl-isothiazolinone (Castelli et al., 1998).

Glucocorticoids

Steroids are well known for their anti-inflammatory and immunosuppressive proper-ties, and formulations containing 0.5% and 1.0% hydrocortisone are standard reme-dies for treating skin irritation and inflammatory responses caused by allergies such as poison oak (Cupps and Fauci, 1982); however, their ability to prevent irritation and the induction of sensitization had not been studied until recently. A series of patents by Alza studied the effect of steroids on the induction phase of sensitization for both intact epidermal skin and mucosal membranes (Amkraut and Shaw, 1991a, 1991b). Coadministration, as well as pretreatment with hydrocortisone, were studied with positive results. Steroids, as well as steroid esters, were found to be suitable; they include hydrocortisone, hydrocortisone ester, betamethasone, betamethasone valerate, fluocinonide, and triamcinolone. In a human test on 80 volunteers (40 sub-jects, each in the control and experimental groups), dex-chlorpheniramine maleate administered transdermally sensitized 16 subjects. When it was codelivered with 2% hydrocortisone, only two subjects were sensitized. In a separate experiment, it was shown that hydrocortisone did not affect the elicitation phase of sensitization (i.e., once sensitized, the presence of hydrocortisone did not prevent a skin reaction in allergic individuals). Two other Alza patents pertain to induction of skin immune tolerance to a sensitizing drug when the drug is delivered both transdermally or through the mucosal membrane (Amkraut, 1991, 1992). Skin immune tolerance is the state of prolonged unresponsiveness to a specific antigen when the antigen is repeatedly applied to the skin. Tolerance to skin sensitization has been induced with repeated doses of UV irradiation and by application to the skin of agents that sup-press the immune system prior to the application of the sensitizer (Cruz et al., 1988; Rheins and Nordlund, 1986). Alza's patents disclose the induction of tolerance in humans to sensitizing drugs such as dex-chlorpheniramine maleate (DCPM) through its coextensive coadministration with hydrocortisone. None of the subjects treated with hydrocortisone experienced sensitization reactions to DCPM, versus 15% for the control group. Drugs that could be delivered transdermally, due to the devel-opment of immune tolerance, included naloxone, clonidine, tetracaine, and scopo-lamine.

Medium potency steroids have been employed to minimize the irritation and sensiti-zation caused by several transdermally administered drugs, such as guanfacine, cloni-dine, fluphenazine, trifluoperazine, and timolol (Franz et al., 1991). Fluocinonide, a medium-potency steroid, was coadministered from polyvinyl chloride patches. From patches containing 10% clonidine, the codelivery of fluocinonide reduced significantly erythema and blistering caused by clonidine to an individual previously sensitized to clonidine. In a separate experiment with patches containing 35% fluphenazine, the codelivery of 1% fluocinonide almost completely blocked the erythema and blistering caused by fluphenazine to an individual previously sensitized to fluphenazine. It is of interest to note that the results from patches codelivering 5% fluocinonide were not as good as those codelivering 1%. This may not be too surprising considering that steroids not only suppress, but also direct and enhance immune functions (Wilckens and DeRijk, 1997). Irritant contact dermatitis was also reduced by the use of fluocinonide. Four polyvinyl chloride patches containing the irritant antihypertensive (5Z, 13E, 15R, 16R)-16-fluoro-15-hydroxy-9-osoprosta-5-dienoic acid (RO 22-1327) and different concentrations of fluocinonide were tested on a human subject not previously exposed to this drug. The incorporation of 1% fluocinonide in the patch resulted in significant reduction of irritant contact dermatitis.

Synthetic glucocorticosteroids, such as glucocorticoid carboxylic acid esters, have been reported to suppress cutaneous delayed hypersensitivity (Ross, 1988, 1990). The advantage of the carboxylic acid esters is that, in contrast to steroids, they do not cause epidermal atrophy and other adverse systemic effects, such as suppression of plasma B. The glucocorticoid carboxylic acid esters and, more specifically, triamcinolone acetonide 21-oic methyl ester at a permeation rate of 100 μg per 24 hours can specifically, temporarily, and locally suppress the cutaneous immune system to the coadministered allergen. The carboxylic acid esters were shown to prevent allergic reactions and also the induction phase of sensitization, which must precede the allergic state. Moreover, their effects were shown to be confined to the site of application so that other areas of the body surface remained armed.

Steroid treatment has been reported for the prevention of pruritis caused by the transdermal administration of Nicotine and the prevention of irritation in the delivery of polypeptides through mucosal membranes (Govil and Kohlman, 1990; Wang et al., 1990).

Immunosuppressive Agents

Recently, favorable clinical results in the treatment of atopic dermatitis were reported for the topical use of macrolide antibiotics (e.g., tacrolimus (FK 506)) (Ruzicka et al., 1997).

The macrolide drugs, such as cyclosporins, rapamycins, FK-506 derivatives, and combinations thereof, are claimed to affect site-specific immune suppression of local inflammatory/immune responses in mammalian tissue and may be useful for the treatment of many autoimmune and T-cell–mediated diseases, including inhibition of contact hypersensitivity (Hewitt and Black, 1996). The major advantage of the topical delivery is that it eliminates the concerns of systemic administration of immunosuppressants with its many side effects (i.e., kidney and liver damage). Cyclosporin A and hydrocortisone produce a synergistic combination to attack multiple disease mechanisms at a local site-specific level.

Local Anesthetics

Topical anesthetics generally are believed to disrupt membrane organization and affect the lipid occupancy by charged molecules. They also could be considered ion channel modulators because their action is probably similar in nature to that of diuretics. Local anesthetics block conduction of the nerve impulse by preventing the large transient increase in the permeability of excitable membranes to Na^+ ions. This action is due to the direct interaction of the anesthetic with the voltage-sensitive Na^+ ion channels. Local anesthetics can also bind to other membrane-bound proteins, and they can block K^+ ion channels as well (Richie and Greene, 1993).

Topical anesthetics, in concentrations up to 25%, have been used in formulations containing capsaicin to inhibit the local topical irritant effect of capsaicin (Bernstein, 1991). The formulations containing up to 1% capsaicin were used to treat superficial pain syndromes. Benzocaine and lidocaine were the local anesthetics of choice. The local anesthetics lidocaine, benzocaine, lignocaine, methocaine, butylaminobenzoate, and procaine have also been disclosed as anti-irritants in transdermal formulations containing nicotine (Bannon et al., 1994). The once-daily transdermal formulations containing nicotine are being used to treat symptoms associated with tobacco smoking cessation.

Emollients and Anti-Irritants

Glycerin has been disclosed as an anti-irritant for reducing the skin irritation of transdermal drug/enhancer compositions (Patel and Ebert, 1989). The method involves treating the skin with glycerin prior to, or concurrently with, the administration of the drug/enhancer composition. It is effective in reducing the irritation caused by drugs such as pindolol and enhancers such as ethanol, propylene glycol, dimethyl sulfoxide, and Azone.

Aluminum chlorhydrate has also been shown to prevent, as well as treat, allergic contact dermatitis caused by urushiol oil, poison ivy, or poison oak (Waali, 1987). Treatment with an aerosol solution of 25% aluminum chlorhydrate was effective in relieving the symptoms in the infected area and preventing further spread of the dermatitis.

A method for preventing or treating irritation or itching of the skin is disclosed where the anti-irritant is 1-[3-4-(diphenylmethyl)-1-piperazinyl propyl]-1H-benzimidazole (Cauwenbergh, 1995). It is recommended for use in cosmetics, including sunscreens, anti-wrinkle products, shampoos, shaving creams, diaper rash, face masks, deodorants, and aftershaves.

Topical compositions containing capsazepine have been shown to be suitable for treating neurogenic skin disorders and diseases (De Lacharriere and Breton, 1997).

In particular, the compositions are useful for preventing and/or controlling skin and eye irritation, itching, and erythema, as well as for reducing the irritancy of an active substance having an irritant side effect. Actives with irritant side effects include alpha and beta hydroxyacids, beta ketoacids, retinoids, anthralins, anthranoids, peroxides, minoxidil, lithium salts, vitamin D, surfactants, reducing and oxidizing agents, and strong acids and bases.

Antagonists of CGRP (calcitonin gene-related peptide) have been reported to prevent and treat irritation and sensitization (Breton, 1996). Compounds disclosed as anti-irritants include CGRP 8–37 and antibodies of anti-CGRP. These compounds have been used successfully in cosmetic, pharmaceutical, and dermatological compositions.

A method has been claimed for the treatment of allergic contact dermatitis, which comprises treating a patient with a formulation capable of inducing oxidative stress and a heat shock response so as to convert the allergic reaction of the allergic contact dermatitis to an irritant reaction (Schmidt, 1996). Oxidative stress is induced, preferably by hydrogen peroxide, through a polymeric precursor such as chitin, gelatin, pectin, and sodium carboxymethylcellulose.

Use of superoxide dismutase (SOD) as an agent for the protection of the skin against inflammatory reactions associated with chemical irritation have also been disclosed in the patent literature (Wilder, 1986). SOD is an oxygen-free radical scavenger that had been previously shown to suppress carrageenan-induced inflammation in the rat (Huber and Saifer, 1977). It is claimed that SOD eliminates the superoxide anion radicals, hydrogen peroxide and hydroxyl radical formation, and thus reduces the damage to cells and irritation from diverse classes of chemical irritants including inorganic and organic peroxides, acids, and bases. The formulation containing the SOD, from 1–1000 CIU/cm^2 (CIU is defined as the cytochrome inhibition unit), is applied topically 5–10 minutes before exposure to chemical irritants. Examples with sodium lauryl sulfate, lauric acid, cinnamaldehyde, benzoyl peroxide, sodium hydroxide, and eugenol are presented.

Arachidonic Acid Cascade Modulators

A number of inhibitors of the arachidonic acid cascade are claimed to prevent contact dermatitis associated with transdermal delivery (Franz et al.,1988). Among the agents that inhibit arachidonic acid cascade are the key enzymes in the process (e.g., phospholipase A_2, 5-lipoxygenase, 12-lipoxygenase, or cyclo-oxygenase) or they block the end response (e.g., receptor antagonists). These are claimed to be effective in treating contact dermatitis, both irritant or allergic. Because the transformation of the arachidonic acid to its metabolic products (leucotrienes, prostaglandins), which are responsible for the inflammatory reaction, is an enzyme-catalyzed free radical oxidative process, free radical scavengers could be used to reduce the inflammatory reactions. Such free radical scavengers disclosed include hindered phenols, such as 2,6-ditertiarybutyl-para-cresol (BHT), p-tertiarybutyl catechol, hydroquinone, benzoquinone, and N,N-diethylhydroxyamine. Materials that compete with the ara-chidonic acid biotransformation are also disclosed as anti-irritants because they allow less arachidonic acid to be transformed into irritating products, such as leucotrienes and prostaglandins with four double bonds. Chemicals that compete with arachi-donic acid, and which bi-products are not irritating, include eicosatetraenoic acid, eicosapentanoic acid, and dihomolinoleic acid.

Omega-9-unsaturated fatty acids and, especially, the cis isomers have been shown to prevent or alleviate the inflammatory symptoms of allergic contact dermatitis (Akimoto et al., 1995). Previous work had shown that synthesis of leucotriene B_4 is inhibited when rats were fed omega-9-unsaturated fatty acids. The inventors stud-ied several unsaturated fatty acids and found the omega-9 gave the best suppression of delayed allergy reactions. The effective omega-9-unsaturated fatty acids had a minimum of 2 double bonds and 18–22 carbon atoms. Specific fatty acids claimed include 6,9-octadecadienoic acid, 8,11-eicosadienoic acid, and 5,8,11-eicosatrienoic acid. Mice fed 5,8,11-eicosatrienoic acid for 1 week showed a 50% reduction of footpad swelling over the control mice sensitized to sheep red blood cells.

Lipid Mediators

Fatty acids have been reported to reduce irritation caused by aqueous detergents, in-cluding anionic, cationic, nonionic, and amphoteric organic detergents. Improvement was obtained by incorporating in the formulation up to 10% of an additive comprising the polymerized product of 2–4 molecules of a monomeric C_{12}–C_{26} fatty acid, said additive containing 2–4 carboxyl, or carboxyl salt groups (Kelly and Ritter, 1970). It was proposed that the fatty acid additives stabilize the keratin layer disrupted by detergent-induced keratin protein denaturation and thereby protect the living tissue. The preferred lipid additives are polymerized ethylenically unsaturated fatty acids into dimers, trimers, and tetramers, which result in a cyclo aliphatic ring structure. The only lipid modifier shown in the examples was the dimer of linoleic acid, and it was tested with several detergents, including sodium lauryl sulphate, alkyl benzene sulfonate, linear alkane sulfonate, and N-coco-β-aminopropionic acid. Similar lipid modifiers were used to reduce the inflammatory response of allergic contact dermati-tis during challenge. Lipid modifiers, which are organic compounds having at least two polar groups (e.g., carboxyl groups) separated by a chain of at least 15 atoms, the majority of which are carbon atoms and containing a cyclic moiety of at least 5 atoms, were applied prior to contact of the skin with the allergenic agent (Willer et al., 1978). Lipid modifiers used in the examples were bis(hydroxyethyl) dimerate (esterification of dimerized linoleic acid with ethylene glycol) and bis(triethanolamine salt) dimerate.

Cis and trans traumatic acid salts have been claimed for the treatment of many skin diseases, including itching and allergic dermatitis (Della Valle and Marcolongo, 1993). Traumatic acid or 10-dodecendioic acid is a linear long chain carboxylic acid having 12 carbon atoms and an unsaturated bond. The anti-irritant activity of traumatic acid is claimed to be due to the carboxylic group distribution along the carbon chain (Goldberg et al., 1977).

The topical application of cis-9-heptadecenoic acid has been shown to be useful in the prophylaxis and treatment of skin diseases, including contact allergies (Degwert et al., 1998). In vitro tests showed that cis-9-heptadecenoic acid suppresses the release of important triggering mediators, such as tumor necrosis factor (TNF)-α, which is important in the prophylaxis and treatment of skin diseases, including allergies. Heptadecenoic acid also showed significant inhibition of mitogen-induced proliferation of lymphocytes, and therefore should also have anti-inflammatory activity by virtue of its immunomodulating potency.

Sphingosine and other aliphatic aminodiol lipids have been shown to provide long-term, anti-irritant activity when formulated into cosmetic compositions containing irritants such as lactic, hydroxybenzoic, or retinoic acids (Herstein, 1995). Comparative data show a beneficial control of long-term irritation induced by several weeks of daily use of a variety of skin-renewal stimulating acids, as mentioned above. In contrast to the aliphatic aminodiol lipids, the incorporation into the formulations of other skin lipids, such as phospholipids, cerebrosides, and ceramides, had little activity in controlling long-term irritation. In addition to sphingosine, phytosphinganine and dihydrosphingosine are claimed as anti-irritants together with their analogues, homologues and derivatives.

Antioxidants

Antioxidants and free radical scavengers have been considered in the patent literature for eliminating or minimizing irritation and sensitization in transdermal delivery. European Patent EP 0314528A1 claims that irritation mechanisms involve inflammatory reaction of skin caused by the metabolic products of arachidonic acid transformations (leucotrienes, 12-hydroxyeicosatetraenoic acid, prostaglandins) (Franz et al., 1988). The enzymatic transformations (lipoxygenase and cyclo-oxygenase) are catalyzed by free radical oxidative processes that can be minimized by the presence in the skin of antioxidants and free radical scavengers. Some free radical scavengers mentioned include hindered phenols such as 2,6-ditertiarybutyl-para-cresol (BHT), p-tertiarybutyl catechol, hydroquinone, benzoquinone, and N,N-diethylhydroxyamine. Vitamin E and nordihydroguaiaretic acid are also disclosed as antioxidants that minimize irritation when used as pretreaments or codelivered with the transdermal drug.

Ascorbic acid (vitamin C) has also been found to decrease skin irritation caused by topical administration of active ingredients. Ascorbic acid, provided in topical application in as low as 5% concentration from a separate solution or by admixture with a cosmetically acceptable vehicle, is effective in reducing the irritation caused by many cosmetic and pharmaceutical ingredients (Duffy, 1996). Examples of such ingredients include alpha hydroxy acids, benzoyl peroxide, retinol (vitamin A), retinoic acid, quaternary ammonium lactates, and salicylic acid.

Vitamin E has been claimed to protect skin and reduce irritation caused by actives and, more specifically, benzoyl peroxide for the treatment of acne and other skin lesions (Hall et al., 1996; Roshdy, 1987). Tocopherol acetate, tocopherol sorbate, tocopherol linoleate, and mixtures thereof are disclosed as useful vitamin E analogues. Alpha tocopherol has also been shown to reduce irritation (Nagy et al., 1993; Roshdy, 1987). Compositions for topical applications containing up to 5% alpha tocopherol reduced irritation caused by retinoic acid. Topical alpha tocopherol was also shown to reduce, in a dose response fashion, the irritation caused by repeated applications of retinoic acid. Inflammatory response was measured by MPO activity (MPO is a neutrophil enzyme marker), which is directly proportional to the increase in neutrophils. Inhibition levels of 20% and 70% were observed with 0.01% and 5% concentrations of alpha tocopherol, respectively.

Methyl nicotinate has also been shown to reduce irritation of transdermally administered drugs, as well as reduce the elicitation response of sensitization (Cormiez et al., 1995). The methyl nicotinate is coadministered with the irritant or sensitizer at preferred concentrations of 0.5% to 5%. Methyl nicotinate was tested against the irritant chloroquine and the sensitizers propranolol, ketoprofen, and tetracaine.

In the case of ketoprofen, 2% hydroxyethylcellulose alcoholic gels were prepared and placed in Finn chambers and applied on the backs of 10 hairless guinea pigs. After 24 hours, the chambers were removed and the intensity of the reaction was scored for edema and erythema at 24 and 48 hours. The average visual score for erythema and edema was greater at 24 and 48 hours for the formulations that did not contain methyl nicotinate.

Panthenol and Derivatives

Panthenol and its analogues and derivatives, such as pantothenic acid, pantetheine, and pantethine, have been claimed as anti-irritants for formulations containing up to 20% of benzoyl peroxide (Hewitt and Black, 1996). The formulations are gels that exhibit low skin irritation and are useful for topical application to human skin for the treatment of acne and other skin lesions. D-panthenol was disclosed as an anti-pruritic in formulations containing nicotine for the treatment of symptoms associated with tobacco smoking cessation (Govil, 1990). Dexpanthenol has also been claimed as an enhancer/anti-irritant useful with many drug families, including opioids, calcium antagonists, beta blockers, and antihypertensives (Kolter, 1990). Specific drugs claimed include hydromorphon, biperiden, and gallopamil. Several formulations containing 15% gallopamil and 15% dexpanthenol were shown to produce substantially less irritation than the control formulations without the dexpanthenol.

Reduction of Electrotransport Irritation

Iontophoresis and electroporation are two processes studied extensively to increase the permeation of drugs through intact skin. By applying electric current through the skin, the irritation issues are amplified in comparison to those of passive transdermal delivery (Ledger, 1992). World Patent WO 9506497 discloses an electrotransport device with a cathodic reservoir containing the agent to be delivered and an anodic reservoir and an electrical power source to apply a voltage across the reservoirs. It is claimed that when the anodic reservoir is maintained at a pH above 4 (4–10) and the cathodic reservoir at a pH below 4 (2–4), the irritation is minimized (Cormiez et al., 1995). Methods have been described for reducing the sensations (pain) during iontophoretic drug delivery by delivering the therapeutic agent with a multivalent ion (Phipps, 1993). Their experimental work indicated that the preferred multivalent ions to mitigate the iontophoretic sensations were calcium, magnesium, phosphate, and zinc. At high iontophoretic current density, $600 \ \mu A/cm^2$, the ranking from best to worst was calcium, phosphate, chloride, acetate, citrate, sulfate, potassium, sodium.

Metabolic Modulators/Lysosome Modifiers

Methods and devices were described for the reduction of sensitization and irritation caused by transdermally delivered drugs, wherein one or more metabolic modulators is coadministered with the sensitizing or irritating drug (Cormiez et al., 1989, 1994). Metabolic modulators affect the enzymes in the skin and thus modify the metabolism of the drug so as to inhibit the formation of reactive or irritating metabolites. An irritant reaction occurs if irritating metabolites are formed that can react adversely with cellular components, and sensitization occurs when the reactive metabolites formed provide a configuration capable of activating the immune system. Enzymes present in the skin include monoamine oxidases, peroxidases, decarboxylases, carboxyl esterases, and alcohol dehydrogenases. Monoamine oxidases are probably the most important, as they can deaminate transdermally delivered amine drugs, such as beta blockers, antihistamines, and local anesthetics. Metabolic modulators that can modify monoamine oxidase activity include harmine, benzyl alcohol, tranylcypromine, phenylhydrazine, 2-phenyl-1-ethanol, and cinnamyl alcohol. Examples showed that 2-phenyl-1-ethanol prevented the induction of sensitization to tetracaine and completely prevented the inflammatory response of already sensitized subjects.

A method for preventing induction of sensitization and treating the inflammatory response upon elicitation have been disclosed that involves the coadministration to the skin or mucosa a sensitizing drug together with an antigen processing-inhibiting agent (Ledger et al., 1992a, 1992b). Events that lead to the association of the antigens with the cell surface of a class II MHC molecule are referred to as *antigen processing*. The above-mentioned association is required for the occurrence of presentation of the antigen by Langerhans cells' to T cells. Presentation is required for both the induction as well as the elicitation step of the sensitization process. Antigen processing inhibitors include ionophores and weak base compounds, such ammonium chloride. The processing inhibitors prevent the processing of the drug in the lysosome due to the increase in pH; thus the proteases in the lysosome are not able to chemically alter the drug into the proper antigenic form for class II MHC association. Excellent data were obtained for two sensitizing drugs: propranolol and tetracaine. Over 90% reduction in inflammatory response to propranolol was obtained with a formulation containing 4% ammonium chloride.

Agents that elevate intralysosomal pH are claimed to reduce irritation caused by drugs that are weak bases, such as the beta-adrenergic antagonist propranolol and the antimalarial chloroquine (Cormiez et al., 1992a, 1992b). Weak bases are known to accumulate in lysosomes due to the low intralysosomal pH (MacIntyre et al., 1988). The weak bases can permeate the lysosomal membranes at their uncharged molecular form; however, the low pH of the lysosomes favors protonation, and the charged weak bases are now relatively membrane impermeable and less able to pass back through the lysosomal membrane. Ionophores have also been shown to accumulate in lysosome membranes and thus increase the pH of the lysosomes (Maxfield, 1982). The invention also pertains to the use of competitive weak bases that would also increase the lysosomal pH. The increased pH would reduce the protonation of the weak base drug and thus reduce its accumulation in the lysosome and reduce irritation. Ionophores suitable for the invention include monensin, nigericin, valinomycin, and gramicidin. Competitive weak bases include ammonia, ammonium chloride, methylamine, ethanolamine, and trimethamine. A formulation containing 5% ammonium chloride reduced 10-fold the irritation caused by 2% chloroquine or 2% propranolol.

MAST CELL DEGRANULATING AGENTS

This section deals with experimental studies that define a critical connection between agents that cause mast cell degranulation and that also inhibit contact hypersensitivity response (CHR) in mice. In particular, it focuses on cis-urocanic acid, as it has an established history and linkage to events that mediate UVB-induced immunosuppression.

The effect of urocanic acid on mast cells and, for that matter, its effect on other targets may depend on whether the effects seen are specific to the cis-isomer or to some indeterminate mixture of the trans- and cis-isomers. This was answered by preparing highly purified (>95%) cis-urocanic acid (Wille et al., 1999a). The solubility of purified cis-urocanic acid (C-UA) was determined to be 25 mg/ml in water, whereas the solubility of the trans-isomer was only 1.1. mg/ml in water at 20°C. This differs from the reported value of 1–2 mg/ml for both the trans- and the cis-isomer (CTFA Final Report on the Safety Assessment of Urocanic acid, 1995). The higher water solubility of the cis-isomer relative to the trans-isomer suggest a greater diffusibility and transport into the dermis following UVB irradiation, where the cis-isomer has greater access to dermal mast cells.

The permeation of purified C-UA through both mouse and human skin was studied. As expected, mouse skin was two to three time more permeable than human skin with its thicker *stratum corneum*. These studies could aid future use of C-UA in transdermal delivery systems, where sustained delivery is required at levels that target dermal mast cell degranulation–induced immunosuppression of CHR. The C-UA concentrations in vehicle that accomplish these goals correspond to skin permeation flux rates in the range of 10–50 $\mu g/cm^2/hr$.

Human skin organ cultures exposed to culture medium containing cis-urocanic acid were reported to induce dermal mast cell degranulation. Depletion of human mast cell–specific chymase enzyme was used as a marker enzyme of mast cells (Wille et al., 1999a). In addition, the cis-isomer was more potent than the trans-isomer and even more potent than morphine sulfate, a standard mast cell secretagogue.

Previous studies had established that ultraviolet irradiation in the B wavelength range induced E-selectin cell adhesion protein surface expression on dermal endothelial cell membranes, and that this effect is mediated through the release of mast cell TNF-α (Klein et al., 1989). Regulation of the induction of these adhesion proteins is, therefore, of importance in leukocyte trafficking during the process of skin inflammation. Neonatal foreskin organ cultures exposed to culture medium containing cis-urocanic acid displayed significantly induced dermal endothelial cell E-selectin expression, and this expression was greater than similar cultures exposed to trans-urocanic acid. The degree of E-selectin expression induced by C-UA was equivalent to the maximal level observed for TNF-α, the positive control. These findings suggest a role for C-UA in mediating UV-induced immunosuppression of allergic contact dermatitis and are supportive of previous work implicating cis-urocanic and TNF-α as mediators of immune suppression (Kurimoto and Streilein, 1992).

Mast cells appear to be the primary and immediate source of TNF-α generated by UVB and C-UA. The question arose as to what the cellular targets are that are transduced by these immunosuppressive signals. The most likely candidates are Langerhans' cells and keratinocytes. More particularly, does C-UA–provoked mast cell cytokine release alter Langerhans' cells cytokine responsitivity and modulate antigen presentation?

Although keratinocytes are known to produce many immunologically active cytokines, C-UA does not affect the secretion of TNF-α, interleukin (IL)-4 or IL-10 (Yarosh et al., 1992). Another possibility is that C-UA–induced release of mast cell cytokines stimulates keratinocyte CSF-1 or inhibits keratinocyte IL-7, both of which could interfere with dendritic cell maturation and antigen presentation (Ullrich, 1999).

Cis-urocanic acid mimics many of the deleterious morphological and biochemical effects induced by UV irradiation on Langerhans' cells, including loss of dendrites, cytoskeletal modifications, and their rate of migration to draining lymph nodes (Bacci et al., 1996; Kurimoto and Streilein, 1992; Moodycliffe et al., 1992). There is uncertainty as to how such alterations result in immunosuppression. The deleterious effects of UVB and cis-urocanic acid can be reversed by antibody to C-UA (El-Ghorr and Norval, 1994). Another suggestion is that mast cell cytokines and TNF-α may upregulate Langerhans' cell IL-4 and IL-10, thereby increasing their immunosuppressive potential and ability to anergize T cells through ablation of their B7 costimulatory signals. In addition, mast cell cytokines and TNF-α induce laminin surface receptors on intraepidermal Langerhans' cells, which either slow down or speed up their transit from the epidermis to the lymph nodes (Iofredda et al., 1993). Alternatively, IL-4 secretion by Langerhans' cells may attract monocytes, which secrete IL-10, leading to deranged costimulatory B7 expression on Langerhans' cells. The latter effect may interfere with Th1 activation through apoptosis induction and at the same time favor the activation of Th2 cells and the production of IL-10 required for the induction of immune tolerance (Lambert and Granstein, 1998).

Cis-urocanic acid inhibits both the induction and elicitation phases of CHR in mice. By contrast, it has no effect on irritation or immediate-type allergic response (Laerma and Maibach, 1995). Previously, De Fabo and Noonan had speculated that C-UA is the chromophore that mediates UVB-induced CHR suppression (De Fabo and Noonan, 1983); however, Reeves et al. found no correlation between suppression of contact hypersensitivity by UV radiation and UV photoisomerization of epidermal urocanic acid, the tran-isomer, in hairless mice (Reeve et al., 1994). Further, C-UA has been implicated in a wide variety of immune functions, including suppression of CHR, induction of a serum suppressor factor, inhibition of epidermal cell cytokine IL-1, defective dendritic cell antigen presentation, prolongation of murine skin allograft survival, and regulation of tumor antigen presentation (Beissert et al., 1997; Gruner et al., 1992; Harriot-Smith and Halliday, 1988; Noonan et al., 1988; Norval et al., 1989; Rasanen et al., 1987; Ross et al., 1986, Ross et al., 1987, Ross et al., 1988a, 1988b).

One of the most compelling indications that C-UA mediates UVB-induced immunosuppression is its reversal by dermal injection of an antibody to TNF-α (Kurimoto and Streilein, 1992). A primary role for C-UA and mast cell–mediated cytokine release in UVB-induced immunosuppression is further supported by its reversal by monoclonal antibody to C-UA and its failure to reverse UVB-induced immunosuppression of DHT responses (Moodycliffe et al., 1993; Moodycliffe et al., 1996). Recently, Yamazaki et al. reported that TNF-α, RANTES, and MCP-1 are potent chemotactic attractants for Langerhans' cells and may influence their migration from the epidermis to regional lymph nodes during contact sensitization (Yamazaki et al., 1998). Also, Udey and Jakob reported that both IL-1 and TNF-α aid in the emigration of Langerhans cell by reducing E-cadherin interaction of Langerhans' cells to epidermal keratinocytes (Jakob and Udey, 1998). Experiments using anti–IL-1 beta antibody administered systemically prior to skin sensitization abolished Langerhans' cell accumulation in the draining lymph nodes and also inhibited TNF-α–induced migration and dendritic cell accumulation (Cumberbatch et al., 1997). Interestingly, Price et al. reported that antibody to alpha 6 integrin (laminin), but not an anti-alpha 4 integrin (fibronectin), blocks Langerhans, emigration and dendritic cell accumulation during the induction phase of contact sensitization (Price et al., 1997).

Other studies have suggested that the immunosuppressive effects of both C-UA and UVB is mediated by DNA damage, as both immunosuppression and pyrimidine dimer formation were reversed in skin exposed to liposome-mediated DNA repair enzymes and reversal of both C-UA and UVB-induced immunosuppression by UVA radiation (Reeve et al. 1998; Yarosh and Kripke, 1996). Because UV-induced DNA damage in keratinocytes mediates IL-6 mRNA and cytokine release, there may be a common pathway underlying cytokine regulation of immunosuppression and its reversal by countervailing cytokines (Petit-Frere et al., 1998).

Purified cis-urocanic acid applied topically to ears of mice was effective in preventing ear swelling in mice sensitized on the abdomen with 2,4-dintrochlorobenzene (DNCB) (Wille et al., 1999a). Nearly complete (>85%) elimination of the sensitization response was attained at or above 10% C-UA, and a dose response relation for prevention was established. In other experiments, codelivery of C-UA was found to be minimally effective relative to 24-hour pretreatment.

Two other mast cell degranulating agents were tested for their ability to suppress contact hypersensitivity in DNCB-sensitized mice. Chloroquine, applied topically to the abdomen prior to DNCB sensitization, significantly inhibited the ear swelling response. Capsaicin, another mast cell secretagogue, reduced ear swelling in DNCB-sensitized mice. These agents were not as effective as C-UA, but no attempt was made to carry out dose optimization studies.

A state of immune unresponsiveness, also known as *immune tolerance*, was shown to exist in mice that had been treated with only a single application of cis-urocanic acid (Wille et al., 1999a). The level of immune tolerance obtained in the absence of a second pretreatment was about half of that achieved by a single C-UA application in the same animal.

To test the hypothesis that degranulation of mast cells prior to skin sensitization is immunosuppressive, it was surmised that agents that block mast release of cytokines should be poor immunosuppressive agents. Sodium cromolyn is a drug that stabilizes mast cell membranes and is used clinically to treat type 1 immediate-type allergic reactions brought about by mast cell release of histamines. It was expected that it might act to suppress the elicitation phase of CHR. In accordance with our hypothesis, sodium cromoyln injected intradermally 1 hour before trinitrochlorobenzene sulfonate (TNCB), a water-soluble sensitizer and analogue, not only failed to prevent ear swelling, it actually enhanced the ear swelling, response, although the enhancement was not statistically significant.

U.S. Patent 5, 843,979 and European Patent Application 0612525A claim the use of mast cell degranulators to abrogate the induction step of delayed hypersensitivity in the dermal or transdermal delivery of drugs (Wille and Kydonieus, 1994, 1998). The authors claim that mast cell degranulators are also capable of inducing a state of immunological tolerance to the skin-sensitizing agent by delivery prior to or at the onset of transdermal administration. Cis-urocanic acid and its analogues, capsaicin, chloroquine, an antihuman IgE antibody, compound 48/80, morphine sulfate, and substance P, are shown to be appropriate mast cell degranulating agents. Cis-urocanic acid was able to reduce in Balb/c mice the sensitization to DNCB in dose response fashion.

In addition, several other agents that affect mast cell release of histamine have been claimed to reduce skin irritation. Topical application of p-substituted phenoxy alkanols jointly with an agent that causes irritation or contact sensitization modify and mitigate such irritation or sensitization (Berger, 1991). The lead compound is chlorphenesin, a phenoxy propanol originally described as a muscle relaxant. Chlorphenesin inhibits histamine release from human leucocytes and inhibits the degranulation of the rat mast cells mediated by IgE/anti-IgE reactions (Kimura et al., 1974; Lichtenstein et al., 1969). It is claimed that p-phenoxy alkanols inhibit irritation and sensitization by inhibiting the release of chemical mediators that are responsible for the symptoms of allergic diseases and inflammation. These mediators are histamine and the IgE-mediated cytokines. In an experiment using Balb/c mice already sensitized to dinitrofluorobenzene, a 2% solution of chlorphenesin reduced by 100% the ear swelling of the mice upon challenge. The use of nitric oxide synthase inhibitors have also been disclosed to reduce the skin irritant effects of topically applied cosmetics and pharmaceutical components (Giacomoni and Andral, 1996). This is in accordance with the observation that nitric oxide directly inhibits the IgE-mediated secretory function of mast cells (Eastmond et al., 1997). Nitric oxide synthase inhibitors useful with the teachings of this patent include N-monomethyl-L-arginine, N-nitro-L-arginine, N-amino-L-arginine, and N,N-dimethyl-arginine. The nitric oxide synthase inhibitors are useful in reducing the irritation caused by retinoic acid, salicylic acid, and vitamin D and its derivatives.

U.S. Patent 5162361 discloses the use of aromatic diamidines to control diseases in which a diminution of IL-1 is beneficial, including skin hypersensitivity (Rosenthal et al., 1992). The inventors have also shown for the first time that aromatic diamidines block the secretion of IL-6 and tumor necrosis factor from cells producing these cytokines. The lead diamidine is 1,5-bis(4-amidinophenoxy) pentane (pentamidine) or its imidazoline-substituted derivatives. In experiments with mice already sensitized to oxazolone, pentamidine was able to reduce the inflammatory response upon challenge in a dose-dependent fashion. Immediate ear pretreatment with 20 μg/ear of pentamidine reduced ear swelling by 50% over the positive control. Application of 80 μg pentamidine per ear reduced ear swelling about 75%.

ION CHANNEL MODULATORS

Ion channel modulators have been studied extensively and they have been shown to modulate both the induction and elicitation steps of the sensitization response. The mechanism of modulation has not been elucidated. One could hypothesize that by modulating the ion channels on the cell surface, you alter the permeation of small molecules into the cells, which then interfere with signal transduction as well as cell-to-cell communication.

The inhibition of contact hypersensitivity by calcium channel antagonists, lanthanium ion, diltiazem in mice, and verapamil in humans has been reported (Diezel et al., 1989; Gruner et al., 1991; McFadden et al., 1992). The mechanism of inhibition has not been determined but may relate the specific inhibitory effect of calcium channel blockers on cytokine-stimulated lymphocyte migration (Bacon et al., 1989). The latter authors showed that DSDZ 202-79, a dihydropyridine analogue but not a stereoisomer of the analogue, which was inactive as a calcium channel antagonist inhibited lymphocyte migration, implicating release of IL-1 alpha, IL-beta, and IL-8 in the mechanism of cutaneous immunity. Recently, Matsunaga et al. reported that skin exposed to haptens but not irritants induced IL-1 beta mRNA, signifying a role for epidermal IL-1 beta in regulating hapten-induced cytokine signalling in the early events of CHR (Matsunaga et al., 1998). These results have been confirmed by Cumberbatch et al., who were able to block dendritic antigen-presenting cell migration to the local lymph nodes following antigenic stimulation by injection of monoclonal antibodies to IL-1 beta (Cumberbatch et al., 1997).

The generality that ion channel modulators are also effective topical immunosuppressants was strengthened by studies of Gallo and Granstein, who reported that amiloride, a sodium-hydrogen channel antagonist, inhibited allergic contact dermatitis (Gallo and Granstein, 1989; Granstein and Gallo, 1990). This notion was later challenged because several analogues of amiloride that lacked Na^+/H^+ ion channel inhibitory activity retained inhibitory activity against contact hypersensitivity response (Lindgren et al., 1995).

To test this hypothesis, we employed the vitamin A hypervitaminosis-enhanced CBA/J mouse model. Previously, it was used to demonstrate the sensitization potential of a number of weakly sensitizing drugs, including nadolol, a beta-blocking cardiovascular drug (Kalsih et al., 1996). Nifedipine and verapamil were shown to inhibit the induction phase of CHR in nadolol-sensitized CBA/J mice (Wille et al., 1999b). Nifedipine was a more potent inhibitor of CHR than verapamil; this agrees with the fact that nifedipine is a more potent calcium channel blocking drug than verapamil and with the fact that nifedipine would have a greater flux through skin than verapamil due to its smaller size (MW = 346.34) relative to verapamil (MW = 454.59).

For the first time it was reported that amiloride was an effective inhibitor of the induction phase of DNCB-induced CHR in mice. A single application of nifedipine, applied 24 hours prior to DNCB sensitization, reduced the ear swelling response by 80% relative to gel vehicle controls. Likewise, a single application of amiloride, applied 24 hours prior to DNCB sensitization, significantly reduced the ear swelling response by 60% relative to vehicle control mice. These studies confirm and extend the concept that interference with cutaneous ion channels inhibits the induction phase of contact hypersensitivity.

Ethacrynic acid (Edecrin), a potassium ion channel blocker and potent loop-diuretic drug, was shown earlier to prevent the induction phase of contact hypersensitivity in both DNCB-sensitized Balb/c mice and in CBA/J mice sensitized with four different cutaneous drug sensitizers, including, albuterol, clonidine, chlorpheniramine, and nadolol (Kalish et al., 1997). Inhibition of nadolol-induced sensitization required that ethacrynic acid be applied to the skin at least a few hours prior to sensitization. These studies provided evidence that ethacrynic acid is a novel topical immunosuppressant. It was proposed that it may permit the transdermal delivery of drugs that otherwise are safe but cause allergic skin reactions in patients that are sensitized by topical delivery of those drugs.

Recently, it was reported that ethacrynic acid (ECA) was effective in inhibiting the elicitation phase of contact hypersensitivity in mice sensitized to DNCB on their bellies and challenged 5 days later on the ear with DNCB (Wille et al., 1998). Ear swelling was reduced 65% relative to control DNCB-sensitized mice.

Interestingly, the ear-swelling response was suppressed to the same extent by topical application of either 1% ECA or 1% hydrocortisone. In addition, ECA was just as effective in suppressing the ear swelling response when applied 1 hour after ear challenge in mice sensitized with oxazolone and challenged on the ear 5 days later with oxazolone.

Of importance to cosmetic applications, ethacrynic acid was reported to be very effective in preventing both the induction and elicitation phases of sensitization induced by the oxidative dye, para-phenylenediamine (PPD). These results indicate that ECA suppression is independent of the specific antigen used to immunize animals and acts generally to inhibit both the efferent and the afferent arm of the skin immune system.

To test the effect of ECA on the prevention of skin irritation caused by exposure to irritants, a rapid and reliable mouse ear swelling test was developed that yielded quantitative assessment of irritant potency of a test substance (Wille et al., 1998). Ethacrynic acid applied topically 5–10 minutes after topical application of six different irritants significantly reduced ear swelling. It completely blocked ear swelling due to a panel of irritants, including phorbol myristate acetate (0.1%), lactic acid (25%), and retinoic acid (5%). It also significantly reduced ear swelling due to irritating concentrations of DNCB (2%), capsaicin (5%), and arachidonic acid (2.5%).

In these studies, the anti-irritant activity of ECA was compared with two topical nonsteroidal anti-inflammatory drugs (NSAIDs): ibuprofen (1%) and indomethacin (1%); neither was as effective as ethacrynic acid in preventing irritation produced by arachidonic acid (5%).

Finally, ethacrynic acid treatment was shown to prevent surfactant-induced cumulative skin irritation. Ethacrynic acid pretreatment was highly effective in preventing ear swelling due to the surfactant di-sodium laureth sulfosuccinate (DSS), which was almost completely prevented by a 1 hour pretreatment preceding each surfactant treatment. Although not as effective, topical ethacrynic acid significantly reduced the extent of ear swelling by 60% when presented simultaneously with 10% sodium lauryl sulfate (SLS).

The aim of another study was to identify a nondrug congener of ethacrynic acid that is as potent an inhibitor of irritant and allergic contact dermatitis as ethacrynic acid. The agent identified was phenoxyacetic acid methyl ester (PAME). PAME has the same phenoxyacetic acid backbone minus the chloride groups and methylene alkyl chain substituents, and is the methyl ester of the phenoxyacetic acid. Unlike ECA, this chemical analog lacked topical diuretic activity.

The anti-irritant topical activity of PAME was demonstrated using the short-term mouse ear swelling test with applications occurring 1–2 hours after painting Balb/c mouse ears with an irritant (Wille et al., 2000). PAME (1%) dissolved in 70% ethanol completely prevented arachidonic acid (2.5%)–induced ear swelling response, when applied either 1 hour before irritant application or 5 minutes after irritant application. In addition, PAME also prevented ear swelling when applied 15 minutes after painting mouse ears with *trans*-retinoic acid (1%); it also prevented DNCB (2%)-induced skin irritation in mice. The anti-irritant activity of PAME was comparable to that observed with ethacrynic acid (1%), whereas topical application of two NSAIDs, indomethacin and ibuprofen, were without appreciable effect. PAME inhibited irritation induced by topical application of phorbol myritatic acid (PMA), a potent inflammatory agent and tumor promoter. By contrast, topical indomethacin was completely without effect on PMA-induced irritation.

In other studies, PAME inhibited surfactant-induced irritation (Wille et al., 2000). In one experimental design, PAME (1%) was applied to mouse ears prior to twice-daily single applications of a detergent, di-sodium sulfosuccinate (15%), for 5 days. It reduced erythema and ear swelling by 58%. In the second design, PAME (1%) was applied to mouse ears simultaneously with sodium lauryl sulfate (10%) twice daily, 1 hour apart, for 2 consecutive days; it reduced ear swelling by 33%. In the third design, PAME (1%) was applied to mouse ears twice daily, 1 hour apart, 5 minutes after each single application of 10% SLS; it reduced ear swelling by 62%. Histological analysis was performed on ear skin samples taken from the same mice used in the above post-treatment protocol. It showed that PAME completely reversed SLS-induced epidermal hyperproliferation, dermal induration, and leukocytic infiltration. ECA treatment was less effective in reducing these histological correlates of SLS-induced inflammation (Wille et al., 2000).

PAME was examined for its ability to inhibit the elicitation phase of contact hypersensitivity response. DNCB-sensitized mice, treated 5 minutes after DNCB ear challenge with PAME (1%), exhibited a more than 60% reduction in ear swelling response. PAME was also effective in inhibiting the elicitation phase in oxazolone-sensitized mice. Topical application of PAME (1%), 5 minutes after oxazolone ear challenge, substantially reduced the ear swelling response by 84%.

PAME was also effective in inhibiting both the induction and elicitation phases of paraphenylenediamine-induced contact hypersensitivity response. PAME (0.5%) was applied one day prior to each of the PPD applications to the shaved abdominal skin of CBA/J mice. It inhibited 70%, 83%, and 98% of the ear swelling response measured at 24, 48, and 72 hours post-challenge, respectively. PAME (1%) was also effective in blocking the elicitation phase when applied 30 minutes prior to ear challenge in paraphenylenediamine-sensitized CBA/J mice. It inhibited the elicitation phase by 80%, 100%, and 100% of the ear swelling response measured at 24, 48, and 72 hours, respectively (Wille et al., 2000).

Bristol-Myers Squibb was recently issued four U.S. patents addressing the abrogation of sensitization, as well as pure irritation by the use of ion channel modulators (Wille and Kydonieus, 1997a, 1997b; Wille et al., 1997a, 1997b). U.S. Patent 5686100 describes loop diuretics, such as ethacrynic acid, for the prevention and/or treatment of adverse reactions of the skin to the presence of a skin-sensitizing or irritating substance. Ethacrynic acid was shown to be effective against the classical sensitizers, such as DNCB and oxazolone, as well as the important sensitizing drugs albuterol, nadolol, clonidine, and chlorpheniramine. U.S. Patent 5716987 pertains to nondiuretic analogues of ethacrynic acid as modulators of irritation and sensitization. The analogues include phenoxyacetic acid and its lower alkyl esters, such as phenoxyacetic acid methyl ester. In general, the analogues were as effective as ethacrynic acid. U.S. Patent 5618557 pertains only to the prophylactic treatment of allergic contact dermatitis by the use of potassium-sparing diuretics, and it is specifically positioned for the prevention of sensitization in the dermal and transdermal delivery of drugs. Granstein and Gallo had previously shown in U.S. Patent WO 9009792 that amiloride, a potassium-sparing diuretic, and its analogues were able to treat the inflammation caused by a sensitizer applied to the skin of previously sensitized subjects. The last Bristol-Myers Squibb patent (WO 9718782) pertains to the use of calcium channel blockers, such as nifedipine and verapamil, to prevent or treat reactions of the skin to the presence of a skin-sensitizing agent. Calcium channel blockers were shown to be effective against DNCB, as well as the beta adrenergic agonist nadolol.

A survey of patent literature on the use of ion channel inhibitors of contact dermatitis revealed several patents. Patents U.S. 5202130 and South African ZA 9006583 also teach the treatment of contact sensitization with calcium flux inhibitors (Grant and Diezel, 1993; Sharpe, 1991). The South African patent teaches the use of nifedipine, which was shown to significantly reduce the inflammation caused by 3% oxazolone to already sensitized mice. Diltiazem, a calcium channel blocker, was shown to effectively treat the inflammation caused by DNCB to already sensitized mice. One significant contribution of this patent pertains to the use of lanthanum ions, from lanthanum chloride or lanthanum citrate, to treat sensitization. Lanthanum ions are competitors to calcium ions for the inhibition of calcium channels. A German (East) patent claims that magnesium ions, optionally together with lanthanum ions, are suitable for the treatment of allergic inflammations (Diezel, 1992). Magnesium chloride (28%) was effective in treating the inflammation caused by 0.5% DNCB.

Several other patents have been issued using metal ions to reduce irritation and sensitization. World Patent WO 9619181 provides formulations containing aqueous-soluble monovalent potassium cations or monovalent lithium ions to reduce irritation and the itch sensation of topically applied formulations (Hahn and Theuson, 1996). These ions affect the sensory nerves and prevent the transmission of sensory impulses and the desire to scratch. French Patent 2740341 pertains to the use of metallic salts, such as lanthanides, tin, cobalt, barium, manganese, strontium, zinc, and others, for the treatment of cutaneous pain and prevention of irritation among other cutaneous disorders (De Lacharriere and Breton, 1996). They claim that the metal salts are substance P antagonists and thus are useful in reducing or preventing irritant effects of cosmetics and dermatological and pharmaceutical compositions. In U.S. Patent 5708023, the use of zinc gluconate is claimed as an irritant-inactivating agent (Modak and Advani, 1998). A method is also disclosed of inactivating irritants in a fluid contacting skin, comprised of applying the composition containing zinc gluconate to the skin. U.S. Patent 5489441 discloses the use of Ruthenium Red as an immunosuppressant, which, among other uses, is useful in alleviating contact dermatitis (Dwyer and Esenther, 1996). Ruthenium Red is an inorganic hexavalent polycationic dye that has been used to stain complex polysaccharides. These dyes have also been shown to affect calcium ion transport in pig stomach cells and in rat liver cells (Kapus et al., 1990; Missaen et al., 1990). Ruthenium Red reduced significantly, and in a dose response fashion, the inflammation caused by trinitrochlorobenzene on previously sensitized mice.

SUMMARY

During the last decade, remarkable progress has been made in identifying a wide variety of potential therapeutic topical agents that have demonstrated activity in reducing skin irritation and skin sensitization. The information we have provided here about the majority of these agents is available in the current patent literature. Some of these proprietary anti-irritants and/or *countersensitizers* have been assigned to large pharmaceutical companies, and their development is likely in the near future.

Our goal here was to highlight recent development of two new categories of countersensitizers: mast cell degranulating agents and ion channel modulators, both of which appear to have unique attributes pertinent to the clinical treatment of dermal and transdermal drug-induced irritation and sensitization.

Mast Cell Degranulating Agents

We believe that the lead candidate among mast cell degranulating agents is cis-urocanic acid. The trans-isomer of urocanic acid is a natural skin metabolite that is located in the upper layers of mammalian epidermis, where it acts as a UV chromophore. It is converted to the cis-isomer upon UV exposure by a UV-dependent photoisomerization reaction. It is believed to play a role in the down-regulation of the skin auto-immune reactions that may occur due to UV-damaging effect on epidermal skin proteins. Our studies and the work of many laboratories have amply demonstrated that it has topical activity in inhibiting contact hypersensitivity. Other perhaps more potent and specific analogues may be developed as second-generation products for future commercial development.

Ion Channel Immune Modulators

Earlier studies provided firm evidence that ion channel antagonists interfere with contact hypersensitivity and skin inflammation. Here we reviewed recent work that confirms and extends this idea to include ethacrynic acid and phenoxyacetic acid methyl ester. Both ethacrynic acid and PAME are effective topical inhibitors of cutaneous irritation and contact hypersensitivity response. ECA and PAME appear to be active in suppressing both the induction and elicitation phases, indicating that they are useful in preventing and treating adverse skin reaction that occur upon single or repeated environmental exposure and are likely to afford protection in the transdermal delivery of commercially important drugs that cause skin sensitization (e.g., antihypertensive (clonidine), beta-blocker (nadolol), antiasthmatic (albuterol), and antihistaminic (chlorpheniramine) drugs.

The prospect for future use of either a mast cell degranulating agents and/or an ion channel modulator to eliminate or treat contact dermatitis now awaits successful validation in prospective randomized human clinical trials.

REFERENCES

Adams, R. M. (1983): Contact dermatitis due to irritation and allergic sensitization. In: *Occupational Skin Disease*. Edited by Adams, R. M. W. B. Saunders, Philadelphia.

Akimoto, K., Kawashimia, H., Hamazacki, T., and Sawazaki, S. (August 1995): European Patent Application 0704211A to Suntory Ltd., Preventive or alleviating agent for medical symptoms caused by delayed allergy reactions.

Amkraut, A., and Shaw, J. (March 1991a): United States Patent 5000596 assigned to Alza., Prevention of contact allergy by coadministration of a corticosteroid with a sensitizing drug.

Amkraut, A., and Shaw, J. (December 1991b): United States Patent 5077054 assigned to Alza., Prevention of contact allergy by coadministration of a corticosteroid with a sensitizing drug.

Amkraut, A. (September 1991): United States Patent 5049387 assigned to Alza., Inducing skin tolerance to a sensitizing drug.

Amkraut, A. (June 1992): United States Patent 5118509 assigned to Alza., Inducing skin tolerance to a sensitizing drug.

Bacci, S., Nakamura, T., and Streilein, J. W. (1996): Failed antigen presentation after UVB radiation correlates with modification of Langerhans cells cytoskeleton. *J. Invest. Dermatol.*, 107:838.

Bacon, K., Westwick, J., and Camp, R. D. R., (1989): Potent and specific inhibition of IL-8, IL-1 alpha, and IL-1 beta-induced in vitro lymphocyte migration by calcium channel antagonists. *Biochem. Biophys. Res. Commun.*, 165:349.

Bannon, Y. B., Corish, J., Corrigan, O., Geoghegan, E., and Masterson, J. (March 1994): United States Patent 5298257 assigned to Elan Transdermal., Method for treatment of withdrawal symptoms associated with smoking cessation and preparations for use in said method.

Beissert, S., Mohammad, T., Torri, H., Lonati, A., Yan, Z., Morrison, H., and Granstein, R. D. (1997): Regulation of tumor antigen presentation by urocanic acid. *J. Immunol.*, 169:92.

Berger, F. M. (April 1991): United States Patent 5008293., Process for the treatment of the skin to alleviate skin diseases arising from contact sensitization or irritation utilizing p-substituted phenoxy alcanols.

Bernstein, J. E. (March 1991): United States Patent 4997853 assigned to Galenpharma., Method and compositions utilizing capsaicin as an external analgesic.

Bos, J., (1989): *Skin Immune System (SIS)*. CRC Press, Boca Ratan, FL.

Breton, L. (January 1996): European Patent Application EP 0723774A assigned to L'Oreal, Use of an antagonist of CGRP in a cosmetic, pharmaceutical or dermatological composition.

Castelli, D., Colin, L., Camel, E., and Ries, G. (1998): Pretreatment of skin with a Ginkgo biloba extract/sodium carboxymethyl-β-1, 3-glucan formulation appears to inhibit the elicitation of allergic contact dermatitis in man. *Contact Dermatitis*, 38:123.

Castro, J. R. (February 1995): United States Patent 5393526 assigned to Elizabeth Arden, Cosmetic Compositions.

Cauwenbergh, G. (December 1995): United States Patent 5476853 to Janssen Pharmaceutica., Agent for use as an anti-irritant.

Cormiez, M., Amkraut, A., and Ledger, P. (September 1995): United States Patent 5451407 assigned to Alza., Reduction or prevention of skin irritation or sensitization during transdermal administration of an irritating or sensitizing drug.

Cormiez, M., Ledger, P., Johnson, J., and Phipps, J. (March 1995): World Patent Application WO 9506497 assigned to Alza., Reduction of skin irritation and resistance during electrotransport.

Cormiez, M., Ledger, P., Amkraut, A., and Marty, J. (December 1989): United States Patent 4885154 assigned to Alza., Method for reducing sensitization in transdermal drug delivery and means thereof.

Cormiez, M., Ledger, P., Amkraut, A., and Marty, J. (April 1994): United States Patent 5304739 assigned to Alza., Method for reducing sensitization in transdermal drug delivery and means thereof.

Cormiez, M., Ledger, P., and Amkraut, A. (July 1992a): United States Patent 5130139 assigned to Alza., Reduction or prevention of skin irritation to drugs.

Cormiez, M., Ledger, P., and Amkraut, A. (November 1992b): United States Patent 5160741 assigned to Alza., Reduction or prevention of skin irritation to drugs.

Cosmetic Ingredient Review Expert Panel. Final Report on the Safety Assessment of Urocanic acid. (1995): *J. Am. College of Toxicology*, 14:388.

Cruz, A., et al. (1988): *Photodermatology*, 5:126.

Cumberbatch, M., Dearman, R. J., and Kimber, I. (1997): Langerhans cells require signals from both tumor necrosis factor-alpha and interleukin-1 beta for migration. *Immunology*, 92:388.

Cupps., T. R., and Fauci, A. S. (1982): *Immunology Rev.*, 65:133.

De Fabo, E. C., and Noonan, F. P. (1983): Mechanism of immune suppression by ultraviolet irradiation in vivo. *J. Exp. Med.*, 157:84.

De Lacharriere, O., and Breton, L. (October 1996): French Patent 2740341 assigned to L'Oreal, Use of metallic salts in the treatment of cutaneous pain and prevention of irritation.

De Lacharriere, O., and Breton, L. (May 1997): World Patent Application WO 9717077 to L'Oreal, Topical composition containing capsazepine.

Degwert, J., Jacob, J., and Steckel, F. (January 1998): United States Patent 5708028 assigned to Beiersdorf A.G., Use of cis-9-heptadecenoic acid for treating psoriasis and allergies.

Della Valle, F., and Marcolongo, G. (November 1993): European Patent Application 0599188A assigned to Lifegroup., Trans and cis traumatic acid salts having cicatrizant activity.

Diezel, W., Gruner, S., Diaz, L. A., and Anhalt, G. J. (1989): Inhibition of cutaneous contact hypersensitivity by calcium transport inhibitors Lanthanum and diltiazem. *J. Invest. Dermatol.*, 93:322.

Diezel, W. (January 1992): East German Patent DD 297062 assigned to TKS Optimum, Magnesium salts as topical inflammation inhibitors of the skin.

Duffy, J. (May 1996): United States Patent 5516793 assigned to Avon Products, Use of ascorbic acid to reduce irritation of topically applied active ingredients.

Dwyer, D. S., and Esenther, K. (February 1996): United States Patent 5489441 assigned to Procept Inc., Method for suppressing immune response associated with psoriasis, contact dermatitis and diabetes mellitus.

Eastmond, N. C., Banks, M., and Coleman, J. (1997): Nitric oxide inhibits IgE-mediated degranulation of mast cells and is the principal intermediate in IFN-gamma-induced suppression and exocytotis. *J. Immunol.*, 159:1444.

El-Ghorr, A. A., and Norval, M. (1994): A monoclonal antibody to *cis*-urocanic acid prevents the UV-induced changes in Langerhans cells and DTH responses in mice, although not preventing dendritic cell accumulation in lymph nodes draining the site of irradiation and contact hypersensitivity responses. *J. Invest. Dermatol.*, 105:264.

Fischer, A. A. (1982): Contact dermatitis from topical medicaments. *Semin. Dermatol.*, 1:49.

Franz, T. J., Shah, K., and Kydonieus, A. (July 1991): United States Patent 5028431 assigned to Hercon Labs, Article for the delivery to animal tissue of a pharmacologically active agent.

Franz, T. J., Shah, K., and Kydonieus, A. (October 1988): European Patent Application 0314528A assigned to Hercon Labs., Article for controlled release and delivery of substances to animal tissues.

Franz, T. J., Shah, K., and Kydonieus, A. (October 1988): European Patent Application EP 0314528A assigned to Hercon Labs., Article for the controlled release and delivery of substances to animal tissue.

Gallo, R. L., and Granstein, R. D. (1989): Inhibition of allergic contact dermatitis and ultraviolet irradiation-induced tissue swelling in the mouse by amiloride. *Arch. Dermatol.*, 125:502.

Giacomoni, P., and Andral C. (September 1996): World Patent Application WO 9626711 assigned to L'Oreal, Nitric oxide synthase inhibitors.

Goldberg, R. L., et al. (1977): Reduction of topical irritation. *J. Chem. Soc.*, 28:667.

Govil, S. K., and Kohlman, P. (March 1990): United States Patent 4908213 assigned to Schering Plough., Transdermal delivery of nicotine.

Govil, S. K., and Kohlman P. (March 1990): United States Patent 4908213 assigned to Schering Plough., Transdermal administration of nicotine.

Govil, S. K. (March 1990): United States Patent 4908213 assigned to Schering Plough., Transdermal delivery of nicotine.

Granstein, R. D., and Gallo, R. L. (September 1990): World Patent Application WO 9009792 assigned to General Hospital Corp., Topical application of amiloride or analogues thereof for treatment of inflammation.

Grant, A., and Diezel, W. (April 1993): United States Patent 5202130 assigned to Johns Hopkins, Suppression of eczematous dermatitis by calcium transport inhibition.

Gruner, S., Diezel, W., Strunk, D., Eckert, R., Siems, W., and Anhalt, G. J. (1991): Inhibition of Langerhans cell ATPase and contact sensitization by lanthanides-role of suppressor cells. *J. Invest. Dermatol.*, 97:478.

Gruner, S., Oesterwitz, H., Stoppe, H., Henke, W., Eckert, R., and Sonnischsen, N. (1992): Cis-urocanic acid as a mediator of ultraviolet light-induced immunosuppression. *Semin. Hematol.*, 1:102.

Hahn, G. S., and Theuson, D. O. (June 1996): World Patent Application WO 9619181 assigned to Cosmederm Technologies, Formulation and method for reducing skin irritation.

Halker-Sorensen, L. (1996): Occupational skin diseases. *Contact Dermatitis*, 35(Suppl. 1), pp. 91–120.

Hall, B. J., Bauer, J. A., and Deckner, G. E. (August 1996): United States Patent 5545407 assigned to Proctor and Gamble, Dermatological compositions and methods of treatment of skin lesions therewith using benzoyl peroxide and tocopherol esters.

Hall, B. J., Bauer, J. A., and Deckner, G. E. (August 1995): United States Patent 5445823 assigned to Proctor and Gamble, Dermatological compositions and methods of treatment of skin lesions therewith.

Harriot-Smith, T. G., and Halliday, W. J. (1988): Suppression of contact hypersensitivity by short-term ultraviolet irradiation.II: The role of cis-urocanic acid. *Clin Exp Immunol.*, 72:174.

Herstein, M. (February 1995): World Patent Application WO 9503028, Cosmetic skin-renewal stimulating composition with long term irritation control.

Hewitt, C. W., and Black, K. S. (January 1996): World Patent Application WO 9600058 assigned to the University of California, Methods for inducing site-specific immunosuppression and compositions of site-specific immunosuppressants.

Huber, W., and Saifer, M. G. (1977): In: *Superoxide and Superoxide Dismutases.* Edited by Michelson, A. M. et al., p. 517 Academic Press.

Iofredda, M. D., Whitaker, D., and Murphy, G. F. (1993): Mast cell degranulation upregulates α6 integrins on epidermal Langerhans cells. *J. Invest. Dermatol.*, 101:749.

Jakob, T., and Udey, M. C. (1998): Regulation of E-cadherin-mediated adhesion in Langerhan cell-like dendritic cells by inflammatory mediators that mobilize Langerhans cell in vivo. *J. Immunol.*, 160:4067.

Kalish, R. S., Wood, J., Kydonieus, A., and Wille, J. (1997): Prevention of contact hypersensitivity to topically applied drugs by ethacrynic acid: Potential application to transdermal drug delivery. *J. Controll Rel.*, 48:79.

Kalish, R.S., Wood, J., Wille, J., and Kydonieus, A. (1996): Sensitization of mice to topically applied drugs: Albuterol, chlorpheniramine, clonidine, and nadolol. *Contact Dermatatitis*, 35:76.

Kapus, A., et al. (1990): Ruthenium Red inhibits mitochondrial Na$^+$ and K$^+$ uniports induced by magnesium removal. *J. Biol. Chem.*, 265:18063.

Kelly, R., and Ritter, E. J. (November 1970): United States Patent 3538009 assigned to Cincinnati Milacron., Method of reducing skin irritation in detergent compositions.

Kimura, J., et al. (1974): *Immunology*, 26:983.

Klein L. M., Lavker, R. M., Matis, W. L., and Murphy, G. F. (1989): Degranulation of human mast cells induces an endothelial antigen central to leukocyte adhesion. *Proc. Natl. Acad. Sci. USA.*, 86:8972.

Kolter, K. (January 1990): European Patent Application 0380989A assigned to Knoll, Pflaster zur transdermalen andwendung.

Kurimoto, I., and Streilein, J. W. (1992): Cis-urocanic acid suppression of contact hypersensitivity induction is mediated via tumor necrosis factor-α. *J. Immunol.*, 148:3072.

Kurimoto I., and Streilein J. W. (1992): Deleterious effects of cis-urocanic acid and UVB radiation on Langerhans cells and on the induction of contact hypersensitivity are mediated by tumor necrosis factor-α. *J. Invest. Dermatol.*, 99:69s.

Kurimoto, I., and Streilein, J. W. (1992): Cis-urocanic acid suppression of contact hypersensitivity induction is mediated via tumor necrosis factor-α. *J. Immunol.*, 148:3072.

Kydonieus, A., and Wille, J. (1999): *Biochemical Modulation of Skin Reactions in Dermal and Transdermal Drug Delivery.* CRC Press, Boca Raton, FL.

Laerma, A. I., and Maibach, H. I. (1995): Topical cis-urocanic acid suppresses both induction and elicitation of contact hypersensitivity in Balb/c mice. *Acta Derm Venereol (Stockh).*, 77:272.

Lambert, R. W., and Granstein, R. D. (1998): Neuropeptides and Langerhans cells. *Exp Dermatol.*, 7:73.

Ledger, P. (1992): Skin biological issues in electronically enhanced transdermal delivery. *Advanced Drug Delivery Reviews*, 9:289.

Ledger, P. W., Cormiez, M., and Amkraut, A. (June 1992a): United States Patent 5120545 assigned to Alza., Reduction or prevention of sensitization to drugs.

Ledger, P. W., Cormiez, M., and Amkraut, A. (September 1992b): United States Patent 5149539 assigned to Alza., Reduction or prevention of sensitization to drugs.

Lichtenstein, J., et al. (1969): *J. Immunol.*, 103:866.

Lindgren, A. M., Granstein R. D., Hosoi, J., and Gallo, R. L. (1995): Structure-function relations in the inhibition of murine contact hypersensitivity by amiloride. *J. Invest. Dermatol.*, 104:38.

MacIntyre, H., et al. (1988): *Biopharm. and Drug Disposition*, 9:513.

Maibach, H. I. (1985): Clonidine: Irritant and contact dermatitis assays. *Contact Dermatitis.* 12:192.

Matsunaga, T., Katayama, H., Nishioka, K. (1998): Epidermal cytokine mRNA expression induced by hapten differs from that induced by primary irritants in human skin organ culture system. *J. Dermatol.*, 25:421.

Maxfield, L. (1982): *J. Cell Biology*, 95:676.

McFadden, J., Bacon, K., and Camp, R. (1992): Topically applied verapamil hydrochloride inhibits tuberculin-induced delayed-type hypersensitivity reactions in human skin. *J. Invest. Dermatol.*, 99:784.

Menne, T., and Maibach, H. I. (1994): *Hand Eczema.* CRC Press, Boca Ratan, FL.

Missaen, J., et al. (1990): Ruthenium Red and compound 48/80 inhibit the smooth-muscle plasma membrane of pig stomach cells., *Biochemica et Biophysica Acta* 1023:449.

Modak, S. M., and Advani, B. H. (January 1998): United States Patent 5708023 assigned to Columbia University, Zinc gluconate gel compositions.

Moodycliffe, A. M., Bucana, C. D., Kripke, M. L., Norval, M., and Ullrich, S. E. (1996): Differential effects of a monoclonal antibody to cis-urocanic acid on the suppression of delayed and contact hypersensitivity following ultraviolet irradiation. *J. Immunol.*, 157:2891.

Moodycliffe, A. M., Kimber, I., and Norval, M. (1992): The effect of ultraviolet B irradiation and urocanic acid isomers on dendritic cell migration. *Immunology*, 77:394.

Moodycliffe, A. M., Norval, M., Kimber, I., and Simpson, T. J. (1993): Characterization of a monoclonal antibody to cis-urocanic acid in the serum of irradiation mice by immunoassay. *Immunology*, 79:667.

Nagy, C. F., Quick, T. W., and Shapiro, S. S. (October 1993): United States Patent 5252604 assigned to Hoffman LaRoche, Compositions of retinoic acids and tocopherol for prevention of Dermatitis.

Noonan, F. P., De Fabo, E. C., and Morrison, H. (1988): Cis urocanic acid, a product formed by ultraviolet B irradiation of the skin, initiates an antigen presentation defect in splenic dendritic cells in vivo. *J. Invest Dermatol.*, 90:92.

Norval, M., Simpson, T. J., Bardshiri, E., and Howie, S. E. M. (1989): Urocanic acid analogues and the suppression of the delayed type hypersensitivity response to Herpes simplex virus. *Photochem Photobiol.*, 9:633.

Parnell, F. (October 1991): World Patent Application WO 9114441 assigned to Parnell Pharma., Eriodictyon drug delivery systems.

Patel, D. C., and Ebert, C. D. (August 1989): United States Patent 4855294 assigned to Theratech., Method for reducing skin irritation associated with drug/penetration enhancer compositions.

Petit-Frere, C., Clingen, P. H., Grewe, M., Krutman, J., Roza, L., Arlett, C. F., and Green M. H. L. (1998): Induction of interleukin-6 production by ultraviolet radiation in normal human epidermal keratinocytes and in a human keratinocyte line is mediated by DNA damage. *J. Invest. Dermatol.*, 111:354.

Phipps, J. B. (June 1993): United States Patent 5221254 assigned to Alza., Method for reducing sensation in iontophoretic drug delivery.

Price, A. A., Cumberbatch, M., Kimber, I., and Ager, A. (1997): Alpha 6 integrins are required for Langerhans cell migration from the epidermis. *J. Exp. Med.*, 186:1725.

Rasanen, L., Jansen, C. T., Reunala, T., and Morrison, H. (1987): Stereospecific inhibition of human epidermal cell interleukin-1 secretion and HLA-DR expression by cis-urocanic acid. *Photodermatol.*, 4:182.

Reeve, V. E., Bosnic, M., et al. (1998): Ultraviolet A radiation (320-400 nm) protects hairless mice from immunosuppression induced by ultraviolet B radiation (280-320 nm) or cis-urocanic acid. *Inter Arch Allergy Immunol.*, 115:316.

Reeve, V. E., Boehm-Wilcox, C., Bosnic, M., Cope, R., and Ley, R. D. (1994): Lack of correlation between suppression of contact hypersensitivity by UV radiation and photosensitization of epidermal urocanic acid in hairless mice. *Photochem. Photobiol.*, 60:268.

Rheins, B., and Nordlund H. (1986): *J. Immunol.*, 136:867.

Richie, J. M., and Greene, N. M. (1993): Local anesthetics. In: *Goodman and Gilman's The Pharmacological Basis of Therapeutics.*, 8th edition. Edited by Gilman, A.G. et al., McGraw Hill.

Rosenthal, G. J., Kouchi, Y., Corsini, E., Blaylock, B., Comment, C., Luster, M., Craig, W., and Taylor, M. (November 1992): United States Patent 5162361 assigned to U.S. Government, Method of treating diseases associated with elevated levels of interleukin 1.

Roshdy, I. (February 1987): German Patent DE 3522572., Mittel zum schutz und behandlung der haut.

Ross, P. M. (January 1990): United States Patent 4897260 assigned to Rockefeller University, Compositions that affect suppression of cutaneous delayed hypersensitivity and products including same.

Ross, P. M. (October 1988): World Patent Application WO 8807371 assigned to Rockefeller University, Prevention and treatment of the deleterious effects of exposing skin to the sun, and compositions thereof.

Ross, J. A., Howie, S. E. M., Norval, M., Maingay, P., and Simpson, T. J. (1986): Ultraviolet irradiated urocanic acid suppresses delayed-type hypersensitivity to Herpes simplex virus in mice. *J. Invest. Dermatol.*, 87:630.

Ross, J. A., Howie, S. E. M., Norval, M., and Maingay, J. (1987): Two phenotypically distinct cells are involved in UV-irradiated urocanic acid-induced suppression of the efferent DTH response to HSV-1 in vivo, *J. Invest. Dermatol.*, 89:230.

Ross, J. A., Howie, S. E. M., Norval, M., and Maingay, J. (1988a): Systemic administration of urocanic acid generates suppression of the delayed type hypersensitivity response to herpes simplex virus in a murine model of infection. *Photodermatol.*, 5:9.

Ross, J. A., Howie, S. E. M., Norval, M., and Maingay, J. P. (1988b): Induction of suppression of delayed type hypersensitivity to Herpes simplex virus by epidermal cells exposed to UV-irradiated urocanic acid in vivo. *Viral Immunol.*, 1:191.

Rowe, A., and Bunker, C. B. (1998): Interleukin-4 and the interleukin-4 receptor in allergic contact dermatitis. *Contact Dermatitis*, 38:36.

Ruzicka, T., Bieber, T., Schopf, E., et al. (1997): A short-term trial of tacrolimus (FK506) ointment for atopic dermatitis. *NEJM*, 337:816.

Schaefer, H., and Rougier, A. (1998): Meeting report: UV and the Immune System. *Eur. J. Dermatol.*, 8:193.

Schmidt, R. J. (November 1996): United States Patent 5,578,300 assigned to U.C.C. Consultants., Treatment of allergic contact dermatitis.

Sharpe, R. J. (September 1991): South African Patent ZA 9006583 assigned to Beth Israel Hospital, Treatment of cutaneous hypersensitivity with topical calcium channel blockers.

Smith, W. P., Pelliccione, N. J., Marenus, K. D., and Maes, D. H. (August 1989): European Patent Application 0354554A assigned to Estee Lauder, Anti-irritant and desensitizing compositions and methods of their use.

Ullrich, S. E. (1999): The effects of ultraviolet radiation on the immune response. In: *Biochemical Modulation of Skin Reactions in Dermal and Transdermal Drug Delivery*. Edited by Kydonieus, A., and Wille, J., CRC Press, Boca Raton, FL.

Waali, E. E. (May 1987): United States Patent 4663151 to Research Corporation., Aluminum chlorhydrate as a prophylactic treatment for poison oak, poison ivy and poison sumac dermatitis.

Wakem, P., and Gaspari, A. (1999): Mechanisms of allergic and irritant contact dermatitis. In: *Biochemical Modulation of Skin Reactions in Dermal and Transdermal Drug Delivery*. Edited by Kydonieus, A., and Wille. J., CRC Press, Boca Raton, FL.

Wang, Y. J., Lee, W. A., and Narog, B. (August 1990): World Patent Application WO 9009167 assigned to California Biotechnology, Composition and method for administration of pharmaceutically active substances.

Wilckens, T., and DeRijk, R. (1997): Glucocorticoids and Immune function: Unknown dimensions and new frontiers. *Immunology Today*, 18:418.

Wilder, M. S. (May 1986): World Patent Application WO 860256 to Centerchem., A method for preventing or alleviating skin irritation using a formulation containing superoxide dismutase.

Wille, J., and Kydonieus, A. (1999): Ion channel modulation of contact dermatitis. In: *Biochemical Modulation of Skin Reactions in Dermal and Transdermal Drug Delivery*. Kydonieus, A., and Wille, J., CRC Press, Boca Raton, FL.

Wille, J., and Kydonieus, A. (April, 1994): European Patent Application 0612525 assigned to Bristol-Myers Squibb, Transdermal treatment with mast cell degranulating agents.

Wille, J., and Kydonieus, A. (December 1998): United States Patent 5843979 assigned to Bristol-Myers Squibb Co., Transdermal treatment with mast cell degranulating agents for drug-induced hypersensitivity.

Wille, J. J., Kydonieus, A., and Kalish, R. S. (1998): Inhibition of irritation and contact hypersensitivity by ethacrynic acid. *Skin Pharm. Appl. Skin Physiol.*

Wille, J. J., Kydonieus, A., and Kalish, R. S. (In press): Inhibition of irritation and contact hypersensitivity by phenoxyacetic acid methyl ester (PAME). *Skin Pharm. Appl. Skin Physiol.*

Wille, J., and Kydonieus, A. (November 1997): United States Patent 5686100 assigned to Bristol-Myers Squibb, Prophylactic and therapeutic treatment of skin sensitization and irritation.

Wille, J., and Kydonieus, A. (December 1997): United States Patent 5716987 assigned to Bristol-Myers Squibb, Prophylactic and therapeutic treatment of skin sensitization and irritation.

Wille, J., Kydonieus, A., and Castellana, F. (April 1997): United States Patent 5618557 assigned to Bristol-Myers Squibb, Prophylactic treatment of allergic contact dermatitis.

Wille, J., Kydonieus, A., and Castellana, F. (May 1997): World Patent Application WO 9718782 assigned to Bristol-Myers Squibb, Treatment with calcium channel blockers for drug induced hypersensitivity.

Wille, J. J., Kydonieus, A. N., and Kalish, R. S. (1999a): Cis-Urocanic acid induces mast cell degranulation and release of preformed TNF-α: A possible mechanism linking UVB and Cis-urocanic acid to inhibition of contact hypersensitivity. *Skin Pharm. Appl. Skin Physiol.*

Wille, J. J., Kydonieus, A., and Kalish, R. S. (1999b): Several different ion channel modulators abrogate contact hypersensitivity in mice. *Skin Pharm. Appl. Skin Physiol.*

Wille, J., and Kydonieus, A. (1999): Mast cell degranulating agents modulate skin immune responses. In: *Biochemical Modulation of Skin Reactions in Dermal and Transdermal Drug Delivery*. Edited by Kydonieus, A. S., and Wille. J., CRC Press, Boca Raton, FL.

Willer, S. G., Yust, P., and Kelley, R. (February 1978): United States Patent 4076799 assigned to Cincinnati Milacron., Method of inhibiting skin irritation.

Yamazaki, S., Yokezeki, H., Sahoh, T., Katayama, I., and Nishioka, K. (1998): TNF-α, RANTES, and MCP 1 are major chemoattractants of murine Langerhans cells to the regional lymph nodes. *Exp. Dermatol.*, 7:35.

Yarosh, D. B., and Kripke, M. L. (1996): DNA repair and cytokines in anti-mutagenesis and anticarcinogenesis. *Mutation Res.*, 350:255.

Yarosh, D. B., Alas, L., Kibital, A. L., and Ullrich, S. E. (1992): Urocanic acid, immunosuppressive cytokines, and the induction of human immunodeficiency virus. *Photodermatol. Photoimmunol. Photomed.*, 9:127.

Yawalkar, N., Brand, C. U., and Braathen, B. R. (1998): Interleukin-12 expression in human afferent lymph derived from the induction phase of allergic contact dermatitis. *Br. J. Dermatol.*, 138:297.

Toxicology of Skin
Edited by Howard I. Maibach
Copyright © 2001 Taylor & Francis

18

Allergic Contact Dermatitis: Clinical Considerations

Yung-Hian Leow

National Skin Centre, Singapore

Howard I. Maibach

Department of Dermatology, University of California, San Francisco, California, USA

INTRODUCTION

Allergic contact dermatitis is the prototypical skin manifestation of type IV delayed-type hypersensitivity reaction upon percutaneous exposure to an allergen. Apart from irritant contact dermatitis, it is probably the next commonest skin reaction to noxious stimuli. As opposed to irritant contact dermatitis, it is a specific immunologic tissue reaction that has been studied for more than 100 years.

CLINICAL PRESENTATION

History

Sensitization is the prerequisite to the manifestation of this immunological response. It does not occur upon first exposure to the allergen, and only manifest clinically after repeated exposure. Hence, there usually is a lag-time between exposure and the detection of dermatitis on the skin. The site of skin involvement usually is confined to the site of exposure to the allergen, but secondary spread can occur occasionally.

Classical history of dermatitis or eczema can occur, namely itch, redness, and blister formation on the affected skin. An important point that differentiates it from other forms of dermatitis is the history of direct skin contact with various environmental agents. They may be daily household products, like toiletries and cosmetics, or items used at work, like solvents and industrial oils; however, very often an occult allergen that may not be apparent to the patient or the physician could be the cause of the patient's dermatitis.

Physical Examination

Classical clinical features of dermatitis or eczema, like erythema, edema, blister formation, and even lichenification, can be detected on the involved skin; however, as opposed to irritant contact dermatitis, allergic contact dermatitis is an immunological response, and secondary spread of the dermatitis with widespread involvement can occur.

During the clinical evaluation of the patient, knowledge of common allergens affecting specific regions of the body will be useful to the attending dermatologist in outlining the differential diagnoses. For example, one needs to consider allergic contact dermatitis to fragrance, colognes, deodorants, and antiperspirants as the likely cause of exogenous eczema occurring at the axillary area.

Unusual clinical presentation of an allergic contact dermatitis or noneczematous contact dermatitis has been reported occasionally. Interesting examples are allergic contact leucoderma from cinnamic aldehyde in toothpaste, purpuric contact dermatitis from rubber antioxidant, and pigmented contact dermatitis (Fisher, 1974; Mathias et al., 1980; Osmundsen, 1970).

EVALUATION/INVESTIGATION

The clinical suspicion of an allergic contact dermatitis has to be ascertained and confirmed. Other common differential diagnoses that must be considered include endogenous eczema, irritant contact dermatitis, and psoriasis. The key investigative tool is patch testing.

Patch Testing

The principle behind patch testing is basically to reproduce, in a controlled clinical setting, an allergic contact dermatitis in a miniscule patch by applying putative allergens in a suitable vehicle dispersed in an appropriate nonirritating concentration.

Indication

The indication for patch testing is mainly to confirm the suspicion of an allergic contact dermatitis. Other indications in the clinical setting includes the exclusion of an irritant contact dermatitis and puzzling clinical situations where contact dermatitis cannot be entirely excluded.

The Procedure

The most commonly adopted system of patch testing in most dermatological institutions is a closed patch test system that involves the use of Finn chambers that are shallow aluminum discs mounted on nonsensitizing and low-irritant acrylate Scanpor tape (Norgesplaster A/S, Norway). The allergens are diluted in various vehicles, of which the most widely used vehicle is petrolatum, as most materials are soluble, stable, nonirritating, and well dispersed in it. Other vehicles include water, ethanol, acetone and olive oil. The appropriate concentration and vehicle at which the allergens are being dispersed are determined through years of clinical experiment in patients. They can be found in referenced textbooks and referred journals (De Groot, 1994); however, in situations where it is a new allergen, the basic principle is to patch test the patient with lowest concentration that does not irritate or sensitize the patient. Concentrations of 0.1% to 1% are preferred. Serial dilutions and testing in adequate controls may be required to establish the validity of testing when handling unfamiliar materials.

The upper back is the most appropriate site for the application of the patches. The outer aspects of the thighs or the arms are used in special circumstances.

Patch Test Material

Through scientific publications and interaction, eminent patch testers have identified and compiled the most commonly occurring allergens into a list of standard series. Allergens are also grouped into special series or batteries to suit the specific clinical situations. Obviously, the same standard series of allergens may not be entirely applicable for physicians practicing in different parts of the world due to differing exposure to environmental agents; however, as a guide, an international standard patch test series is often cited as the starting point for the development of a regional patch test series (Lachapelle et al., 1997).

Patch Test Reading

As allergic contact dermatitis is a delayed-type IV hypersensitivity reaction, the patches are applied on the skin for 48 hours. The reaction will then be evaluated after 48 hours, 72 hours, 96 hours, or 1 week later, depending on the set-up of the patch test clinic (Marks and DeLeo, 1992). Some positive patch test reactions may be delayed, like neomycin or corticosteroid. In such situations, the patient should be advised to return for reading after 1 week.

The following scoring system for patch test reaction is universally accepted (Wilkinson, 1970):

±: Erythema
+: Erythema, with papules
++: Oedematous or vesicular reaction
+ + +: Spreading bullous or ulcerative reaction

++ and + + + reactions are convincing positive reactions. + reaction is often being accepted as positive; however ± reaction usually is interpreted as negative reaction.

Relevance of Patch Testing

It is often said that doing a patch test is very simple, but the interpretation of patch test reaction requires skill and experience. The elicitation of a positive patch test reaction does not equate finding the allergen, as it had to be correlated with the patient's presenting clinical problem.

There are four possible interpretations to a positive patch test reaction:

a. Present or current relevance—this refers to positive reaction that is relevant to the patient's current presenting clinical problem.
b. Past relevance—this refers to a positive patch test reaction that does not explain the patient's current clinical problem but uncovers an allergen that caused dermatitis in the past (but of no current relevance).
c. Exposed—the elicited positive reaction has no clinical significance, but unmasked an allergen that the patient has previously encountered in the past.
d. Unknown relevance—there is no clinical relevance or the patient cannot recall previous encounter with the allergen that has caused a positive reaction.

Strict criteria have been proposed for proper interpretation of patch test reaction (Tables 18.1, 18.2, and 18.3) (Ale and Maibach, 1995; Marrakchi and Maibach, 1994).

Pitfalls of Patch Testing

As with all biological tests, patch testing is fraught with the same problem of false-positive and false-negative reactions.

TABLE 18.1. *Assessment of clinical relevance*

History of exposure to the sensitizer
  Occupational exposure
  Nonoccupational exposure (homework, hobbies)
  Use of pharmaceutical products (over-the-counter and prescription), cosmetics,
    clothing, jewelry, bandages (especially when intolerance is referred)
  Type of exposure: dose, frequency, site
  Environmental conditions: humidity, temperature, occlusion, vapors, powders,
    photo-exposure
Clinical characteristics of the dermatitis
  Dermatitis area corresponding to the exposure site
  Some morphologies suggest specific allergens
  Clinical course (caused or aggravated by the exposure)
Correlation with other tests
  Positive patch test to a product
  Positive patch test to a product's extract
  Positive Provocation Use Test (PUT), or Repeated Open Application Test (ROAT):
    twice daily application to small test area for 7 days
  Positive use test (typical product use)
Reproduction or aggravation of dermatitis by the patch test or use tests

Adapted from (Ale and Maibach, 1995).

False-positive reaction is often interpreted as an irritant reaction. This arises when the putative allergen is being patch tested at too high a concentration or it is dispersed in an incorrect vehicle. This can be overcome by patch testing the allergen at a lower concentration and in the appropriate nonirritating vehicle. The other important cause of false-positive patch test reaction arises from the *angry back syndrome* (Bruynzeel and Maibach, 1986). The latter refers to the resultant hyperirritability of the skin surrounding any eczema. In the case of patch testing, this occurs in the vicinity of strong positive reactions, thus resulting in multiple weak positive reactions that may not be of clinical relevance. This can be overcome by patch testing the patient to each suspected allergen separately at different occasions.

False-negative reaction arises when the allergen is being patch tested at too low a concentration, in an inappropriate vehicle, too small an amount, or from inadequate occlusion onto the skin. Other possible causes of a false-negative reaction can arise from reading the patch test reaction too early or patch testing the patient during the

TABLE 18.2. *Suggested guidelines for the assessment of relevance*

Requestion the patient in light of the test results
Show the patient a list of products containing the allergen (sometimes, he/she will
  be able to recognize the source of exposure)
Seek cross-reacting substances
Consider concomitant and simultaneous sensitization
Repeat the patch test with dilution, test with extracts, Use test, PUT, ROAT, test
  in vitro
Use "lists" of allergens for specific occupations
Obtain information from the product's manufacturer
Perform chemical analysis of products

Adapted from (Ale and Maibach, 1995).

TABLE 18.3. *Operational definition of allergic contact dermatitis*

Appropriate morphology
Positive patch test
Repatch test, when appropriate, to rule out Excited Skin Syndrome and "rogue" reactions
Provocative Use Test (PUT), Repeated Open Application Test (ROAT) or Use test
Serial dilution patch testing (when indicated)
Review reliability of irritant control for uncommonly used allergens

Adapted from (Marrakchi & Maibach, 1994).

refractory phase of the skin while it is still recovering from an episode of severe eczema.

OTHER TEST METHODS

Provocative Usage Test (Repeated Open Application Test)

After the patient has been patch tested in the usual way, Provocative Usage Test (PUT) or Repeated Open Application Test (ROAT) is sometimes required in certain clinical situations to further substantiate and confirm the relevance of a particular positive or negative patch test reaction (Hannuksela, 1991). This is especially true for the evaluation of products that are being used on a daily basis, like cosmetics and skin care products. This test basically requires the patient to apply the actual product twice daily to a small site of the antecubital fossa or forearm for 7 days. Like patch testing, it basically constitutes reproducing the eczema on the test area.

TRUE Test

Although patch testing is the preferred method to evaluate a suspected case of allergic contact dermatitis in most dermatological institutions, it is not the perfect test system, as outlined in other segments of this chapter. *TRUE Test, or Thin Layer Rapid Use Epicutaneous Test*, is a well-standardized, easy-to-use test system that overcomes some of the inaccuracies of the traditional patch test system.

The allergens are incorporated in hydrophilic gels, which are then coated onto water-impermeable sheets of polyester. They are mounted on nonwoven cellulose tape with acrylic adhesive, which are covered with siliconized plastic. The commercially available test system is packed in airtight and light-impermeable envelope. Upon direct contact with the skin, perspiration will hydrate the film and transforms it into a gel, which then releases the allergens in measured amounts.

The advantages of this system include ease of handling by the attending physicians, comfort to the patients, exact standardized dosage, stability of test materials, consistent panel-to-panel location, and reproducibility of test results (Fischer and Maibach,

1985, 1989; Lachapelle et al., 1988); however, at this point in time, the TRUE Test system is essentially a general basic screening tool for dermatologists who do not subspecialized in the field of contact dermatitis, as it only evaluates a limited commercial standard series of allergens. Thus, it does not supersede the conventional patch test system with Finn chamber and Scanpor Tape, as the latter is more flexible with regards to the choice of test material and it is still the gold standard by which investigative work on allergic contact dermatitis is based on.

TREATMENT

Management of a clinical case of allergic contact dermatitis goes beyond making a diagnosis and the search for a putative allergen. Allergic contact eczema should be treated in the same manner as with any other form of dermatitis or eczema, like the use of topical corticosteroid to suppress the inflammation.

Armed with the knowledge of the patch test results, clinicians can give specific advice to patients with regards to the avoidance of the incriminating allergens for the rash; however, if avoidance is not possible, other strategies should be considered. These include the use of personal protective equipment (e.g., impervious gloves) to prevent direct contact with the allergen and substitution of the allergen with an alternative compound. This is of paramount importance, especially in industrial or occupational dermatitis.

CONCLUSION

There had been recent proliferation of published literature that emphasizes the benefits of objective assessment of the skin with noninvasive bioengineering methods, which probably are of greater significance in the investigative aspect of contact dermatitis. In the clinical practice of dermatology, the logarithmic approach (viz. history, physical examination, patch testing, treatment) is still the most practical and logical approach to solve a suspected case of allergic contact dermatitis, with other objective scientific method–like chemical analysis and bioengineering serving more as adjuncts to problem solving.

REFERENCES

Ale, S. I., Maibach, H. I. (1995): Clinical relevance in allergic contact dermatitis: An algorithmic approach. *Derm. Beruf. Umwelt.*, 43:119.
Bruynzeel, D. P., Maibach, H. I. (1986): Excited skin syndrome ("angry back"). *Arch. Dermatol.*, 122:323.
De Groot, A. C. (1994): *Patch Testing: Test Concentrations for 3700 Chemicals.* Elsevier, The Netherlands.
Fischer, T., Maibach, H. I. (1985): The thin layer rapid use epicutaneous test (TRUE-test): A new patch test method with high accuracy. *Br. J. Dermatol.*, 112:63.
Fischer, T., Maibach, H. I. (1989): Easier patch testing with TRUE test. *J. Am. Acad Dermatol.*, 20:447.
Fisher, A. A. (1974): Allergic petechial and purpuric rubber dermatitis: The PPP syndrome. *Cutis*, 14:25.

Hannuksela, M. (1991): Sensitivity of various skin sites in the repeated open application test. *Am. J. Contact Dermatitis.*, 2:102.

Lachapelle, J. M., Bruynzeel, D. P., Ducombs, G, Hannuksela, M., Ring, J., White, I. R., Wilkinson, J., Fischer, T., Billberg, K. (1988): European multicenter study of the TRUE test. *Contact Dermatitis*, 19:91.

Lachapelle, J.-M., Ale, S. I., Freeman, S, Frosch, P. J., Goh, C. L., Hannuksela, M, Hayakawa, R, Maibach, H. I., Wahlberg, J. E. (1997): Proposal for a revised international standard series of patch tests. *Contact Dermatitis.* 36:121.

Marks, J. G., DeLeo, V. A. (1992). *Contact and Occupational Dermatology.* Mosby, St. Louis.

Marrakchi, S, Maibach, H. I. (1994): What is occupational contact dermatitis? An operational definition. *Dermatol Clin.* 12:477.

Mathias, C. G. T., Maibach, H. I., Conant, M. A. (1980): Perioral leukoderma simulating vitiligo from use of a toothpaste containing cinnamic aldehyde. *Arch. Dermatol.*, 116:1172.

Osmundsen P. E. (1970): Pigmented contact dermatitis. *Br. J. Dermatol.*, 83:296.

Wilkinson, D. S., Fregert, S., Magnusson, B., Bandmann, H. J., Calnon, C. D., Cronin, E., Hjorth, N., Maibach, H. I., Malalten, K. E., Meneghini, C. L., Pirila, V. (1970): Terminology of contact dermatitis. *Acta. Dermato-Venereol.*, 50:287.

Toxicology of Skin
Edited by Howard I. Maibach
Copyright © 2001 Taylor & Francis

19

Pharmacology, Toxicology, and Immunology of Experimental Contact Sensitizers

Whitney Hannon, Tim Chartier, and J. Richard Taylor

School of Medicine, University of Miami, Miami, Florida, USA

INTRODUCTION

Contact sensitizers are substances that stimulate an allergic type IV or delayed–type hypersensitivity (DTH) reaction through topical administration. A thorough under-standing of most commonly used contact sensitizers is essential for any clinician seeking to employ contact sensitizers in diagnosis, and treatment. This review covers the basic pharmacology, toxicology, and immunology of the three most well character-ized three-contact sensitizers: dinitrochlorobenzene (DNCB), diphenylcyclopropra-none (DPCP), and squaric acid dibutylester (SADBE). All are potent sensitizers with known chemical structures and studied pharmacological profiles, permitting repro-ducible conditions during experimentation (Catalona et al., 1972). In addition, none of them rely on previous exposure to the antigen (Catalona et al., 1972). Although DNCB, DPCP, and SADBE appear to stimulate DTH by fairly similar mechanisms, clinicians should be aware of the properties that make each chemical unique.

DNCB has the longest and most extensive history of the three. Its sensitizing properties were recognized during the late 19th century as a result of the industrial exposures. Laborers with a severe contact dermatitis caused by DNCB were first described in the literature several decades later (Bernstein, 1912; Wedroff, 1928). Although industrial exposures to DNCB were controlled by storing DNCB in closed containers away from human contact, it was noticed that DNCB had other interesting properties that warranted investigation by the scientific and medical community. For example, DNCB was useful in characterizing the cellular immune response in healthy individuals and those with a variety of diseases. It was also found to have an intrinsic therapeutic value (Stricker et al., 1997) in various types of cancer, alopecia areata (Rokhsar et al., 1988), recalcitrant warts, and, most recently, HIV (Mills, 1986). DNCB has been employed extensively by a diverse group of medical professionals and investigators: 886 studies from 1966–1999 mention DNCB in their Medline database abstracts. In 1982, out of 1756 practicing dermatologists who responded to a survey, 30% had used DNCB at some point in their careers for evaluation of cell-mediated immunity or therapy (Millikan, 1982). When DNCB was found to be mutagenic in the 1980s, investigators began to seek alternative contact sensitizers.

SADBE, a potent sensitizer that is nonmutagenic in the Ames test, was the first proposed alternative to DNCB. A similar compound, squaric acid diethyl ester (SADEE), was first discovered in 1959 (Cohen et al., 1959) and recognized to be a strong sensitizer in 1976 (Noster et al., 1976). SADBE was later found to be superior to SADEE in sensitization properties (Happle et al., 1983). The instability of SADBE, the difficulty finding commercial sources are disadvantages to its use. In addition, squaric acid esters are not approved by the Food and Drug Administration (FDA), nor is a pharmaceutical grade available for clinical use (Wilkerson et al., 1985). Once again investigators began looking for alternative contact sensitizers.

The newest addition to this group of sensitizers is DPCP. First synthesized in 1959 (Breslow et al., 1959; Volpin et al., 1959), DPCP's ability to provoke contact hypersensitivity was recognized considerably later, in the 1970s (Rosen et al., 1976; Whittaker, 1972) and dermatologists started using DPCP in the 1980s (Hausen and Stute, 1980). A European patent was granted for its use in 1982 (Hausen and Happel, 1982). DPCP has the advantage that it is nonmutagenic in the Ames Test and more stable than squaric acid, but the disadvantage is that it is photoreactive.

The purpose of this review is to (1) put contact sensitizing agents into historical perspective, (2) examine the important properties of contact sensitizing agents in general as well as the unique properties of each compound specifically, and (3) discuss the advantages and disadvantages of each particular agent.

DINITROCHLOROBENZENE

Pharmacology

Formula: 1-chloro-2,4 dinitrobenzene, $C_6H_3(NO_2)_2Cl$.
MW: 202.55 g/mol.

Solubility: DNCB dissolves well in acetone, alcohol, ether, benzene; insoluble in water.

Properties: Yellow to brownish needles, m.p. = 27-53 C, b.p. = 315 C.

Stability: DNCB is stable at room temperature. The active ingredient in DNCB decays with time, so it is recommended that fresh solution be made every two months (Dunagin and Millikan, 1982). In addition, more dilute solution may be less stable than more concentrated solution secondary to hydrolysis (H. I. Maibach, personal communication).

Synthesis: DNCB is synthesized commercially by nitration of orthochlorobenzene (Winholz et al., 1976).

Absorption and Excretion: DNCB is absorbed very readily by human skin in comparison with other compounds (Feldmann and Maibach, 1979). The minimal contact time was 12 hours for DNCB in acetone and 6 hours for DNCB in DMSO (Vakilzadeh et al., 1973).

The half-life of DNCB is 4 hours. With intravenous administration, 64% of the total dose was recovered in urine. After topical administration, 53% ± 12% was absorbed. The fastest absorption rate (3.4% dose/hr) was in the first 0–12 hours, with a dramatic decrease in absorption rate after 12 hrs: 0.57% dose/hr between 12–24 hrs, 0.13% dose/hr between 24–48 hours, and 0.05% dose/hr after 48 hours. In 12 patients receiving the highest doses of DNCB, DNCB was found in the urine in one patient and traces of aromatic acetyl acid metabolites of DNCB were found in the urine 5 other patients (De Prost et al., 1982).

Toxicology

Mutagenicity studies: DNCB is mutagenic in the Ames Test (Black et al., 1985; Kratka et al., 1979; Strobel and Rohrborn, 1980; Summer and Goeggelmann, 1980) that is used to screen chemicals to undergo carcinogenicity testing in animals (Ames, 1984; Dunagin, 1986). DNCB was inherently mutagenic even in the presense of potentially mutagenic contaminants (Black et al., 1985; Wilkerson et al., 1983) in all studies but one (Drill and Lazar, 1977). In addition, DNCB transforms BHK cells (Strobel and Rohrborn, 1980).

Carcinogenicity studies: Tests in mice and rats revealed that DNCB is not a carcinogen (Weigburger et al., 1978). The lab rats and mice in this National Cancer Institute study were fed DNCB as a large part of their diet, 3000 mg of DNCB per kilogram of food, for most of their life span (18 months) and had no increased risk of tumors. It is estimated that the amount tested in this study were 5 million times greater than standard human dosages and yet carcinogenic effect was detected (Dunagin, 1986).

Biological studies: DNCB appears to rapidly deplete the content of glutathione in rat skin (Summer and Goeggelmann, 1980), inducing an increase in glutathione synthesis (Schmidt and Chung, 1992); however, it doesn't significantly alter the activity of rat skin glutathione-S-transferases (Summer and Goeggelmann, 1980). In addition, when applied to rat skin, DNCB did not influence the cytochrome P-450 system or the activities of some correlated enzymes. Post-mitochondrial liver supernatant from rats treated locally with DNCB was not mutagenic in the Ames Test (Kratka et al., 1979). DNCB induced synthesis of heat shock proteins in murine splenocytes in vivo (Albers et al., 1998). DNCB combines with methionine at physiological pH, forming 1-S-methyldinitrobenzene, a compound that readily attacts DNA directly through nucleophilic substitution of the halogen atom (Gupta et al., 1997).

In human studies, routine biological parameters remained normal throughout the course of treatment (Daman et al., 1978; De Prost et al., 1982; Happle et al., 1978); however, DNCB is genotoxic by sister chromatid exchange in human skin fibroblasts at concentrations well below those used clinically (De Leve, 1996).

Toxicity: Lab animals have been fed very large doses of the compound without any deleterious effects (Weisburger et al., 1978). Human research subjects have also consumed a total dose of up to 60 mg of DNCB without any toxic effects (Lowney, 1973). No clinical syndromes related to chronic exposure to nitrochlorinated benzene derivatives were observed (Heid et al., 1984). The toxicity of dinitrophenol, one of the metabolites of DNCB, has toxicity characterized by hepatic and renal cellular changes, convulsive seizures, and hyperthermia (De Prost et al., 1982).

Contaminants: No pharmaceutical grade of DNCB is available (Wilkerson et al., 1983), The synthesis process creates other compounds that have been shown to contaminate commercial samples of DNCB (Wilkerson et al., 1983). Contaminants caused no significant increase in mutagenicity (Wilkerson et al., 1988). One contaminant orthochloronitrobenzene, however, may be the unchanged precursor that causes cancer in rats (Weisburger et al., 1978).

Occupational exposures: DNCB has industrial applications in color photography (Stricker et al., 1991) as an algicide used in air conditioners, refrigeration equipment, and other water systems (Adams et al., 1971; Zimmerman, 1970), and in the manufacture of sulfur blacks, rubber chemicals, explosives (Adams et al., 1971), textile dyes (Dorn, 1959; Raijka et al., 1956), and chloroamphenicol (Foussereau and Benezra, 1970; Gutschmidt, 1979). Because DNCB's potential for sensitization is well recognized, the use of DNCB has mainly been confined to closed systems. In the first decades of the century, industrial exposure to DNCB were reported (Bernstein, 1912; Wedrow, 1928). In the 1970s, four cases of contact dermatitis in air-conditioning repairmen exposed to DNCB were reported. Biochemistry workers who synthesize DNCB have also been sensitized (Pye and Burton, 1976).

Adverse events: Painful lymphadenopathy in the nodes draining the treated area has been noted (De Prost et al., 1982). Widespread exfoliative dermatitis, also known as generalized autosensitization (Carrel and Davidson, 1973; De Prost et al., 1982; De Prost et al., 1979; Lee et al., 1984; Lewis, 1973; Young, 1976) or erythema multiforme-like syndrome (Viraben et al., 1990), may develop during treatment. These episodes usually can be treated adequately by a short course of corticosteroids (Bukner and Price, 1978; Lewis, 1973). Bacterial superinfections (Bukner and Price, 1978; Czarnecki, 1979) and cellulitis (Grayson et al., 1982) have been a complication of eczema. Generalized urticaria and pruritis have been reported

(De Prost et al., 1982; Erikson, 1980; McDaniel et al., 1982). A case of general malaise, with chills, arthragias and insomnia evolving in a transitory fashion, was associated with DNCB treatment (De Prost et al., 1982). Depigmentation of short and/or longer duration is also reported (Bukner and Price, 1978; De Prost et al., 1982; Grosshans and Foussereau, 1982). Dyshidrodiform dermatitis and unexplained pain in the foot was described in one patient treated with DNCB for warts on the foot (Dunagin and Millikan, 1982). Local neck reaction of near tracheotomy severity has also been reported (Lewis, 1973).

Immunology

Mechanism of action: DNCB may act as a prohapten that is activated enzymatically by NADPH-dependent reductases in the skin with the concomitant generation of superoxide and hydrogen peroxide to form potentially protein-reactive free radicals and other metabolites (Schmidt and Chung, 1992). DNCB then combines covalently with proteins to form an immunogenic hapten-protein conjugate (Basketter et al., 1997). DNCB specifically binds to lysine groups intradermally (Catalona et al., 1972). DNCB also acts as an irreversible inhibitor of human thioredoxin reductase that is accompanied by a large increase in NADPH oxidase activity (Arner et al., 1995) and the stimulation of NAPDH-dependent oxygen utilization (Schmidt and Chung, 1992). Production of reactive oxygen species and inhibited reduction of thioredoxin by modified thioredoxin reductase after reactions with DNCB may play a major role in the inflammatory reaction (Nordberg et al., 1998).

Sensitization features: 95% of normal patients can be sensitized to DNCB (Catalona et al., 1972). DNCB sensitization properties and kinetics have been reviewed extensively (Friedmann, 1990). The ED50 or dose of DNCB required to sensitize 50% of patients is 116 μg; 100% sensitization occurs at 500 μg. The proportion of subjects sensitized increases with the log of the sensitizing dose (Friedmann, 1990). The main determinant of sensitivity is the antigen dose per unit area rather than the total dose or total area (Friedmann, 1990). The sensitization potential of DNCB depends on the vehicle used, possibly due to the effect of the vehicle on absorption and deposition (Heylings et al., 1996). For example, DNCB in acetone induces more lymph node proliferation than DNCB in propylene glycol (Heylings et al., 1996).

DNCB applied to skin causes a significant increase in tumor necrosis factor (TNF)-alpha and interleukin (IL)-6 production (Holliday et al., 1997). Dendritic cells upregulate CD86 expression (Aiba et al., 1997). DNCB has not been reported to induce respiratory sensitization (Dearman and Kimber, 1991) or produce circulating antibodies (Catalona et al., 1972; Chakravorty et al., 1973).

Factors modulating sensitization (Menne and Wilkinson, 1995): The majority of authors report that ability to sensitize decreases with age (Bleumink et al., 1974; Catalona et al., 1972; Gross, 1965; Waldorf et al., 1969); however, one study found no effect (Schwartz, 1953).

Sex may be a modulating factor. A large, well-controlled study found that men were more susceptible to DNCB sensitization than women (Walker et al., 1967). Another study found that women had an increased reactivity to challenge than men (Rees et al., 1989). Two studies found that women receiving progestogens through combined oral contraceptive pills or intramuscular progesterone had increased reactions to DNCB (Gerretsen et al., 1975; Sadoff et al., 1972).

Race may be a modulating factor as well. Blacks are reported to be less susceptible to DNCB contact sensitization than whites (Andersen, 1979; Kligman, 1966).

Diet may also play a role. A low-protein diet may reduce the density and alter the function of Langerhans' cells such that there is less sensitization to DNCB (Ghaznawie et al., 1989). Vitamin E deficiency decreases the contact sensitization responses to DNCB in mice (Ikarashi et al., 1998).

Certain diseases appear to cause a decrease in sensitizing ability: stage III malignant melanoma (Eilber et al., 1975), rosacea (Manna et al., 1982), squamous cell carcinomas in the head and neck region (Bier et al., 1983), chronic renal failure (Maxwell et al., 1981), advanced cutaneous T-cell lymphoma (Vonderheid et al., 1998), and postmastectomy lymphedemous limb (Mallon et al., 1997). Active Hodgkin's disease is associated with complete anergy (Aisenberg, 1962). Most studies find a decreased sensitization ability in atopics (Menne and Wilkinson, 1995). In melanoma patients, age (Camacho et al., 1981) and sex (Bleumink et al., 1974; Camacho et al., 1981) did not appear to have an effect on sensitization.

Decreased reactions to DNCB have been noted for the following treatments: UVB light exposure (Mommaas et al., 1993; Tseng et al., 1992), PUVA therapy (Moss et al., 1981), X-rays, systemic prednisone, topical corticosteroids (Menne and Wilkinson, 1995). Hypnotic suggestion may increase or decrease the response to DNCB during sensitization (Zachariae and Bjerring, 1993).

PAR: The PAR (primary allergic reaction), also known as spontaneous flare, is of clinical and theoretical interest (Catalona et al., 1972). The PAR is an allergen-specific reaction that occurs if allergen or T cells are retained at a skin site (Scheper and Von Blomberg, 1995). The PAR corresponds to the time when sufficient numbers of effector T cells have entered the circulation and come in contact with allergen retained in the skin from the first exposure. PAR-associated local allergen retention usually is of short duration only (Scheper and Von Blomberg, 1995). In contrast, if a PAR is due to T-cell retention, the locally increased antigen-specific hyperreactivity can last for several months (Scheper and Von Blomberg, 1995). Only a few specific T cells seem necessary for a macroscopic effect (Scheper and Von Blomberg, 1995). The intensity of a PAR can be a reliable predictor of the intensity of the elicitation response; however, the absense of a PAR does not necessarily indicate a failure to sensitize (Kelly et al., 1998). A PAR is characterized histologically by intracellular and intercellular edema, exocytosis, and intrafiltration of the dermis by basophils, monocytes, large lymphocytes, macrophages, and eosinophils.

One study observed a PAR in approximately 50% of normal individuals 10–14 days after the first DNCB exposure (Waksman, 1960). Another reported a higher percentage with a 2-mg sensitization: 97.2% of 143 normal subjects had a PAR within 9–10 days (Catalona, 1972). DNCB appears to be only capable of staying in the skin and inducing such a response for 2 weeks, explaining in part why the PAR disappears over time.

Tolerance: Tolerance is a state of immunologic unresponsiveness to a typically potent antigen (Tron and Sunder, 1991). There at least three ways to induce tolerance to a contact sensitizer (Tron and Sunder, 1991): (1) administer the antigen first via the oral intravenous, intramuscular, or intraperitoneal route instead of epicutaneously or cutaneously (Manrer, 1983; Waksman, 1960); (2) apply to a cutaneous site that has somehow been altered (e.g., UV irradiated) (Elmets et al., 1983; Toews et al., 1980); (3) first administer an antigen similar to the sensitizer but modify to some degree (Sommer et al., 1975). One theory on the phenomenon of tolerance is that it reflects suppressor T-cell induction.

The ultimate level of sensitivity decreases significantly when humans consume more than 20 mg of DNCB by mouth (total dose) (Lowney, 1973). In mice, high-dose oral/parenteral DNCB induces systemic tolerance, as evidenced by depressed DTH responses and reduced serum IgG responses (Oliver and Silbart, 1998), whereas low-dose DNCB induces only local tolerance (Oliver and Silbart, 1998). Cholera toxin was incapable of breaking this preexisting tolerance (Oliver and Silbart, 1998). Cimetidine modulates immune responsiveness (Avella et al., 1978; Rocklin, 1976) and has been shown to inhibit suppressor T-cell activity (Nair and Schwartz, 1983). Sensitivity to DNCB can sometimes be induced in unresponsive patients by the administration of cimetidine (Breuillard and Szapiro, 1978; Daman and Rosenberg, 1977; Daman et al., 1978; De Prost et al., 1982; Swanson et al., 1981; Thestrup-Pederson et al., 1982).

Desensitization: Desensitization, also known as hyposensitization, is defined as the phenomenon of losing responsiveness to a chemical during a course of therapy, and is believed to be due to receptor down-regulation. Typically, desensitization is temporary. If treatment is stopped, sensitivity will return spontaneously within 6–10 months (Maibach et al., 1985). The terms "desensitization" and "tolerance" often are interchanged in the literature. Cimetidine sometimes reverses clinical desensitization. In 25 patients treated for alopecia areata (AA), 2 desensitized and both reversed their desensitization with cimetidine (Swanson, 1981). In another study, out of 30 patients treated for AA, one desensitized and was successfully reversed with cimedine (Temmerman et al., 1984). (De Prost et al., 1982) had 5/42 AA patients desensitize and 4/5 cases reversed with cimetidine. Breuillard reports that in 30 patients treated for AA, 15 became "tolerant" (i.e., desensitized to DNCB), but not all attempts to treat with cimetidine were successful (Breuillard and Szapiro, 1978). Another agent that has been useful in reversing desensitization is BCG (Klein et al., 1975).

Cross-sensitization: Cross-sensitization to nitrobenzene chemicals used in labs, agriculture, and industry is theoretically possible. DNCB has been reported to cross-sensitize to pyrrolnitrin (Micotrin®), an antifungal cream used in Europe and Japan, and chloramphenicol (Meneghini and Angelini, 1975; Schwank and Jirasek, 1963); however, in the case of chloramphenicol, subsequent studies failed to confirm this finding (Djawari et al., 1980; Eriksen, 1978; Lowney, 1973; Palacios et al., 1968; Strick, 1983). Therefore, it is hypothesized that (Strick, 1983) an irritation reaction may have been misinterpreted as DTH in the first study (Schwank and Jirasek, 1963). Most likely, therefore, DNCB does not cross-sensitize to chloramphenicol. After administration of DNCB to 82 patients, 2 of those patients became sensitized to 1% picric acid but not to other derivatives with structures related to DNCB (Djawari et al., 1980). Other possible cross-sensitizers are DNFB, the Sanger reagent used by biochemists to mark the terminal amino acid of polypeptides and protein (Garcia-Perez, 1978), and the herbicide DNA-nitraline (Nishioka et al., 1983).

Repeated DNCB exposure has been reported to cause sensitization to nonrelated contact allergens, such as dithranol (De Groot et al., 1981); however, unfortunately these tests were not followed up sufficiently to distinguish between irritant and true contact hypersensitivity (Heid et al., 1984). Sensitization to nickel after treatment with DNCB has also been reported (De Prost et al., 1982). It is unclear how many of these observations may be false-positives due to the excited skin syndrome.

SQUARIC ACID DIBUTYL ESTER

Pharmacology

Alternate names: 3,4 dihydroxy-3-cyclobutene-1,2 dione or 1,2dibutoxycyclobutene.
Formula: $C_{12}H_{18}O_4$.
MW: 226.28 g/mol.
Solubility: Acetone, butanol, ethanol, isopropanol.
Properties: d = 1.047.
Stability: The stability of SADBE depends on the storage solution. SADBE is sensitive to hydrolysis in a few hours (Noster et al., 1976) by atmospheric water, forming squaric acid. A half squarate ester is formed prior to complete hydrolysis. SADBE stored in alcohol undergoes significant hydrolysis in molar excess of water. If SABDE is stored in 100% isopropanol, it undergoes a transesterification reaction, resulting in possible changes in sensitization properties. Hydrophylic petrolatum also is an acceptable storage medium. If SADBE is stored in ethanol, it undergoes transesterification to SADEE. SADBE is best stored in butanol or acetone solution. It is important to add molecular sieves in order to prevent hydrolysis and maintain a consistent concentration, especially with very dilute solutions of squarate esters (Wilkerson et al., 1985). There was slight hydrolysis over a 3-week period even if optimal conditions were followed (i.e., molecular sieves and acetone) (Wilkerson et al., 1985).
Synthesis: Wilkerson describes a protocol to synthesize SADBE from squaric acid or 1,2 dihydroxy-cyclobutene-dione (Arner et al., 1995; Basketter et al.,

1997; Catalona et al., 1972). Because squaric acid is produced by cyclization of the carcinogen hexachlorobutadiene and also has a carcinogenic intermediate tetrachloro-2-cyclobutene-1-one, commercial samples could potentially contain these chemicals as contaminants (Wilkerson et al., 1985). One study screened commercial SADBE with gas chromatography and mass spectrometry and found no contaminants (Wilkerson et al., 1985).

Absorption: In mouse skin (Sherertz and Sloan, 1988), there is a linear relationship between cumulative milligrams delivered and time for the first 24 hours. Between 24 and 48 hours, the amount delivered reaches a plateau of 3 mg cumulative dose. The flux of 2% squaric acid through human skin in vitro was measured to be $400 \pm 200 \frac{mg}{cm^2 h}$ (SD) with a lag time of 2.1 hours (Sherertz and Sloan, 1988).

Toxicology

Mutagenicity studies: SADBE was not mutagenic in the Ames Test (Happle et al., 1983; Strobel and Rohrborn, 1980).

Carcinogenicity studies: No data were found on carcinogenicity testing of SADBE. Breakdown product squaric acid causes a low incidence of tumors at the injection site in rats (Van Duuren et al., 1971). The SADBE precursor, Hexochlorobutadiene, is nonmutagenic in Ames test but causes cancer in rats (Kochiba et al., 1977). Another intermediate in the synthesis process, tetrachloro-2-cyclobutene-1-one, was found to induce tumors in mice (Kochiba et al., 1977). Lifetime subcutaneous injections of tetrachloro-2-cyclobutene-1-one, an intermediate in the synthesis of SADBE, caused rhabdomyosarcoma (Van Duuren et al., 1971).

Biological studies: SADBE does not deplete epidermal glutathione in rat skin at a concentration of 2%. The observation that 0.2 mM of SADBE depletes glutathione in isolated rat hepatocytes (K. H. Summer, unpublished data) suggests that SADBE has a reactive electrophilic nature (Happle et al., 1983).

(Case, 1984) monitored complete blood count (cbc), urinalysis, liver, and renal function tests every 2–3 months while patients were on SADBE therapy and observed no changes.

Toxicity: There are no data on the toxicity of SADBE.

Contaminants: No carcinogenic contaminants were found when SADBE studied with gas chromatography-mass spectrometry (Wilkerson et al., 1985).

Occupational exposures: No data were found.

Adverse events: Foley reported 71% (10/14) patients with severe eczematous, reactions that were enough to withdraw from the study (Foley et al., 1996). She also reports disseminated eczematous reactions in 9/14 patients (Foley et al., 1996). Barth also had 12/17 patients (70%) with severe eczematous reactions (Barth et al., 1986). He also describes a burning erythema lasting from 2–5 days, believed to be an irritant reaction, as well as an edematous weeping eczema on day 7 (Barth et al., 1986). Fowler describes a case of persistent contact dermatitis recalcitrant to all therapies but excision (Fowler et al., 1993). Valsecchi reports depigmentation at treatment sites (Valsecchi and Caintelli, 1984). Foley also reported temporary hypopigmentation at treatment sites (Foley et al., 1996), and posterior cervical lymphadenopathy (Flowers et al., 1982), painful at times (Case et al., 1984; Caserio, 1987).

Immunology

Mechanism of action: The portion of the squarate molecule that produces sensitization is not known. The dicarbonyl portion of squarate esters reacts with arginine and lysine residues (Whitfield and Friedman, 1972), secondary and tertiary amines, and thiols to produce stable derivatives. This multiplicity of possible reaction sites may explain its excellent sensitizing ability. SADBE appears to only allergenically alter the skin at site of contact (Daman et al., 1978; Fowler et al., 1993; Noster et al., 1976).

Sensitization features: (Perret et al, 1990) reported that about 1%–2% of patients could not be sensitized with SADBE. Sensitization declines over time in some patients, but gradual hydrolysis of the ester over time does not explain this observation (Claudy and Roche, 1978; Wilkerson et al., 1985).

SADEE, formed when SADBE is stored in ethanol, has strong sensitizing activity (Noster et al., 1976), but SADBE produces a more pronounced 24-hour effect in guinea pigs than SADEE (Happle et al., 1983). Thus, SADBE has been preferentially used (Happle et al., 1983). The sensitization properties of the intermediate breakdown product, a squarate half ester, are not known (Wilkerson et al., 1985). The final hydrolysis breakdown product, squaric acid, presumably does not sensitize because it cannot form bonds with proteins to form an antigen as can the two esters.

Factors modulating sensitization: No data were found.

Cross-sensitization: SADBE exposure has been reported to cause increased nickel sensitivity (Valsecchi and cainelli, 1984) and persistent patch test reactions to substances such gold (Fowler et al., 1993), but excited skin syndrome was not excluded in these cases. The patch test reaction usually disappears after 3–6 weeks, but in one case the dermatitis was so intense and persistent that it only resolved after excision (Fowler et al., 1993).

Desensitization: (Case et al., 1984) reported that 2/26 patients became desensitized to SADBE. After treatment with cimetidine, 1/2 regained sensitization.

DIPHENYLCYCLOPROPENONE

Pharmacology

Formula: $C_{15}H_{10}O$.
Other names: Diphencyprone (Happle et al., 1983), 2,3-diphenyl cyclopropenone-1.

MW: 206.24 g/mol.

Solubility: DPCP is soluble in ethanol, cyclohexane, and acetone (Wilkerson et al., 1984); however, it is not very soluble in cyclohexane, making it difficult to prepare a 1% solution (Wilkerson et al., 1984). It is stable in darkness at room temperature and −70°C, but undergoes pyrolysis upon melting (melting point is 121–121.5°C) or exposure to sunlight, incandescent or fluorescent light to yield diphenylacetylene (DPA), and carbon monoxide (CO) (Wilkerson et al., 1984).

Properties: White powder, m.p = 119–120°C.

Stability: DPCP breaks down into DPA and CO in the presense of light. All solutions exposed to sunlight or fluorescent light had completely decomposed (>98%) after 2 weeks. After 1 hour of exposure to sunlight, DPCP underwent at least a 60% conversion rate. Fluorescent light at a distance of 0.5 m produced a 4% conversion in 6 hours, whereas an incandescent light bulb produced 2% conversion in 6 hours (Wilkerson et al., 1984). Storage in acetone may protect DPCP from breaking down because acetone absorbs UV light (Wilkerson et al., 1987). Wilkerson (Wilkerson et al., 1987) wrapped bottles in aluminum foil to exclude light. Solutions that were protected from light and stored at either −70°C or room temperature did not change spectral absorption patterns over the 4-week period (Wilkerson et al., 1984). No one has studied the shelf stability of DPCP protected from light beyond 4 weeks. Some authors assert that the stable half-life of the drug is at least 6 months, probably longer (Van der Steen et al., 1991). The greater photolability of DPCP brings up the increased theoretical possibility of phototoxicity reactions.

Storage: Because of its lability, DPCP requires rigorous storage conditions. It must be protected from light, either by storing in amber bottles or covering with aluminum foil, and replaced every 2–4 weeks.

Synthesis: DPCP is synthesized by bromination of dibenzylketone to yield a, a′-dibromodibenzylketone followed by cyclization with base to yield DPCP (Wilkerson et al., 1984).

Absorption: DPCP is not detectable in urine or serum following topical application (Berth-Jones and McBurney, 1994), and therefore does not appear to accumulate in the body unchanged. It is still possible, however, that DPCP is absorbed and metabolized exceptionally rapidly or that its breakdown products are absorbed instead. Further work is needed to look for breakdown products or metabolites in blood and urine.

Toxicology

Mutagenicity studies: DPCP is nonmutagenic in Ames test at concentrations of 50 µg/ml and 100 µg/ml, and also in the host-mediated assay, a modification of the Ames test using metabolic activation with rat liver (Stuli et al., 1981; Wilkerson et al., 1984). Precursor dibenzlketone is nonmutagenic, but the intermediate, a, a′ dibromodibenzylketone, is a potent mutagen (Wilkerson et al., 1987). As a synthetic precursor to DPCP, this mutagen could potentially contaminate commercially obtained samples of DPCP (Wilkerson et al., 1987), and some clinicians using DPCP have protocols to screen for this contaminant.

Although the photolysis product, DPA, is nonmutagenic in the Ames test (Wilkerson et al., 1987), UV light causes a detectable, twofold over background, mutagenic response to DPCP in Ames bacterial test strain UTH8413 (Wilkerson et al., 1987). It is hypothesized, therefore, that S-9 microsomal enzymes may be trapping at least one photochemically produced, short-lived, high-energy intermediate that is mutagenic (Wilkerson et al., 1987).

Carcinogenicity studies: There are no data.

Biological studies: DPCP had no teratogenic or organ toxicity in hen's egg test (Happle, 1985) or mouse teratogenicity assay (Happle, 1986). No alteration in blood count, renal, or hepatic function was observed in treated patients checked at 3 (Van der Steen et al., 1991) or 4–6 month intervals (Hull and Cunliffe, 1992).

Toxicity: There are no data.

Contaminants: There are no data.

Occupational Exposures: Industrial uses of DPCP (e.g., as a synthetic intermediate) may increase in the future because of the potential for new products (Quinkert et al., 1963; Rosen et al., 1976; Toda, 1976; Wilkerson et al., 1984). Dermatologists (Shah et al., 1996), chemists (Sansom et al., 1995), pharmacy workers, and nurses (Adisesh et al., 1997) have become sensitized. Appropriate precautions should be taken to avoid exposure: restrict use to trained staff, fully inform of risks, perform procedures in designated area with adequate ventilation, wear gloves and protective garments, change garments between patients, and assure safe storage and disposal of drug and contaminated applicators/clothing (Shah et al., 1996).

Adverse events: Some of the reported adverse events associated with therapy were severe eczema, both localized and generalized (Pericin and Trueb, 1988), frequent headaches (Van der Steen et al., 1991); sleep disturbances due to pruritis (Van der Steen et al., 1991); erythema multiforme (Kim et al., 1996; Oh et al., 1998; Pericin and Trueb, 1988; Perret et al., 1990; Puigand Allegre, 1994; Van der Steen et al., 1991); dyschromia in confetti (Van der Steen et al., 1991); generalized rash (Lane and Hogan, 1988); extrasystoles (Lane and Hogan, 1988); generalized urticaria (Lane and Hogan, 1988; Tosti, 1989); vitiligo (Duhra and Foulds, 1990; Hatzi et al., 1988; Henderson and Ilchyshyn, 1995; MacDonald-Hull et al., 1989); transient or longer-lasting pigmentation in treated areas (Lautenschlager and Itin, 1998; Orecchia et al., 1988); regional lymphadenopathy (Berth-Jones and McBurney, 1994); generalized dermatitis (Hatzi et al., 1988); high fever, chills, and headache (Hatzi et al., 1988); serum sickness–like disease with fever, arthralgias, and elevated serum IgG levels (Pericin and Trueb, 1988); fainting, probably due to anaphylactic reaction in one woman who was positive to a prick test with .01% DPCP (Hatzi et al., 1988). Berth acertains that such side effects are rare (Berth-Jones and McBurney, 1994).

Immunology

Mechanism of action: It has been hypothesized that DPCP could be acting as a contact sensitizer, a photoallergic sensitizer, or both (Wilkerson et al., 1984). In addition, DPCP could act as a prosensitizer, with the actual sensitizer liberated by a photoactivated stable, metastable, or unstable intermediate (Wilkerson et al., 1984).

Sensitization features: Mayerhausen sensitized 10 patient with malignant melanoma simulaneously with DNCB and DPCP and found that reactions correlated within individuals but that doses of DPCP were lower than DNCB to obtain a skin reaction of the same intensity (Mayerhausen, 1987). Other authors observed a linear dose-dependent increase in sensitization response with DPCP and DNCB and also found DPCP to be a more potent sensitizer than DNCB (Kelly et al., 1998). For example, patients sensitized with twofold lower doses and challenged with 10-fold lower doses developed stronger elicitation responses with DPCP than with DNCB (Kelly et al., 1998). The mean slope of the elicitation responses to DPCP was 18 times greater than the slope for DNCB (Kelly et al., 1998). There was no difference in DPCP's ability to sensitize people with skin type I/II vs. III/IV (Kelly et al., 1998). To our knowledge, the ED50 of DPCP has not been reported. There are no data on the ability of breakdown product DPA to sensitize. The mutagenic precursor to DPCP, a,a'dibromodibenzylketone, elicts a reaction in guinea pigs sensitized to DPCP.

Factors modulating factors: Hypnotic suggestion is reported to be capable of both increasing or decreasing response to DPCP during sensitization (Zachariae and Bjerring, 1993). More highly hypnotizable subjects experienced more relaxation and greater reactions during sensitization (Zachariae et al., 1997). In general, relaxation during sensitization appears to increase the erythema and induration (Zachariae et al., 1997).

Cross-sensitization: None were reported.

Desensitization: (Van der Steen et al., 1992) reports that 15/139 patients became de-sensitized. Seo applied low concentrations (0.001) of DPCP weekly for 4 months and observed systemic and local hyposensitization (Seo and Eun, 1996). According to (Hatzi et al., 1988), desensitization was observed in 4 to 7 months in 12/45 patients (26.7%), a higher rate of desensitization than with DNCB or SADBE. The same phenomenon was observed by (Orecchia and Rabbiosi, 1985).

CONCLUSIONS

DNCB Mutagenicity/Carcinogenicity Debate

There has been considerable debate in the literature over the mutagenicity of DNCB. (Kratka et al., 1979) was the first to raise concerns about mutagenicity. Driven by these concerns, in the 1980s there was a shift away from DNCB to DPCP and SADBE.

Several authors recommended abandoning the use of DNCB for benign conditions (Happle, 1985; Heid et al., 1984). In 1979, however, DNCB had been rigorously tested for carcinogenicity and found not to be a carcinogen. Several authors continue to defend the safety of DNCB (Dunagin, 1986; Stricker et al., 1991). Some authors argue that for ordinary therapeutic doses, DNCB is safer than a large number of other drugs currently in use for benign conditions as well as many common foods (Dunagin, 1986). The safety of DPCP and SADBE remain unknown, as animal and sister chromatid experiments have not yet been completed (Happle, 1986).

Advantages/Disadvantages of the Respective Agents

The ideal contact sensitizer would be a substance that is available, able to sensitize at least 95% of normal individuals, chemically stable, free of adverse effects, and not found in human environments (Naylor et al., 1988).

DNCB, DPCP, and SADBE are all potent sensitizers. DPCP is significantly more potent than DNCB for a given dose. There is a good correlation between the intensity of responses to DNCB and DPCP within a given individual.

SADBE and DPCP are nonmutagenic in the Ames Test, whereas DNCB is mutagenic. Carcinogenicity studies on DNCB were negative, but more recent genotoxicity studies were positive. SADBE and DPCP have not undergone carcinogenicity or genotoxicity testing . They carry approximately a 17% risk of being carcinogenic if the sensitivity of the Ames test is 83% (Ames, 1984). DNCB is nontoxic at high doses in mice and human subjects, whereas DPCP and SADBE have not undergone such rigorous testing.

DNCB is the most stable of all three compounds. Squaric acid requires refrigeration and possibly special solvents and additives to prevent hydrolysis and loss of potency (Naylor et al., 1988). DPCP must be stored in dark glass in a dark place because it readily undergoes photodecomposition.

All three compounds can have severe adverse side effects, and it is unclear which has less side effects than the others.

All three compounds can be found in human environments, albeit at low levels in specialized populations. DNCB, given its longer history, has more applications and therefore a potentially greater chance of exposure; however, in the future, the other sensitizers may become equally prevalent.

DNCB and DPCP are more easily available because they can be ordered from well-known supply companies. SADBE, however, has no consistent commercial source according to (Wilkin, 1986). Thus, concerns can be raised about solvent, hydration of solution, type and number of impurities, type of container employed, contaminants, storage factors, and shelf-life.

Summary

Given the current state of knowledge, DNCB is the best characterized chemical and has superior stability, but is genotoxic and therefore should be used only with careful consideration of clinical risks and benefits. DPCP and SADBE are less well characterized and more unstable.

The presence of rare but serious side effects associated with the use of these agents should be investigated further. Patients need to be made fully aware of the potential adverse effects and give their informed consent before starting therapy. Some of the most severe side effects occurred when patients were allowed to handle the chemicals themselves. Contact sensitizers should be regarded as investigational drugs (Wilkin, 1986) and carefully monitored by an experienced physician. In summary, currently there is no ideal contact sensitizer. Investigative efforts in the future should be focused on better characterizing the safety of the known sensitizers, DNCB, DPCP, and squaric acid, and searching for new sensitizers with better safety profiles and less side effects.

REFERENCES

Adams, R. M., Zimmerman, M. C., Bartlett, J. B., and Preston, J. F. (1971): 1-Chloro-2,4-Dinitrobenzene as a algicide. *Arch. Derm.*, 103:191–193.

Adisesh, A., Beck, M., and Cherry, N. M. (1997): Hazards in the use of diphencyprone. *British Journal of Dermatology*, 136:465–479.

Aiba, S., Terunuma, J., Manome, H., and Tagami, H. (1997): Dendritic cells respond differently to haptens and irritants by their production of cytokines and expression of co-stimulatory molecules. *Eur. Journal. Immunol.*, 27:3031–3038.

Aisenberg, A. (1962): Studies on delayed hypersensitivity in Hodgkin's disease. *J. Clin. Invest.*, 41:1964–1970.

Albers, R., Van der Pijl, A., Bol, M., Seinen, W., and Pieters, R. (1998): Stress proteins (HSP) and chemical-induced autoimmunity. *Toxicol Appl. Pharmacol.*, 140:70–76.

Ames, B. N. (1984): Cancer and diet. *Science*, 224;668–670; 757–760.

Ames, B. N. (1984): The detection of environmental mutagens and potential carcinogens. Cancer 53:2034–1040.

Andersen, K. E., and Maibach, H. I. (1979): Black and white skin differences. *J. Am. Acad. Dermatol.*, 1:276–282.

Arner, E. S. J., Bjornstedt, M., and Holmgren, A. (1995): 1-chloro2,4-nitrobenzene is a irreversible inhibitor of human thioredoxin reductase. *J. Biological Chemistry.*, 270:3479–3482.

Avella, J., Binder, H. J., Madsen, J. E., Binder, H. J., and Askenase, P. W. (1978): Effect of histamine H2: Receptor antagonists on delayed hypersensitivity. Lancet, 1:624–626.

Barth, J. H., Darley, C. R., and Gibson, J. R., (1985): Squaric acid esters in the treatment of alopecia areata. *Dermatologica.*, 170:40–42.

Barth, J. H., Darley, C. R., and Gibson, J. R., (1986): Squaric acid esters in the treatment of alopecia areata. *J. Am. Acad. Dermatol.*, 5:846.

Basketter, D. A., Dearman, R. J., Hilton, J., Kimber, I. (1997): Dinitrohalobenzenes: evaluation of relative skin sensitization potential using the local lymph node assay. *Contact Dermatitis*, 36:97–100.

Bernstein, M. J. (1912): A dermatitis caused by "di-nitro-chlor benzol." Lancet, 1:534.

Berth-Jones, J., McBurney, A., and Hutchinson, P. E. (1994): Diphencyprone is not detectable is serum or urine following topical application. *Acta. Derm. Veneol.*, 74:312–313.

Bier, J., Nicklisch, U., and Platz, H. (1983): The doubtful relevance of nonspecific immune reactivity in patients with squamous cell carcinoma of the head and neck region. Cancer, 52:1165–1172.

Black, H. S., Castrow, F. F., and Gerguis, J. (1985): The mutagenicity of Dinitrochlorobenzene. *Arch. Dermatol.* Vol. 121.

Bleumink, E., Nater, J. P., Koops, H. S., The T. H. (1974): A standard method for DNCB sensitization in patients with neoplasms. Cancer, 33:911–915.

Breslow, R., Haynie, R., and Mirra, J. (1959): Synthesis of diphenylcylcopropenone. *J. Am. Chem. Soc.*, 81:247–248.

Breuillard, F., and Szapiro, E. (1978): Cimetidine in acquired tolerance to dinitrochlorobenzene. Lancet, 726.

Buckner, D., and Price, N. M. (1978): Immunotherapy of verrucae vulgares with dinitrochlorobenzene. *Br. J. Derm.*, 98:451–455.

Camacho, E. S., Pinsky, C. M., Braun, D. W., Golbey, R. D., Fortner, J. G., Wanebo, H. J., and Oettgen, H. F. (1981): DNCB reactivity and prognosis in 419 patients with malignant melanoma. *Cancer* 47:2446–2450.

Carrel, J. M., and Davidson, D. M. (1973): Topical Immunotherapy of Recalcitrant verrucae: A new therapeutic approach. *Journal American Podiatric Association.*, 63:293–296.

Case, P, Mitchell, A., Swanson, N., Vanderveen, E. E., Ellis, C. N., and Headington, J. T. (1984): Topical therapy of alopecia areata with squaric acid dibutylester. Topical therapy of alopecia areata with squaric acid dibutylester. *J. Am. Acad. Dermatol.*, 10:447–451.

Caserio, R. (1987): Treatment of alopecia areata with squaric acid dibutylester. *Arch. Dermatol.*, 123:1036–1041.

Catalona, W. J., Taylor, P. T., and Chretien, P. B. (1972): Quantitative dinitrochlorobenzene contact sensitization in a normal population. *Clin. Exp. Immunol.*, 12:325–333.

Catalona, W. J., Taylor, P. T., Rabson, A. S., and Chretien, P. B. (1972): A method for dinitrochlorobenzene contact sensitization. *N. Engl. J. Med.*, 286:399–402.

Chakravorty, R. C., Curutchet, H. P., Copolla, F. S., Park, C. M., Blaylock, W. K., and Lawrence, W. (1973): The delayed hypersensitivity reaction in the cancer patient: Observations on senstization by DNCB. *Surgery*, 73:730–735.

Claudy, A. L., and Roche, H. (1978): Immunotherapy of multiple recurring warts. II: Reassessment of the use of squaric acid dibutyl ester (SADBE). *Ann. Dermatol. Venereol.*, (Paris). 108:765–767.

Cohen, S., Lacher, J. R., and Park, J. D. (1959): Diketocyclobutenediol. *Journal of the American Chemical Society*, 81:3480.

Czarnecki, N., and Hintner, H. (1979): Verrucae vulgaris. Lokalbehandlung mit DNCB. *Wien. Klin. Wochenschr.*, 91:822–825.

Daman, L. A., and Rosenberg, E. W. (1977): Acquired tolerance to dinitrobenzene reversed by cimetidine. *Lancet*, 2:1087.

Daman, L. A., Rosenberg, E. W., and Drake, L. (1978): Treatment of alopecia areata with dinitrochlorobenzene. *Arch. Dermatol.*, 114:1036–1038.

De Groot, A. C., Nater, J. P., Bleumink, E., and De Jong, M. C. J. M. (1981): Does DNCB Therapy potentiate epicutaneous sensitization to non-related contact allergens? *Clinical and Experimental Dermatology*, 6:139–144.

De Leve, L. D. (1996): Dinitrochlorobenzene is genotoxic by sister chromatid exchange in human skin fibroblasts. *Mutat. Res.*, 371:105–108.

De Prost, Y., Paquez, F., and Touraine, R. (1982): Dinitrochlorobenzene treatment of alopecia areata. *Arch. Dermatol.*, 118:542–545.

De Prost, Y., Paquez, F., and Touraine, R. (1979): Traitement de la pelade par applications locale de DNCB. *Ann. Dermatol. Venerol*, 106:437–440.

Dearman, R. J., and Kimber, I. (1991): Differential stimulation of immune function by respiratory and contact chemical allergens. *Immunology*, 72:563–570.

Djawari, D., Koerner, E., and Haneke, E. (1980): Krenzallergien nach Sensibiliserung mit Dinitrochlorobenzol (DNCB). *Hautarzt*, 31:198–202.

Dorn, H. (1959): Chronische Ekzeme durch Kleiderfarben und Textilhilfsprodukte. *Z. Hautkr.*, 26:4–8.

Drill, V. A., Lazar, P. (1977): Cutaneous Toxicity. Academic Press, New York, Inc. pp. 189–201.

Duhra, P., and Foulds, I. S. (1990): Persistant Vitiligo induced by diphencyprone. *Br. J. Derm.*, 123:415–416.

Dunagin, W. G., and Millikan, L. E. (1982): Dinitrochlorobenzene immunotherapy for verrucae resistant to standard treatment modalities. *J. Am. Acad. Derm.*, 6:40–45.

Dunagin, W. G. (1986): Potential hazards of DNCB disputed. *Arch. Dermatol.*, 122:12–13.

Eilber, F. R., Nizze, J. A., and Morton, D. L. (1975): Sequential evaluation of general immune competence in cancer patients: Correlation with clinical course. *Cancer* 35:660–665.

Elmets, C. A., Bergstresser, P. R., Tigelar, R. E., Wood, P. J., and Streilein, J. W. (1983): Analysis of the mechanism of unresponsiveness produced by haptens painted on skin exposed to low dose ultraviolet radiation. *J. Exp. Med.*, 158:781–794.

Eriksen, K. (1978): Cross allergy between paranitro compounds with special reference to DNCB and chloramphenicol. *Contact Dermatitis.*, 4:29–32.

Erikson, K. (1980): Treatment of the common wart by induced allergic inflammation. *Dermatologica*, 160:161–166.

Feldmann, R. J., and Maibach, H. I. (1979): Absorption of some organic compounds through the skin in man. *J. Invest. Dermatol.*, 54:399–404.

Flowers, F. P., Slazinski, L., Fenske, N. A., and Pullara, T. J. (1982): Topical squaric dibutyleester therapy for Alopecia Areata. Cutis 30:433–436.

Foley, S., Blattel, S. A., and Martin, A. G. (1996): Clinical sequelae associated with squaric acid dibutylester topical sensitization. *Am. J. Contact Dermatitis*, 7:104–108.

Foussereau, J., and Benezra, C. (1970): *Les eczemas allergiques professionnels*. Masson, Paris, pp. 347–351.

Fowler, J. F., Hodge, S. J., and Tobin, G. R. (1993): Persistent allergic contact dermatitis from squaric acid dibutyl ester. *J. Am. Acad. Dermatol.*, 28:259–260.

Friedmann, P. (1990): The immunology of allergic contact dermatitis: The DNCB Story. *Adv. Dermatology*, 5:175–196.

Garcia-Perez. (1978): Occupational dermatitis from DNCB with cross-sensitivity to DNCB. *Contact Dermatitis.*, 4:125–127.

Gerretsen, G., Bleumink, E., Kremer, J., and Nater, J. P. (1975): Dinitrochlorobenzene sensitization test in women on hormonal contraceptives. Lancet, 23:347–349.

Ghaznawie, M., Papadimitriou, J. M., and Heenan, P. J. (1989): Reduced density and morphologic alteration of 1a and ADP Positive Langerhans cells after low protein diet. *Br. J. Dermatol.*, 120–341.

Grayson, R. J., Ratner, S. W., and Shaps, R. S. (1982): Topical DNCB Therapy for recalcitrant verruca plantaris. *Journal of the American Podiatry Association*, 72:557–559.

Gross, L. (1965): Immunological defect in aged population and its relation to cancer. Cancer 18:201–204.

Grosshans, E., and Foussereau, J. (1982): Complications et complexites de lectures des tests epicutanes. *Ann. Dermatol. Venerol.*, 110:259–268.

Gupta, R. L., Saini, B. H., and Juneja, T. R. (1997): Nitroreductase independent mutagenicity of 1-halogenated-2,4-dinitrobenzenes. *Mutat. Res.*, 381:41–47.

Gutschmidt, E. (1979): Beitrag zur DNCB. Therapie der Alopecia areata. *Z. Hautkr.*, 54:430–435.

Happle, R., Hausen, B. M., and Wiesner-Menzel, L. (1983): Diphencyprone in the treatment of alopecia areata. *Acta. Derma. Venereol.* (Stockh), 63:49–52.

Happel, R. (1985): The potential hazards of DNCB. *Arch. Dermatol.*, 121:330–331.

Happle, R., Kalveram, K. J., Buechner, U., Echternacht-Happle, K., Goeggelmann, W., and Sumner, K. H. (1980): Contact allergy as a therapeutic tool for alopecia areata: Application of squaric acid dibutylester. *Dermatologica*, 161:289–297.

Happle, R. (1985): Immunologische Behandlung der Alopecia areata. *Hautarzt*, 36(suppl 7):108–111.

Happle, R. (1986): Alopecia areata und alopecia totalis: Pathogenese und topische Immuntherapie mit Diphencyprone. *Therapiewoche*, 36:2706–2716.

Happle, R. (1986): In reply to Potential hazards of DNCB disputed. *Arch. Derm.*, 122:12–13.

Happle, R., Cebulla, K., and Echternacht-Happle K. (1978): Dinitrochlorobenzene therapy for alopecia areata. *Arch. Dermatol.*, 14:1629–1631.

Hatzi, J., Gourgiotou, A., Tosca, A., Varelzidis, A., and Straigos, J. (1988): Vitiligo as a reaction to topical treatment with diphencyprone. *Dermatologica*, 177:146–148.

Hausen, B. M., and Stute, J. (1980): Diphenylcyclopropenone: A strong contact sensitizer. *Chem. and Industry*, 17:699–700.

Hausen, B., and Happel, G. (1982): Cyclopropenenones for the local treatment of alopecia areata. *European Patent Application EP*, 62:157, Granted, CA98:59919h.

Heid, E., Grosshans, E., Laplanche, G., and Caussade, P. (1984): Dinitrochlorobenzene (DNCB): Usage therapeutique. *Ann. Dermatol. Venereol.*, 111:841–849.

Henderson, C. A., and Ilchyshyn, A. (1995): Vitiligo complicating diphencyprone sensitization therapy for alopecia universalis (letter). *Br. J. Dermatol.*, 133:496–497.

Heylings, J. R., Clowes, H. M., Cumberbatch, M., Dearman, R. J., Fielding, I., Hilton, J., and Kimber, I. (1996): Sensitization to 2,4-dinitrochlorobenzene: Influence of vehicle on absorption and lymph node activation. *Toxicology*, 109:57–65.

Holliday, M. R., Corsini, E., Smith, S., Basketter, D., Dearman, R., and Kimber, I. (1997): Differential induction of cutaneous TNF and IL-6 by topically induced chemicals. *Am. J. Contact Dermatitis*, 8:158–164.

Hull, S. M., and Cunliffe, W. J. (1992): Treatment of alopecia areata with diphenylcyclopropenone. *J. Am. Acad. Dermatol.*, 26:276–277.

Ikarashi, Y., Tsuchiya, T., Nakamura, A., Beppu, M., and Kikugawa, K. (1998): Effect of vitamin E on contact sensitization responses induced by 2,4 dinitrochlorobenzene in mice. *J. Nutr. Sci. Vitaminol.*, 44:225–236.

Kelly, D. A., Walker, S. L., McGregor, J. M., and Young, A. R. (1988): A single exposure of solar simulated radiation suppresses contact hypersensitivity responses both locally and systemically in humans: Quantitative studies with high frequency ultrasound. Journal of Photochemistry and Photobiology B: Biology 44:130–142.

Kim, J. K., Lee, H. S., Yoon, T. J., Oh, C. W., and Kim, T. H. (1996): Erythema multiforme due to diphenylcyclopropenone. *Ann. Dermatol.*, 250–252.

Klein, E., Holtermann, O. A., Helm, F., Rosner, D., Milgrom, H., Adler, S., Stoll, H. L., Case, R. W., Prior, R. L., and Murphy, G. P. (1975): Immunological approaches to the management of primary and secondary tumors involving the skin and soft tissues: review of a ten-year program. *Transplantation Proceedings*, VII:297–315.

Kligman, A. M. (1966): The identification of contact human assay. II: Factors influencing the induction and measurements of allergic contact dermatitis. *J. Invest. Dermatol.*, 47:375–392.

Kochiba, R. J., Keyes, D. G., Jersey, J. J. (1977): Results of a two year chronic toxicity study with hexachlorobutadiene in rats. *Am. Ind. Hyg. Assoc. J.*, 38:589–602.

Kratka, J., Goerz, G., Vizethum, W., and Strobel, R. (1979): Ditrochlorobenzene: Influence of the Cytochrome P-450 System and Mutagenic Effects. *Arch. Dermatol. Res.*, 266:315–318.

Lane, P. R., and Hogan, D. J. (1988): Diphencyprone. *J. Am. Acad. Dermatol.*, 19:364–365.

Lautenschlager, S., and Itin, P. H. (1998): Reticulate, patchy and mottled pigmentation of the neck. Acquired forms. Dermatology (Switzerland), 197:291–296.

Lee, S., Cho, C. K., Kim, J. G., and Chun, S. I. (1984): Therapeutic effect of dinitrochlorobenzene (DNCB) on verruca plana and verruca vulgaris. *International J. Dermatol.*, 23:624–626.

Lewis, H. M. (1973): Topical immunotherapy in refractory warts. Cutis, 12:863–867.

Lowney, E. D. (1973): Suppression of contact sensitization in man by prior feeding of antigen. *J. Invest. Dermatology.*, 61:90–93.

Maahs, G. (1985): Herstellung und Reaktionen des Perchlorcyclobutenons. *Ann. Chem.*, 686:55–63.

MacDonald-Hull, S. P., Cotterill, J. A. C., and Norris, J. F. B. (1989): Vitiligo following diphencyprone dermatitis. *Br. J. Derm.*, 120:323.

Maibach, H. I., Epstein, E. (1985): Allergic contact dermatitis. In: *Dermatologic Immunology and Allergy*. Edited by Stone Jr, CV Mosby, St. Louis, pp. 355–390.

Mallon, E., Powell, S., Mortimer, P., and Ryan, T. J. (1997): Evidence for altered cell-mediated immunity in postmastectomy lymphoedema. *Br. J. Dermatol.*, 137:928–933.

Manna, V., Marks, R., and Holt, P. (1982): Involvement of immune mechanisms in the pathogenesis of rosacea. *Br. J. Dermatol.*, 107:203–208.

Maurer, T. (1983): Contact and Photocontact Allergens. Marcel Dekker, New York.

Maxwell, J. G., Lawrence, P. F., and Warden, G. W. (1981): Dinitrochlorobenzene skin reactivity as a predictor of outcome in transplantation of juvenile onset diabetes. Transplantation. 32:381–384.

Mayerhausen, W., and Remy, W. (1987): Testung der zellularen Immunreaktivitat bei Melanompatienten mit Diphenylcyclopropenon (DPCP) im Vergleich mit Dinitrochlorobenzol (DNCB). Hausarzt. 38:449–452.

McDaniel, D. H., Blatchley, D. M., and Welton, W. A. (1982): Adverse systemic reaction to dinitrochlorobenzene. *Arch. Dermatol.*, 118:371.

Meneghini, C. L., and Angelini, G. (1975): Contact dermatitis from pyrrolnitrin (an antimycotic agent). *Contact Dermatitis*, 1:288–292.

Menne, T., and Wilkinson, J. D. (1995): Individual Predisposition to Contact Dermatitis. In: *Textbook of Contact Dermatitis*, 2nd ed., edited by Rycroft, R. J. D., Menne, T., Frosch, P. J. pp. 123–130, Springer Verlag, Munich.

Millikan, L. (1982): Task Force on DNCB reports survey results. *J. Am. Acad. Derm.*, 7:91–93.

Mills, L. B. (1986): Stimulation of T-cellular immunity by cutaneous application of dinitrochlorobenzene. *J. Am. Acad. Dermatol.*, 14:1089–1090.

Mommaas, A. M., Mulder, A. A., Vermeer, M., Boom, B. W., Tseng, C., Taylor, J. R., and Streilein, J. W. (1993): Ultrastructural studies bearing on mechanism of UVB-impaired induction of contact hypersensitivity to DNCB in man. *Clin. Exp. Immunol.*, 92:487–493.

Moss, C., Friedmann, P. S., and Shuster, S. (1981): Impaired contact hypersensitivity in untreated psoriasis and the effects of photochemotherapy and dithranol UV-B *Br. J. Dermatology*, 105:503–508.

Nair, M. P. N., and Schwartz, S. A. (1983): Effect of histamine and histamine antagonists in natural and antibody dependent cellular cytotoxicity in vitro. *Cell. Immunol.*, 81:45–60.

Naylor, M. F., Nelder, K. H., Yarbrough, G. K., Rosio, T. J., Iriondo, M., and Yeary, J. (1988): Contact immunotherapy of resistant warts. *J. Am. Acad. Dermatol.*, 19:679–683.

Nishioka, K., Asagami, C., Kurata, M., and Fujita, H. (1983): Sensitivity to the weed-killer DNA nitralin and cross-sensitivity to dinitrochlorobenzene. *Arch. Dermatol.*, 119:304–306.

Nordberg, J., Zhong, L., Holmgren, A., and Arner, E. (1998): Mammalian Thioredoxin reductase is irreversibly inhibited by dinitrohalobenzenes by alkylation of both the redox active selenocysteine and its neighboring cysteine residue. *Journal Bio. Chem.*, 273:10835–10842.

Noster, U., Hausen, B. M., Krische, B., and Schultz, K. H. (1976): Squaric acid diethylester: A strong sensitizer. *Contact Dermatitis*, 2:99–104.

Oh, C. W., Han, K. D., and Kim, T. H. (1998): Bullous erythema multiforme following topical diphenyl-cyclopropenone application. *Contact Dermatitis*, 38:220–221.

Oliver, A. R., and Silbart, L. K. (1998): Local and systemic tolerance to orally administered DNCB is not broken by cholera toxin. *Int. Arch. Allergy Immunol.*, 116:318–324.

Orecchia, G., Douville, H., Santagostino, L., and Rabbiosi, G. (1988): Treatment of multiple relapsing warts with diphenciprone. *Dermatologica*, 177:225–231.

Orecchia, G., and Rabbiosi, G. (1985): Treatment of alopecia areata with diphencyprone. *Dermatologica*, 171:193–196.

Palacios, J. S., Nemuth, M. G., and Blaylock, W. K. (1968): Lack of cross sensitization between 2,4 dinitrobenzene and chloramphenicol. *South Med. J.*, 51:243–245.

Pericin, M., and Trueb, R. M. (1998): Topical immunotherapy of severe alopecia areata with diphenylcyclopropenone: Evaluation of 68 cases. *Dermatology*, 196:418–421.

Perret, C. M., Steijlen, P. M., and Happle, R. (1990): Alopecia areata, pathogenesis and topical immunotherapy. *Int. J. Dermatol.*, 29:83–88.

Perret, C. M., Steijlen, P. M., Zaun, H., and Happle, R. (1990): Erythema multiforme-like eruptions: A rare side effect of topical immunotherapy with diphenylcyclopropenone. *Dermatologica*, 180:5–7.

Puig, L., and Allegre, M. (1994): Erythema multiforme-like reaction following diphencyprone treatment of plane warts. *International Journal of Dermatology*, 33:201–203.

Pye, R. J., and Burton, D. L. (1976): DNCB chemical lab worker and chloromycetin (letter). *Br. Med. J.*, 6044:1130–1131.

Quinkert, G., Opitz, K., Wiersdorff, W. W., and Weinlich, J. (1963): Light induced decarbonylation of dissolved ketones. *Tetra Letters*, 1868.

Raijka, G., and Vincze, E. (1956): Durch Textilefarbstoffe verursachte Dermatosen. *Berufsdermatosen*, 4:169–172.

Rees, J. L., Friedmann, P. S., and Matthews, J. N. S. (1989): Sex differences in susceptibility to development of contact hypersensitivity with DNCB. *Br. J. Dermatol.*, 120:371–374.

Rocklin, R. E. (1976): Modulation of cellular-immune responses in vivo or in vitro by histamine receptor binding lymphocytes. *J. Clin. Invest.*, 57:1051–1058.

Rokhsar, C. K., Shupack, J. L., Vafai, J. J., and Washenik, K. (1998): Efficacy of contact sensitizers in the treatment of alopecia areata. *J. Am. Acad. Derm.*, 39:751–761.

Rosen, M. H., Fengler, I., Bonet, G., Giannina, T., Butler, M. C., Poplck, F. R., and Steinetz, B. G. (1976): Contraceptive agents from cycloaddition reactions of diarylcyclopropanones and diarylthiirene 1,1 dioxides. *J. Med. Chem.*, 19:414–419.

Sadoff, L., Glovsky, M., Alenky, A. (1972): DNCB test in cancer patients. *NEJM.* 287:47–48.

Sansom, J. E., Molloy, K. C., and Lovell, C. R. (1995): Occupational sensitization to diphencyprone in a chemist. *Contact Dermatitis.*, 32:363.

Scheper, R. J., and Von Blomberg (1995): Cellular mechanisms in allergic contact dermatisis: In: *the Textbook of Contact Dermatitis*, 2nd ed., edited by Rycroft, R. J. D., Menne, T., Frosch, P. J. pp. 11–27, Springer Verlag, Munich.

Schmidt, R. J., and Chung, L. Y. (1992): Biochemical responses to allergenic and nonallergenic nitrohalobenzenes: Evidence that an NADPH-dependent reductase in skin may act as a prohapten-activating enzyme. *Arch. Dermatol. Res.*, 284:400–408.

Schwank, V. R., and Jirasek, L. (1963): Kontaktallergie gegen chloramphenicol mit besonderer berucksichtigung der gruppensensibilisierung. *Hautarzt*, 14:24–30.

Schwartz, M. (1953): Eczematous sensitization in various age groups. *J. Allergy*, 24:143–148.

Seo, K. I., and Eun, H. C. (1996): Loss of contact sensitization evaluated by laser Doppler blood flowmetry and transepidermal water loss measurement. *Contact Dermatitis*, 34:233–236.

Shah, M., Lewis, F. M., and Messenger, A. G. (1996): Hazards in the use of diphencyprone. *British Journal of Dermatology*, 134:1151–1165.

Sherertz, E. F., and Sloan, K. B. (1988): Percutaneous penetration of squaric acid and its esters in hairless mouse and human skin in vitro. *Arch. Dermatol. Res.*, 280:57–60.

Sommer, G., Parker, D., and Turk, J. L. (1975): Epicutaneous induction of hyporeactivity in contact sensitization. Demonstration of suppressor cells inducted by contact with 2,4 dinitrothiocyanatebenzene. *Immunology*, 29:517–525.

Strick, R. (1983): Lack of cross reactivity between DNCB and chloramphenicol. *Contact dermatitis*, 9:484–487.

Stricker, R. B., Goldberg, B., and Epstein, W. L. (1997): Topical immune modulation (TIM): A novel approach to the immunotherapy of systemic disease. *Immunology Letters*, 59:145–150.

Stricker, R. B., Elswood, B. F., and Abrams, D. I. (1991): Dendritic cells and dinitrochlorobenzene (DNCB): A new treatment approach to AIDS. *Immunology Letters*, 29:191–196.

Strobel, R., and Rohrborn, G. (1980): Mutagenic and cell-transforming activities of 1-chloro-2,4-dinitrobenzene and squaric acid dibutyl ester (SADBE). *Arch. Toxicol.*, 45:307–314.

Stute, J., Hausen, B. M., and Schulz, K. H. (1981): Diphenylcyclopropenon-ein stark wirksames Kontaktallergen. *Dermatosen.*, 29:12–15.

Summer, K. H., and Goeggelmann, W. (1980): 1-Chloro-2,4-Dinitrobenzene depletes glutathione in rat skin and is mutagenic in *Salmonella typhimurium*. Mutation Research, 77:91–93.

Swanson, N. A., Mitchell, A. J., Leany, M. S., Headington, J. T., and Diaz, L. A. (1981): Topical treatment of alopecia areata. *Arch Dermatol.*, 17:384–387.

Temmerman, L., De Weert, J., De Keyser, L., and Kint, A. (1984): Treatment of alopecia areata with dinitrochlorobenzene. *Acta. Derm. Venereol.*, (Stockh), 64:441–443.

Thestrup-Pedersen, K., Bisballe, S., Jensen, J. R., and Zachariae, H.: Immunological studies in patients with alopecia receiving dinitrochlorobenzene and cimetidine therapy. *Arch. Dermatol. Res.*, 273:261–266.

Toda, R. (1976): Diarylcyclopropenones and diarylthiirene 1,1-dioxides. *J. Med. Chem.*, 19:414–419.

Toews, G. B., Bergstresser, P. R., and Streilein, J. W. (1980): Epidermal Langerhans cell density determines whether contact sensitivity or unresponsiveness follows skin painting with DNFB. *J. Immunol.*, 124:445–453.

Tosti, A., Guerra, L., and Bardazzi, F. (1989): Contact urticaria during topical immunotherapy. *Contact Dermatitis*, 21:196–197.

Tron, V. A., and Sauder, D. N. (1991): Allergic contact dermatitis. In: ed. *Immunologic Diseases of the skin*. Edited by Jordon R. E., Connecticut: Appleton and Lange, pp. 253–261.

Tseng, C., Hoffman, B., Kurimoto, I., Shimizu, T., Schmieder, G. J., Taylor, J. R., and Streilein, J. W. (1992): Analysis of the effects of ultraviolet B radiation on induction of primary allergic reactions. *J. Invest. Dermatol.*, 98:871–875.

Vakilzadeh, F., Bruss, R., and Rupec, M. (1973): Die Beeinflussung des experimentellen Kontaktekzems durch Dimethylsulfoxid. *Arch. Dermatol. Forsch.*, 246:125–130.

Valsecchi, R., and Caintelli, T. (1984): Depigmentation from squaric acid dibutylester. *Contact Dermatitis*, 10:109.

Valsecchi, R., and Cainelli, T. (1984): Nickel sensitivity as a complication of squaric acid dibutylester treatment of alopecia areata. *Contact Dermatitis*, 234

Van der Steen, P. H. M., Van Baar, H. M. J., Perret, C. M., and Happle, R. (1991): Treatment of alopecia areata with diphenylcyclopropenone. *J. Am. Acad. Dermatol.*, 24:253–257.

Van der Steen, P., and Happle, R. (1992): "Dyschromia in confetti" as a side effect of topical immunotherapy with diphenylcyclopropenone. *Arch. Dermatol.*, 128:518–520.

Van der Steen, P. H. M., Boezeman, J. B. M., and Happle, R. (1992): Topical immunotherapy for alopecia areata: Reevaluation of 139 cases after an additional follow-up period of 19 months. *Dermatology*, 184:198–201.

Van Duuren, V., Benjamin, L., Methionne, S., Blair, R., Goldschmidt, B. M., and Katz, C. (1971): Carcinogenicity of isosters of epoxides and lactones: Aziridine ethanol, propane sultone and related compounds. *Journal of the National Cancer Institute*, 46:143–149.

Viraben, R., Labrousse, and J. L., Bazex, J. (1990). Erythema multiforme due to DNCB. *Contact Dermatitis*, 22:179.

Volpin, M. E., Koreshkov, Y. D., and Kursanov, D. N. (1959): Diphenylcyclopropenone: A three membered analog of tropone. Izvest Akad Naak SSSR, Otdel Khim 3:560.

Vonderheid, E. D., Ekbote, S. K., Kerrigan, K., Kalmanson, J. D., Van Scott, E. J, Rook, A. H., and Abrams, J. T. (1998): The prognostic significance of delayed type hypersensitivity to dinitrochlorobenzene and mechlorethamine hydrochloride in cutaneous T cell lymphoma. *J. Invest. Dermatol.*, 110:946–950.

Waksman, B. H. (1960): Delayed hypersensitivity reactions: A growing class of immunologic phenomenon. *J. Allergy*, 31:468–475.

Waldorf, D. S., Willkens, R. F., and Decker, J. L. (1969): Impaired delayed hypersensitivity in an aging population associated with anitnuclear reactivity and rheumatic fever. JAMA, 203:831.

Walker, F. B., Smith, P. D., and Maibach, H. I. (1967): Genetic factors in human contact dermatitis. *Int. Arch. Allergy*, 32:453–462.

Wedroff, N. (1928): Zur Frage der Dinitrochlorbenzoldermatosen. *Arch. Derm. Syph.*, 154:143–145.

Wedrow, N. (1928): Zur Frage der Dinitrobenzoldematosis. *Arch. Derm. Syph.*, 154:143–153.

Weisburger, E. K., Russfield, A. B., Homberger, F., Weisburger, J. H., Roger, E., Van Dongen, C. G., and Chu, K. C. (1978): Testing of twenty-one environmental aromatic amines or derivatives for long term toxicity or carcinogenicity. *J. Environ. Pathol. Toxicol.*, 2:325.

Whitfield, R. E., and Friedman, M. (1972): Chemical modification of wool with dicarbonyl compounds in dimethyl sulfoxide. *Textile Research Journal*, 42:344–346.

Whittaker, M. (1972): Severe dermatitis caused by diphenylcyclopropenone. *Contact Dermatitis News Letter*, 11:264.

Wilkerson, M. G., Henkin, J., Wilkin, J. K., and Smith. R. G. (1985): Squaric acid and esters: Analysis for contaminants and stability in solvents. *J. Amer. Acad. Derm.*, 13:229–234.

Wilkerson, M. G., Henkin, J., and Wilkin, J. K. (1984): Diphenylcyclopropenone: Examination for potential contaminants, mechanisms of sensitization and photochemical stability. *J. Am. Acad. Dermatol.*, 11:802–807.

Wilkerson, M. G., Conner, T. H. Henkin, J., Wilkin, J., and Matney, T. (1987): Assessment of diphenyl-cyclopropenone for photochemically induced mutagenicity in the Ames assay. *J. Am. Acad. Dermatol.*, 17:606–611.

Wilkerson, M. G., Wilkin, J. K., and Smith, R. G. (1983): Contaminants of DNCB. *J. Am. Acad. Dermatol.*, 9:554–557.

Wilkerson, M. G., Connor, T. H., and Wilkin, J. K. (1988): Dinitrochlorobenzene is inherently mutagenic in the presence of trace mutagenic contaminants. *Arch. Dermatol.*, 124:396–398.

Wilkin, J. K. (1986): Reply to Squaric acid esters in the treatment of alopecia areata. *J. Am. Acad. Derm.*, 14:486.

Winholz, M., Budavari, S., Stroumtsos, L. Y., Fertig, and M. N. (1976): The Merck Index, ed. 9. Merck & Co., Inc., Rahwau, NJ, 2113.

Young, H. R. (1976): DNCB reported effective for recalcitrant warts. *Skin and Allergy News*, 7:19.

Zachariae, R., and Bjerring, P. (1993): Increase and decrease of delayed cutaneous reactions obtained by hypnotic suggestions during sensitization. *Allergy*, 48:6–11.

Zachariae, R., Jorgensen, M. M., Chistensen, S., and Bjerring, P. (1997): Effects of relaxation on the delayed-type hypersensitivity (DTH) reaction to diphenylcyclopropenone (DCP). *Allergy*, 52:760–764.

Zimmerman, M. C. (1970): Dinitrochlorobenzene in water systems. *Contact Dermatitis*, Newsl. 7:165–167.

Toxicology of Skin
Edited by Howard I. Maibach
Copyright © 2001 Taylor & Francis

20

Modulation of the Dermal Immune Response by Acute Restraint

Sally S. Tinkle and Melanie S. Flint

Centers for Disease Control and Prevention, National Institute for Occupational Safety and Health, Morgantown, West Virginia, USA

INTRODUCTION

Immunity is an important multicellular mechanism used by organisms to recognize and eliminate foreign substances. The stress response is a multisystem mechanism designed to recognize and ameliorate emotional and physical perturbations in homeostasis. These two mechanisms frequently act in concert, through feedforward and feedback pathways, to restore physiologic equilibrium (Misery, 1997; Tomaszewska and Przekop, 1997). The stress response exerts its influence through two pathways, the neuroendocrine pathway, mediated through the hypothalamus-pituitary-adrenal axis (HPAA), and the catecholamine pathway of the sympathetic nervous system (Figure 20.1) (Chrousos, 1995). Glucocorticoids, a major activation product of the HPAA, affect multiple steps in the immune response, such as leukocyte proliferation and trafficking, adhesion molecule and cytokine expression, B and T lymphocyte ratios, and CD4/CD8 ratios (Belsito et al., 1982; Snyder and Unanue, 1982; Chrousos and Gold, 1992; Steer et al., 1998; Wiegers and Reul, 1998). The immune response to antigen involves the complex interplay of humoral and cellular mediators, many of which also regulate HPAA function (Boumpas et al., 1993). For example, tumor necrosis factor (TNF)-α, interleukin (IL)-1, and IL-6, important proinflammatory cytokines, activate the hypothalamus alone or in synergy with each other to

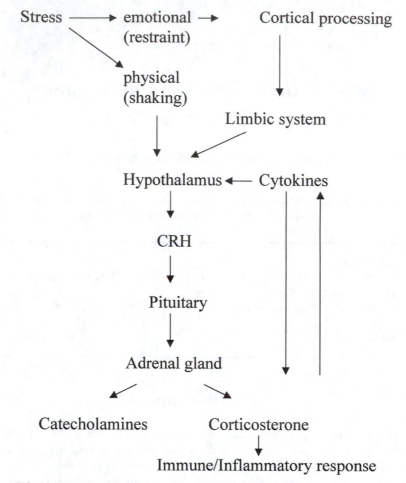

FIG. 20.1. The interplay between the stress response and the immune response occurs through multiple feedforward and feedback loops.

increase serum levels of adrenocorticotropin hormone (ACTH) and glucocorticoids (van Deventer et al., 1990; Zhou et al., 1993; Van Dam et al., 1998), and may also stimulate directly pituitary ACTH and adrenal cortisol secretion (Bernardini et al., 1990; Salas et al., 1990; Imura et al., 1991). Historically, glucocorticoids were thought to counter-regulate cytokine responses and protect the host from excessive catabolic effects of unregulated cytokine production (Schleimer et al., 1989). More recent studies have documented glucocorticoid enhancement of the immune response (Dhabhar and McEwen, 1999). The interaction between the immune system and the HPAA is complex and depends on the type of immune response and the type of stressor.

The skin provides an important, direct interface between external stimuli and the internal milieu, whereas the HPAA receives integrated physical and emotional stimuli and coordinates the neurohormone response. Therefore, when stress occurs simultaneously with dermal exposure to antigen, the cutaneous cell-mediated immune response is the sum of the dermal immune response and the HPAA-mediated response to stress.

Allergic contact dermatitis (ACD) is characterized as a delayed-type IV hypersensitivity (DTH) response requiring two exposures to antigen: a sensitization exposure and a challenge exposure. Several studies have documented crosstalk between the HPAA and the dermal cell-mediated immune response. Using changes in rodent pinnae thickness as an indirect measure of the DTH response to chemical challenge, (Dhabhar et al., 1996) demonstrated that mild, acute-restraint induced a significant and persistent increase in pinnae thickness and in the number of leukocytes at the site of the DTH response. Adrenalectomy negated this stress-induced amplification, and the acute administration of low doses of corticosterone restored the augmented response in a dose-dependent manner (Dhabhar and McEwen, 1999). Conversely, high doses of corticosterone, or administration of the synthetic glucocorticoid, dexamethasone, to adrenalectomized mice suppressed significantly the DTH response. Murine studies have demonstrated an immobilization time-dependent decrease in the dermal DTH when restraint is applied prior to sensitization and no change in the response when mice are restrained prior to chemical challenge (Kawaguchi et al., 1997). Langerhans' cells are the professional antigen-presenting cell in the dermal DTH response. Several reports demonstrate that, coupled to chemical exposure, the topical and systemic application of glucocorticoids to guinea pigs or the application of acute restraint stress to Balb/c mice influences Langerhans' cell morphology and antigen presentation capabilities (Belsito et al., 1982; Kawaguchi et al., 1997). These data underscore the importance of the HPAA in mediating stress-induced changes in the dermal immune response, however, the effects of restraint stress and the multiple factors that impinge on the stress response, and thus the development of ACD, are largely unknown.

MULTIPLE FACTORS MODIFY THE IMMUNOLOGIC RESPONSE AND THE STRESS RESPONSE IN THE DEVELOPMENT OF CONTACT DERMATITIS

Immunologic Predisposition

The skin allergic response is considered a predominantly Th1 DTH response (Kimber, 1994); however, several studies suggest that psychological stress shifts the Th1/Th2 cytokine ratio to a Th2-predominant profile. (Iwakabe et al., 1998) demonstrated a restraint-induced decrease in Con A-stimulated BALB/c spleen cell production of interferon (IFN)-γ and no change in IL-4 production. Dexamethasone-induced decreases in IFN-γ and modest increases in IL-4 and IL-10 were documented in phytohemagglutinin (PHA) and anti-CD3 antibody-stimulated human peripheral blood mononuclear cells (PBMCs) (Agarwal and Marshall, 1998). Utilizing the Th2-predisposed BALB/c and Th1-predisposed C57BL/6 mice, (Brenner and

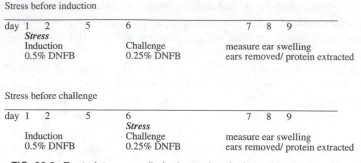

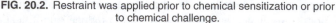

FIG. 20.2. Restraint was applied prior to chemical sensitization or prior to chemical challenge.

Moynihan, 1997) combined repeated foot shock and viral exposure to demonstrate strain-dependent changes in PBMC cytokine response but no shift toward Th2-dominant immunity.

To investigate modulation of Th1-predominant cutaneous hypersensitivity by restraint stress, we applied 100 μl of 0.25% 2,4-dinitrofluorobenzene (DNFB) in acetone-olive oil vehicle (4:1) to the flank of male BALB/c and C57BL/6 mice on days 1 and 2 (Figure 20.2). Each experimental animal was restrained in an adequately ventilated 50 ml conical tube for 2 hours on day 6. Mice were able to assume a prone or supine position but were not able to turn head to tail. Control animals were left undisturbed in their home cages. Immediately following the restraint period, 50 μl of 0.25% DNFB was applied to the right pinnae, and the left pinnae was treated with vehicle only. The concentration of serum corticosterone immediately following restraint was measured by RIA. Pinnae thickness was measured on days 7 through 9. The concentration of cytokines in dorsal ear tissue homogenates was measured by ELISA and normalized by tissue weight.

Consistent with previous studies, employing this restraint technique, 2-hour restraint stress increased serum corticosterone levels 10-fold in BALB/c mice and 16-fold in C57BL/6 mice (Table 20.1) (Dhabhar and McEwen, 1996).

At 24 hours post-treatment, DNFB-induced increases in pinnae thickness were not changed by the application of restraint for either strain of mice (Figure 20.3). Restraint-enhanced ear swelling, with respect to the nonrestrained control group, was noted at 48 and 72 hours post-challenge in BALB/c mice, whereas the measurements

TABLE 20.1. *Stress-induced changes in plasma corticosterone levels*

Mouse strain	Control (ng/ml)	Restrained (ng/ml)
BALB/c	90 ± 27.0	931 ± 56.0*
C57BL/6	32 ± 10.4	533 ± 28.6*
B6,129 (Males)	96 ± 35.7	436 ± 65.7*
B6,129 (Females)	110.5 ± 39	539 ± 96*

*Data (n = 4 per group) are represented as the mean ± SEM. Statistically significant differences were measured by the unpaired t-test (p < 0.01).

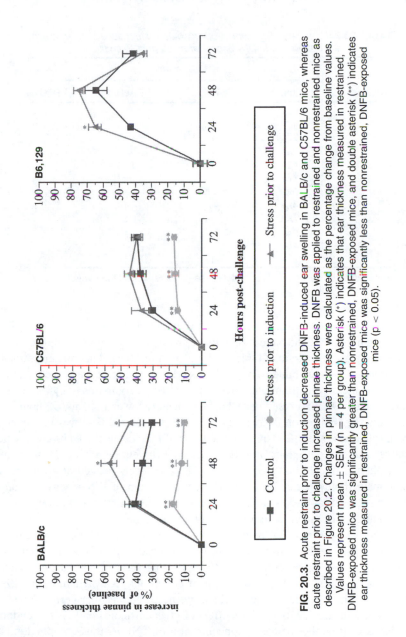

FIG. 20.3. Acute restraint prior to induction decreased DNFB-induced ear swelling in BALB/c and C57BL/6 mice, whereas acute restraint prior to challenge increased pinnae thickness. DNFB was applied to restrained and nonrestrained mice as described in Figure 20.2. Changes in pinnae thickness were calculated as the percentage change from baseline values. Values represent mean ± SEM (n = 4 per group). Asterisk (*) indicates that ear thickness measured in restrained, DNFB-exposed mice was significantly greater than nonrestrained, DNFB-exposed mice, and double asterisk (**) indicates ear thickness measured in restrained, DNFB-exposed mice was significantly less than nonrestrained, DNFB-exposed mice ($p < 0.05$).

TABLE 20.2. *Summary of cytokine profiles analyzed in mouse skin, sensitized and challenged to DNFB, and stressed prior to challenge*

Cytokine	Time	Strain		
		C57BL/6	BALB/c	B6,129
TNF-α	24	↓	↔	↔
	48	↑	(—)	
	72	↔	(—)	
IL-1β	24	↓	↔	(—)
	48	↔	(—)	
	72	↑	(—)	
IFN-γ	24	↔	↓	(—)
	48	↔	↓	
	72	↔	↓	
IL-4	24	↔	↑	↓
	48	↓	↔	
	72	↑	↑	

Arrows indicate increase (↑) or decrease (↓) or remained the same (↔) compared to control.

in C57BL/6 mice displayed an increase in the mean thickness that did not achieve statistical significance. The pattern of cytokine response was more complex and differed between the two strains (Table 20.2 and Figure 20.4). The concentrations of TNF-α and IL-1 β in restrained Th2-predisposed BALB/c mice were unchanged 24, 48, and 72 hours after restraint plus chemical challenge; compared to cytokine levels in nonrestrained, DNFB-treated mice, but significantly lower at 48 and 72 hours. The concentration of IFN-γ was decreased at all time points, and IL-4 was increased. In contrast, cytokine levels in Th1-predisposed C57BL/6 mice displayed no clear pattern of change. The concentrations of TNF-α and IL-1 β were depressed 24 hours after restraint plus chemical challenge; however, TNF-α levels peaked at 48 hours and IL-1 β remained elevated at 72 hours. The concentration of IFN-γ was unchanged at all time points. IL-4 production, maximal at 0.16 ng/g tissue at 48 hours in nonrestrained mice, was increased in restrained mice at 72 hours.

These data demonstrate that restraint stress prior to chemical challenge causes a moderate increase in the murine ear swelling response in BALB/c mice, but not in C57BL/6 mice, and agree with a BALB/c study that showed no change in pinnae thickness at 24 hours under similar treatment conditions (Kawaguchi et al., 1997). Changes in pinnae thickness, characterized by increased vascular permeability and cell extravasation, are due to the direct effect of glucocorticoids on vascular endothelial cells, as well as the increased concentration of TNF-α, thus complicating the biological interpretation of the ear swelling response (Falaschi et al., 1999).

C57BL/6 mice display a higher IFN-γ/IL-4 ratio in the cutaneous cytokine response than BALB/c mice. The restraint-induced decrease in the BALB/c ratio results from simultaneous decrease in IFN-γ and increase in IL-4. In contrast, IFN-γ levels in C57BL/6 mice display minimal change whereas increases in the IL-4 concentration drive the shift in the ratio.

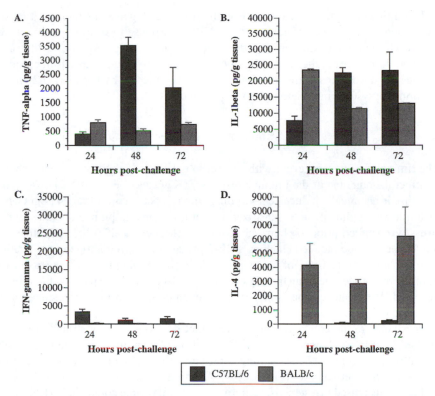

FIG. 20.4. Acute restraint prior to chemical challenge alters DNFB-induced cutaneous cytokine production in a strain-dependent manner. (A) TNF-α; (B) IL-1β; (C) IFN-γ; (D) IL-4. The C57BL/6 Th1 and the BALB/c Th2 bias are evident in the restraint before challenge modulation of cytokine production. Data are expressed as mean $\pm$ SEM (n = 4 per group).

The complexity of the stress response mediated by the HPAA is underscored by studies in rodents that have demonstrated a relationship between a blunted HPAA and increased susceptibility to inflammation and autoimmunity (Sternberg and Licinio, 1995). Other researchers have shown that injection of corticosterone increased pinnae thickness in rodents sensitized and challenged with DNFB (Dhabhar and McEwen, 1999). To evaluate the strain-dependent sensitivity of the immune response to glucocorticoid manipulation, we injected 2 mg of corticosterone per kilogram of body weight subcutaneously into C57BL/6 mice 2 hours prior to chemical challenge in restrained and control mice to determine if increased corticosterone would shift the C57BL/6 Th1 response to a Th2-predominant response (data not shown). Exogenous corticosterone did not alter the concentrations of TNF-α and IL-1β; however, the concentration of IL-4 increased at 24 hours, with no accompanying change in IFN-γ. The rise in the concentration of IL-4 decreased the IFN-γ/IL-4 ratio eight-fold, shifting this immune parameter in the Th2 direction at 24 hours only.

The differential effects of restraint on the immune response, as measured in this study, demonstrate that differences in the HPAA response to restraint stress, as reflected by the sixfold difference in the serum corticosterone response to the same stressor, differences in the sensitivity of each cytokine to changes in serum corticosterone, and differences in the immune response, as reflected by the Th1-Th2 bias of the immune system, modulate the development of ACD.

Timing of Acute Stress

The timing of a stressful event with respect to exposure to antigen has been shown to affect the outcome of the immune response. (Kawaguchi et al., 1997) observed a decrease in chemically induced ear swelling with respect to controls at 24 hours when restraint was applied prior to sensitization, but no change in pinnae thickness when stress was applied prior to chemical challenge. (Munck et al., 1984) hypothesized that corticosterone-induced changes in IL-2 production during antigen sensitization may explain the effectivity of glucocorticoids in suppressing the early events of the immune response rather than the late events when T-cell clonal expansion has already occurred. We evaluated the application of restraint stress prior to sensitization in BALB/c and C57BL/6 mice.

In contrast to the enhanced ear swelling that develops when restraint stress was applied prior to challenge (day 6), the application of restraint stress prior to sensitization (day 1) suppresses ear swelling approximately twofold to threefold at all time points in both strains of mice (Figure 20.3). Additionally, the concentrations of TNF-α and IL-1β increased in a stepwise pattern over the 3-day time course in BALB/c mice (Figure 20.5). Restraint applied prior to sensitization also induced significantly higher levels of IFN-γ and lower levels of IL-4 than observed when stress was applied prior to challenge. In C57BL/6 mice, the concentration of TNF-α increased more slowly than observed for restraint prior to challenge, and the concentrations of IL-1β, IFN-γ, and IL-4 match those measured when stress was applied before challenge. These data demonstrate that the timing of restraint with respect to antigen exposure affects important parameters of the immune response and suggest that the BALB/c mice may be more sensitive to the timing of stress than the C57BL/6 mice.

Gender

Gender differences in HPAA structure and function and the impact of these differences on immunity are well documented (Lesniewska et al., 1990; Viau and Meaney, 1991; Da Silva, 1995). To investigate the impact of gender on cutaneous hypersensitivity, male and female B6,129 mice were sensitized and challenged with DNFB according to the treatment paradigm for restraint before chemical challenge. The constitutive concentration of serum corticosterone matched that of BALB/c mice, and, immediately following restraint, was approximately four-fold higher than the concentration measured in nonrestrained mice (Table 20.1). Consistent with our previous data, restraint

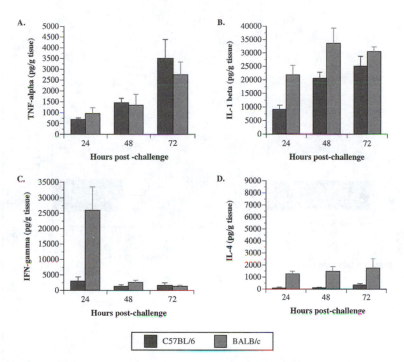

FIG. 20.5. Acute restraint prior to sensitization alters the cutaneous cytokine response to chemical challenge in a strain-dependent manner. (A) TNF-α; (B) IL-1β; (C) IFN-γ; (D) IL-4. Restraint before sensitization enhanced IFN-γ production by BALB/c mice more significantly than did restraint prior to chemical challenge. Data are expressed as mean ± SEM (n = 4 per group).

plus chemical challenge increased ear swelling for both male and female B6,129 mice to a comparable degree along a time course similar to that observed for nonrestrained B6,129 mice and for BALB/c and C57BL/6 mice (data not shown). The B6,129 male ear swelling response was greater at 24 and 48 hours after restraint plus chemical challenge, whereas the female response was further elevated at 48 hours only.

The cytokine profile for the B6,129 males and females at 24 hours differed in the magnitude, but not the pattern, of the response (Figure 20.6), a finding consistent with data demonstrating that the female immune response is more sensitive to fluctuations in glucocorticoids than the male response (Da Silva, 1995). In particular, DNFB-induced TNF-α, IL-1β, and IFN-1 concentrations in non-restrained mice were comparable in males and females in the same treatment paradigm, whereas levels of IL-4 were lower in females. However, restraint stress decreased IL-4 in both sexes, and IL-1β and TNF-α were unchanged. Furthermore, this cytokine profile is distinct from the BALB/c levels, with the concentration of TNF-α corresponding to the BALB/c levels, and the concentration of IL-1β, IFN-γ, and IL-4 corresponding to that of C57BL/6 mice.

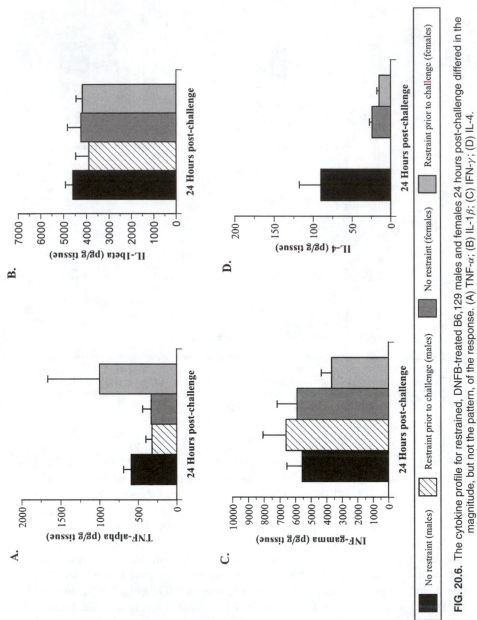

FIG. 20.6. The cytokine profile for restrained, DNFB-treated B6,129 males and females 24 hours post-challenge differed in the magnitude, but not the pattern, of the response. (A) TNF-α; (B) IL-1β; (C) IFN-γ; (D) IL-4.

Types of Stress

Stressors generally are classified as physical or psychological. The hypothalamus receives physically stressful signals directly and psychologically stressful signals through the limbic system (Figure 20.1) (Khansari et al., 1990). These different pathways provide qualitatively different signals to the HPAA (Borysenko and Borysenko, 1982). Restraint stress is considered a psychological stress because of its anxiety-generating component. To examine the cytokine response to physical stress, we employed the same experimental paradigm; however, B6,129 mice, four per cage, were placed on an orbital shaker revolving at two oscillations per minute for 30 minutes (Dhabhar and McEwen, 1996).

We observed significant differences in the ear swelling response and cytokine profile between restraint stress and shaking stress. DNFB induced a 40%–70% increase in pinnae thickness for restrained mice and a 6%–12% increase for mice subjected to shaking stress (data not shown). This physical stress increased IFN-γ production 10-fold compared to restrained mice, IL-4 levels increased twofold, and TNF-α concentrations were halved. The concentration of IL-1β was unchanged.

SUMMARY

The ability to recognize adverse stimuli and to initiate a cascade of cellular and molecular events that restore homeostasis is imperative for survival of an organism. The immune response has evolved to protect against foreign substances, and the HPAA and the sympathetic nervous system respond to perturbations in physical and psychological homeostasis. These systems act independently and in concert to restore and maintain physiologic equilibrium systemically and in the skin. Cumulative data confirm the cellular and molecular complexity of each system and the complexity of the interaction between them. The cutaneous immune response is modulated by stressors in a complex manner that depends, in part, on the type of stressor, the type of immune response, immunologic predisposition, gender, and the timing of the acute stress with respect to antigen exposure. The nonlinearity of the cytokine response and the lack of correlation with the ear swelling response suggest that measurements of vascular changes underlying increased pinnae thickness provide an incomplete picture of cutaneous hypersensitivity and that the cutaneous immune response is highly sensitive to environmental changes mediated through the hypothalamic-pituitary-adrenal axis.

REFERENCES

Agarwal, S. K., and Marshall, G. D., Jr. (1998): Glucocorticoid-induced type 1/type 2 cytokine alterations in humans: A model for stress-related immune dysfunction. *J. Interferon Cytokine Res.*, 18:1059–1068.

Belsito, D. V., Flotte, T. J., Lim, H. W., Baer, R. L., Thorbecke, G. J., and Gigli, I. (1982): Effect of glucocorticosteroids on epidermal Langerhans cells. *J. Exp. Med.*, 155:291–302.

Bernardini, R., Kamilaris, T. C., Calogero, A. E., Johnson, E. O., Gomez, M. T., Gold, P. W., and Chrousos, G. P. (1990): Interactions between tumor necrosis factor-alpha, hypothalamic corticotropin-releasing hormone, and adrenocorticotropin secretion in the rat. *Endocrinology*, 126:2876–2881.

Borysenko, M., and Borysenko, J. (1982): Stress, behavior, and immunity: Animal models and mediating mechanisms. *Gen. Hosp. Psychiatry*, 4:59–67.

Boumpas, D. T., Chrousos, G. P., Wilder, R. L., Cupps, T. R., and Balow, J. E. (1993): Glucocorticoid therapy for immune-mediated diseases: Basic and clinical correlates. *Ann. Intern. Med.*, 119:1198–1208.

Brenner, G. J., and Moynihan, J. A. (1997): Stressor-induced alterations in immune response and viral clearance following infection with herpes simplex virus-type 1 in BALB/c and C57B1/6 mice. *Brain. Behav. Immun.*, 11:9–23.

Chrousos, G. P. (1995): The hypothalamic-pituitary-adrenal axis and immune-mediated inflammation (see comments). *N. Engl. J. Med.*, 332:1351–1362.

Chrousos, G. P., and Gold, P. W. (1992): The concepts of stress and stress system disorders: Overview of physical and behavioral homeostasis (published erratum appears in JAMA 1992, 268:200). *JAMA*, 267:1244–1252.

Da Silva, J. A. (1995): Sex hormones, glucocorticoids and autoimmunity: Facts and hypotheses. *Ann. Rheum. Dis.*, 54:6–16.

Dhabhar, F. S., and McEwen, B. S. (1996): Stress-induced enhancement of antigen-specific cell-mediated immunity. *J. Immunol.*, 156:2608–2615.

Dhabhar, F. S., and McEwen, B. S. (1999): Enhancing versus suppressive effects of stress hormones on skin immune function. *Proc. Natl. Acad. Sci. USA*, 96:1059–1064.

Dhabhar, F. S., Miller, A. H., McEwen, B. S., and Spencer, R. L. (1996): Stress-induced changes in blood leukocyte distribution: Role of adrenal steroid hormones. *J. Immunol.*, 157:1638–1644.

Falaschi, P., Martocchia, A., Proietti, A., and D'Urso, R. (1999): The immune system and the hypothalamus-pituitary-adrenal (HPA) axis. In: *Cytokines: Stress and Immunity*. Edited by N. P. Plotnikoff, R. E. Faith, A. J. Murgo, and R. A. Good, pp. 325–337, Boca Raton, CRC.

Imura, H., Fukata, J., and Mori, T. (1991): Cytokines and endocrine function: An interaction between the immune and neuroendocrine systems. *Clin. Endocrinol. (Oxf)*, 35:107–115.

Iwakabe, K., Shimada, M. et al. (1998): The restraint stress drives a shift in Th1/Th2 balance toward Th2-dominant immunity in mice (In Process Citation). *Immunol. Lett.*, 62:39–43.

Kawaguchi, Y., Okada, T., Konishi, H., Fujino, M., Asai, J., and Ito, M. (1997): Reduction of the DTH response is related to morphological changes of Langerhans cells in mice exposed to acute immobilization stress. *Clin. Exp. Immunol.*, 109:397–401.

Khansari, D. N., Murgo, A. J., and Faith, R. E. (1990): Effects of stress on the immune system (see comments). *Immunol. Today*, 11:170–175.

Kimber, I. (1994): Cytokines and regulation of allergic sensitization to chemicals. *Toxicology*, 93:1–11.

Lesniewska, B., Nowak, M., and Malendowicz, L. K. (1990): Sex differences in adrenocortical structure and function. XXVIII: ACTH and corticosterone in intact, gonadectomised and gonadal hormone replaced rats. *Horm. Metab. Res.*, 22:378–381.

Misery, L. (1997): Skin, immunity and the nervous system. *Br. J. Dermatol.*, 137:843–850.

Munck, A., Guyre, P. M., and Holbrook, N. J. (1984): Physiological functions of glucocorticoids in stress and their relation to pharmacological actions. *Endocr. Rev.*, 5:25–44.

Salas, M. A., and Evans, S. W. (1990): Interleukin-6 and ACTH act synergistically to stimulate the release of corticosterone from adrenal gland cells. *Clin. Exp. Immunol.*, 79:470–473.

Schleimer, R. P., Claman, H. N. and Oronsky, A. (1989): *Anti-Inflammatory Steroid Action*. San Diego, Academic Press.

Snyder, D. S., and Unanue, E. R. (1982): Corticosteroids inhibit murine macrophage Ia expression and interleukin 1 production. *J. Immunol.*, 129:1803–1805.

Steer, J. H., and Ma, D. T. (1998): Altered leucocyte trafficking and suppressed tumor necrosis factor alpha release from peripheral blood monocytes after intra-articular glucocorticoid treatment. *Ann. Rheum. Dis.*, 57:732–737.

Sternberg, E. M., and Licinio, J. (1995): Overview of neuroimmune stress interactions: Implications for susceptibility to inflammatory disease. *Ann. N. Y. Acad. Sci.*, 771:364–371.

Tomaszewska, D., and Przekop, F. (1997): The immune-neuro-endocrine interactions. *J. Physiol. Pharmacol.*, 48:139–158.

Van Dam, A. M., and Malinowsky, D. (1998): Interleukin 1 (IL-1) type I receptors mediate activation of rat hypothalamus-pituitary-adrenal axis and interleukin 6 production as shown by receptor type selective deletion mutants of IL-1beta. *Cytokine*, 10:413–417.

van Deventer, S. J., and Buller, H. R. (1990): Experimental endotoxemia in humans: Analysis of cytokine release and coagulation, fibrinolytic, and complement pathways. *Blood*, 76:2520–2526.

Viau, V., and Meaney, M. J. (1991): Variations in the hypothalamic-pituitary-adrenal response to stress during the estrous cycle in the rat. *Endocrinology*, 129:2503–2511.

Wiegers, G. J., and Reul, J. M. (1998): Induction of cytokine receptors by glucocorticoids: Functional and pathological significance. *Trends. Pharmacol. Sci.*, 19:317–321.

Zhou, D., and Kusnecov, A. W. (1993): Exposure to physical and psychological stressors elevates plasma interleukin 6: Relationship to the activation of hypothalamic-pituitary-adrenal axis. *Endocrinology*, 133: 2523–2530.

Toxicology of Skin
Edited by Howard I. Maibach
Copyright © 2001 Taylor & Francis

21

Textile-Dye Allergic Contact Dermatitis Prevalence*

Kathryn L. Hatch

College of Agriculture, The University of Arizona, Tucson, Arizona, USA

Howard I. Maibach

University of California, San Francisco, California, USA

- **Introduction**
- **Methods**
- **Findings**
  - · Research Methods · Study Prevalence · Disperse Dye Prevalence · Non-disperse Dye Prevalence
- **Discussion**
- **Recommendations**

INTRODUCTION

Although dermatologists have reported cases of skin reactions suspected to be caused by textile dyes since 1869 (Wilson, 1869), they have reported textile-dye prevalence results only during the last decade (Balato et al., 1990; Dooms-Goossens, 1992; Gasperini et al., 1989; Goncalo et al., 1992; Lodi et al., 1998; Manzini et al., 1991; Manzini et al., 1996; Manzini et al., 1996; Seidenari et al., 1991; Seidenari et al., 1995; Soni and Sherertz, 1996). This chapter summarizes the textile-dye prevalence studies (Balato et al., 1990; Dooms-Goossens, 1992; Gasperini et al., 1989; Goncalo et al., 1992; Lodi et al., 1998; Manzini et al., 1991; Manzini et al., 1996; Manzini et al., 1996; Seidenari et al., 1991; Seidenari et al., 1995; Soni and Sherertz, 1996) and makes recommendations for advancing knowledge about textile-dye sensitization.

---

*Originally published in *Contact Dermatitis*, 2000, volume 42, pages 187–195. Reprinted with permission.

METHODS

A database was created to allow ready comparison of prevalence data by study and by specific dye. Three tables were developed to record information about each study, with records including (a) journal article designation, (b) patients tested, (c) name and origin of patch test allergens, (d) patch test details, and (e) study prevalence. In Table 21.1, records are placed into two categories based on type of patient patch tested. Table 21.1 contains the records of those studies in which patients appeared for routine patch testing which included textile dyes, Table 21.2 contains the records of those studies in which patients were suspected to have textile-dye allergic contact dermatitis (ACD), and Table 21.3 contains the records of those studies in which patients were known or were very likely to have textile-dye ACD.

Additionally, two tables were constructed to record patch test results about each dye used in testing, one table for disperse dyes (Table 21.4), and one for nondisperse dyes (Table 21.5). Each record in these tables consists of dye name, number of patch test–positive patients in the study, number of patients who were patch tested, study code (1 for studies in Table 21.1, 2 for studies in Table 21.2, and 3 for studies in Table 21.3), prevalence, and reference. Dyes are arranged in order of amount of prevalence information available about them.

FINDINGS

Eleven textile-dye prevalence articles were published from 1989 to 1999 (Balato et al., 1990; Dooms-Goossens, 1992; Gasperini et al., 1989; Goncalo et al., 1992; Lodi et al., 1998; Manzini et al., 1991; Manzini et al., 1996; Manzini et al., 1996; Seidenari et al., 1991; Seidenari et al., 1995; Soni and Sherertz, 1996), and an additional article provides data from which textile-dye prevalence can be obtained (Seidenari et al., 1997) (Tables 21.1, 21.2, 21.3). One article reports 4 studies (Seidenari et al., 1991), 1 article reports 3 studies (Seidenari et al., 1997), 2 articles report 2 studies each (Balato et al., 1990; Dooms-Goossens, 1992), and the remaining 8 articles report 1 study each for a total of 19 studies. The focus of 2 articles (Gasperini et al., 1989; Soni and Sherertz, 1996) is on textile-dye prevalence among textile and garment manufacturing employees. The remaining 10 focus on prevalence among people without occupational exposure to dyes.

Research Methods

Three different patient populations were selected for testing with textile dyes. Ten studies were conducted by patch testing eczematous patients to determine the cause of ACD (Table 21.1), 3 studies were conducted by patch testing eczematous patients who were suspected to be textile-dye sensitive (Table 21.2), and 6 studies were conducted by patch testing eczematous patients who were or probably were textile-dye sensitive (Table 21.3). The objectives in the latter 2 types of studies were to either determine specific dyes to which a patient was sensitive or to determine the best set of dyes to use in routine patch testing to detect textile-dye sensitivity. Further, 15 of the 19 studies were performed in Italian clinics (4 in reference Seidenari et al., 1991, 3 in reference Seidenari et al., 1997, 2 in reference Balato et al., 1990, and 1 each in references Gasperini et al., 1989; Lodi et al., 1998; Manzini et al., 1991; Manzini et al., 1996; Manzini et al., 1996; Seidenari et al., 1995). The remaining studies were done in Belgian (Dooms-Goossens, 1992), United States, (Soni and Sherertz, 1996), and Portuguese (Goncalo et al., 1992) clinics. Patients were adults except in the (Goncalo et al., 1992) and Seidenari et al. investigations (Seidenari et al., 1997, study 2), where patients were children.

Numbers of eczematous patients who were patch tested in Table 21.1 studies varied from 6203 (Seidenari et al., 1997, Study 1) to 312 (Manzini et al., 1996), and in Table 21.3 studies varied from 145 (Balato et al., 1990) to 23 (Seidenari et al., 1997, Study 2). Often, gender distribution was not provided; however, when reported (Balato et al., 1990; Dooms-Goossens, 1992; Gasperini et al., 1989; Lodi et al., 1998), more women than men were in the study population.

Most of the studies included disperse and nondisperse dyes, but far more disperse than nondisperse dyes were used. Three studies focused exclusively on nondisperse dyes (Manzini et al., 1996; Manzini et al., 1996; Seidenari et al., 1995). Dyes for patch testing were obtained from F.I.R.M.A. (Fabbrica Italiana Ritrovati Medicinali Affini) (Firenze, Italy), Hermal/Trolab (Hamburg, Germany), Chemotechnique Diagnostics (Malmo, Sweden), or directly from dye manufacturers. Usually, the dyes were not checked for identity or purity. Number of dyes used in any study varied from 27 (Manzini et al., 1991) to 4 (Seidenari et al., 1997, study 2). In all, 18 disperse, 17 reactive, 4 acid, 4 basic, 1 direct, and 10 nondisperse dyes of unidentified application class were used.

Patch test application materials included use of Finn Chambers (Epitest, Helsinski, Finland) on Scanpor tape (Norgesplaster, Oslo, Norway), A1-test strips, and Van de Bend square chambers with Micropore (3M, St. Paul, MN) and covered with Mefix (Smith & Nephew, Largo, FL). Techniques mentioned were those of (Wilkinson et al., 1970) and (Seroli et al., 1987). Application times were either 2 days or 2 and 4 days (3 days exceptionally). Times given for interval between removal of patch to reading the skin response were 1 hours and 24 hours, 48 hours and 72 hours, 2–3 hours, and 3 hours and 4 days. Not all authors reported application method and reading procedure.

Study Prevalence

In those studies in which patients appeared for routine patch testing and disperse dyes were included (Balato et al., 1990; Lodi et al., 1998; Manzini et al., 1991; Seidenari et al., 1991; Seidenari et al., 1997 in Table 21.1), prevalence values range from 5.8% (Seidenari et al., 1991, study 3) to 1.4% (Manzini et al., 1991). In those Table 21.1 studies in which only nondisperse dyes were used for patch testing (Manzini et al., 1996;

TABLE 21.1. *Studies in which patients appeared for routine patch testing that included textile dyes*

Article	Patients	Name and origin of dye allergens	Patch test details	Study prevalance
Balato et al., 1990 Study 2	576 consecutive patients with various eczemas; no time period stated	Disperse dyes Blue 124, Red 1, Orange 3, and Yellow 3 and GIRDA standard series	A1-test strips on Scanpor tape, removed after 48 hours and read 1 hour and 24 hours later; removed after 48 hours and read 1 hour and 24 hours later	3.3% (19 of 576—6 men and 13 women)
Manzini et al., 1991	569 patients aged 20–70 years who were referred to Italian allergological service with dermatitis suspected of being allergic in nature; no time period stated	GIRDCA standard series (Hermal/Trolab), Textile dye series (Firma, Italy)*, 17 nondisperse dyes directly provided by industry applied at 2% pet.	Finn chambers on Scanpor tape to healthy back skin; removed 2 days and read 1 day later	1.4% (6 of 569)
Seidenari et al., 1991 Study 2	861 consecutive Italian patients seen between 12/1/1988 and 11/18/1989	GIRDCA standard series supplemented with Disperse dyes Blue 124, Red 1, Yellow 3, Orange 3	Finn chambers with Scanpor tape using Wilkinson et al.'s procedure;	4.8% (41 of 861)
Seidenari et al., 1991 Study 3	746 Italian patients seen between 11/19/1990 and 4/30/1990	GIRDCA standard series, 4 disperse dyes above and 12 other dyes: disperse dyes Red 17, Black 1, Orange 76, Yellows 9 and 54, and Blues 3 and 35, Acid Black 48, Basic Black 1, aminoazobenzene, and p-dimethylaminoazobenzene	Finn chambers with Scanpor tape using Wilkinson et al.'s procedure	5.8% (43 of 746)
Dooms-Goossens, 1992 Study 1	3336 contact dermatitis patients in Belgium clinic seen between 1987 and 1991	European standard series (Hermal/Trolab) which included PPD as a textile dye indicator	Van de Bend sq. chambers fixed with Micropore and covered with Mefix 2 and 4 (exceptionally 3) days, following ICDRG guidelines; excluded doubtful reactions	0.2% (8 of the 3336 were positive to PPD)

Reference	Population	Dyes tested	Application method	Prevalence
Seidenari et al., 1995	1814 consecutive patients attending Italian patch test clinic between 2/1993 and 6/1994	GIRDCA standard series augmented by 5 nondisperse dyes frequently used to dye natural fibers at 5% pet.: Direct Orange 34, Direct Yellow 169, Acid Yellow 61, Acid Reds 118 and 359; origin was dye manufacturer	Applied on back; removal at 3 days and read 2–3 hours later according to ICDRG guidelines	0.88% (16 of 1814)
Manzini et al., 1996	1813 consecutive patients seen between 2/93 and 6/94	GIRDCA standard series and 12 commonly used textile reactive dyes: Reds 123, 238 and 244, Orange 107, Yellow 17, Blues 18, 21, 122 and 238, Violet 5, Black 5, Green 12. Origin was Ciba, Hoechst, and Sandoz dye manufacturers	As powder at 5% pet. to healthy back skin with Finn chambers on Scanpor tape; removed after 3 days and read 3 hours later and at 4 days; evaluated according to ICDRG guidelines	0.99% (18 of 1813 to reactive dyes and allergens in GIRDCA series and 4 to reactive dyes only)
Manzini et al., 1996	312 consecutive Italian patients admitted to allergological service for suspected ACD; no time period stated	5 reactive dyes: Reds 180 and 198, Oranges 16 and 96, and Blue 220 at 5% and 10% in pet; Origin was Hoechst dye manufacturer	Applied to healthy back skin with Finn chambers on Scanpor tape; removal at 3 days and read after 3 hours and 4 days	0%
Seidenari et al., 1997[†] Study 1	6203 Italian patients during 1990–1995 period	Disperse Dyes Blue 124, Orange 3, Yellow 3, and Red 1 from FIRMA*; purity of dyes checked with another working group	No information provided	3.8% (236 of 6203)
Lodi et al., 1998	Reviewed 1012 Italian patients (311 men and 701 women) with suspected ACD; no time period stated	GIRDCA standard series augmented by Disperse Dyes Black 1 (1% pet.) and Blue 3 (1% pet.) and a mix composed of Disperse dyes Blue 124, Red 1, Yellow 3, and Orange 3; patch tests with suspected fabric in cases with a history of wearing colored garments	No information provided	3.1% (31 of 1012—26 women and 5 men)

*F.I.R.M.A patch test series of 9 dyes: Disperse Reds 1 and 17; Disperse Blues 3 and 124; Disperse Orange 3; Disperse Yellows 3 and 9, Disperse Black 1, and Basic Brown 1.

[†] Not strictly a prevalence study, but rather a study of cross-sensitization of dyes. Included because prevalence can be obtained from the data presented.

TABLE 21.2. Studies in which patients were suspected to have textile-dye ACD

Article	Patients	Name and origin of dye allergens	Patch test details	Study prevalence
Seidenari et al., 1991 Study 1	1145 Italian patients, some of whom were suspected of having textile-dye ACD, seen between 10/1/1987 and 11/30/1988	Textile industry patch test series (specific source not given)	Finn chambers with Scanpor tape using Wilkinson et al.'s procedure	?% (16 patients with positive test to dyes but unclear how many of the 1145 were patch tested with the dyes)
Dooms-Goossens, 1992 Study 2	164 contact dermatitis patients in Belgium clinic seen between 1987 and 1991 who were suspected of having textile-dye ACD	159 tested with 14 Chemotechnique textile dyes* with p-aminoazobenzene (0.25% pet.) added, and 5 tested with only Disperse dyes Orange 3, Yellow 3, and Blue 106	Same as Study 1 described in Table 21.1 of this chapter	17.1% (28 of 164; 21 women and 7 men)
Goncalo et al., 1992	329 children 14 years and younger seen between 1985 and 1989 at a Portuguese clinic, some of whom were suspected of having textile-dye ACD	GPEDC (Hermal/Trolab) series† and additional series indicated by clinical data including the following dyes: Disperse dyes Orange 3, Yellow 3, Red 1, and Bismarck Brown	Read at 2 and 3 or 4 days according to ICDRG guidelines	?% (4 children with positive test to dyes but unclear how many children were patch tested with the dyes)

*Chemotechnique patch test series of 15 dyes: Disperse Reds 1 and 17; Disperse Blues 3, 35, 85, 106, 124, and 153; Disperse Oranges 1, 3, and 13; Disperse Yellows 3 and 9, Disperse Brown 1; and Basic Red 46.
†Hermal/Trolab patch test series of 5 dyes: Disperse Dyes Reds 1 and 17, Orange 3, Yellow 3, and Blue 3.

Manzini et al., 1996; Seidenari et al., 1995), prevalence values range from 1.0%–0.0%. Only 1 prevalence value could be calculated from the information provided in those studies in which patients suspected of having textile-dye ACD were tested with a series of textile dyes (Table 21.2 studies); that prevalence was 17.1%. Prevalence values in patient populations known or most likely sensitized to dyes ranged from 72.7%–15.9%.

Prevalence for women appears to be higher than for men. Specifically, in (Balato et al., 1990, Study 2, Table 21.1), 13 women and 6 men were patch test positive to at least 1 dye, in (Dooms-Goossens, 1992, study 2, Table 21.2), 21 women and 7 men were patch test positive, and in (Lodi et al., 1998, Table 21.1), 26 women and 5 men were patch test positive. The percentage of patch test positive subjects who were women in each study is, therefore, 75%, 83%, and 68%, respectively. Whether women are more susceptible to sensitization by textile dyes is not known, as these percentages may be a reflection of the proportion of men to women in the study patient population and not necessarily reflective of the total population. The only data available to compare the proportion of women patch test positive to the proportion in the study population is that of (Lodi et al., 1998); there were 701 women and 311 men, making the proportion of females 69.3%. This proportion approximates the proportion of women who were patch test positive to at least 1 dye (68%).

Disperse Dye Prevalence

The amount of prevalence data collected for various disperse dyes varies considerably (Table 21.4) with the greatest amount being available for Disperse dyes Orange 3, Yellow 3, Red 1, and Blue 124. For these 4 dyes, patch testing has occurred in all three study categories or patient population types (Balato et al., 1990; Bide and McConnell, 1996 and Dooms-Goossens, 1992). For these 4 dyes the highest prevalence value in the Study 1 category is 2.2% in 1 of the Disperse Blue 124 studies and 0.2% in 1 of the Disperse Orange 3 studies. Average prevalence values calculated for these 4 dyes using the Study 1 data are 2.4% for Disperse Blue 124, 1.0% for Orange 3 and Red 1, and 0.7% for Yellow 3. In contrast, prevalence values for these same 4 dyes in type 3 studies were as high as 60.9%. There is considerable variation in the prevalence data for the Category 3 studies, but the lowest value is usually greater than the lowest prevalence value for the dye in a Category 1 study.

Nondisperse Dye Prevalence

As shown in Table 21.5, only 6 of the nondisperse dyes—Acid Black 48, Basic Red 46, Basic Brown 1, Acid Yellow 61, Acid Red 118, and Acid Red 359—have been used in more than 1 prevalence study. Interestingly, most of the studies are category 1 studies, those in which the patients are *not* suspected, known, or very likely to be sensitive to textile dyes. Prevalence is consistently less than 0.4% in the Category 1

TABLE 21.3. *Studies in which patients were known or most likely sensitized to textile dyes*

Article	Patients	Name and origin of dye allergens	Patch test details	Study prevalence
Gasperini et al., 1989	104 garment-industry workers (11 men and 93 women) plus 57 textile-industry workers (42 men and 5 women) with occupational contact dermatitis seen during 1986 and 1987	Disperse dyes Blue 124, Yellow 3, Black 1, Orange 76, Yellow 54, Red 17, Blue 35, and Orange 3, and Acid Black 48 in 1% pet.; origin/series not stated	Prepared haptens according to Cronin, 1980[*] and Fisher, 1986[†]; used Rapid Patch Test (RPT) technique	31.7% ACD in garment workers 49.2% in textile workers
Balato et al., 1990 Study 1	145 Italian patients suspected of having ACD from textile chemicals seen during 1989 and 1989	Unnamed textile series: Disperse dyes Blue 124, Red 1, Orange 3, Yellow 3, Red 17, Orange76; Basic Brown 1 and Acid Black 48 in 1% pet. plus the GIRDCA standard series	A1-test strips on Scanpor tape; removed after 48 hours and read 1 hour and 24 hours later	15.9% (23 of 145)
Seidenari et al., 1991 Study 4	98 Italian patients identified as having dye allergy by patch testing with dyes described in Table 21.1 of this chapter	4 dyes: Disperse Blue 124, Red 1, Orange 3, and Yellow 3; 9 textile dyes: Disperse Red 17, Black 1, Orange 76, Yellow 9 and 54, Blue 3 and 35; Acid Black 48, Black Base 1	Finn chambers with Scanpor tape using Wilkinson et al.'s procedure	72% and 63.3% (72 of 100 to at least one dye in the 4 dye series and 62 of 98 to at least one dye in the 9 dye series)

314

Soni and Sherertz, 1996	72 U.S./NC textile industry workers seen between 1989 and 1994 for evaluation of possible work-related contact dermatitis; age range 19–62 years with average age of 44	Standard screening series (Hermal or North American Contact Dermatitis Group series); 70% underwent testing with dye and chemical series (Chemotechnique*). Patient-specific allergens including individual dyed fabrics	Finn chamber method; North American Contact Dermatitis group grading scheme	6.9% (5 of 72)
Seidenari et al., 1997 Study 2	23 children identified as being azo-dye sensitive by patch testing with dyes described in Table 21.1 of this chapter	Disperse dyes Red 1, Orange 3, Yellow 3, and Blue 124	See Seidenari et al., 1997, Study 1, Table 21.1 of this chapter	100% (23 of 23)
Seidenari et al., 1997 Study 3	33 patients identified as being dye sensitive by patch testing with dyes described in Table 21.1 of this chapter	Disperse dyes Orange 76, Blue 3, Red 17, Black 1, Yellow 54, Yellow, Blue 35, and Basic Brown	See Seidenari et al., 1997, Study 1, Table 21.1 of this chapter	72.7% (24 of 33)

*Chemotechnique patch test series of 15 dyes: Disperse Reds 1 and 17; Disperse Blues 3, 35, 85, 106, 124, and 153; Disperse Oranges 1, 3, and 13; Disperse Yellows 3 and 9, Disperse Brown 1; and Basic Red 46.

†Fisher, A. A. (1986): *Contact Dermatitis*, 3rd ed. Lea and Febiger, Philadelphia.

The original publication did not provide the number for the dye.

TABLE 21.4. *Prevalence of each disperse dye by study*

Dye	No. of +s	No. of patients	Study category*	Prevalence	Reference	Study
Disperse	5	576	1	0.8%	Balato et al., 1990	2
Orange 3	1	569	1	0.2%	Manzini et al., 1991	
	107	6203	1	1.7%	Seidenari et al., 1997	1
	5	164	2	3.0%	Dooms-Goossens, 1992	2
	6	145	3	4.1%	Balato et al., 1990	1
	2	72	3	2.8%	Soni and Sherertz, 1996	
	28	100	3	28.0%	Seidenari et al., 1991	4
	12	23	3	52.2%	Seidenari et al., 1997	2
	1	62	3	1.6%	Gasperini et al., 1989	
Disperse	42	6203	1	0.7%	Seidenari et al., 1997	1
Yellow 3	8	1012	1	0.8%	Lodi et al., 1998	
	3	576	1	0.5%	Balato et al., 1990	2
	1	164	2	0.6%	Dooms-Goossens, 1992	2
	5	145	3	3.4%	Balato et al., 1990	1
	24	100	3	24.0%	Seidenari et al., 1991	4
	4	23	3	17.4%	Seidenari et al., 1997	2
	1	72	3	1.4%	Soni and Sherertz, 1996	
	4	62	3	6.5%	Gasperini et al., 1989	
Disperse	67	6203	1	1.1%	Seidenari et al., 1997	1
Red 1	6	576	1	1.0%	Balato et al., 1990	2
	5	1012	1	0.5%	Lodi et al., 1998	
	2	159	2	1.3%	Dooms-Goossens, 1992	2
	7	145	3	4.8%	Balato et al., 1990	1
	14	23	3	60.9%	Seidenari et al., 1997	2
	29	100	3	29.0%	Seidenari et al., 1991	4
	1	72	3	1.4%	Soni and Sherertz, 1996	
Disperse	11	576	1	1.9%	Balato et al., 1990	2
Blue 124	104	6203	1	1.7%	Seidenari et al., 1997	1
	22	1012	1	2.2%	Lodi et al., 1998	
	6	159	2	3.8%	Dooms-Goossens, 1992	2
	12	145	3	8.3%	Balato et al., 1990	1
	1	72	3	1.4%	Goncalo et al., 1992	
	36	100	3	36.0%	Seidenari et al., 1991	4
	4	23	3	17.4%	Seidenari et al., 1997	2
	8	62	3	12.9%	Gasperini et al., 1989	
Disperse	5	159	2	3.1%	Dooms-Goossens, 1992	2
Red 17	3	145	3	2.1%	Balato et al., 1990	1
	20	98	3	20.4%	Seidenari et al., 1991	4
	3	33	3	9.1%	Seidenari et al., 1997	3
	1	62	3	1.6%	Gasperini et al., 1989	
Disperse	6	159	2	3.8%	Dooms-Goossens, 1992	2
Blue 35	2	145	3	1.4%	Balato et al., 1990	1
	5	98	3	5.1%	Seidenari et al., 1991	4
	1	33	3	3.0%	Seidenari et al., 1997	3
	1	62	3	1.6%	Gasperini et al., 1989	
Disperse	3	145	3	2.1%	Balato et al., 1990	1
Orange 76	11	98	3	11.2%	Seidenari et al., 1991	4
	12	33	3	36.4%	Seidenari et al., 1997	3
	2	62	3	3.2%	Gasperini et al., 1989	
Disperse	1	569	1	0.2%	Manzini et al., 1991	
Black 1	12	98	3	12.2%	Seidenari et al., 1991	4
	3	33	3	9.1%	Seidenari et al., 1997	3
	2	62	3	3.2%	Gasperini et al., 1989	

TABLE 21.4. (*Continued*)

Dye	No. of +s	No. of patients	Study category*	Prevalence	Reference	Study
Disperse Yellow 54	3	98	3	3.1%	Seidenari et al., 1991	4
	2	33	3	6.1%	Seidenari et al., 1997	3
	1	62	3	1.6%	Gasperini et al., 1989	
Disperse Blue 3	0	159	2	0.0%	Dooms-Goossens, 1992	2
	4	98	3	4.1%	Seidenari et al., 1991	4
	5	33	3	15.2%	Seidenari et al., 1997	3
Disperse Yellow 9	2	159	2	1.3%	Dooms-Goossens, 1992	2
	11	98	3	11.2%	Seidenari et al., 1991	4
Disperse Orange 1	5	159	2	3.1%	Dooms-Goossens, 1992	2
Disperse Orange 13	2	159	2	1.3%	Dooms-Goossens, 1992	2
Disperse Yellow	2	33	3	6.1%	Seidenari et al., 1997	3
Disperse Blue 85	1	159	2	0.6%	Dooms-Goossens, 1992	2
Disperse Blue 106	16	159	2	9.8%	Dooms-Goossens, 1992	2
Disperse Blue 153	3	159	2	1.9%	Dooms-Goossens, 1992	2
Disperse Brown 1	4	159	2	2.5%	Dooms-Goossens, 1992	2

*1 refers to studies recorded in Table 21.1; 2 refers to studies recorded in Table 21.2; 3 refers to studies recorded in Table 21.3.

studies. No positive patch tests were obtained for 18 of the dyes listed when they were used in Category 1 studies. Prevalence values range from 9.2% for the 1 Category 3 for Basic Black 1 to 1.4% for the 1 Category 3 study for Basic Brown 1.

DISCUSSION

The findings reported in textile-dye prevalence papers are valuable in defining the extent of human sensitivity to textile dyes as well as to specific dyes. Most have been done to determine sensitivity to disperse than to nondisperse dyes. Available data indicate lower prevalence to nondisperse than to disperse dyes, that prevalence is probably less than 6% of eczematous patient populations patch tested with a number of different dyes to determine possible textile-dye sensitivity, and that sensitivity to any one dye usually is less than 2% in eczematous patient groups.

The data assembled in this article must be placed in perspective. First, the data are largely reflective of prevalence in Europe, primarily Italy. Only one study was conducted outside of Europe (i.e., North America). Prevalence values, therefore, should not be generalized to represent sensitivity in all countries.

Secondly, prevalence values presented reflect sensitivity to a dye placed directly on the skin and not sensitivity to a dyed textile. Prevalence would be expected to be lower if a dyed textile were placed on the skin rather than the dye itself because the dye would have to transfer from the fabric to skin. Reactive dyes properly applied to fabric are not readily available for transfer from fabric to skin, as they are covalently bonded to the cellulose polymers of fibers (cotton, rayon, flax/linen) dyed with them; however, excess reactive dye on fabric is readily available for transfer. Dyes in the

TABLE 21.5. *Prevalence of each nondisperse dye by study*

Dye	No. of +s	No. of patients	Study Category*	Prevalence	Reference	Study
Acid Black 48	1	569	1	0.2%	Manzini et al., 1991	
	1	62	3	1.6%	Gasperini et al., 1989	
	3	145	3	2.1%	Balato et al., 1990	1
	4	98	3	4.1%	Seidenari et al., 1991	4
Basic Red 46	1	569	1	0.2%	Manzini et al., 1991	
	1	159	2	0.6%	Dooms-Goossens, 1992	2
	2	72	3	2.8%	Soni and Sherertz, 1996	
Basic Brown 1	1	159	2	0.6%	Dooms-Goossens, 1992	2
	2	145	3	1.4%	Balato et al., 1990	1
Acid Yellow 61	1	569	1	0.2%	Manzini et al., 1991	
	5	1814	1	0.3%	Seidenari et al., 1995	
Acid Red 118	1	569	1	0.2%	Manzini et al., 1991	
	1	1814	1	<0.1%	Seidenari et al., 1995	
Acid Red 359	1	569	1	0.2%	Manzini et al., 1991	
	2	1814	1	0.1%	Seidenari et al., 1995	
Basic Black 1	9	100	3	9.2%	Seidenari et al., 1991	4
Basic Brown ?	1	33	3	3.0%	Seidenari et al., 1997	3
Direct Orange 34	8	1814	1	0.4%	Seidenari et al., 1995	
Direct Yellow 169	0	1814	1	0.0%	Seidenari et al., 1995	
Reactive Red 123	1	1813	1	<0.1%	Manzini et al., 1996	
Reactive Red 180	0	312	1	0.0%	Manzini et al., 1996	
Reactive Red 198	0	1813	1	0.0%	Manzini et al., 1996	
Reactive Red 238	1	1813	1	<0.1%	Manzini et al., 1996	
Reactive Red 244	1	1813	1	<0.1%	Manzini et al., 1996	
Reactive Orange 16	0	312	1	0.0%	Manzini et al., 1996	
Reactive Orange 96	0	312	1	0.0%	Manzini et al., 1996	
Reactive Orange 107	2	1813	1	<0.1%	Manzini et al., 1996	
Reactive Yellow 17	1	1813	1	<0.1%	Manzini et al., 1996	
Reactive Green 12	0	1813	1	0.0%	Manzini et al., 1996	
Reactive Blue 18	0	1813	1	0.0%	Manzini et al., 1996	
Reactive Blue 21	1	1813	1	<0.1%	Manzini et al., 1996	
Reactive Blue 122	3	1813	1	0.2%	Manzini et al., 1996	
Reactive Blue 220	0	312	1	0.0%	Manzini et al., 1996	
Reactive Blue 238	2	1813	1	<0.1%	Manzini et al., 1996	
Reactive Violet 5	5	1813	1	0.3%	Manzini et al., 1996	
Reactive Black 5	2	1813	1	<0.1%	Manzini et al., 1996	
Yorkshire Red	0	569	1	0.0%	Manzini et al., 1991	
Yorkshire Joracril Yellow	0	569	1	0.0%	Manzini et al., 1991	
Yorkshire Joracril Blue	0	569	1	0.0%	Manzini et al., 1991	
Sandolan Blue	0	569	1	0.0%	Manzini et al., 1991	
Diazol Black	0	69	1	0.0%	Manzini et al., 1991	
Brilliant Red	0	569	1	0.0%	Manzini et al., 1991	
Marine Foron Blue	0	569	1	0.0%	Manzini et al., 1991	
Supramine Blue	0	569	1	0.0%	Manzini et al., 1991	
Astrazon MS Green	0	569	1	0.0%	Manzini et al., 1991	
Astrazon M Green	0	569	1	0.0%	Manzini et al., 1991	

*1 refers to studies recorded in Table 21.1; 2 refers to studies recorded in Table 21.2; 3 refers to studies recorded in Table 21.3.

acid, basic, and disperse groups, although not covalently bonded to textile polymers, usually are strongly held within fibers due to current methods of dye application.

Thirdly, one should not infer that a person showing sensitivity to a dye in a patch test setting has developed sensitivity to the dye by wearing fabric colored with it. The primary sensitizer may be another product or be a cross-sensitization (Seidenari et al., 1997).

Fourth, certain dyes named as contact allergens may not be the contact allergens, especially those used in only 1 study and/or obtained directly from a manufacturer rather than a firm specializing in producing patch test allergen series. Dermatology teams usually did not check the purity nor identify the dyes used in patch testing, which opens questions about each dye's purity and identity. In 1986, (Foussereau and Dallara, 1986) found a problem with purity of several dyes in a commercially available patch test series. When dye products contain more than one compound, it may not be the dye molecule that causes the skin reaction, but one of the other chemicals in the dye product. Although the purity problem identified in the series has possibly been rectified, no published research was found to confirm that dyes in those series are now pure and that the composition of same-named dyes in various patch testing series are chemically identical. Further, a study by (Bide and McConnell, 1996) established that dye manufacturers do incorrectly label dyes and same-name dyes differ in formulation. Differences occur from batch to batch as well as from one manufacture to another. To illustrate purity and labeling problems, (Bide and McConnell, 1996; Seroli et al., 1987) show thin-layer chromatograms in which (a) 5 dyes marketed as Disperse Blue 3 are all chemically different, (b) 1 dye labeled as Disperse Violet 27 was actually Disperse Violet 1, and (c) 2 dyes labeled as Disperse Violet 1 were not chemically identical. Only by testing with the dyes themselves and the contaminants will we learn the real rather than the presumed allergen.

In summary, findings should not be inferred

to apply to the general population, as people tested were referrals to dermatology clinics;
to apply to world-wide populations, as patients usually were European;
to reveal sensitivity to dyed textiles, as dyes were applied directly to skin;
to mean that textile dyes are primary sensitizers.

Further, it should not be concluded that all dyes named are the allergens, as dye purity and identity usually were not verified.

RECOMMENDATIONS

Additional prevalence studies are encouraged as one key in understanding sensitization of human populations to textile dyes. Such studies need to be done in North America, European countries other than Italy, and other world countries. Studies that are performed with people from the general population, rather than dermatologic referral populations, would add greatly to knowing sensitization to textile dyes. To assist in comparison of data collected in various countries and to track prevalence over time in any locale, a standard procedure or protocol would be invaluable.

Researchers are encouraged to include dyes for which little prevalence data have been collected. Those dyes listed in Tables 21.4 and 21.5, which have been used in only one prevalence study to date, are prime candidates for use in future studies. An investigation to determine which dyes are used most for coloring apparel items, those to which there is greatest exposure, would be another source of candidates. Further, researchers are encouraged to include Disperse Blue 14 in one or more prevalence studies, as Disperse Blue 14 was identified to have the properties of an ACD dye in a quantitative structure activity relationship (QSAR) study undertaken by (Hatch and Magee, 1998).

When future prevalence studies are designed and implemented, researchers are encouraged to verify the identity and purity of the dyes used, especially when the dyes are obtained directly from dye manufacturers. When reporting these studies, authors are encouraged to include gender distribution as well as a clear presentation of the number of patients who are patch test positive to each dye, and a prevalence for each dye as well as a prevalence for the study. Collaboration with industry and dye trade associations should expedite identification of the less sensitizing dye formulations.

The reviewed publications offer our best current insight into how frequent the textile dye ACD problem is. Even with this growing corpus, we still remain unclear as to how often the positive patch tests biologically mean clinically relevant intolerance, and cross-reactivity, nor do we know threshold levels and clinical outcome.

REFERENCES

Balato, N., Lembo, G., Patruno, C., Avala, F. (1990): Prevalence of textile dye contact sensitization. *Contact Dermatitis*, 23:111–126.

Bide, M. J., McConnell, B. L. (1996): In-house testing of dyes. *Textile Chemist and Colorist*, 28(3): 11–15.

Cronin, E. (1980): *Contact Dermatitis*, 2nd ed. Churchill Livingston, Edinburgh.

Dooms-Goossens, A. (1992): Textile dye dermatitis. *Contact Dermatitis*, 27:321–323.

Foussereau, J., Dallara, M. (1986): Purity of standardized textile dye allergens: A thin layer chromatography study. *Contact Dermatitis*, 14:303–306.

Gasperini, M., Farli, M., Lombardi, P., Sertoli, A. (1989): Contact dermatitis in the textile and garment industry. In: *Current topics in contact dermatitis*. Edited by Frosch, P., pp. 326–329. Springer-Verlag, Berlin.

Goncalo, S., Goncalo, M., Azenha, M. A., Barros, A., Bastos, A. S., Brandão, F. M., Faria, A., Marques, M. S. J., Pecegueiro, M., Rodriques, J. B., Salgueiro, E., Torres, V. (1992): Allergic contact dermatitis in children. *Contact Dermatitis*, 26:112–115.

Hatch, K. L., Magee, P. (1998): A discriminant model for allergic contact dermatitis in anthraquinone disperse dyes. *Quantitative Structure-Activity Relationships*, 17:20–26.

Lodi, A., Ambonati, M., Coassini, A., Chirarelli, G., Mancini, L. L., Crosti, C. (1998): Textile dye contact dermatitis in an allergic population. *Contact Dermatitis*, 39:314–315.

Manzini, B. M., Seidenari, S., Danese, P., Motolese, A. (1991): Contact sensitization to newly patch tested non-disperse textile dyes. *Contact Dermatitis*, 25:331–332.

Manzini, B. M., Motolese, A., Conti, A., Ferdani, G., Seidenari, S. (1996): Sensitization to reactive textile dyes in patients with contact dermatitis. *Contact Dermatitis*, 34:172–175.

Manzini, B. M., Donini, M., Motolese, A., Seidenari, S. (1996): A study of 5 newly patch-tested reactive textile dyes. *Contact Dermatitis*, 35:313.

Seidenari, S., Manzini, B. M., Danese, P. (1991): Contact sensitization to textile dyes: description of 100 subjects. *Contact Dermatitis*, 24:253–258.

Seidenari, S., Manzini, M., Schiavi, M. E., Motolese, A. (1995): Prevalence of contact allergy to non-disperse azo dyes for natural fibers: A study in 1814 consecutive patients. *Contact Dermatitis*, 33:118–122.

Seidenari, S., Mantovani, L., Manzini, B. M., Pignatti, M. (1997): Cross-sensitization between azo dyes and para-amino compound. *Contact Dermatitis*, 36:91–96.

Seroli, A., Farli, M., Francalanci, S., Gola, M., Lombardi, P. (1987): Vantaggi di un nuovo metodo (Rapido) per l'esecuzione di patch test. 8a Riunione G.I.D.C.A. Riassunto 57.

Soni, B. P., Sherertz, E. F. (1996): Contact dermatitis in the textile industry: A review of 72 patients, *American Journal of Contact Dermatitis*, 7:226–230.

Wilkinson, D. S., Fregert, S., Magnusson, B., Bandman, H.-J., Calnan, C. D., Cronin, E., Hjorth, N., Maibach, H. I., Malten, K. E., Meneghini, C. L., Pirila, V. (1970): Terminology of contact dermatitis. *Acta. Dermato-venereologica*, 50:287–292.

Wilson, E. (1869): Des stries et des macules atrophiques au fausses cicatrices de la peau. *Annual of Dermatology and Syphology*, 141.

Toxicology of Skin
Edited by Howard I. Maibach
Copyright © 2001 Taylor & Francis

22

Dose-Response Studies in Guinea Pig Allergy Tests

Søren Frankild

Department of Dermatology, Odense University Hospital, Denmark

INTRODUCTION

Contact allergy is a delayed, T-cell–mediated immune response to a sensitizing chemical, which in some patients progresses to incapacitating allergic contact dermatitis. Therefore, proper identification and labeling of environmental contact allergens is a request from governmental agencies, consumer organizations, industry, and unions. So far, more than 3700 substances have been reported as potential contact sensitizers (de Groot, 1994). Predictive testing of new chemicals for allergenicity is mandatory prior to marketing in the European Union (EU) in order to reduce or omit exposure to sensitizing chemicals. Predictive allergy testing in animals is the current methods of choice (Andersen and Maibach, 1985c; Botham et al., 1991; Klecak, 1991; Maurer, 1996; Andersen and Frankild, 1997).

Guidelines for predictive allergenicity testing in animals have been adopted by the Organisation for Economic Cooperation and Development (OECD) and by the

EU in 1992, commonly known as OECD guideline #406 and the Annex V methods, respectively (OECD guideline #406, 1992; Annex V methods 1994). The guidelines recommend two guinea pig assays: the Guinea Pig Maximization Test (GPMT) and the Buehler test. (Magnusson and Kligman, 1970) developed the Guinea Pig Maximization Test in the late 1960s by determining and maximizing each single factor in the procedure in favor of sensitization and combining them into a single protocol. The test procedure was designed to be as simple as possible to reduce the need for manpower; therefore, only two induction treatments are prescribed. The GPMT comprises intradermal induction with the chemical in question combined with Freund's complete adjuvant (FCA) and occluded topical application of the compound (Figure 22.1). To decide if the chemical is a sensitizer, an occluded topical challenge is performed 2 weeks later.

Buehler developed the Buehler test in the 1960s as a test solely dependent, on topical occluded induction without the use of FCA (Buehler 1964, 1965, Ritz and Buehler, 1980). The Buehler test employs three 6-hour topical induction patches, one patch/week, while the animals are restrained (Figure 22.1). Challenge is similar to the induction procedure and performed 2 weeks after induction. The guidelines contain detailed description of the test procedure. The standard GPMT and Buehler test is done in groups of guinea pigs divided into a test and a control group. It is recommended that at least 10 test animals and 5 control animals are used. If the chemical does not sensitize 15% of the animals in the GPMT or 30% in the Buehler

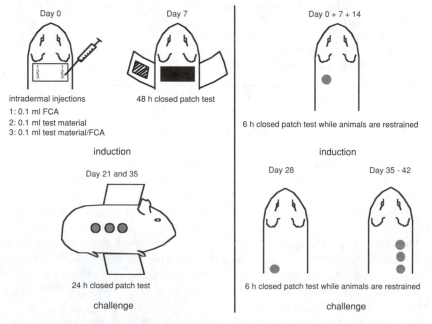

FIG. 22.1. Schematic presentation of the Guinea Pig Maximization Test and the Buehler test procedure.

test, additionally 10 test and 5 control animals should be tested (i.e., 30 animals all together). The OECD guideline #406 does not require that test results are analyzed statistically. According to the EU Commission Directive any chemical inducing 30% or more sensitized animals in the GPMT and 15% or more, sensitized animals in the Buehler test should be labeled R43, indicating potential allergenicity (Annex VI, 1993). Three chemicals are suggested to be used as positive controls: benzocaine, cinnamic aldehyde, and mercaptobenzothiazole to assure the sensitivity of the test in the laboratory.

VARIATION IN GPMT RESULTS

According to the present guidelines, one single induction concentration producing mild to moderate skin irritation should be used for each induction treatment in the GPMT (OECD guideline #406, 1992); however, what is a mild to moderate induction concentration? This is not defined in the guidelines, and the literature shows great variation in choice of induction concentration (Wahlberg and Boman, 1985). This is unfortunate because contact allergy is not an all-or-none phenomenon, but an adverse reaction dependent upon several factors, including the induction concentration (Kligman, 1966b; Marzulli and Maibach, 1974; Andersen et al., 1995; Johansen et al., 1996; Bacci et al., 1997). The other factors that influence the test results in predictive testing are challenge concentration, choice of vehicle, technical details as quality of occlusion, different batches of test chemical, patch test reading, source of animals, and ambient test conditions (Andersen et al. 1994).

Consequently, variation in the induction concentration in the standard GPMT is a significant factor giving variation in test results and poor reproducibility (Basketter et al., 1995b). The use of multiple doses for induction may improve standardization of the GPMT and thereby reduce interlaboratory variation (Carmichael and Foulds, 1990).

THE INFLUENCE OF CONCENTRATION ON SENSITIZATION

The dose-response relationship is a fundamental concept in toxicology. There are at least three quantitative parameters encountered in the study of skin sensitization: (1) the concentration (dose of chemical per unit area of skin), (2) area over which the chemical is delivered, and (3) the total dose of chemical. Quantitative characteristics of skin sensitization in human volunteres have been studied with 2,4-dinitrochlorobenzene (DNCB) (Friedmann et al., 1983a, 1983b). When humans with no history of contact dermatitis received various sensitizing doses of DNCB, it was found that the proportion of subjects sensitized showed a positive sigmoid relationship with the log of the induction concentration (Friedmann, 1990). In contrast, when increasing amounts of DNCB was given to proportionally larger areas of skin (the concentration per unit of skin area remained constant), there was no increase in sensitization frequency. Thus, the induction of skin sensitization is, within limits, dependent upon the concentration of the allergen (dose of chemical per unit area of skin), not the total dose. In corollary it was found that with a constant dose of DNCB and varying the area of application, the sensitization frequency was highest in the groups who received the dose of DNCB on the smallest area (White et al., 1986). (Upadhye and Maibach, 1992) concluded that a higher ratio of allergen molecules per Langerhans' cell dramatically increased the sensitization frequency; however, in diagnostic patch tests in humans, the total amount of allergen applied may also be important for the frequency and degree of contact allergy (Fowler et al., 1995).

MULTIPLE-DOSE RELATIONSHIP IN THE GUINEA PIG

Two dose-response relationships may be considered when predictive allergenicity testing is done. First, the test group may be divided into smaller groups, where each group is treated with different induction concentrations to obtain the relationship between the induction concentration and the response rate. Secondly, for the challenge reaction, the relationship between challenge concentrations and frequency and degree of sensitization may be studied.

The impact of different induction doses on the sensitization frequency has been studied in humans. A dose-response relationship was demonstrated for neomycin, penicillin G, paraphenylenediamine, and tetrachlorosalicylanilide (Kligman, 1966b). A dose-response relationship was also demonstrated with the modified human Draize test using the following chemicals: benzocaine, bronopol, formaldehyde, glutaraldehyde, mafenide, and paraphenylenediamine (Marzulli and Maibach, 1974).

(Magnusson and Kligman, 1969) already described the correlation between induction concentration and sensitization rate in the GPMT in the original paper. This information, however, was not studied systematically in the following years, and only scanty information was published on the dose-response concept (Ziegler et al., 1988; Jayjock and Lewis, 1992). The importance of using dose-response in toxicological testing may not have been fully appreciated, partly due to inadequate statistical tools for dose-response analysis and the requirement of a large number of animals that has to be used to show dose-response relationships.

(Bronaugh et al., 1994) used more than 200 guinea pigs in each dose group to show dose-response relationship with the open epicutaneous test and the Buehler test with dinitrochlorobenzene (DNCB) and paraphenylenediamine.

A GPMT dose-response study with the fungicide Maneb, covering an induction dose range of 1000 intradermally and 10,000 topically, showed a dose-response relationship between sensitization and intradermal as well as topical induction (Matsushita and Aoyama, 1980). The challenge concentration 0.05% gave negative test reactions, though the animals were sensitized documented by positive challenge to high concentration. (Koschier et al., 1983) investigated the dose-response relationship for 2,4-toluene diisocyanate and showed that the sensitization response was dependent upon the dose used both at induction and challenge. With the preservative Kathon CG and a dose-response protocol in the Buehler assay, (Chan et al., 1983) demonstrated a 3-dimensional dose-response relationship between sensitization frequency, induction, and challenge concentration. The same relationship was recently shown for the GPMT with 2-4-DNCB, phenylisocyanate, and phenylisothiocyanate (Nakamura et al., 1994). They found the test response dependent upon both intradermal induction and challenge for most compounds; however, they omitted the impact of the topical induction in the analysis because using both intradermal and topical concentration for analysis complicates the analysis. This is unfortunate because topical induction may contribute to the sensitization rate.

(Andersen et al., 1985a, 1995) used the GPMT with a dose-response design comprising both different intradermal and topical concentrations and analyzed the test results with logistic regression methods. The studies were based on the assumption that 30 animals could be divided into 5 test groups of 5 animals and 1 control group without losing statistical power. A number of chemicals were studied: nickel sulfate, lidocaine, Kathon CG, formaldehyde, cinnamic aldehyde, mercaptobenzothiazole, propylparaben, chloroacetamide, benzocaine, methyl metacrylate, potassium dichromate, and methyl acrylic acid (Rohold et al., 1991; Andersen, 1986; Andersen, 1993; Andersen et al., 1995; Boman et al., 1996). A dose relationship could be demonstrated for some of the chemicals tested; however, with very potent allergens (Kathon CG) covering an intradermal dose range of 30 or weak allergens (lidocaine, propyl paraben) covering an intradermal dose range of 30 and 100, respectively, the dose range seemed to be too narrow. Consequently, the curves were either fixed at the top or bottom of the response limits. Another key point was that the intradermal induction, not the topical dose, was decisive for the sensitization rate for formaldehyde, cinnamic aldehyde, and mercaptobenzothiazole. The two reports were recently reviewed by (Basketter, 1998), who concluded that the GPMT is excellent for identifying weak allergens and that the standard GPMT protocol should be used when the ultimate hazard identification is needed, but he recommended newer assays, as the local lymph node assay should be used rather than modifying the old assays.

DOSE-RESPONSE RELATIONSHIP AND STATISTICS

The dose-response models considered in the following describe the probability of a positive response as a function of the dose. The most widely used models are lognormal and logistic distributions, though there are other posibilities (Finney, 1971). The logistic regression model was chosen because it is mathematically simple, has desirable statistical properties, and the existence of sufficient estimators (Ashton 1972; Vølund 1978, 1982). Furthermore, this model has previously been used for analysis of dose response GPMT data (Andersen et al., 1995) using Magnusson and Kligman's (1969) ordinal grading scale with four response levels (0, 1+, 2+ and 3+); however, the distance between two levels on the scale is unknown, in contrast to an interval scale assigned with numerical values 0, 1, 2, and 3 (Siegel, 1956; Andersen et al., 1994). Binomial data were obtained from the ordinal four-grade ranking scale (Table 22.1) by grading $(0 \cup 1+)$ as negative and $(2+ \cup 3+)$ as positive

TABLE 22.1. *The 4 grade ranking scale for the evaluation of challenge patch test reactions (Magnusson and Kligman, 1969)*

0	=	no visible change
1+	=	discrete or patchy erythema
2+	=	moderate and confluent erythema
3+	=	intense erythema and swelling

responses. Weak-positive (1+) reactions were chosen to represent a negative response to eliminate false-positive reactions due to irritation from bandaging (Andersen et al., 1995, 1996); however, if all control animals show 0 response and the test animals show a 1+ response, the reactions may be relevant and an evidence of contact allergy. Therefore, analysis of ordinal data is desirable, and they may also be analyzed with logistic regression methods (Agresti, 1984).

In previous GPMT studies, 30 animals were divided into groups of 5 animals (Andersen et al., 1995); however, (Basketter, 1998) has questioned the consequences of using only five animals in each group in the multiple-dose design. With logistic regression analysis of the dose-response data, it is possible to maintain the statistical precision in analysis of multiple-dose groups without increasing the total number of animals required for each test. This is possible because logistic regression analysis includes all animals used for the experiment in a simultaneous statistical analysis. Thus, the larger number of groups will counterbalance the smaller group size. This correlation has been demonstrated by (Finney, 1978), who obtained the same statistical precision when doses of one drug were given to 2 groups of 12 animals, 3 groups of 8 animals or 4 groups of 6 animals. (Vølund, 1993) further confirmed this with computerized stochastic simulations on experimentally obtained dose-response data; however, too many test groups may for practical reasons be difficult. From a statistical point of view, however, the statistical power of the test is retained with logistic regression analysis (Finney 1978; Vølund 1993).

Based on logistic regression analysis, dose-response curves may be estimated. The shape of the dose-response curve may be monotone or nonmonotone (bell-shaped curve) and is not only related to the induction concentration, but also the challenge concentration (Roberts and Williams, 1982). The nonmonotonous dose-response relationship is of toxicological interest because it shows that false low sensitization rate may be obtained in the standard GPMT, not only by using too low an induction concentration, but also when a too-high induction concentration is used. The nonmonotone curve is used when it gives a statistically superior curve fit compared to the monotone curve; however, estimating the concentration sensitizing 50% of the animals (ED_{50}) for nonmonotone curves is questionable because two ED_{50}'s can be estimated from a bell-shaped curve. For the nonmonotone curves the ED_{50} is read before the maximal sensitisation is reached.

The nonmonotone shaped curve has been reproduced with several chemicals in the multiple-dose GPMT (i.e., formaldehyde and mercaptobenzothiazole), but it is of cause dependent on the dose design and concentrations used. A nonmonotonous dose-response relationship may be explained by a local toxic effect of the chemical to the antigen-presenting cells or to systemic toxicity. A toxic effect on the Langerhans' cells is questionable because sensitization may then be induced via the dermal antigen-presenting cells (Bacci et al., 1997). Systemic toxicity is an alternative possibility, as the animals may lose weight (Stadler and Karol, 1985); however, no weight loss was observed in the highest induction groups in our experiments. Another explanation may be a downregulation of the immune response caused by an overload phenomenon related to the use of FCA (Andersen, 1985b). The nonmonotonous dose response relationship was, also seen in the Buehler test with formaldehyde, and in the MEST with TNCB, dicyclohexylmethane-4,4'-diisocyanate, and picryl chloride (Sullivan et al., 1990; Stadler and Karol 1985), and even in humans with Furacin and ammoni-ated mercury (Kligman, 1966b).

DOSE-RESPONSE STUDIES IN THE GPMT AND BUEHLER TEST

The standard GPMT using only one moderately irritating intradermal and topical induction dose gives a qualitative assessment of allergenicity of the chemical—hazard identification. Refining the standard GPMT with a dose-response design is an attempt to provide quantitative GPMT data that may be used for risk estimation in the risk assessment process instead of only hazard identification (Canada Health Protection Branch, 1990; Derelanko, 1995; Neumann, 1995). Dose-response data may give an estimation of the maximal sensitization frequency (EMS), an ED_{xx} (the concentration that sensitizes xx% (e.g., 50% = ED_{50}) of the animals), a dose-response relationship, and an estimate of a threshold concentration, below which the allergen does not sensitize the guinea pigs.

Dose-response studies with the GPMT using well-defined chemicals with varying allergenicity are presented: benzocaine, chlorhexidine, chlorocresol, (α-hexyl cin-namic aldehyde, hydroxycitronellal, Kathon CG, and parthenolide and dose-response studies in the Buehler test with formaldehyde and neomycin. The GPMT and Buehler test procedure is illustrated in Figure 22.1, and details are described elsewhere (Buehler, 1964, 1965; Magnusson and Kligman, 1969). The test procedures recom-mended in the OECD guideline were followed. The animals were restrained during treatment in the Buehler test (Buehler, 1965, 1985). In both tests, 30 animals were used for each test and divided into six groups of five animals—one control group and five test groups treated with different induction concentrations. The dose-response design for the GPMT is shown in Table 22.2. Simultaneously increasing intradermal and topical induction doses were chosen for the GPMT. The advantage of using simulta-neous increasing intradermal and topical induction is that the impact of both induc-tion modes is tested simultaneously is one statistical analysis. This gives a simplified analysis compared to nonparallel dose designs. The dose-response curve will be the same for either induction mode, but parallel displaced according to the induction concentrations used. For the Buehler test, increasing topical induction concentra-tions were used. The choice of induction and challenge concentrations is based on

TABLE 22.2. *The concentrations shown here are an example to illustrate the distribution of doses (the concentrations in brackets indicate the dose combination covering a dose range of 100)*

	1	X	X	X	X	X	[X]
	0.3	X	X	X	X	[X]	X
Intradermal	0.01	X	X	X	[X]	X	X
Dose (%)	0.03	X	X	[X]	X	X	X
	0.01	X	[X]	X	X	X	X
	0	[X]	X	X	X	X	X
		0	0.1	0.3	1	3	10
				Topical dose (%)			

dose-finding experiments. A dose range between 50 and 10,000 was used, which, from a biological point of view, seems reasonable; however, the choice of dose range for induction may be difficult because some chemicals are strong allergens with a steep dose-response curve whereas other chemicals are weaker allergens with a rather flat curve. Choosing test concentrations for chemicals with unknown allergenic potential may be even more difficult in a narrow dose design, and therefore a broader dose range may be preferable. The concentration range chosen and the distribution of concentrations within the dose range should preferably encompass concentrations giving negative to maximal sensitization, thus making it possible to estimate the dose response curve and the three parameters: threshold concentration, ED_{50}, and the maximal sensitization rate (EMS).

GPMT RESULTS

Benzocaine

Benzocaine is a local anesthetic and a fairly weak sensitizer in humans (Kligman, 1966a). Benzocaine is included in the European standard patch test series and used in a concentration of 5% in petrolatum. The results after increasing intradermal and topical induction with a factor 10,000 is shown in Table 22.3. The estimated maximal sensitization rate was EMS = 0.9 and the ED_{50} = 0.2%, respectively (Figure 22.2).

TABLE 22.3. *Test results with benzocaine, id. Induction in propylene glycol (PG), topical induction in petrolatum (pet.), challenge concentration 10% in pet. Induction dose range of 10000. No. positive/no. tested. Concentrations in percentage*

Benzocaine							
Challenge conc.	Day	Control	id 0.0005 top 0.005	id 0.005 top 0.05	id 0.05 top 0.5	id 0.5 top 5	id 5 top 50
10	2	0/5	0/5	0/5	3/5	2/5	3/5
	3	0/5	0/5	0/5	3/5	3/5	4/5

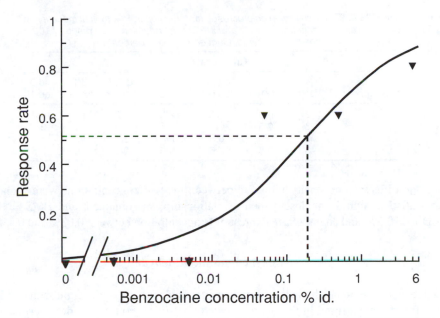

FIG. 22.2. Dose-response GPMT. The dose response curves for benzocaine. Day 3 readings after challenge with 10% benzocaine. ▼ is the observed responses after increasing id. and topical induction with a factor 10,000; —— is the monotone logistic curve fit. The ED$_{50}$ is shown.

There was a significant dose-response relationship and an intradermal threshold concentration between 0.005% and 0.05%.

Chlorhexidine

Chlorhexidine is a disinfectant with a weak sensitizing potential, commonly causing irritant contact dermatitis (Angelini, 1995). No animals were sensitized in the dose-response GPMT with chlorhexidine in a dose range of 1000 (id.: 0.001%–1%).

Chlorocresol

Chlorocresol is a biocide with widespread use in industry and pharmaceutical products. It is an occasional human contact sensitizer (Andersen, 1986) and an experimental sensitizer in the standard GPMT, sensitizing between 15%–84% of the animals, depending on the choice of dose and vehicle (Wahlberg and Boman, 1985). Table 22.4 and Figure 22.3 show the results and fitted curve after simultaneous increase in intradermal and topical induction concentrations with a dose range of 100. In this experiment, two challenge concentrations were used at 1% and 0.1%,

TABLE 22.4. *Test results with chlorocresol, id. Induction in PG, topical induction in pet., challenge concentration 0.1% and 1% in pet. Induction dose range of 100. No. positive/no. tested. Concentrations in percentage*

| | | | Chlorocresol | | | | |
| | | | id 0.03 | id 0.1 | id 0.3 | id 1 | id 3 |
Challenge conc.	Day	Control	top 0.1	top 0.3	top 1	top 3	top 10
0.1	2	0/5	0/5	2/5	4/5	3/5	4/5
	3	0/5	0/5	3/5	3/5	3/5	3/5
1	2	0/5	0/5	2/5	3/5	3/5	5/5
	3	0/5	1/5	4/5	3/5	4/5	5/5

and no difference in sensitization rate between the two concentrations were found. A statistical significant dose-response relationship was obtained, and EMS = 1, $ED_{50} = 0.3\%$, and an intradermal threshold concentration between 0.03% and 0.1%.

α-Hexyl Cinnamic Aldehyde

α-Hexyl cinnamic aldehyde is a fragrance chemical and one of the eight included in the fragrance mix used in the standard patch test series. Four series of experiments were performed with α-hexyl cinnamic aldehyde. Figure 22.4 and Table 22.5 show

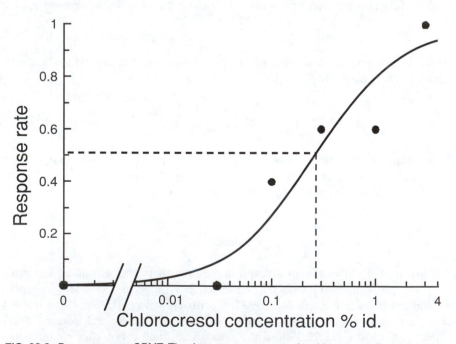

FIG. 22.3. Dose-response GPMT. The dose response curves for chlorocresol. Day 2 readings after challenge with 1% chlorocresol. • is the observed responses with increasing intradermal and topical induction with a dose range of 100; —— is the monotone logistic curve fit. The ED_{50} is shown.

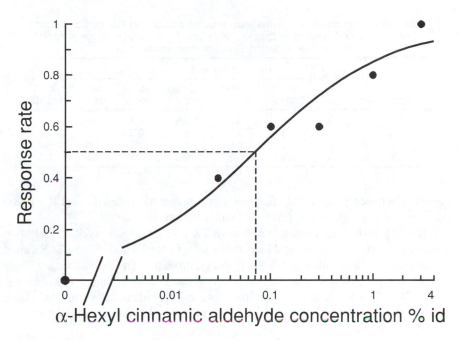

FIG. 22.4. Dose-response GPMT. The dose response curve for α-hexyl cinnamic aldehyde. Day 3 readings after increasing intradermal and topical induction with a dose range of 100 and challenge with 3% α-hexyl cinnamic aldehyde. • is the observed responses; —— is the monotone logistic curve fit. The ED_{50} is shown.

the results after combined increasing intradermal and topical induction of α-hexyl cinnamic aldehyde with a dose range of 100. The monotone curve fit after day 3 readings was used. α-Hexyl cinnamic aldehyde is a strong sensitizer in the multiple-dose GPMT with an EMS = 0.9 and $ED_{50} = 0.08\%$. A statistical significant dose-response relationship was obtained and the threshold concentration was less than 0.03%.

Hydroxycitronellal

Hydroxycitronellal is one of the eight constituents in the fragrance mix used in the standard patch test series. In the human maximization test, there is a sensitization

TABLE 22.5. Test results with α-hexyl cinnamic aldehyde, id. Induction in PG, topical induction in pet., challenge concentration 3% in pet. Induction dose range of 100. No. positive/no. tested. Concentrations in percentage

			α-hexyl cinnamic aldehyde				
Challenge conc.	Day	Control	id 0.03 top 0.3	Id 0.1 top 1	Id 0.3 top 3	id 1 top 10	id 3 top 10
3	2	0/5	3/5	4/5	2/5	3/5	3/5
	3	0/5	2/5	3/5	3/5	4/5	5/5

TABLE 22.6. *Test results with hydroxycitronellal, id. Induction in PG, topical induction in pet., challenge concentration 2% and 20% in pet. Induction dose range of 10,000. No. positive/no. tested. Concentrations in percentage*

			Hydroxycitronellal				
Challenge conc.	Day	Control	Id 0.001 Top 0.01	id 0.01 top 0.1	id 0.1 top 1	id 1 top 10	id 10 top 100
	2	0/5	0/5	0/5	0/5	3/5	3/5
2	3	0/5	0/5	0/5	0/5	3/5	4/5
	2	0/5	0/5	0/5	0/5	4/5	5/5
20	3	0/5	0/5	0/5	0/5	4/5	5/5

frequency between 8% and 70%, depending on concentration used (Ford et al., 1988). One series of experiment was performed with hydroxycitronellal.

Table 22.6 shows the challenge data with 2% and 20% hydroxycitronellal after combined intradermal and topical induction with a factor 10,000. Figure 22.5 shows the fitted curve after challenge with 2% hydroxycitronellal. The 2% challenge results were chosen because only one intermediate reaction between 0 and 1 was obtained with 20% challenge concentration, and therefore no curve fit could be estimated. After challenge or rechallenge, no animals were sensitized in the lowest induction groups.

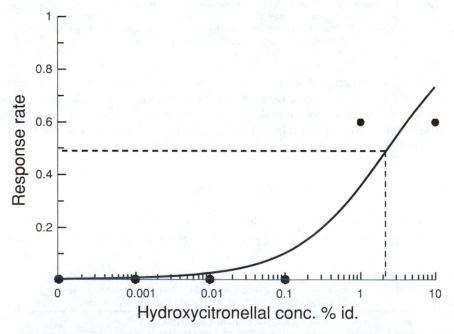

FIG. 22.5. Dose-response GPMT. The dose response curve for hydroxycitronellal. Day 2 readings after challenge with 2% hydroxycitronellal. • is the observed responses; —— is the monotone logistic curve fit after combined increasing intradermal and topical induction with a factor 10,000. The ED_{50} is shown.

TABLE 22.7. *Test results with Kathon CG, id. and topical induction in saline, challenge and rechallenge concentration 0.01% in saline. Induction dose range of 300. No. positive/no. tested. Concentrations in percentage*

			Kathon CG				
	Day	Control	id 3×10^{-5} top 0.0003	id 0.0001 top 0.001	id 0.0003 top 0.003	id 0.001 top 0.01	Id 0.01 top 0.1
Challenge conc. 0.01	2	0/5	3/5	2/5	3/5	3/5	4/5
	3	0/5	3/5	2/5	3/5	4/5	5/5
Rechallenge conc. 0.01	2	0/5	4/5	3/5	4/5	5/5	3/5
	3	0/5	4/5	3/5	4/5	5/5	4/5

The EMS $= 0.7$, ED$_{50} = 2\%$, intradermal threshold concentration between 0.1% and 1%, and a significant dose relationship was obtained.

Kathon CG

Isothiazolinones (5-chloro-2-methyl-4-isothiazolin-3-one and 2-methyl-4-isothiazolin-3-ono, 3:1 w/w) are the active ingredients in Kathon CG, a cosmetic preservative. Kathon CG is an extreme potent sensitizer in man and guinea pig, as demonstrated in a previous multiple-dose GPMT. Using intradermal doses ranging from 0.0003%–0.01% and topical doses ranging from 0.003% and 0.3%, all test animals were sensitized using a challenge concentration of 0.03% (Andersen et al., 1995). In the present study, animals were treated with lower induction and challenge concentrations compared to the previous study. The results are shown in Table 22.7 and Figure 22.6.

Increasing intradermal and topical induction with a dose range of 300 was used. In this study, three challenge concentrations were used; 0.01, 0.003, and 0.001%. With the highest challenge concentration, the EMS $= 1$, ED$_{50} = 0.00003\%$, and intradermal threshold concentration was <0.00003. None of the Kathon CG-sensitized animals reacted to the 0.003% or 0.001% challenge concentration; however, three animals were positive after rechallenge with 0.003%, but none after rechallenge with 0.001%.

Parthenolide

Parthenolide is a sesquiterpene lactone found in fewerfew and is known as a potent human sensitizer. Table 22.8 shows the results after combined increasing intradermal and topical induction of parthenolide.

Figure 22.7 shows the fitted dose-response curve after challenge with 0.03% parthenolide. Parthenolide is a very strong sensitizer in the multiple-dose GPMT, with an estimated maximal sensitization rate of EMS $= 1$, ED$_{50} = 0.002\%$, and intradermal threshold concentration between 0.0001% and 0.001%. After challenge with 0.3%, nearly all animals became positive, resulting in only one group showing a sensitization frequency between 0 and 1, and therefore no dose-response curve could be estimated.

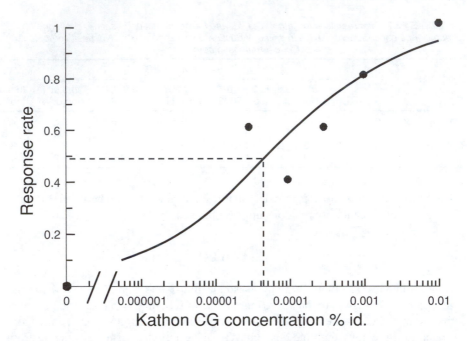

FIG. 22.6. Dose-response GPMT. Dose response GPMT for Kathon CG. Day 3 readings after challenge with 0.01% Kathon CG. • is the observed responses; —— is the monotone logistic curve fit after combined increasing intradermal induction with a factor 300. The ED_{50} is shown.

BUEHLER TEST RESULTS

Formaldehyde

Formaldehyde is a common contact allergen present in the European standard patch test series. Formaldehyde is used as a preservative in a large variety of products in industry and domestic environment. Formaldehyde exposure is not limited to contact with the chemical as such, but also from several formaldehyde releasers. Previous reported standard Buehler test results in the literature with formaldehyde have

TABLE 22.8. *Test results with parthenolide, id. induction in PG and topical induction in pet., challenge concentration 0.03% and 0.3% in pet. Induction dose range of 10,000. No. positive/no. tested. Concentrations in percentage*

			Parthenolide				
Challenge conc.	Day	Control	Id 0.0001 top 0.0003	id 0.001 top 0.003	id 0.01 top 0.03	id 0.1 top 0.3	id 1 top 3
0.03	2	0/5	0/5	1/5	5/5	5/5	5/5
	3	0/5	0/5	3/5	4/5	5/5	5/5
0.3	2	0/5	0/5	2/5	5/5	5/5	5/5
	3	0/5	0/5	3/5	5/5	5/5	5/5

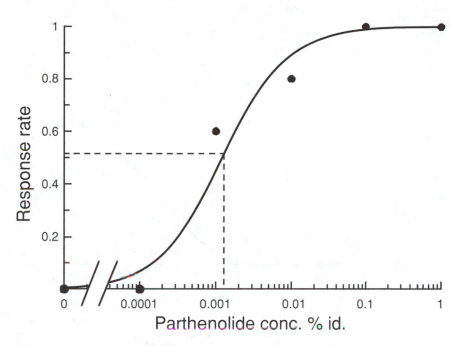

FIG. 22.7. Dose-response GPMT. The dose response curve for parthenolide. Day 3 reading after challenge with 0.03% parthenolide. • is the observed responses; —— is the monotone logistic curve fit after combined increasing intradermal and topical induction with a factor 10,000. The ED_{50} is shown.

given varying sensitization rates between 0–70% (Hilton et al., 1996; Buehler, 1985; Marzulli and Maguire, 1982).

Table 22.9 shows data with formaldehyde, and Figure 22.8 shows the fitted dose response curves for the readings day 2 with the Buehler test.

After the first topical induction with the highest concentration of formaldehyde, severe erythema was observed in the animals. For the second induction patch, this group was treated with a lower induction of 10%; however, after the second induction, both groups treated with 10% formaldehyde reacted with severe erythema. The highest induction concentration applied for the third induction was therefore reduced to 3%. Challenge was performed with 0.3% and 1% formaldehyde, but 3 of 5 control animals responded with a 1+ to 3+ after challenge with 1% formaldehyde. In the initial

TABLE 22.9. *Buehler test results with formaldehyde , topical induction in saline, challenge concentration 0.3% in saline. Induction dose range of 100. No. positive/no. tested. Concentrations in percentage*

		Formaldehyde					
Challenge conc.	Day	Control	top 0.3	top 1	top 3	top 10 (3)	top 30 (3)
0.3	2	1/5	1/5	3/5	2/5	2/5	1/5
	3	0/5	1/5	3/5	2/5	3/5	1/5

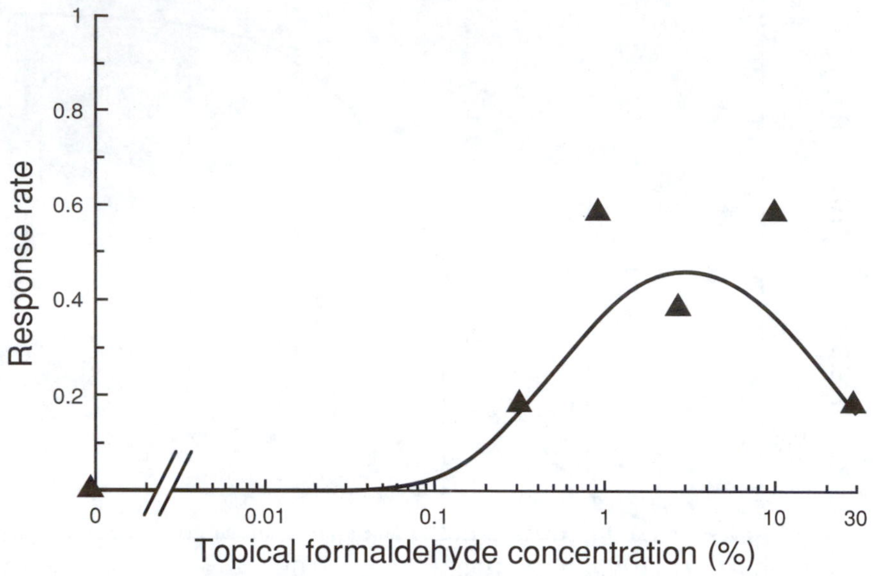

FIG. 22.8. Dose-response Buehler test. The dose response curves for formaldehyde. Buehler test readings day 2 after challenge with 0.3% formaldehyde. ▲ is the observed responses and; —— is the monotone logistic curve fit.

dose-finding studies, 1% formaldehyde was not irritant. One control animal reacted to the 0.3% formaldehyde patch at the day 2 reading after challenge, but the reaction disappeared at the day 3 reading. The data after challenge and rechallenge with 0.3% is used for statistical analysis. The EMS = 0.6, ED_{50} = 1%, and topical threshold concentration was <0.3%.

Neomycin Sulfate

Neomycin is an antibiotic used in topical drugs and included in the standard patch test series and a common sensitizer (Lipozencic et al., 1993).

Table 22.10 and Figure 22.9 show the neomycin challenge data and the fitted dose-response curves for the day 3 readings with neomycin.

TABLE 22.10. *Buehler test results with neomycin, topical induction in saline, challenge concentration 10% and 30% in saline. Induction dose range of 50. No. positive/no. tested. Concentrations in percentage*

		Neomycin					
Challenge conc.	Day	Control	top 1	top 3	top 10	top 30	top 50
10	2	0/5	0/5	0/5	0/5	0/5	0/5
	3	0/5	0/5	0/5	0/5	0/5	0/5
30	2	0/5	0/5	0/5	1/5	0/5	0/5
	3	0/5	0/5	0/5	1/5	2/5	2/5

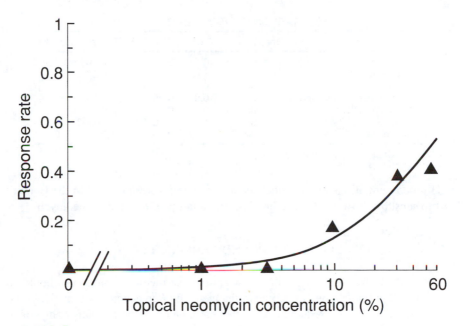

FIG. 22.9. Dose-response Buehler test. The dose response curves for neomycin. Buehler test readings day 3 after challenge with 30% neomycin. ▲ is the observed responses; —— is the monotone logistic curve fit.

Sensitized animals were observed at the day 3 readings after challenge with 30% neomycin. No reactions was observed after challenge with 10% concentration, but after rechallenge with the same concentration, neomycin sensitization was observed. The ED_{50} could not be estimated, EMS = 0.4, and topical induction threshold concentration was between 3% and 10%. The sensitization rate was higher after rechallenge, compared to challenge with a maximal sensitization rate of EMS = 1.

SUMMARY OF TEST RESULTS

Tables 22.11 and 22.12 give an overview of the estimated maximal sensitization rate (EMS), the intradermal concentration sensitizing 50% of the animals (ED_{50}), and the concentration below which the allergen does not sensitize the guinea pigs (threshold concentration).

GPMT and Buehler test results are dependent upon the induction concentration. It is seen that Kathon CG is a very strong sensitizer, even at very low concentrations (ED_{50} (id.) of 0.00003%). Benzocaine and chlorocresol were moderate sensitizers with EMS of 0.8–1 and ED_{50} of 0.2–0.4. The ED_{50} of α-hexyl cinnamic aldehyde and parthenolide was 10-fold less but with the same EMS as benzocaine and chlorocresol, whereas hydroxycitronellal had a higher ED_{50}. No animals were positive with chlorhexidine in the GPMT, as could be expected from human studies.

TABLE 22.11. *GPMT results*

Chemical	Dose range	EMS	ED_{50} (%)	Threshold id. (%)
Benzocaine	1000	0.9	0.2	>0.005 < 0.05
Chlorocresol	100	1	0.3	>0.03 < 0.1
Chlorhexidine	1000	0	—	not sensitising
α-Hexyl cinnamic aldehyde	100	0.9	0.08	<0.03
Hydroxycitronellal	1000	0.7	2	>0.1 < 1
Kathon CG	300	1	0.00003	<00003
Parthenolide	1000	1	0.03	>0.0001 < 0.001

— = Could not be estimated

High concentrations of formaldehyde and neomycin were needed in the Buehler test to sensitize the animals, and the ED_{50} could not be estimated for neomycin.

CONCLUSION AND PERSPECTIVES

(Russell and Burch, 1959) described how animal test methods constantly should be subject to development by *R*eduction, *R*efinement and *R*eplacement alternatives - the three *R*'s. They defined *R*eduction as implementation of methods giving comparable levels of information with the use of fewer animals, or methods giving more information from a given number of animals, so that, in the long run, fewer animals would be needed. *R*efinement was obtained by a decrease in use of stressful procedures applied to the animals used, and *R*eplacement was defined as alternative methods using nonsentient material instead of living animals.

Modifying the standard GPMT and Buehler test with a dose-response design and analyzing the test results with logistic regression analysis is an approach of a *R*efinement and *R*eduction alternative for better animal welfare. It has been argued that the use of inappropriate statistical analysis of experimental results may not only misguide interpretation of test results, but also cause inefficient use of animals in biomedical research (Festing, 1992, 1994). (Balls et al., 1995; Holzhütter et al., 1996) have specifically advocated further investigation of statistics in predictive toxicological methods in favor of better test design and more efficient data handling. Thus, focusing future studies on improved statistics for predictive allergy tests seems most relevant. Guinea pig data usually are presented as binomial data, negative or positive reactions; however, all four levels in the Kligman grading scale may be used as ordinal data for logistic regression analysis. The advantage of using ranked ordinal data is that all reactions are used for statistical analysis, which may narrow the confidence

TABLE 22.12. *Buehler test results*

Chemicals	Dose range	EMS	ED_{50} (%)	Threshold id. (%)
Formaldehyde	100	0.6	1	<0.3
Neomycin Sulfate	50	0.4	—	>3 < 10

— = Could not be estimated

TABLE 22.13. *Benzocaine rechallenge data after 2 days*

Concentration id. (%)	Ordinal data				Binomial data # ≥ 2+ No. positive/no. tested.
	# = 0	# = 1+	# = 2+	#= 3+	
0	5	0	0	0	0/5
0.1	2	2	1	0	1/5
0.3	1	1	1	2	3/5
1	1	0	1	3	4/5
3	0	1	0	4	4/5
5	0	2	3	0	3/5

intervals and statistically demonstrate a true dose-response relationship. Benzocaine rechallenge results are shown in Table 22.13 and Figure 22.10 as binomial and ranked ordinal data. Rechallenge data for benzocaine were chosen because there were many intermediate reactions (grade 1+ and 2+).

There is a significant dose-response relationship when the ranked ordinal data are used for analysis ($p = 0.023$). This was not the case with the binomial data ($p = 0.08$); however the curve shape and EMS was very similar (Figure 22.10). Using ordinal dose-response data may be especially valuable if many intermediate responses (1+ and 2+) are obtained (Frankild et al., 1997).

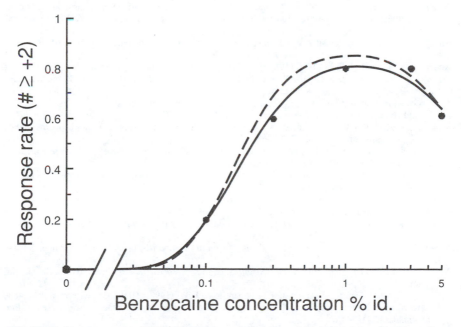

FIG. 22.10. Dose-response GPMT. The dose response curves for benzocaine. Day 2 readings after rechallenge with 10% benzocaine. Induction with increasing id. and constant low topical induction 1% benzocaine. • is the observed responses (# ≥ 2+); ---- is the monotone logistic curve fit for the ordinal data; —— is the monotone logistic curve fit for the binomial data.

Refining the standard guinea pig protocol with a dose-response design and analyzing data with logistic dose regression may lead to improved characterization of the allergenic potential and provide quantitative data that can be used for risk estimation in the risk assessment process. This is a prerequisite for better regulation of contact allergens in the EU. Furthermore, it may improve animal welfare by obtaining more reproducible and reliable test results from the animals used and thus reduce the need for repetitive tests, thereby reducing the total number of animals needed for the purpose of allergenicity testing.

REFERENCES

Canada Health Protection Branch (1990): Risk Management in Health Protection Branch. Ottawa: Minister of Supply and Services.

OECD Guideline #406 (1992): Organisation for Economic Cooperation and Development. Guideline 406: Skin sensitization.

Annex VI to Dangerous Substances Directive 92/32/EC (1993): Annex VI to the European Commision Directive 93/21/EEC of April 1993 adapting to technical progress for the 18th time Council Directive 67/548/EEC on the Aproximation of the Laws, Regulation and Administrative Provisions relating to the Classification, Packing and Labelling of Dangerous Substances. *Off. J. Eur. Commun., L110A.*

Annex V methods EEC (1994): Commission Directive of 29 December 1992 adapting to technical progress for the 17th council directive on laws, regulations and administrative provisions relating to classification, packaging and labelling of dangerous substances. *Off. J. Eur. Commun., L383 A* 131.

Agresti, A. (1984): *Analysis of Ordinal Categorical Data.* J. Wiley, New York.

Andersen, K. E., Boman, A., Vølund, A., and Wahlberg, J. E. (1985a): Induction of formaldehyde contact sensitivity: Dose response relationship in the guinea pig maximization test. *Acta. Derm.Venereol. (Stockh.)*, 65:472–478.

Andersen, K. E. (1985b): Sensitivity and subsequent "down regulation" of sensitivity induced by chlorocresol in guinea pigs. *Arch. Dermatol. Res.*, 277: 84–87.

Andersen, K. E., and Maibach, H. I. (1985c): Guinea pig sensitization assays: An overview: In *Contact Allergy: Predictive Tests in Guinea Pigs.* Edited by Andersen, K. E., and Maibach, H. I., pp. 263–290. Karger, Basel.

Andersen, K. E. (1986): Contact allergy to chlorocresol, formaldehyde, and other biocides. Guinea pig tests and clinical studies. *Acta. Derm.Venereol. (Stockh.)*, 66:1–21.

Andersen, K. E. (1993): Potency evaluation of contact allergens. I: Dose-response studies using the guinea pig maximization test. TemaNord 570:1–63.

Andersen, K. E., Bruze, M., Karlberg, A. T., Wahlberg, J. E., and Vølund, A. (1994): How to do sensitization tests in guinea pigs (letter). *Contact Dermatitis*, 31:278–279.

Andersen, K. E., Vølund, A., and Frankild, S. (1995): The guinea pig maximization test—with a multiple dose design. *Acta. Derm. Venereol.*, 75:463–469.

Andersen, K. E., Frankild, S., Wahlberg, J. E., and Boman, A. (1996): Non-specific hyperreactivity related to the use of Freund's complete adjuvant. *Contact Dermatitis*, 34:127.

Andersen, K. E., and Frankild, S. (1997): Allergic contact dermatitis. *Clinics in Dermatology* 15:645–654.

Angelini, G. (1995): Topical drugs. In: *Textbook of Contact Dermatitis*, pp. 477–503. Springer Verlag, Berlin.

Ashton, W. D. 1972. *The Logit Transformation with Special Reference to its Uses in Bioassays.* Ch. Griffin, London.

Bacci, S., Alard, P., Dai, R., Nakamura, T., and Streilein, J. W. (1997): High and low doses of haptens dictate whether dermal or epidermal antigen-presenting cells promote contact hypersensitivity. *Eur. J. Immunol.*, 27:442–448.

Back, O., and Larsen, A. (1982): Contact sensitivity in mice evaluated by means of ear swelling and a radiometric test. *J. Invest. Dermatol.*, 78:309–312.

Balls, M., Goldberg, A. M., Fentem, J. H., and Broadwell, D. C. (1995): The three Rs: The way forward. *Atla*, 23:838.

Basketter, D. A., and Chamberlain, M. (1995b): Validation of skin sensitization assays. *Food Chem. Toxicol.*, 33:1057–1059.

Basketter, D. (1998): Book review. *Am. J. Cont. Derm.*, 8:252–253.

Boman, A., Dyhring-Jacobsen, S., Klastrup, S., Svendsen, O., and Wahlberg, J. E. (1996): Potency evaluation of contact allergens. II: Dose-response studies using the guinea pig maximization test. *TemaNord*, 570:9–58.

Botham, P. A., Basketter, D. A., Maurer, T., Mueller, D., Potokar, M., and Bontinck, W. J. (1991): Skin sensitization: A critical review of predictive test methods in animals and man. *Food Chem. Toxicol.*, 29:275–286.

Bronaugh, R. L., Roberts, C. D., and McCoy, J. L. (1994): Dose-response relationship in skin sensitization. *Food Chem. Toxicol.*, 32:113–117.

Buehler, E. V. (1964): A new method for detecting potential sensitizers using the guinea pig. *Toxicol. Appl. Pharmacol.*, 6:341.

Buehler, E. V. (1965): Delayed contact hypersensitivity in the guinea pig. *Arch. Dermatol.*, 91:171–177.

Buehler, E. V. (1985): A rationale for the selection of occlusion to induce and elicit delayed contact hypersensitivity in the guinea pig: A prospective test. *Curr. Probl. Dermatol.*, 14:39–58.

Chan, P. K., Baldwin, R. C., Parsons, R. D., Moss, J. N., Stiratelli, R., Smith, J. M., and Hayes, A. W. (1983): Kathon biocide: Manifestation of delayed contact dermatitis in guinea pigs is dependent on the concentration for induction and challenge. *J. Invest. Dermatol.*, 81:409–411.

de Groot, A. C. (1994): *Patch Testing: Test Concentrations and Vehicles for 3700 Chemicals*. Elsevier, Amsterdam.

Derelanko, M. J. (1995): Risk assessment. In: *CRC Handbook of Toxicology*. Edited by Derelanko, M. J., and Hollinger, M. A., pp. 591–676. CRC Press, Boca Raton.

Festing, M. F. (1992): The scope for improving the design of laboratory animal experiments. *Lab. Anim.*, 26:256.

Festing, M. F. (1994): Reduction of animal use: Experimental design and quality of experiments. *Lab. Anim.*, 28:212.

Finney, D. J. (1971): *Probit Analysis.* Cambridge University Press, London.

Finney, D. J. (1978): *Statistical Method in Biological Assay*, 3rd. Edition. Griffin, C. H., London.

Ford, R. A., Api, A. M., and Suskind, R. R. (1988): Allergic contact sensitization potential of hydroxycitronellal in humans. *Food Chem. Toxicol.*, 26:921–926.

Fowler, J. F., Jr., and Finley, B. (1995): Quantity of allergen per unit area is more important than concentration for effective patch testing. *Am. J. Cont. Derm.*, 6:157–159.

Frankild, S. (1997): Logistic regression analysis of ranked (ordinal) GPMT data. ERGECD meeting in Odense, Denmark, May 22–24, 1997.

Fregert, S. (1978): Molecular weights of haptens. *Acta. Derm. Venereol. Suppl. Stockh.*, 79:71–72.

Friedmann, P. S., Moss, C., Shuster, S., and Simpson, J. M. (1983a): Quantitation of sensitization and responsiveness to dinitrochlorobenzene in normal subjects. *Br. J. Dermatol.*, 109 Suppl. 25:86–88.

Friedmann, P. S., Moss, C., Shuster, S., and Simpson, J. M. (1983b): Quantitative relationships between sensitizing dose of DNCB and reactivity in normal subjects. *Clin. Exp. Immunol.*, 53:709–715.

Friedmann, P. S. (1990): The immunology of allergic contact dermatitis: The DNCB story. *Adv. Dermatol.*, 5:175–195.

Hilton, J., Dearman, R. J., Basketter, D. A., Scholes, E. W., and Kimber, I. (1996): Experimental assessment of the sensitizing properties of formaldehyde. *Food Chem. Toxicol.*, 34:571–578.

Holzhütter, H.-G., Archer, G., Dami, N., Lovell, D. P., Saltelli, A., and Sjöström, M. (1996): Recommendations for the application of biostatistical methods during the development and validation of alternative toxicological methods: ECVAM Biostatistic Task Force Report 1. *ATLA*, 24:511–530.

Jayjock, M. A., and Lewis, P. G. (1992): Low-applied-dose extrapolation of indiction and elicitation of contact allergy in the evaluation and management of sensitization risk from kathon CG isothiazolone in products. *Am. J. Cont. Derm.*, 3:86–91.

Johansen, J. D., Andersen, K. E., and Menne, T. (1996): Quantitative aspects of isoeugonol contact allergy assessd by use and patch tests. *Contact Dermatitis*, 34:414–418.

Kimber, I., and Basketter, D. A. (1992): The murine local lymph node assay: A commentary on collaborative studies and new directions. *Food Chem. Toxicol.*, 30:165–169.

Klecak, G. (1991): Identification of contact allergens: Predictive tests in animals. In: *Dermatotoxicology*. Edited by Marzulli, F. N., and Maibach, H. I., pp. 363–413. Hemisphere Publishing Corpration, New York.

Kligman, A. M. (1966a): The identification of contact allergens by human assay. III: The Maximization Test: A procedure for screening and rating contact sensitizers. *J. Invest. Dermatol.*, 47:393–409.

Kligman, A. M. (1966b): The identification of contact allergens by human assay. II: Factors influencing the induction and measurement of allergic contact dermatitis. *J. Invest. Dermatol.*, 47:5,375–392.

Koschier, F. J., Burden, E. J., Brunkhorst, C. S., and Friedman, M. A. (1983): Concentration-dependent elicitation of dermal sensitization in guinea pigs treated with 2,4-toluene diisocyanate. *Toxicol. Appl. Pharmacol.*, 67:401–407.

Lipozencic, J., Milavec Puretic, V., and Trajkovic, S. (1993): Neomycin: A frequent contact allergen. *Arh. Hig. Rada. Toksikol.*, 44:173–180.

Magnusson, B., and Kligman, A. M. (1969): The identification of contact allergens by animal assay: The guinea pig maximization test. *J. Invest. Dermatol.*, 52:268–276.

Magnusson, B., and Kligman, A. M. (1970): *Allergic Contact Dermatitis in the Guinea Pig: Identification of Contact Allergens*, pp. 102–123. Charles C. Thomas, Springfield.

Marzulli, F. N., and Maibach, H. I. (1974): The use of graded concentrations in studying skin sensitizers: Experimental contact sensitization in man. *Food Cosmet. Toxicol.*, 12:219–227.

Marzulli, F., and Maguire, H. C. J. (1982): Usefulness and limitations of various guinea-pig test methods in detecting human skin sensitizers-validation of guinea-pig tests for skin hypersensitivity. *Food Chem. Toxicol.*, 20:67–74.

Matsushita, T., and Aoyama, K. (1980): Dose-response relationship in delayed type contact sensitivity with maneb: An experimental model of "simple chemicals". *Ind. Health*, 18:31–39.

Maurer, T. (1996): Predictive testing for skin allergy. In: *Allergic Hypersensitivities Induced by Chemicals: WHO*, pp. 237–259. Edited by Vos, J. G., Young M. and Smith. E. CRC Press.

Nakamura, A., Momma, J., Sekiguchi, H., Noda, T., Yamano, T., Kaniwa, M., Kojima, S., Tsuda, M., and Kurokawa, Y. (1994): A new protocol and criteria for quantitative determination of sensitization potencies of chemicals by guinea pig maximization test. *Contact Dermatitis*, 31:72–85.

Neumann, D. A. (1995): Immunotoxicity testing and risk assessment summary of a 1994 workshop. *Food Chem. Toxicol.*, 10:887–894.

Ritz, H. L. and Buehler, E. V. (1980): Planning, conduct and interpretation of guinea pig sensitization patch tests. In: *Current Concept in Cutaneous Toxocity*, pp. 25–40. Edited by Drill, V. A., and Lazar, P. Academic, New York.

Roberts, D. W., and Williams, D. L. (1982): The derivation of quantitative correlations between skin sensitisation and physico-chemical parameters for alkylating agents, and their application to experimental data for sultones. *J. Theor. Biol.*, 99:807–825.

Rohold, A. E., Nielsen, G. D., and Andersen, K. E. (1991): Nickel-sulfate-induced contact dermatitis in the guinea pig maximization test: A dose-response study. *Contact Dermatitis*, 24:35–39.

Russell, W. M. S., and Burch, R. L. (1959): *The Principles of Human Experimental Technique*. Meuthen, London.

Siegel, S. (1956): *Nonparametric Statistics for the Behavioral Sciences*. McGraw-Hill, Kogakusha Ltd., Tokyo.

Stadler, J. C., and Karol, M. H. (1985): Use of dose-response data to compare the skin sensitizing abilities of dicyclohexylmethane-4,4'-diisocyanate and picryl chloride in two animal species. *Toxicol. Appl. Pharmacol.*, 78: 445–450.

Sullivan, S., Bergstresser, P. R., and Streilein, J. W. (1990): Analysis of dose response of trinitrochlorobenzene contact hypersensitivity induction in mice: Pretreatment with cyclophosphamide reveals an optimal sensitising dose. *J. Invest. Dermatol.*, 94:711–716.

Upadhye, M. R., and Maibach, H. I. (1992): Influence of area of application of allergen on sensitisation in contact dermatitis. *Contact Dermatitis*, 27:281–286.

Vølund, A. (1978): Application of the four-parameter logistic model to bioassay: Comparison with slope ratio and parallel line models. *Biometrics*, 34:357–365.

Vølund, A. (1982): Combination of multivariate bioassay results. *Biometrics*, 38:181–190.

Vølund, A. (1993): Statistical part. Potency evaluation of contact allergens. *TemaNord*, 570.

Wahlberg, J. E., and Boman, A. (1985): Guinea Pig Maximization Test. In: *Contact Allergy: Predictive Tests in Guinea Pigs*. Edited by Andersen, K. E., and Maibach, H. I., pp. 59–106. Karger, Basel.

White, S. I., Friedmann, P. S., Moss, C., and Simpson, J. M. (1986): The effect of altering area of aplication and dose per unit area on sensitisation by DNCB. *Br. J. Dermatol.*, 115:663–668.

Ziegler, V., Ziegler, B., and Kipping, D. (1988): Dose-response sensitization experiments with imidazolidinyl urea. *Contact Dermatitis*, 19:236–237.

Toxicology of Skin
Edited by Howard I. Maibach
Copyright © 2001 Taylor & Francis

23

Operational Definition of Allergic Contact Dermatitis

S. Iris Ale

Department of Dermatology, University Hospital, Republic University, Montevideo, Uruguay

Howard I. Maibach

Department of Dermatology, University of California, San Francisco, California, USA

INTRODUCTION

When establishing the diagnosis of clinical allergic contact dermatitis (ACD), two prerequisites should be fulfilled: (1) demonstrating the existence of delayed hypersensitivity to one or several allergens, and (2) proving their clinical relevance. To accomplish this goal, the physician relies upon a detailed history, complete physical examination, and comprehensive skin testing. Although skin testing is conducted according to the clinical history and physical examination, the diagnostic process is bidirectional and skin test results will guide further questioning and investigation.

DIAGNOSIS OF ALLERGIC CONTACT DERMATITIS

A correct diagnosis and characterization of the causative agent(s) of the patient's dermatitis constitute essential prerequisites for both therapeutic and preventive measures to be adopted. Although exposure to the responsible allergen is the substantial factor, whether or not it will actually result in clinical dermatitis, depends on additional circumstances; thus, the diagnostic approach should also include evaluating all predisposing and contributory factors, and providing insights in differentiating contact dermatitis from endogenous dermatoses, such as atopic dermatitis and psoriasis.

Clinical History

The assessment starts with a comprehensive dermatological history (Table 23.1). The physician should investigate all probable sources of allergenic exposure in the patient's environment, and determine whether or not this exposure may be responsible for the patient's dermatitis. In addition, history should provide insights in differentiating ACD from other exogenous or endogenous dermatoses. This is crucial when dealing with multifactorial dermatitis. Previous episodes of dermatitis, presence of other skin diseases and personal or family atopy must also be recorded.

Assessment of Exposure

The history must determine which potential allergens the subject is exposed to, including occupational and nonoccupational agents. This should include a meticulous

TABLE 23.1. *Clinical data for the assessment of ACD*

History of exposure to the sensitizer (present or past), specially seeking for intolerance.
   Occupational exposure
      complete job description and materials
      personal protective measures at work (gloves, masks, barrier creams)
      other materials present in the working environment
   Non-occupational exposure
      homework, hobbies
      skin care products, nail and hair products, fragrances
      pharmaceutical products (by prescription and over the counter)
      personal protective measures. Use of gloves, detergents, etc.
      jewelry and clothing
   Indirect contact (Skin care and other products of partner, fomites, etc.)
   Seasonal related contact (plants and other environmental agents)
   Photoexposure
   Type of exposure: dose, frequency, site
   Environmental conditions: humidity, temperature, occlusion, vapors, powders, mechanical
      trauma, friction, etc.
Clinical characteristics of the present dermatitis
   Time of onset and characteristics of the initial lesions
   Dermatitis area corresponding to the exposure site
   Some morphologies suggest specific allergens
   Clinical course (caused or aggravated by the exposure)
   Time relationship to work. Effect of holidays and time-off work
History of previous dermatitis and other clinical events
   Past exogenous dermatitis with similar or different characteristics
   previous patch testing
   other endogenous skin diseases (psoriasis, atopic dermatitis, stasis, etc.)
Personal and family atopy and history of other family skin diseases

study of the working activities of the patient. If occupation appears causative, an occupational history must be obtained and a workplace visit may be required. Nonoccupational sources of exposure comprising hobbies and home environment, personal skin care products, clothing and accessories, current and previous medications, including self-treatments, etc., have to be thoroughly investigated. The environmental evaluation involves the determination of the intrinsec allergenic potential and other physicochemical characteristics of the suspected agent. Quantitative data regarding exposure (i.e., dose, frequency, area of exposure, etc.) should also be determined. Delayed sensitivity is a dose-related phenomenon and there is a threshold surface concentration of the allergen required to induce sensitization and/or elicitation of the response (Marzulli and Maibach, 1976; Upadhye and Maibach, 1992). Sensitization and consequent clinical dermatitis may result from a single exposure to a strong allergen; however, several exposures usually are required before sensitization and dermatitis result. Along with the intrinsic sensitizing properties of the substance, we have to consider concomitant exposure factors that might enhance the percutaneous penetration (i.e., irritation, occlusion, heat, mechanical trauma, etc.) (Andersen, 1994; Meneghini, 1985; Shmunes, 1988; Skog, 1958). As a consequence of repeated exposure, the allergen progressively accumulates in the epidermis; chemical and mechanical injuries increase its percutaneous absorption. Therefore, the clinical scenario will often show that an allergic contact dermatitis is preceded by an irritant contact dermatitis.

Occasionally, even when a careful clinical history has been taken, the source of exposure remains unidentified. This can be due to multiple causes: the exposure can be indirect (i.e., produced through a contaminated item, airborne, ectopic, or connubial), infrequent, or sporadic. Patients usually disregard as potential causes of dermatitis those substances to which they have limited exposure, and therefore do not disclose them in the history. Furthermore, some sources of exposure remain occult due to insufficient chemical identification in many household or industrial products (Table 23.2).

Time Course of the Dermatitis

The recognition of exposure to one or many allergens is not enough to establish the diagnosis of ACD; the clinical history should determine whether there is a time relationship between the attributable exposure and the clinical course of the dermatitis. The time between first contact with the causative agent and the clinical onset of dermatitis partly depends on the sensitizer and the conditions of exposure and partly on constitutional susceptibility. Sometimes the causative agent is a substance of recent exposure (weeks or months); however, sensitization often occurs only after several years of repeated contact with the allergen, although patients are seldom aware of this fact.

Location and clinical aspect of the earliest lesions and subsequent course of the dermatitis frequently provide diagnostic clues to assessing etiology. Secondary dissemination or infection may obscure the original pattern of the dermatitis afterwards.

TABLE 23.2. *Sources of information on exposure and drawbacks of different sources*

Product labeling
Labeling depends on regulatory policies that can vary in different countries
Substances used in the manufacturing process usually are not included
Substances added to raw materials may be not declared
Information from manufacters or suppliers
Time consuming
Many times specialized information is not available
Depends on the manufacturer's cooperation
Product data bases
Sometimes inadequate or information is not updated
Material Safety Data Sheets
Sometimes inadequate or information is not updated
Information from textbooks
Sometimes not updated
Chemical analysis of the products
Time consuming
Difficult to perform in complex products
Methodology is still not available or not validated for certain substances

The elicitation time depends on the characteristics of the sensitizer, intensity of exposure, and degree of sensitivity. Lesions usually appear 24–72 hours after the last exposure to the causative agent, but they may develop as early as 5 hours or as late as 7 days after exposure. Occupational origin should be suspected if the dermatitis improves during weekends or holidays and flares when work is resumed. It should be remembered that chronic dermatitis might require 3–4 weeks away from work before any improvement occurs. Besides, failure to improve during work leaves does not necessarily invalidate a causal relationship, due to the eventual existence of other nonoccupational sources of exposure. Often, ACD improves more slowly than irritant contact dermatitis (ICD) when exposure ceases and recurs faster (in few days) when exposure is restored. In contrast, cumulative irritant contact dermatitis to several weak irritants, usually requires many days or even weeks to develop when the exposure is reestablished. A clinical course characterized by iterative, sudden flares of dermatitis usually indicates ACD.

Physical Examination

When studying a probable ACD, it is essential to examine the entire skin surface. Morphology, location, and pattern of distribution frequently suggest the diagnosis and may point out the causative allergen or rather a range of allergens (Edman, 1985). The clinical appearance of ACD usually is characterized by eczematous inflammation; however, the lesions sometimes are noneczematous (Batscharov and Minkov, 1968; Holst et al., 1976) (i.e., erythema multiforme-like, purpuric, lichenoid, etc.). The primary signs in acute ACD are erythema, edema, papules, vesicles, and weeping, reflecting early inflammatory changes in the dermis and spongiosis with rupture of the intercellular attachments. The cardinal symptom is pruritus. In this stage, ACD may be differentiated from other forms of dermatitis, such as ICD, by the presence of vesicles; however, some irritants may also induce vesiculation.

Several circumstances, such as persistence of allergen within the skin and continued or repeated exposure to the allergen or concurrent effect of additional factors (i.e., irritation, mechanical trauma, secondary allergens), may determine the persistence of the dermatitis. Chronic ACD is characterized by xerosis, scaling, lichenification, and fissures; reflecting acanthosis, hyper- and parakeratosis, and cellular infiltration in the dermis. Subacute ACD may show clinical and histological features of both stages. In these stages it is impossible to differentiate ACD from cumulative irritant contact dermatitis on the exclusive basis of morphological criteria. Furthermore, in some dermatitis, especially hand dermatitis, constitutional, irritant, and allergic factors frequently coexist.

The pattern of distribution of ACD usually is the single most important diagnostic clue, and usually the area of greatest disease is the area of maximum contact with the offending allergen. Sometimes the most severe dermatitis appears in an area distant to the apparent site of contact. This can be due to the transfer of the antigen to other body sites by the hands, which spread the allergen to different areas of the body, such as the face, neck, and genitalia. Other confounding factors should be considered, i.e., the clinical pattern of the dermatitis may have been modified by previous treatment or secondary infection, the dermatitis has a multifactorial background, etc.

Patch Testing

Once the history and the physical examination are completed and contact allergy is suspected, skin tests should be performed. Patch testing constitutes at present the most important tool for the study of delayed hypersensitivity and the only "scientific proof" of allergic contact dermatitis (Fischer, 1986; Lachapelle, 1997), however, it still confronts several inherent methodological and interpretative problems and requires strict observation of the technical aspects and critical assessment of the results. Some of the most difficult aspects include determining which patients should be tested, selecting which allergens to use in a particular patient, interpreting patch test results, including false-positive and false-negative responses, deciding when to retest or perform special tests, and assessing the clinical relevance of positive responses.

The major indication for patch testing is the investigation of a probable ACD. It will confirm the clinical diagnosis and identify the responsible allergen(s), many times unexpected by history. Besides, providing an objective proof of the allergic condition is essential for patient cooperation in allergen avoidance. In addition, patients diagnosed as having skin diseases other than ACD may benefit from patch testing. Chronic or nonresponsive dermatitis are good candidates, esspecially hand or hand and foot dermatitis (irritant, dyshidrotic, hyperkeratotic, and even psoriasis and pustulosis palmaris et plantaris), stasis dermatitis, atopic dermatitis, and numular eczema. Not infrequently, ACD proves to be the actual diagnosis or may arise as a complicating factor. In the latter case the offending agent usually is present in the topical agents prescribed or self-administered for treatment of the primary disease. Other patients in whom patch testing may be considered are those with unclassified eczema, "endogenous" eczema, essential pruritus, possible drug eruptions, and suspected systemic contact dermatitis.

All patients should be tested to a standard screening series and additional series or individual allergens will be selected depending on the history and the distribution of the dermatitis. Ideally, allergens should be chemically defined and have high purity. Standardized, commercially available allergens should be used whenever possible. Testing with nonstandard antigens should be undertaken with caution. These may be chemically pure substances but usually are compound products and may contain unknown components. Moreover, some components can be irritants; such is the case for many industrial products. Information about the materials to be tested should be procured from labels of ingredients for household or skin care products or the material safety data sheets for industrial products. When not enough information is available, the product's manufacturer can be consulted. Specialized textbooks regarding test's concentration for many nonstandardized materials are currently available (de Groot, 1986) (Table 23.2). As a general rule, it can be assumed that products that are intended to remain on the skin ("leave on" products) can be tested "as is," whereas those products that have to be rinsed out ("wash off" products) should be diluted for testing. Should any substance be considered potentially irritant, an open test may be envisaged. It should be performed with diluted substances, whose concentration can be progressively increased as far as no response, either allergic or irritant, appears.

Several factors may influence patch test results, and sources of variability still persist, namely patch test system, allergen concentration and vehicle, amount of test substance applied, proper technique, etc. (Fischer and Maibach, 1984; Fisher and Rystedt, 1985; Kim et al., 1987; Marzulli and Maibach, 1976; Tanglertsampan and Maibach, 1993; Upadhye and Maibach, 1992; Van Ketel, 1979; Vanneste et al., 1980; Wahlberg, 1980). Many of these problems seem to be overcome with the "ready-to-use" delivery systems, (such as the TRUE Test), which have been pharmaceutically optimized, and the allergen dosage, as well as the amount of allergen per unit area, have been standardized (Fischer and Maibach, 1984; Fischer et al., 1989; Kreilgard and Hansen, 1989; Lachapelle et al., 1988; Rietschel et al., 1989).

Timing of patch testing should also be considered; testing is not recommended when the dermatitis is acute or when the patient is receiving systemic corticosteroids. Such circumstances can lead to false-positive or false-negative responses.

Even when the methodology and technique of patch testing are appropriate, accurate reading and interpretation of patch test responses still represents a challenge. Reading of patch test responses is subjective based on inspection and palpation of the test sites- and therefore subjected to interindividual and intraindividual variations (Bruze et al., 1995).

The crucial point in assessing a positive patch test response is ascertaining whether it represents an allergic reaction or a false-positive reaction. The most common cause of false-positive responses is the use of irritant materials for patch testing. It is imperative to follow the established guidelines: materials of unknown composition are never to be used under occlusion and open tests with diluted substances should be preferred as an initial approach. Some standard allergens are marginal irritants, which means that the concentration necessary to induce positive allergic responses is close to the irritant concentration. Allergens such as nickel, chromate, formaldehyde, etc., frequently produce irritant reactions that require careful interpretation.

Morphological features of irritant reactions are sometimes distinct from allergic responses; they are sharply demarcated and confined to the test area, have a glazed or "burned" appearance or may be entirely blistered, and tend to decrease in severity between readings. Pustular reactions and the "edge effect" (positive response limited to the edge of the disk area) usually represent irritancy. When the test reaction is weakly positive, it is not easy to make a clear-cut distinction on a morphological basis. Such reactions can be demonstrated to be false-positive irritant reactions by applying serial dilutions of the chemical, as irritant reactions tend to stop abruptly below a certain concentration, whereas allergic reactions tend to persist, although progressively weaker, with decreasing concentrations.

Skin hyperreactivity ("excited skin syndrome," "angry back") (Maibach, 1981; Mitchell and Maibach, 1997), has to be considered when multiple positive reactions appear. This may be the result of strongly positive responses that induce non-specific reactions in contiguous test sites. Besides, several weak or doubtful positive reactions may occur in patients who have active dermatitis at the time of testing. The presence of simultaneous positive reactions does not necessarily indicate an excited skin syndrome (ESS); they can represent true positive allergic responses to different test substances. Assessment of the significance of several simultaneous patch test reactions requires individual, sequential testing of the involved substances.

False-negative patch test reactions represent a further complication of patch testing and an even more complicated problem, as the cause of the dermatitis may be missed. When suspicion of contact allergy persists, it is necessary to revise the entire testing procedure seeking for methodological or technical pitfalls. The test should be repeated or special tests should be performed: (1) perform the patch test on a previous dermatitic area, (2) perform a Provocative Use Test (PUT) or a Repeated Open Application Test (ROAT) (Hannuksela, 1997). The former has limitations, since appropriate controls are seldom available.

In addition, it should be remembered that contact dermatitis may occur in association with immediate contact reactions. Therefore, immediate-type hypersensitivity testing may prove sometimes necessary.

Assessment of Relevance

After arriving at an interpretation indicating sensitivity to one or many substances, an even more important and complicated task to accomplish is to determine the clinical relevance of the positive response(s). Little or no data on clinical relevance of the positive reactions is provided in many clinical studies. Moreover, there is no consensus as to its definition, how should it be scored and how it should be assessed. In assessing the clinical relevance of a positive patch test reaction, we must establish whether the responsible allergen either represents the primary cause or an aggravating factor of the patient's dermatitis. An allergen is clinically relevant if (1) we can establish the existence of an exposure and (2) the patient's dermatitis is explainable with regard to that exposure (Ale and Maibach, 1995; Lachapelle, 1997). It must also be determined, whether the positive patch test represents the cause of the current dermatitis ("current" or "present" relevance) or a previous one ("past" relevance). In addition, current or past relevance can be scored according to the degree of clinical certainty, using terms such as: "relevance possible", "relevance probable" and "relevance definite" (Tanglertsampan and Maibach, 1993; Upadhye and Maibach, 1992). Furthermore, the exposure to the incriminated allergen may explain the dermatitis entirely, i.e. "complete" relevance; or may represent a contributory or complicating factor in a multifactorial dermatitis, i.e. "partial" relevance. It is complicated, and often unattainable, to assess the relative influence of the different exogenous and endogenous factors on a given clinical dermatitis. Finally, if the source of the positive response is not traced, we consider the response as being of "unknown" relevance. The greater the physician's knowledge of environmental allergens, the higher the relevance scores and the accuracy assessment. A judgment of unknown relevance often reflects physicians' ignorance of the chemical environment and the pattern of cross-sensitivity rather than actual irrelevancy. A reaction of unknown relevance may afterwards assume primary significance if new information is available and therefore should always be recorded.

If the clinical and historical facts are consistent with the patch test result, the physician can be reasonably certain for its significance. Often, however, he or she must reverse the process and attempt to detect in the patient's environment a chemical unexpectedly found reactive on patch testing.

In some circumstances it may be difficult to substantiate the presence of the allergen in the patient's environment. This may be due to the difficulty in detecting certain allergens or to our restricted knowledge about the composition of many products. Industrial and household products often are not labeled with ingredient content, whereas cosmetic products still do not have ingredient labeling in many countries. By chemical analysis of environmental or industrial products, it is possible to demonstrate the presence or absence of some particular allergens. For certain allergens there are simple methods or spot tests that can be useful, such as the dimethylglyoxime test for nickel (Menne et al., 1987), diphenylcarbazide test for chromium (Fregert, 1988), lutidine test for formaldehyde (Fregert et al., 1984), and the filter paper test for epoxy resin (Fregert and Trulsson, 1978). More often, however, complicated chemical methods must be applied (Fregert, 1988). Close collaboration between chemists and dermatologists is essential for optimum management of the problem (Foussereau et al., 1982).

TABLE 23.3. *Testing procedures for the assessment of relevance in ACD*

Testing with the suspected allergen(s)
  Sequential patch testing
  Repeated open application test (ROAT) or Provocative Use Test (PUT)
    on normal skin
    on slightly damaged or previously dermatitic skin
Testing with products suspected to contain the responsible allergen
  Patch testing (using suitable vehicle and appropriate concentration, frequently
    starting with highly dilute substances)
  ROAT (similar as above stated, using proper vehicle and adequate concentration)
  Use test (typical product use)
  Testing in normal controls (if necessary)
Testing with product's extracts
  Similar than 2
Testing with cross-reacting allergens and products suspected to contain them
  Similar than 1

Additional skin tests are frequently necessary for relevance assessment (Table 23.3). Testing with products that may contain the causative allergen and to which the patient refers being exposed can be of some help; however, patch testing with the unmodified product may give false-negative results, usually because of insufficient concentration of the allergen in the product. This may be partially overcome by performing the test with the product's extracts or performing use tests, such as the PUT or ROAT.

Once the source or sources of the allergen have been identified, the patient should be given clear suggestions for alternatives and methods of minimizing future contact with the allergen.

As seen above, diagnosis in allergic contact dermatitis is complex. Eight steps in the assessment of ACD can be considered:

1. Relevant clinical history
2. Appropriate morphology
3. Positive diagnostic patch test with appropriate vehicle and concentration
4. Repeat patch test, when appropriate, to exclude ESS and "rogue" reactions
5. Perform serial dilution (when necessary)
6. Reviewing controls for nonirritating concentrations and perform special test for uncommonly utilized allergens
7. Perform a PUT or ROAT
8. Clearing of the dermatitis when allergen is removed or exposure significantly decreased

With a better knowledge of the chemical environment and more advanced test techniques, it is probable that still more cases of contact dermatitis will eventually be proven to have an allergic component. The above guidelines can provide a simplified rational approach for an operational definition of allergic contact dermatitis that may represent a useful tool until more information becomes available.

REFERENCES

Ale S. I., and Maibach H. I. (1995): Clinical relevance in allergic contact dermatitis: An algorithmic approach, *Dermatosen*, 43:119–121.

Andersen, K. E. (1994): Mechanical trauma and hand eczema. In: *Hand Eczema*, edited by Menné, T. and Maibach H. I., pp 31–34, CRC Press, Boca Raton, FL.

Batscharov, B., and Minkov, D. M. (1968): Dermatitis and purpura from rubber in clothing, *Trans. St. John's Hosp. Derm. Soc.*, 54:178–182.

Bruze, M., Isaksson, M., Edman, B., Björkner, B., Fregert, S., and Möller, H. (1995): A study on expert reading of patch test reactions: inter-individual accordance, *Contact Dermatitis*, 32:331–337.

de Groot, A. C. (1986): *Patch Testing Concentrations and Vehicles for 2800 Allergens.*, Elsevier, Amsterdam.

Edman, B. (1985): Sites on contact dermatitis in relationship to particular allergens, *Contact Dermatitis*, 13:120–135.

Fischer, T., and Maibach, H. I. (1984): Patch test allergens in petrolatum: A reappraisal, *Contact Dermatitis*, 11:224–228.

Fischer, T., and Maibach, H. I. (1984): Amount of nickel applied with a standard patch test, *Contact Dermatitis*, 11:285–287.

Fischer, T., and Maibach, H. I. (1984): The thin layer rapid use epicutaneous test (TRUE-test), a new patch test method with high accuracy, *Br. J. Dermatol.*, 112:63–68.

Fischer, T., Hansen, J., Kreilgard, B., and Maibach, H. I. (1989): The science of patch test standardization, *Immunol. and Allergy Clin.*, 9:417

Fisher, A. A. (1986): *Contact Dermatitis*, 3rd. ed. Lea & Febiger, Philadelphia.

Fisher, T., and Rystedt I. (1985): False-positive, follicular and irritant patch test reactions to metal salts, *Contact Dermatitis*, 12:93–98.

Foussereau, J., Benezra, C., and Maibach, H. I. (1982): *Occupational Contact Dermatitis: Clinical and Chemical Aspects*, 1st ed., pp 80–89, W. B. Saunders, Philadelphia.

Fregert, S. (1988): Physicochemical methods for detection of contact allergens, *Dermatol. Clin.*, 6:97–104.

Fregert, S., Dahlquist, I., and Gruvberger, B. (1984): A simple method for the detection of formaldehyde, *Contact Dermatitis*, 10:132–134.

Fregert, S., and Trulsson, L. (1978): Simple methods for demonstration of epoxy resins of bisphenol A type, *Contact Dermatitis*, 4:69–72.

Hannuksela, M. (1997): The repeated open application test (Roat), *Contact Dermatitis*, 36:39–43.

Holst, R., Kirby, J., and Magnusson, B. (1976): Sensitization to tropical woods giving erythema multiforme-like eruptions, *Contact Dermatitis*, 2:295.

Kim, H. O., Wester, R. C., Mc Master, J. A., Bucks, D. A., and Maibach, H. I. (1987): Skin absorption from patch test systems, *Contact Dermatitis*, 17:178.

Kreilgard, B., and Hansen, J. (1989): Aspects of pharmaceutical and chemical standardization of patch test material,. *J. Am. Acad. Dermatol.*, 21:836–838.

Lachapelle, J-M., Bruynzeel, D. P., Ducombs, G., Hannuksela, M., Ring, J., White, I. R., Wilkinson, J., Fischer, T., and Billberg, K. (1988): European multicenter study of the TRUE Test, *Contact Dermatitis*, 19:91–97.

Lachapelle, J-M. (1997): A proposed relevance scoring system for positive allergic patch test reactions: Practical implications and limitations, *Contact Dermatitis*, 36:39–43.

Maibach, H. I. (1981): The E.S.S.: Excited Skin Syndrome (Alias the "Angry Back"). In: *New Trends in Allergy*, edited by Ring, J., and Burg, G., 208–221, Springer-Verlag, Berlin, Heidelerg, New York.

Marks, J. G. Jr., Belsito, D. V., De Leo, V. A., Fowler, J. F., Jr., Fransway, A. F., Maibach, H. I., Mathias, C. G., Nethercott, J. R., Rictschel, R. L., Sherertz, E. F., Storrs, F. J., and Taylor, J. S. (1998): North American Contact Dermatitis Group patch test results for the detection of delayed-type hypersensitivity to topical allergens, *J. Am. Acad. Dermatol.*, 38:911–918.

Marzulli, F. N., and Maibach, H. I. (1976): Effects of vehicles and elicitation concentration in contact dermatitis testing I: Experimental contact sensitization in humans. *Contact Dermatitis*, 2:325–329.

Meneghini, C. L. (1985): Sensitization in traumatized skin, *Am. J. Ind. Med.*, 8:319–321.

Menne, T., Andersen, K. E., Kaaber, K., Osmundsen, P. E., Andersen, J. R., Yding, F., and Valeur, G. (1987): Evaluation of the dimethylglyoxime stick test for the detection of nickel, *Dermatosen/Occup. Environ.*, 35:128–130.

Mitchell, J., and Maibach, H. I. (1997): Managing the excited skin syndrome: Patch testing hyperirritable skin, *Contact Dermatitis*, 37:193–199.

Rietschel, R. L., Marks, J. G., Adams, R. M., et al. (1989): Preliminary studies of the TRUE Test patch test system in the United States, *J. Am. Acad. Dermatol.*, 21:841–843.

Shmunes, E. (1988): Predisposing factors in occupational skin diseases, *Dermatol. Clin.*, 6:7–13.

Skog, E. (1958): The influence of pre-exposure to alkyl benzene sulphonate detergent, soap and acetone on primary irritant and allergic eczematous reactions, *Acta. Derm. Venereol.*, 38:1–14.

Tanglertsampan, C., and Maibach, H. I. (1993): The role of vehicles in diagnostic patch testing: A reappraisal, *Contact Dermatitis*, 29, 169–174.

Upadhye, M. R., and Maibach, H. I. (1992): Influence of area of application of allergen on sensitization in contact dermatitis, *Contact Dermatitis*, 27:281–286.

Van Ketel, W. G. (1979): Petrolatum again: An adequate vehicle in cases of metal allergy?, *Contact Dermatitis*, 5:192–193.

Vanneste, D., Martin, P., and Lachapelle, J-M. (1980): Comparative study of the density of particles in suspensions for patch testing, *Contact Dermatitis*, 6:197–203.

Wahlberg, J. E. (1980): Petrolatum-A reliable vehicle for metal allergens? *Contact Dermatitis*, 6:134–136.

Wester, R., and Maibach, H. I. (1976): Relationship of topical dose and percutaneous absorption in rhesus monkey and man. *J. Invest. Dermatol.*, 67:518.

Toxicology of Skin
Edited by Howard I. Maibach
Copyright © 2001 Taylor & Francis

24

Predictive Immunotoxicological Risk Assessment

Recent Developments and Applications

B. Homey and P. Lehmann

Department of Dermatology, Heinrich-Heine University, Düsseldorf, Germany

H. W. Vohr*

Research Toxicology, Bayer AG, Wuppertal, Germany

INTRODUCTION

Chemical-induced adverse effects, such as contact and photocontact allergy as well as irritancy and photoirritancy, are of major importance in clinical dermatology and during the development of new pharmaceuticals and industrial chemicals. Allergic and irritant contact dermatitis represent the majority of occupational skin disorders, which show an incidence of 76 cases per 100,000 United States workers (Bureau of Labor Statisitics, 1993). Therefore, occupational skin disorders are the most common

---

*This chapter was part of the post-doctoral thesis of H. W. V.

nontrauma-related occupational disease affecting workers in many different occupations.

Because clinical and histological features of allergic and irritant contact dermatitis are similar, the differentiation between both types of dermatitis in the preclinical and clinical evaluation of chemicals remains difficult; however, the underlying immunological mechanisms are thought to be fundamentally different. For allergic skin immune responses, it is well established that antigen presentation induces T-cell activation and the formation of antigen-specific memory T cells (Enk and Katz, 1995). In contrast, irritant skin reactions are believed to activate the immune cascade independent of the antigen presentation pathway by inducing proinflammatory mediators and cytokines that directly recruit and activate T cells (Baadsgaard and Wang, 1991; Hunziker et al., 1992; Kondo et al., 1994). Thus, irritant skin reactions are defined to be "nonimmunological" (nonspecific) reactions which do not result in the induction of antigen-specific memory T cells. Furthermore, it is well known that patients with irritant contact dermatitis easily aquire hapten-specific sensitization. This clinical observation has been confirmed in several experimental settings in mice; however, the immunological mechanisms of action how irritants may facilitate hapten-specific sensitization remain unlear (Elsner and Burg, 1993; Frantzen et al., 1993).

The majority of tests for predicting allergenicity or photoallergenicity of chemicals use guinea pigs or mice with biphasic, long-term protocols comprising a sensitization phase (induction) and an elicitation phase (challenge) (Buehler et al., 1985; Buehler, 1982; Gad, 1994; Kaidbey and Kligman, 1980; Magnusson and Kligman, 1969; Maurer et al., 1980; Scholes et al., 1991; 1992). In guinea pig models, such as the guinea pig maximization test, Buehler's occlusive patch test, or the optimization test, contact reactivity is assessed by a subjective local erythema score and determined from the frequency of animals exhibiting a positive response (Buehler et al., 1985; Buehler, 1982; Magnusson and Kligman, 1969; Maurer et al., 1980; Maurer, 1985). Together with the challenge-induced mouse ear swelling test (MEST) (Gad, 1994), these models are of proven value for immunotoxicology (Andersen and Maibach, 1985a, 1985b; Oliver et al., 1986); however, they are time consuming, expensive, and may require the injection of adjuvant or other skin pretreatments. Furthermore, the majority of animal models for predicting photoreactivity are heterogeneous concerning administration of test compounds and UV irradiation regimen (Gerberick and Ryan, 1990; Giudici and Maguire, 1985; Harber, 1981; Maguire et al., 1982; Maurer, 1984; Miyachi and Takigawa, 1982, 1983; Scholes et al., 1991). As a consequence, results concerning photoreactivity of chemicals in such preclinical models are often controversial and difficult to interpret.

Recently, a local lymph node assay (LLNA) was described as a predictive assay for identifying the contact sensitizing potential of chemicals in mice (Kimber and Weisenberger, 1989; Kimber et al., 1989). In comparison to the widely used Buehler's occluded patch test and other guinea pig test data (Basketter and Scholes, 1992; Kimber and Weisenberger, 1989; Kimber et al., 1989), the LLNA represents a rapid and cost-effective alternative method for the detection of at least moderate to strong contact sensitizers. In contrast to guinea pig models and the MEST, the LLNA is based on the detection of a primary immune response as a function of auricular lymph node activation following topical application of chemicals on the dorsal surface of ears (Basketter and Scholes, 1992; Kimber and Weisenberger, 1989; Kimber et al., 1989).

In previous experiments with mice, we modified the LLNA by avoiding radioactive labeling of lymph nodes cells with ^{3}H-thymidine in vivo (Homey et al., 1997; Homey et al., 1997; Homey et al., 1998; Ulrich et al., 1998; Vohr et al., 1994). Mainly lymph node weight and cell counts were established as reliable alternative read-out parameters. In addition, ultraviolet A (UVA) irradiation within the same basic protocol allowed to the identification of photoreactive compounds (Ulrich et al., 1998; Vohr et al., 1994).

Although the LLNA was first described to selectively detect allergic skin immune responses, recent studies showed that irritant and photoirritant test compounds also induce lymph node cell proliferation in vivo (Homey et al., 1995; Ikarashi et al., 1993; Scholes et al., 1991; Vohr et al., 1994). For immunotoxicological risk assessment of drugs and industrial chemicals, however, the differentiation between allergic and irritant responses is of major importance.

Based on the current understanding of the mechanisms underlying both irritant skin reactions and the induction of contact allergy, we hypothesized that irritants predominantly induce skin inflammation, which in turn may elicit draining lymph node proliferation. In contrast, the induction of contact allergy should comprise only marginal skin inflammation, whereas activation of skin-draining lymph node cells is predominant.

The aim of this overview is to summarize recent modifications of the original LLNA which may offer a broader application of this predictive test, as well as the differentiation of allergic versus irritant chemical-induced skin responses by measuring immunologically relevant end points within the skin and skin-draining lymph node cells. As a prerequisite, the chosen parameters should be easy to determine and suitable for routine preclinical testing.

PREDICTIVE TESTS

Guinea Pig Assays

More than 30 years ago, E. V. Buehler first published an in vivo model for the screening of chemical-induced contact hypersensitivity in guinea pigs (Buehler et al., 1985; Buehler et al., 1985; Buehler, 1982). He used an occluded patch test technique on the preshaved animal skin. The recommended number of animals was n = 20 in the sensitized group, n = 10 naive animals for challenge, and n = 10 naive control animals for rechallenge. To define the concentration of the test compound, the minimal irritative concentration should be determined in a preliminary irritation study. Therefore, the whole procedure could easily expand to a 6–8 week period.

The interpretation of the Buehler test is based on the subjective assessment of skin erythema 24 and 48 hours after patch removal using erythema scores. Thus, the assessment of positive reactions, as well as the reliability of this test, strongly depends on well-trained technicians, the irritative potential, and the color of the test

TABLE 24.1. *Predictive tests for allergic contact dermatitis in guinea pigs recommended in current guidelines*

Test	Induction	Challenge
Buehler patch test*	Epidermal (occlusive)	Epidermal (occlusive)
Open epidermal test	Epidermal	Epidermal
Split adjuvant test	Epidermal (occlusive); adjuvant	Epidermal (occlusive)
Freund's complete adjuvant test	Intradermal (+/− adjuvant)	Epidermal
Maximization test*	Intradermal (+/− adjuvant); epidermal (occlusive)	Epidermal (occlusive)
Optimization test	Intradermal (+/− adjuvant);	Intradermal
Draize test	Intradermal	Epidermal (occlusive); Intradermal

*favored by OECD guideline 406.

compound (Andersen and Maibach, 1985a, 1985b; Botham et al., 1991; Robinson et al., 1990).

To increase the sensitivity of the Buehler test, many modifications were adopted to this basic principle: (a) pretreatment of the skin with moderate irritants for nonirritative test substances; (b) additional intradermal injection of the test compound during the induction phase; (c) usage of Freund's complete adjuvant (FCA) to increase the immune reaction to the test material. Several of these modifications are currently accepted by regulatory agencies (Table 24.1); however, the tests described by Buehler, as well as by Magnusson and Kligman ("Guinea Pig Maximization Test"), are currently the preferred screening procedures for the "classical" evaluation of chemical-induced contact hypersensitivity.

In addition, the different guinea pig assays were further modified for the detection of phototoxic properties of compounds after topical, dermal, and oral applications. A well-accepted modification with respect to the screening for photosensitization is the one described by Maurer, which represents a sophisticated but time-consuming and labor-intensive system (Maurer, 1984). This is also true for other modifications of guinea pig assays by other authors (Andersen and Maibach, 1985b; Kashima et al., 1993a, 1993b; Maurer, 1985; Ziegler and Suss, 1985).

During recent years it has become more and more apparent that the use of immunological endpoints like proliferation of immune competent cells instead of the assessment of skin erythema and edema may represent appropriate alternative end points for the screening of chemical-induced skin reactions (i. e., measuring the immune reactions in the draining lymph nodes).

Popliteal Lymph Node Assay

The popliteal lymph node assay (PLNA) had been originally developed to measure genetic mismatches in rats as well as in mice. At the beginning, increases in the weight of the popliteal lymph nodes were used as simple read-out parameters. On the basis of increasing understanding of the fundamental mechanisms of immune reactions, it was logical to use the PLNA for other indications than intraspecies mismatch reactions. The first applications of this kind were screenings of pharmaceuticals, which are known to lead to undesired side-effects in humans, like diphenylhydantoin (Gleichmann, 1981) or zimeldine (Thomas et al., 1989).

Over time many modifications regarding the endpoints measured and technical procedures of the PLNA had been described by different authors. Major modifications are (a) the assessment of lymph node cell proliferation instead of lymph node weight increases (Hurtenbach et al., 1987; Kubicka-Muranyi et al., 1993; Thomas et al., 1990), (b) calculation of a stimulation index (SI) for individual animals by injecting the disolved compound vehicle into one foot pad and the vehicle only into the contralateral foot pad (Gleichmann, 1981), (c) further analysis of the lymph node cell suspensions by flow cytometry (Brouland et al., 1994; Descotes, 1992; Descotes et al., 1997; Krzystyniak et al., 1992; Krzystyniak et al., 1992; Vial et al., 1997), cytokine releases, or macrophage activity, (d) adoptive transfer of activated (antigen-specific) cells (Klinkhammer et al., 1988; Kubicka-Muranyi et al., 1993). These latter modifications opened the door for applying the PLNA also for mechanistic investigations of local immune reactions.

The broad use of the PLNA as a screening assay for immunotoxicological effects of low molecular weight chemicals was due to statements of some groups that the PLNA is exclusively suitable for screening autoimmunogenetic effects of chemicals (De Bakker et al., 1990; Ford et al., 1970; Kubicka-Muranyi et al., 1993; Schuhmann et al., 1990). Despite compelling evidence that the PLNA will also detect the sensitizing (allergenic) or irritating properties of a compound, the international acceptance of this test suffered from the controversial discussion. In addition, some false-negative as well as false-positive responses compared to human data aggregated the discussion about the application of the PLNA for predictive testing (Kammuller et al., 1989; Thomas et al., 1990; Verdier et al., 1990). To increase the reliability of the PLNA, some groups included investigations of the tissue by standard histology or by immunohistochemistry (Brouland et al., 1994; De Bakker et al., 1990; Krzystyniak et al., 1992). Although it had been possible to identify skin sensitizers by using standard histology, it was impossible to discriminate the effect of a primary irritant chemical. Furthermore, neither standard histology nor immunohistochemistry could provide a clear differentiation from autoimmune responses (Brouland et al., 1994). To avoid false-negative responses, some authors increased the concentration of the test substance; however, this approach may induce systemic toxic effects, influencing the overall response in the PLN (Descotes et al., 1997).

In contrast to the LLNA (see below), the PLNA had not been consequently validated in intra- or interlaboratory trials. Thus, to date the PLNA predominantly represents a tool for research investigating the mechanisms of chemical-induced reactions rather than a widely accepted instrument for immunotoxical screening.

In 1995, Bloksma et al. stated that the PLNA is the only assay which is able to screen for sensitizing properties of chemicals entering the body by other routes than skin (Bloksma et al., 1995). Among other aspects, this publication provided the basis to start new interlaboratory studies to validate the PLNA for the screening of immunomodulating potentials of chemicals (Descotes et al., 1997; Vial et al., 1997).

Local Lymph Node Assay

In contrast to the PLNA, the LLNA had been published and validated in a consequent manner from the very beginning by I. Kimber et al. (Kimber and Weisenberger, 1989; Kimber et al., 1989, 1991, 1995; Scholes et al., 1992). This assay has been developed not only to complement but also to replace the guinea pig assays (Basketter et al., 1996). The LLNA is based on the assessment of immunological parameters activated during the induction phase of local immune responses. While the guinea pig assays make use of the pathological symptoms for the evaluation of sensitizers, the LLNA investigates the initiation phase of a primary immune response in the skin-draining lymph nodes.

During recent years, many modifications of the LLNA have been published. One adaptation represents the treatement of rats instead of mice (Arts et al., 1996). To avoid the use of radioactive labeling as described for the original LLNA, some authors used the assessment of increases in lymph node cell counts instead of ^{3}H-thymidin incorporation. This modification provides the advantage that further investigations of immune competent cells of the lymph node, as well as the local skin tissue, is applicable (i.e., cell separation, flowcytometry, macrophage activity, in vitro restimulation, etc.). Moreover, an important modification of the LLNA was the inclusion of UV irradiation after topical application of the test compound and the assessment of chemical-induced photoreactions.

Furthermore, it should be mentioned here that beside mechanistical studies, the LLNA was also used for investigating structure-activity relationships of compounds (Basketter et al., 1992; Hostynek et al., 1995; Ikarashi et al., 1993). Although the problem of measuring false-positive responses due to the irritative potential of a compound is not as high as for the PLNA, the original LLNA cannot provide a clear differentiation between chemical-induced allergic and irritant skin responses (Ikarashi et al., 1993).

AN INTEGRATED MODEL FOR THE DETECTION AND DIFFERENTIATION OF CHEMICAL-INDUCED ALLERGIC AND IRRITANT SKIN RESPONSES

During recent years we have been interested in the skin immune sytem and the characterization of contact allergic and irritant responses in the skin and skin-draining lymph nodes. The observations that (a) irritant and allergic skin responses show different molecular and cellular events in the skin immune system, (b) irritant skin responses can be described as nonspecific but not as nonimmunological reactions, (c) irritants predominantly induced skin inflammation, which in turn stimulated draining lymph node cell proliferation, (d) in contrast, the induction phase of contact or photocontact allergy was characterized by marginal skin inflammation, but a marked activation and proliferation of skin-draining lymph node cells have led to the development of a novel integrated model to detect and differentiate chemical-induced irritant and allergic skin responses (Homey et al., 1997; Homey et al., 1997; Homey et al., 1998; Homey et al., 1995; Homey et al., 1998; Ulrich et al., 1998; Vohr et al., 1994).

In adaptation to both the LLNA and the MEST, mice were topically treated on the dorsal surfaces of both ears with the test chemical or vehicle alone on three consecutive days (Figure 24.1). For the induction of photocontact allergy and photoirritancy, mice were irradiated with 10 J/cm² UVA immediately after topical treatment. On day

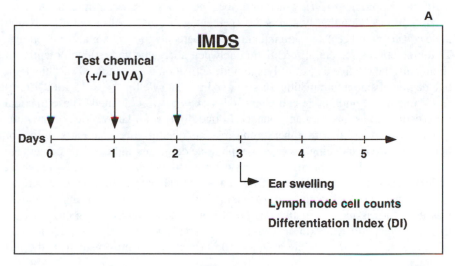

FIG. 24.1a.

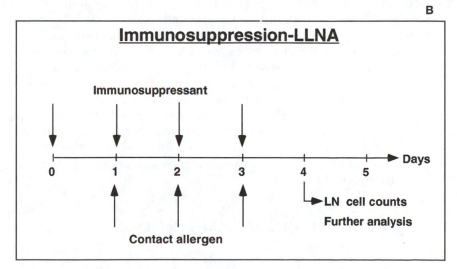

FIG. 24.1b.

0 and day 3, ear thickness was measured and mean ear swelling was calculated. Furthermore, auricular lymph nodes were removed, lymph node cell counts per mouse were determined, and lymph node cell count indices were calculated. Lymph node cell count indices were defined as the ratio of mean lymph node cell counts from mice treated with the test compound to corresponding results of vehicle-treated control groups. Positive test reactions were defined as either significant ear swelling or significant increase in lymph node cell counts.

To quantitatively distinguish between irritant and allergic reactions, we defined a differentiation index (*DI*), which describes the relation between inflammatory responses in local draining lymph nodes (percentage of maximal increases in lymph node cell count indices) and skin inflammation (percentage of maximal ear swelling). To define criteria for the differentiation between allergenic and irritant potential of chemicals, the relative degree of lymph node activation was compared with the relative degree of skin inflammation. First, a maximal ear swelling ($x = 15 \times 0.01$ mm) and a maximal lymph node cell count index ($x = 5$) were defined. These maxima were estimated as mean value (rounded to one digit) from large series of previous experiments ($n = 50$) with either strong irritants eliciting maximal ear swelling or potent sensitizers inducing maximal lymph node cell count indices. The percentage of each test reaction of the maximal ear swelling and the maximal increase in lymph node cell count index was calculated. For mathematical reasons, the maximal increase in the lymph node cell count index (maximal lymph node cell count index$-1 = 4$) was used to classify the relative degree of lymph node activation. In case of positive reactions either in skin or skin-draining lymph nodes, a $DI > 1$ indicated an allergic reaction pattern while a $DI < 1$ described an irritant potential of the test chemical.

Results of the initial study regarding the development of this integrated model for the differentiation of chemical-induced (photo-)allergic and (photo-)irritant skin responses (IMDS) confirmed that irritants predominantly induced skin inflammation, which in turn stimulated draining lymph node cell proliferation. In contrast, the induction phase of contact or photocontact allergy was characterized by marginal skin inflammation, but a marked activation and proliferation of skin-draining lymph node cells (Homey et al., 1998). Furthermore, experiments with the contact allergen oxazolone, the photocontact allergen 3',4'-tetrachlorosalicylanilide (TCSA) + UVA, the irritant croton oil and the photoirritant 8-methoxypsoralen + UVA confirmed the predictive value of the differentiation index (Homey et al., 1998). Furthermore, flow cytometric analysis of lymph node–derived T- and B-cell subpopulations revealed that contact sensitzer, but not irritant, induced the expression of CD69 on the surface of I-A+ cells (Homey et al., 1998).

IMMUNOSUPPRESSION-LOCAL LYMPH NODE ASSAY

We studied immunomodulatory mechanisms of action of topically applied immunosuppressant during primary skin immune responses in epidermal and skin-draining

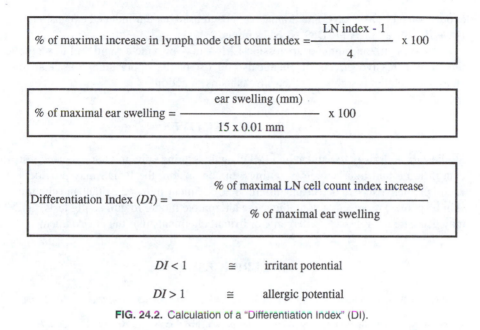

$$\text{\% of maximal increase in lymph node cell count index} = \frac{\text{LN index - 1}}{4} \times 100$$

$$\text{\% of maximal ear swelling} = \frac{\text{ear swelling (mm)}}{15 \times 0.01 \text{ mm}} \times 100$$

$$\text{Differentiation Index } (DI) = \frac{\text{\% of maximal LN cell count index increase}}{\text{\% of maximal ear swelling}}$$

$DI < 1 \quad \cong \quad$ irritant potential

$DI > 1 \quad \cong \quad$ allergic potential

FIG. 24.2. Calculation of a "Differentiation Index" (DI).

lymph node cells which finally led to the development of an "Immunosuppression-LLNA" (Figure 24.1B) (Homey et al., 1997; 1998; Homey et al., 1997; 1998). Analogous to the LLNA, mice were treated with a standard contact allergen (e.g., oxazolone) for three consecutive days; however, mice received pretreatments with putative immunosuppressants.

Topical application of the novel macrolide immunosuppressant FK506 resulted in a dose-dependent suppression of oxazolone-induced increases in lymph node cell counts. Furthermore, this Immunosuppression-LLNA was used to compare immunosuppressive effects of novel macrolids with standard glucocorticosteroids (Homey et al., 1997). Moreover, this assay provided the opportunity to study the immunosuppressive mechanisms of action of topical FK506 in more detail. Topical FK506-induced immunosuppression of contact allergen-induced increases in lymph node counts was accompanied by a depression of interleutrint (IL)-1α, IL-1β, granulocyte-macrophage-colony stimulating factor (GM-CSF), tumor necrosis factor (TNF)-α, macrophage inflammatory protein-2 (MIP-2), and interferon (IFN)-γ mRNA expression within the epidermis and impaired IL-12 p35 and p40 mRNA expression in local draining lymph nodes. Further analysis of the lymph node cell cytokine pattern revealed that the production of both Th1 (IFN-γ, IL-2) and Th2 (IL-4) cytokines were dramatically impaired after topical FK506 treatment. Flow cytometric analysis depicted that topical FK506 decreased the population of epidermis-infiltrating CD4+ T cells and suppressed the expression of CD54 and CD80 on I-A+ epidermis and lymph node cells during hapten-induced contact hypersensitivity. Furthermore, topical FK506 profoundly impaired oxazolone-induced upregulation of CD25 expression on CD4+ LNC and dramatically decreased hapten-induced expansion of I-A+/B220+ and I-A+/CD69+ LNC subsets.

In conclusion, the Immunosuppression-LLNA may provide a rapid tool to screen immunosuppressive effects of chemicals and offers the opportunity for detailed analyses to obtain insights into their mechanisms of action.

FUTURE PERSPECTIVES

Studies are underway to validate the IMDS against guinea pig and human patch test data in intra- and interlaboratory settings. In conclusion, the IMDS may provide a useful predictive tool to detect and differentiate chemical-induced irritant and allergic skin responses, represent a nonradioactive alternative model to guinea pig tests, and may decrease the release of allergenic or irritant chemicals into the environment.

REFERENCES

Andersen, K. E., and Maibach, H. I. (1985a): Guinea pig allergy tests: An overview. *Toxicol. Ind. Health.*, 1:43.

Andersen, K. E., and Maibach, H. I. (1985b): Guinea pig sensitization assays: An overview. *Curr. Probl. Dermatol.*, 14:263.

Arts, J. H., Droge, S. C., Bloksma, N., and Kuper, C. F. (1996): Local lymph node activation in rats after dermal application of the sensitizers 2,4-dinitrochlorobenzene and trimellitic anhydride. *Food Chem. Toxicol.*, 34:55.

Baadsgaard, O., and Wang, T. (1991): Immune regulation in allergic and irritant skin reactions. *Int. J. Dermatol.*, 30:161.

Basketter, D. A., and Scholes, E. W. (1992): Comparison of the local lymph node assay with the guinea-pig maximization test for the detection of a range of contact allergens. *Food Chem. Toxicol.*, 30:65.

Basketter, D. A., Gerberick, G. F., Kimber, I., and Loveless, S. E., (1996): The local lymph node assay: A viable alternative to currently accepted skin sensitization tests. *Food Chem. Toxicol.*, 34:985.

Basketter, D. A., Roberts, D. W., Cronin, M., and Scholes, E. W. (1992): The value of the local lymph node assay in quantitative structure-activity investigations. *Contact Dermatitis*, 27:137.

Bloksma, N., Kubicka-Muranyi, M., Schuppe, H. C., Gleichmann, E., and Gleichmann, H. (1995): Predictive immunotoxicological test systems: Suitability of the popliteal lymph node assay in mice and rats. *Crit. Rev. Toxicol.*, 25:369.

Botham, P. A., Basketter, D. A., Maurer, T., Mueller, D., Potokar, M., and Bontinck, W. J. (1991): Skin sensitization: A critical review of predictive test methods in animals and man (see comments). *Food Chem. Toxicol.*, 29:275.

Brouland, J. P., Verdier, F., Patriarca, C., Vial, T., and Descotes, J. (1994): Morphology of popliteal lymph node responses in Brown-Norway rats. *J. Toxicol. Environ. Health.*, 41:95.

Buehler, E. V., Newmann, E. A., and Parker, R. D. (1985): Use of the occlusive patch to evaluate the photosensitive properties of chemicals in guinea-pigs. *Food Chem. Toxicol.*, 23:689.

Buehler, E. V., Ritz, H. L., and Newmann, E. A. (1985): A proposed plan for the detection and identification of potential sensitizers. *Regul. Toxicol. Pharmacol.*, 5:46.

Buehler, E. V. (1982): Comment on guinea-pig test methods [letter]. *Food Chem. Toxicol.*, 20:494.

De Bakker, J. M., Kammuller, M. E., Muller, E. S., Lam, A. W., Seinen, W., and Bloksma, N. (1990): Kinetics and morphology of chemically induced popliteal lymph node reactions compared with antigen-, mitogen-, and graft-versus-host- reaction-induced responses. *Virchows Arch. B. Cell. Pathol. Incl. Mol. Pathol.*, 58:279.

Descotes, J. (1992): The popliteal lymph node assay: A tool for studying the mechanisms of drug-induced autoimmune disorders. *Toxicol. Lett.*, 64–65, Spec No:101.

Descotes, J., Patriarca, C., Vial, T., and Verdier, F. (1997): The popliteal lymph node assay in 1996. *Toxicology*, 119:45.

Elsner, P., and Burg, G. (1993): Irritant reactivity is a better risk marker for nickel sensitization than atopy [see comments]. *Acta. Derm. Venereol.*, 73:214.

Enk, A. H., and Katz, S. I. (1995): Contact sensitivity as a model for T-cell activation in skin. *J. Invest. Dermatol.*, 105:80S.

Ford, W. L., Burr, W., and Simonsen, M. (1970): A lymph node weight assay for the graft-versus-host activity of rat lymphoid cells. *Transplantation*, 10:258.

Frantzen, E., Goerdt, S., Goebeler, M., and Sorg, C. (1993): Changes in phenotypically distinct phagocyte subpopulations during nonspecific modulation of contact sensitization. *Int. Arch. Allergy. Immunol.*, 101:182.

Gad, S. C. (1994): The mouse ear swelling test (MEST) in the 1990s. *Toxicology*, 93:33.

Gerberick, G. F., and Ryan, C. A. (1990): A predictive mouse ear-swelling model for investigating topical photoallergy. *Food Chem. Toxicol.*, 28:361.

Giudici, P. A., and Maguire, H. C., Jr. (1985): Experimental photoallergy to systemic drugs. *J. Invest. Dermatol.*, 85:207.

Gleichmann, H. (1981): Studies on the mechanism of drug sensitization: T-cell-dependent popliteal lymph node reaction to diphenylhydantoin. *Clin. Immunol. Immunopathol.*, 18:203.

Harber, L. C. (1981): Current status of mammalian and human models for predicting drug photosensitivity. *J. Invest. Dermatol.*, 77:65.

Homey, B., Neubert, T., Arens, A., Schuppe, H. C., Ruzicka, T., Lehmann, P., and Vohr, H. W. (1997): Sunscreen and immunosuppression [letter; comment] (published erratum appears in *J. Invest. Dermatol.*, 1997, 109:820). *J. Invest. Dermatol.*, 109:395.

Homey, B., Schuppe, H. C., Assmann, T., Vohr, H. W., Lauerma, A. I., Ruzicka, T., and Lehmann, P. (1997): A local lymph node assay to analyse immunosuppressive effects of topically applied drugs. *Eur. J. Pharmacol.*, 325:199.

Homey, B., Assmann, T., Vohr, H. W., Ulrich, P., Lauerma, A. I., Ruzicka, T., Lehmann, P., and Schuppe, H. C. (1998): Topical FK506 suppresses cytokine and costimulatory molecule expression in epidermal and local draining lymph node cells during primary skin immune responses. *J. Immunol.*, 160:5331.

Homey, B., Vohr, H. W., Schuppe, H. C., and Kind, P. (1995): UV-dependent local lymph node reactions: Photoallergy and phototoxicity testing. *Curr. Probl. Dermatol.*, 22:44.

Homey, B., von Schilling, C., Blumel, J., Schuppe, H. C., Ruzicka, T., Ahr, H. J., Lehmann, P., and Vohr, H. W. (1998): An integrated model for the differentiation of chemical-induced allergic and irritant skin reactions. *Toxicol. Appl. Pharmacol.*, 153:83.

Hostynek, J. J., Lauerma, A. I., Magee, P. S., Bloom, E., and Maibach, H. I. (1995): A local lymph-node assay validation study of a structure-activity relationship model for contact allergens (published erratum appears in *Arch. Dermatol. Res.*, 1995, 287:767). *Arch. Dermatol. Res.*, 287:567.

Hunziker, T., Brand, C. U., Kapp, A., Waelti, E. R., and Braathen, L. R. (1992): Increased levels of inflammatory cytokines in human skin lymph derived from sodium lauryl sulphate-induced contact dermatitis. *Br. J. Dermatol.*, 127:254.

Hurtenbach, U., Gleichmann, H., Nagata, N., and Gleichmann, E. (1987): Immunity to D-penicillamine: Genetic, cellular, and chemical requirements for induction of popliteal lymph node enlargement in the mouse. *J. Immunol.*, 139:411.

Ikarashi, Y., Tsukamoto, Y., Tsuchiya, T., and Nakamura, A. (1993): Influence of irritants on lymph node cell proliferation and the detection of contact sensitivity to metal salts in the murine local lymph node assay. *Contact Dermatitis*, 29:128.

Kaidbey, K. H., and Kligman, A. M. (1980): Photomaximization test for identifying photoallergic contact sensitizers. *Contact Dermatitis*, 6:161.

Kammuller, M. E., Thomas, C., De Bakker, J. M., Bloksma, N., and Seinen, W. (1989): The popliteal lymph node assay in mice to screen for the immune disregulating potential of chemicals: A preliminary study. *Int. J. Immunopharmacol.*, 11:293.

Kashima, R., Oyake, Y., Okada, J., and Ikeda, Y. (1993a): Studies of new short-period method for delayed contact hypersensitivity assay in the guinea pig. I: Development and comparison with other methods (published erratum appears in *Contact Dermatitis*, 1993, 29:288). *Contact Dermatitis*, 28:235.

Kashima, R., Oyake, Y., Okada, J., and Ikeda, Y. (1993b): Studies of new short-period method for delayed contact hypersensitivity assay in the guinea pig. 2: Studies of the enhancement effect of cyclophosphamide. *Contact Dermatitis*, 29:26.

Kimber, I., and Weisenberger, C. (1989): A murine local lymph node assay for the identification of contact allergens: Assay development and results of an initial validation study. *Arch. Toxicol.*, 63:274.

Kimber, I., Hilton, J., and Weisenberger, C. (1989): The murine local lymph node assay for identification of contact allergens: A preliminary evaluation of in situ measurement of lymphocyte proliferation. *Contact Dermatitis*, 21:215.

Kimber, I., Hilton, J., Botham, P. A., Basketter, D. A., Scholes, E. W., Miller, K., Robbins, M. C., Harrison, P. T., Gray, T. J., and Waite, S. J. (1991): The murine local lymph node assay: Results of an inter-laboratory trial. *Toxicol. Lett.*, 55:203.

Kimber, I., Hilton, J., Dearman, R. J., Gerberick, G. F., Ryan, C. A., Basketter, D. A., Scholes, E. W., Ladics, G. S., Loveless, S. E., House, R. V., Guy, A., Ladies, G. S., and Loveless, S. E. (1995): An international evaluation of the murine local lymph node assay and comparison of modified procedures. *Toxicology*, 103:63.

Klinkhammer, C., Popowa, P., and Gleichmann, H. (1988): Specific immunity to streptozocin: Cellular requirements for induction of lymphoproliferation. *Diabetes*, 37:74.

Kondo, S., Pastore, S., Shivji, G. M., McKenzie, R. C., and Sauder, D. N. (1994): Characterization of epidermal cytokine profiles in sensitization and elicitation phases of allergic contact dermatitis as well as irritant contact dermatitis in mouse skin. *Lymphokine Cytokine Res.*, 13:367.

Krzystyniak, K., Brouland, J. P., Panaye, G., Patriarca, C., Verdier, F., Descotes, J., and Revillard, J. P. (1992): Activation of CD4+ and CD8+ lymphocyte subsets by streptozotocin in murine popliteal lymph node (PLN) test. *J. Autoimmun.*, 5:183.

Krzystyniak, K., Panaye, G., Descotes, J., and Revillard, J. P. (1992): Activation of CD4+ and CD8+ lymphocyte subsets by streptozotocin in the popliteal lymph node assay. II: Comparison with acute graft-vs-host reaction in H-2 incompatible F1 mouse hybrids. *Immunopharmacol. Immunotoxicol.*, 14:865.

Kubicka-Muranyi, M., Goebels, R., Goebel, C., Uetrecht, J., and Gleichmann, E. (1993): T lymphocytes ignore procainamide, but respond to its reactive metabolites in peritoneal cells: Demonstration by the adoptive transfer popliteal lymph node assay. *Toxicol. Appl. Pharmacol.*, 122:88.

Kubicka-Muranyi, M., Behmer, O., Uhrberg, M., Klonowski, H., Bister, J., and Gleichmann, E. (1993): Murine systemic autoimmune disease induced by mercuric chloride (HgCl2): Hg-specific helper T-cells react to antigen stored in macrophages. *Int. J. Immunopharmacol.*, 15:151.

Magnusson, B., and Kligman, A. M. (1969): The identification of contact allergens by animal assay: The guinea pig maximization test. *J. Invest. Dermatol.*, 52:268.

Maguire, H. C., Jr., and Kaidbey, K. (1982): Experimental photoallergic contact dermatitis: A mouse model. *J. Invest. Dermatol.*, 79:147.

Maurer, T., Weirich, E. G., and Hess, R. (1980): The optimization test in the guinea pig in relation to other predictive sensitization methods. *Toxicology*, 15:163.

Maurer, T. (1985): The optimization test. *Curr. Probl. Dermatol.*, 14:114.

Maurer, T. (1984): Experimental contact photoallergenicity: Guinea pig models. *Photodermatol.*, 1:221.

Miyachi, Y., and Takigawa, M. (1983): Mechanisms of contact photosensitivity in mice. III: Predictive testing of chemicals with photoallergenic potential in mice. *Arch. Dermatol.*, 119:736.

Miyachi, Y., and Takigawa, M. (1982): Mechanisms of contact photosensitivity in mice. II: Langerhans cells are required for successful induction of contact photosensitivity to TCSA. *J. Invest. Dermatol.*, 78:363.

Oliver, G. J., Botham, P. A., and Kimber, I. (1986): Models for contact sensitization: Novel approaches and future developments. *Br. J. Dermatol.*, 115 (Suppl 31):53.

Robinson, M. K., Nusair, T. L., Fletcher, E. R., and Ritz, H. L. (1990): A review of the Buehler guinea pig skin sensitization test and its use in a risk assessment process for human skin sensitization. *Toxicology*, 61:91.

Scholes, E. W., Basketter, D. A., Lovell, W. W., Sarll, A. E., and Pendlington, R. U. (1991): The identification of photoallergic potential in the local lymph node assay. *Photodermatol. Photoimmunol. Photomed.*, 8:249.

Scholes, E. W., Basketter, D. A., Sarll, A. E., Kimber, I., Evans, C. D., Miller, K., Robbins, M. C., Harrison, P. T., and Waite, S. J. (1992): The local lymph node assay: Results of a final inter-laboratory validation under field conditions. *J. Appl. Toxicol.*, 12:217.

Schuhmann, D., Kubicka-Muranyi, M., Mirtschewa, J., Gunther, J., Kind, P., and Gleichmann, E. (1990): Adverse immune reactions to gold. I: Chronic treatment with an Au(I) drug sensitizes mouse T cells not to Au(I), but to Au(III) and induces autoantibody formation. *J. Immunol.*, 145:2132.

Thomas, C., Groten, J., Kammuller, M. E., De Bakker, J. M., Seinen, W., and Bloksma, N. (1989): Popliteal lymph node reactions in mice induced by the drug zimeldine. *Int. J. Immunopharmacol.*, 11:693.

Thomas, C., Lippe, W., Hogberg, T., Seinen, W., and Bloksma, N. (1990): Induction of popliteal lymph node enlargement and antibody production in the mouse by pyridylallylamines related to zimeldine. *Int. J. Immunopharmacol.*, 12:569.

Thomas, C., Punt, P., Warringa, R., Hogberg, T., Seinen, W., and Bloksma, N. (1990): Popliteal lymph node enlargement and antibody production in the mouse induced by zimeldine and related compounds with varying side chains. *Int. J. Immunopharmacol.*, 12:561.

Ulrich, P., Homey, B., and Vohr, H. W. (1998): A modified murine local lymph node assay for the differentiation of contact photoallergy from phototoxicity by analysis of cytokine expression in skin-draining lymph node cells. *Toxicology*, 125:149.

Verdier, F., Virat, M., and Descotes, J. (1990): Applicability of the popliteal lymph node assay in the Brown-Norway rat. *Immunopharmacol. Immunotoxicol.*, 12:669.

Vial, T., Carleer, J., Legrain, B., Verdier, F., and Descotes, J. (1997): The popliteal lymph node assay: Results of a preliminary interlaboratory validation study. *Toxicology*, 122:213.

Vohr, H. W., Homey, B., Schuppe, H. C., and Kind, P. (1994): Detection of photoreactivity demonstrated in a modified local lymph node assay in mice. *Photodermatol. Photoimmunol. Photomed.*, 10:57.

Ziegler, V., and Suss, E. (1985): The TINA test. *Curr. Probl. Dermatol.*, 14:172.

Toxicology of Skin
Edited by Howard I. Maibach
Copyright © 2001 Taylor & Francis

25

Allergic Contact Dermatitis from Transdermal Systems

Cheryl Levin and Howard I. Maibach

Department of Dermatology, University of California, San Francisco, California, USA

INTRODUCTION

Transdermal therapeutic systems (TTS) provide an effective method to deliver certain drugs through skin. Lipophilic drugs with relatively high potency and high volume of distribution are promising candidates for transcutaneous delivery. Currently, eight TTS are marketed (scopolamine, clonidine, nitroglycerin, nicotine, estradiol, estradiol/progesterone, fentanyl) (Hogan et al., 1991). Transdermal delivery provides a constant, controlled amount of drug and eliminates the gastrointestinal and hepatic "first-pass" metabolism experienced with oral medication. Other advantages include the avoidance of high peak levels in achieving a steady state concentration and increased patient compliance. Although beneficial, these devices have adverse effects. A limitation in transdermal delivery is allergic contact dermatitis. This delayed hypersensitivity reaction may result from various components of the TTS, namely the drug, its vehicle, or the adhesive used to secure the TTS to the skin (Ademola et al., 1997).

DATA MANAGEMENT

The following reviews ACD in transdermal delivery from 1989 to 1999. Sources searched include Medline/Melvyl, Science Citation Index, and a hand search of all contact dermatitis journals from the years 1989 to 1999.

Scopolamine

The antimotion sickness drug, scopolamine, was the first to be delivered transdermally (Brown et al., 1988). There have been few reported cases of allergic contact dermatitis with this TTS (Fisher, 1984; Trozak, 1985; Van der Willigen et al., 1988). (Gordon et al., 1989) reported delayed-type hypersensitivity reactions in an unexpectedly high rate (10%) of users; however, the report's high results may reflect the subjects' wearing the patch for a prolonged period. Patch tests have confirmed that scopolamine is the sensitizing agent in most cases. Allergic contact dermatitis may be confirmed by patch testing 1% scopolamine in petrolatum or water (Fisher, 1984).

Clonidine

Clonidine is a centrally acting alpha-adrenoreceptor agonist used for the treatment of hypertension. Although early allergencity tests in guinea pigs failed to identify clonidine as a sensitizer, once accessible, many users developed allergic contact dermatitis (Shaw et al., 1987). Therefore, some claim that weak allergens occluding the skin may not be identified with routine guinea pig models (Schper, 1990). Others suggest that the immunosuppressive effects of clonidine on the induction of sensitization in unrelated allergens may account for the inability to detect clonidine sensitization with acute exposure (Robinson et al., 1990). (Kalish et al., 1996) has developed another predictive model of clonidine sensitization (among other weak allergens) using CBA/J mice.

Adverse skin reactions, including erythema, pruritis, and allergic contact dermatitis, have been associated with clonidine utilization in up to 50% of cases (Grattan et al., 1985; Groth et al., 1985; Weber et al., 1984; Boekhorst, 1983; Horning et al., 1988; Maibach, 1987; Fillingim et al., 1989; Corazza et al., 1995; Polster et al., 1999); however, only 2%–3% of topically sensitized patients experience systemic reactions when challenged with oral clonidine (Corazza et al., 1995; Maibach, 1985). Therefore, the current recommendation is to continue oral treatment when topical application is no longer feasible. Clinical studies suggest that the incidence of sensitization to allergic contact dermatitis is higher in females and lower in blacks (Holdiness, 1989).

A majority of studies record sensitization from clonidine itself; however, (Corazza et al., 1995) reported a patient with allergy to the TTS *in toto* and not to the recommended concentrations of the drug itself. This hypersensitivity reaction may be due to an impure compound formation resulting from the interaction of clonidine and acetaldehyde, a component of the TTS (Holdiness, 1989). Dermatitis from other components of the transdermal patch has rarely been reported. One patient with concomitant mycosis fungiosis developed allergic contact dermatitis from a clonidine patch (Polster et al., 1999). The etiology of the ACD could not be determined, as patch testing was not performed. The current recommendation for patch testing is 9% clonidine in petrolatum (Hogan et al., 1991). The dermatotoxicologic experience with clonidine led to the observation that predictive assays (Draize Repeat Insult

Patch test) for transdermals may need to be 12 weeks rather than the typical 6 weeks (unpublished data).

Unsubstantiated reports suggest that skin pretreatment with hydrocortisone, an aluminum-magnesia suspension, or the anti-inflammatory beclomethasone dipropionate reduces the clonidine-induced contact dermatitis in sensitized patients (McChesney, 1991). A study by (Silva and Berman, 1992) tested a 0.1 mL alumina, magnesia, and simethicone suspension (of alumina, magnesia and simethicone). Thin application of the pretreatment solution to back skin prior to application of the clonidine TTS reduced dermatitis in one patient and prevented the type IV hypersensitivity reaction in the other two patients. Clonidine's hypertensive effect was maintained (Silva and Berman, 1992). Hydrocortisone preapplication has been found to increase the absorption of clonidine, although the exact mechanism has yet to be elucidated (Ito et al., 1991). Further studies must be performed to investigate the effects of various skin pretreatment drugs on the development of ACD from clonidine TTS.

TABLE 25.1. *Incidence of ACD in transdermal patches*

Drug	Incidence of allergic contact dermatitis	Specific etiology	Reference
Scopolamine	Rare	The active drug in most cases	Fisher, 1984; Trozak, 1985; Van der Willigen et al., 1988; Gordon et al., 1989
Clonidine	Relatively common; reported in up to 50% of cases	The active drug in most cases	Grattan et al., 1985; Groth et al., 1985; Weber et al., 1984; Boekhorst, 1983; Horning et al., 1988; Maibach, 1987; Fillingim et al., 1989; Corazza et al., 1995; Polster et al., 1999 (Maibach, 1987)
Nitroglycerin	Rare	The active drug in most cases	Harari et al., 1987; Carmichael, 1994; Carmichael et al., 1989; Torres et al., 1992; Landro et al., 1989; Machet et al., 1999; Kounis et al., 1996; Torres et al., 1992; Wainright et al., 1993
Nicotine	Somewhat common	The active drug in most cases	Eichelberg et al., 1989; Abelin et al., 1989; Jordan, 1992; Bircher et al., 1991; Vincenzi et al., 1993; Farm, 1993; Von Bahr et al., 1997; Dwyer et al., 1994; Fiore et al., 1993
Estradiol	Relatively common	A component of the TTS other than the active drug in most cases	McBurney et al., 1989; Grebe et al., 1993; Pecquet et al., 1992; Torres et al., 1992; Schwartz et al., 1988; Corazza et al., 1993; Boehnicke et al., 1996; El-Sayed et al., 1996; Goncalo et al., 1999
Testosterone	Rare	The active drug	Buckley et al., 1997
Fentanyl	No reported cases		

Nitroglycerin

The topically applied prophylactic anti-anginal drug nitroglycerin has rarely caused allergic contact dermatitis (Valliant et al., 1990; de la Fuente et al., 1994). Irritant reactions have been observed in about 15% of cases, although erythema has often been attributed to nitroglycerin's vasodilatory effect and not to allergy or irritation (Valliant et al., 1990).

None of the 3273 patients participating in a large cohort study presented with a nitroglycerin-induced hypersensitivity reaction (Harari et al., 1987; Carmichael, 1994). Similarly, a prospective study of 33 patients found that none experienced ACD (Valliant et al., 1990). (Kounis et al., 1996) reported 4 of 320 (1.2%) patients with an allergic reaction from a nitroglycerin patch, including one generalized anaphylactic response.

In the past decade, only four isolated cases of ACD from nitroglycerin TTS have been reported (Carmichael et al., 1989; Torres et al., 1992; Landro et al., 1989; Machet et al., 1999). Nitroglycerin was responsible for the dermatitis in two of these cases (Carmichael et al., 1989; Torres et al., 1992), whereas one case reported allergy to the TTS *in toto* but not to a placebo patch or nitroglycerin (Landro et al., 1989). Patch testing was not performed in the final case report, although the dermatitis was observed in two transdermal systems (Machet et al., 1999).

Reaction to the TTS system, but not to nitroglycerin, is rare (Torres et al., 1992; Carmichael, 1994; Wainright et al., 1993). It has been hypothesized that the complex composition of the patch could be responsible for the dermatitis. The nitroglycerin patch consists of nitroglycerin adsorbed on lactose and bathed in a medical silicone fluid. The rate-controlling membrane is made of an ethylene vinyl acetate copolymer. Individual components of the TTS should be tested to determine the etiology of the hypersensitivity reaction. Finnish occupational ACD reports suggest patch testing with 0.5%–2% nitroglycerin in petrolatum (Kanerva et al., 1991).

Nicotine

The nicotine patch has been implicated in adverse skin reactions, including allergic contact dermatitis. Twenty-five percent to 50% of patients have experienced localized skin reactions, such as erythema, burning, or itching, upon application of the patch (Abelin et al., 1989; Transdermal Nicotine Study Group, 1991). The cutaneous pharmacologic effects of nicotine are extensively reviewed by (Smith et al., 1992). In a majority of hypersensitivity reactions from this TTS, the active drug nicotine has been responsible. One study involving 183 heavy smokers reported a 2.6% incidence of ACD from nicotine among patients using transdermal patches (Eichelberg et al., 1989). Similar results in a *JAMA* study found 16 of 664 patients (2.4%) with nicotine allergy to the TTS, 11 of whom were confirmed by rechallenge (Abelin et al., 1989). Finally, (Jordan et al., 1992) reported 3 of 186 subjects exhibiting evidence of delayed contact sensititzation to ACD; however, in each of these studies, patch testing was not performed, and therefore allergy to other components of the TTS cannot be ruled

TABLE 25.2. *Typical formulations in transdermal delivery (Physician's Desk Reference 2000)*

Drug	Typical formulation
Scopolamine	Scopolamine 1.5 mg base
Clonidine	Clonidine 0.1 mg, 0.2 mg, or 0.3 mg tablets
Nitroglycerin	Nitroglycerin 0.1 mg, 0.2 mg, 0.3 mg, 0.4 mg, 0.6 mg, or 0.8 mg per hour
Nicotine	Nicotine—three steps of 21 mg, 14 mg, and 7 mg per hour *Or* Nicotine 15 mg in 16 hours
Estradiol	Estradiol 0.025 mg, 0.05 mg, 0.075 mg, or 0.1 mg per day
Estradiol/Progestin	Estradiol/Progestin 0.05 / 0.14 mg per day *Or* Estradiol/Progestin 0.05/0.25 mg per day
Testosterone	Testosterone 2.5 mg, 4 mg, 5 mg, or 6 mg per day
Fentanyl	Fentanyl 2.5 mg, 5 mg, 7.5 mg, or 10 mg per hour

out. In contrast, 5 of 14 volunteers with a history of adverse skin reactions to TTS experienced sensitization to nicotine, as confirmed by patch testing. One individual reacted to the patch itself (Bircher et al., 1991).

There have also been several case reports of ACD from nicotine patches (Vincenzi et al., 1993; Farm, 1993; Von Bahr et al., 1997; Dwyer et al., 1994, Fiore et al., 1993). Patch testing confirmed nicotine as the responsible agent in a majority of these cases (Vincenzi et al., 1993; Farm, 1993). One report involved systemic reactions, including papulovesicular rashes in places of high pressure from underwear. The switch to nicotine gum induced further systemic spread to the legs and feet (Farm, 1993). Two reports suggest that components of the patch itself are cause for the allergic response; however, patch testing was performed only in one of these cases, which found methacrylates to be the causative agent (Dwyer et al., 1994).

Patients displaying symptoms of ACD should be patch tested to confirm the source of the allergy. The observed erythema may not be an allergic reaction at all, but may result from the vasodilatory effect of nicotine, as was reported in one case (Von Bahr et al., 1997). The recommended test agent and concentration for patch testing is an aqueous solution of 10% nicotine base (Vincenzi et al., 1993; Bircher et al., 1991).

Estradiol

Useful in suppressing hot flashes among postmenopausal women, estradiol patches are a source of skin irritation among 2%–25% of users (McCarthy et al., 1992; The Transdermal HRT Investigators Group, 1993; Utiah, 1987; de Cetina et al., 1999). Adverse skin reactions are more prevalent in tropical climates (de Cetina et al., 1999; Bhathena et al., 1998), probably due to the increased occlusive effects of the warm, humid environment. Two types of estradiol TTS are currently marketed. The drug reservoir system is composed of a reservoir containing the drug, a rate-limiting membrane, and an adhesive. The newer matrix system eliminates the alcohol-based reservoir and incorporates the estrogen into the adhesive. Matrix patches may cause fewer skin reactions than reservoir patches according to both unsubstantiated and published reports (The Transdermal HRT Investigators Group, 1993; Howie et al., 1995; Ross et al., 1997).

There have been several reported cases of allergic contact dermatitis from transdermally applied estradiol. Most studies implicate a component of the TTS and not estradiol as the sensitizing agent. Several reports of allergic contact dermatitis from the receptacle constituents (McBurney et al., 1989), ethanol (Grebe et al., 1993; Pecquet et al., 1992), or a component of the drug reservoir (Torres et al., 1992), such as hydroxypropyl cellulose (Schwartz et al., 1988), have been recorded. In one patient there was a putative cross-reactivity between the propylene glycol vehicle used in acyclovir cream 5% used to treat herpes simplex labialis infection and the hydroxypropyl cellulose in an estradiol patch (Corazza et al., 1993). The two compounds share a chemical similarity, and propylene glycol has been indicated in inducing sensitization by other allergens (Hannuksela, 1987). ACD from estradiol TTS has been confirmed by patch testing in only three reported cases (Boehnicke et al., 1996; El-Sayed et al., 1996; Goncalo et al., 1999), with one reporting systemic effects as well (Goncalo et al., 1999); however, estradiol was the probable sensitizing agent in other unsubstantiated reports (McGurney et al., 1989; Carmichael et al., 1992; Quince et al., 1996). Thus far there is no declared vehicle and concentration for patch testing, though two studies have found testing in 96% ethanol to be appropriate (Boehnicke et al., 1996; Goncalo et al., 1999).

Testosterone

The testosterone transdermal patch may be applied to scrotal or glabrous skin. Scrotal skin allows the highest rate of percutaneous absorption for steroids and thus can be utilized to deliver higher levels of drug (Bals-Pratsch et al., 1986). There is only one confirmed case of allergic contact dermatitis from TTS testosterone, a system utilized to treat hypogonadal men (Buckley et al., 1997). A man with Kleinefelter's syndrome became allergic to the testosterone within the nonscrotal TTS (Andropatch), as validated by patch testing. A large cohort study by (Jordan, 1997) compared irritation and allergy associated with nonscrotal and scrotal transdermal systems in 60 healthy adult males. The incidence of irritation and allergy were significantly higher (p < .0001) in the nonscrotal system, although patch testing was not performed. Further studies are necessary to understand the adverse effects of various components of both scrotal and nonscrotal TTS on skin.

Fentanyl

Fentanyl is used as a preanesthetic analgesic. Currently there are no reported cases of ACD from fentanyl.

Prophylactic Measures

Improved predictive tests of sensitization should be developed. Though standard predictive tests indicated clonidine was not an allergen, when marketed, many users developed allergic contact dermatitis. Currently, mouse strains are being developed as an alternative to the standard guinea pig model (Kalish et al., 1996). (Robinson and Cruze, 1996) used guinea pig models and a local lymph node assay (LLNA) in mouse models to aid in the detection of weak allergens. Through vitamin A supplementation and the introduction of chronic conditions, they were able to detect contact sensitization of clonidine in both mice and guinea pigs. Other predictive models have analyzed the electrophiles within various haptens, such as scopolamine and clonidine, and their relationship with nucleophilic groups of skin protein to form antigens (Benezra, 1991).

Contact sensitization from transdermal d-chlorpheniramine and benzoyl peroxide were reduced when hydrocortisone was coadministered (Amkraut et al., 1996). Additionally, when applied prior to the transdermal patch, ion channel modulators, such as ethacrynic acid, have been found to prevent sensitization in mice (Kalish et al., 1997; Wille et al., 1999). The potential role of corticosteroids and ion channel modulators in the prevention of contact sensitization from TTS should be further investigated and better defined.

Additionally, minimizing reapplication to the same application site (Hogan et al., 1991) and maintaining caution when performing oral provocation tests (Vermeer, 1991) would help to reduce the induction of allergic contact dermatitis among TTS users.

With the exception of clonidine ACD and estradiol irritant dermatitis, the first generations of transdermals have had a high level of chemical tolerance. The next generation of investigators have three areas requiring improvement:

1. Prediction of ACD in new systems. Although the 12-week Draize Repeat Insult Patch test is sensitive and specific, shorter assays would be welcome.
2. The epidemiology of transdermal ACD is qualitative rather than quantitative. Future studies providing both numerator and denominator combined with definite product component testing (i.e., each ingredient) will provide valuable benchmarks for future formulations.
3. Lastly, effective anti-irritant and antiallergens, as discussed in references (Amkraut, 1996; Kalish et al., 1997; Wille et al., 1999) would aid the intolerant patient.

REFERENCES

Abelin, T., Buehler, A., Muller, P., Vesanen, K., and Imhof, P. R. (1989): Controlled trial of transdermal nicotine patch in tobacco withdrawal. *Lancet*, 1:7–10.
Ademola, J., Maibach, H. I. (1997): Safety assessment of transdermal and topical dermatological products. In: *Transdermal and Topical Drug Delivery Systems*. Edited by Ghosh, T. K., Pfister, W. R., Yum, S., Interpharm Press, Buffalo Grove, IL.

Amkraut, A. A., Jordan, W., Taskovich, L. (1996): Effect of coadministration of corticosteroids on the development of contact sensitization. *J. Am. Acad. Dermatol.*, 35:27–31.

Bals-Pratsch, M., Yoon, Y. D., Knuth, U., Nieschlag, E. (1986): Transdermal testosterone substitution therapy for male hypogonadism. *The Lancet*, 2:943–946

Benezra, C. (1991): Structure-activity relationships of skin haptens with a closer look at compounds used in transdermal devices. *Journal of Controlled Release*, 15:267–270.

Bhathena, R. K., Anklesaria, B. S., Ganatra, A. M. (1998): Skin reactions with transdermal estradiol therapy in a tropical environment. *Int. J. Gynecol Obstet.*, 60:177–179.

Bircher, A. J., Howald, H., Rufli, T. (1991): Adverse skin reactions to nicotine in a transdermal therapeutic system. *Contact Dermatitis*, 25:230–236.

Boehnicke, W.-H., Gall, H. (1996): Type IV hypersensitivity to topical estradiol in a patient tolerant to it orally. *Contact Dermatitis*, 35:187–188.

Boekhorst, J. C. (1983). Allergic contact dermatitis with transdermal clonidine. *Lancet*, 1:1031–1032.

Brown, L. and Langer, R. (1988): Transdermal delivery of drugs. *Ann. Rev. Med.*, 39:221–29.

Buckley, D. A., Wilkinson, S. M., Higgins, E. M. (1997): Contact allergy to a testosterone patch. *Contact Dermatitis*, 39:91–92.

Carmichael, A. J. (1994): Skin sensitivity and transdermal drug delivery: A review of the problem. *Drug Safety*, 10:151–159.

Carmichael, A. J., Foulds, I. S. (1989): Allergic contact dermatitis from transdermal nitroglycerin. *Contact Dermatitis*, 21:113–114.

Carmichael, A. J., Foulds, I. S. (1992). Allergic contact dermatitis from estradiol in oestrogen patches. *Contact Dermatitis*, 26:194–195.

Corazza, M., Mantovani, L., Virgili, A., Strumia, R. (1995): Allergic contact dermatitis from a clonidine transdermal delivery system. *Contact Dermatitis*, 32:246.

Corazza, M., Virgili, A., Mantovani, L., Malfa, W. L. (1993): Propylene glycol allergy from acyclovir cream with cross-reactivity to hydroxypropyl cellulose in a transdermal estradiol system. *Contact Dermatitis*, 29:283–284.

de Cetina, T. C., Reyes, L. P. (1999): Skin reactions to transdermal estrogen replacement therapy in a tropical climate. *Int. J. Gynecol Obstet.*, 64:71–72.

de la Fuente, Prieto, R., Medina, A. A., Perez, J. M. D. (1994): Contact dermatitis from nitroglycerin. *Annals of Allergy*, 72:344–346.

Dwyer, C. M., Forsyth, A. (1994): Allergic contact dermatitis from methacrylates in a nicotine transdermal patch. *Contact Dermatitis*, 30:309–310.

Eichelberg, D., Stolze, P., Block, M., et al., (1989): Contact allergy induced by TTS-treatment. *Methods and Finding in Experimental Clinical Pharmacology*, 11:223–225.

El-Sayed, F., Bayle-Lebey, P., Marguery, M. C., Basex, J. (1996): Sensibilisation systemique au 17-B-oestradiol induite par voie transcutanee. *Ann. Dermatol. Venereol.*, 123:26–28.

Farm, G. (1993): Contact allergy to nicotine from a nicotine patch. *Contact Dermatitis*, 29:214–215.

Fillingim, J. M., Matzek, K. M., Hughes, E. M., Johnson, P. A., Sharon, G. S. (1989): Long-term treatment with transdermal clonidine in mild hypertension. *Clinical Therapeutics*, 11:398–408.

Fiore, M. C., Hartman, M. J. (1993): Side effects of nicotine patches. *JAMA*, 270:2735.

Fisher, A. A. (1984): Dermatitis due to transdermal therapeutic systems. *Cutis*, 34:526–531.

Goncalo, M., Oliveira, H. S., Monteiro, C., Clerins, I., Figueiredo, A. (1999): Allergic and systemic contact dermatitis from estradiol. *Contact Dermatitis*, 40:58–59.

Gordon, C. R., Shupak, A., Doweck, I., Spitzer, O. (1989): Allergic contact dermatitis caused by transdermal hyoscine. *BMJ*, 298:1220–1221

Grattan, C. E. H. and Kennedy, C. T. C. (1985): Allergic contact dermatitis to transdermal clonidine. *Contact Dermatitis*, 19:225–226

Grebe, S. K. G., Adams, J. D., Feek, C. M. (1993): Systemic sensitization to ethanol by transdermal estrogen patches. *Arch. Dermatol.*, 129:379–380.

Groth, H., Vetter, H., Knuesel, J., Better, W. (1985): Allergic skin reactions to transdermal clonidine. *The Lancet*, 2:850–851.

Hannuksela, M. (1987): Propylene glycol promotes allergic patch test reactions. *Bollettino di Dermatologia Allergologica e Professionale*, 2:40–44.

Harari, Z., Sommer, I., Knobel, B. (1987): Multifocal contact dermatitis to nitroderm TTS 5 with extensive postinflammatory hypermelanosis. *Dermatologica*, 174:249–252.

Hogan, D. J., Maibach, H. I. (1991): Transdermal drug delivery systems: Adverse reactions—dermatologic overview. In: *Exogenous Dermatoses: Environmental Dermatitis*. Edited by Menne, T., Maibach, H. I., CRC Press, Boca Raton, FL.

Holdiness, M. R. (1989): A review of contact dermatitis associated with transdermal therapeutic systems. *Contact Dermatitis*, 20:3–9.

Horning, J. R., Zawada, E. T., Simmons, J. L., Williams, L., McNutly, R. (1988). Efficacy and safety of two year therapy with transdermal clonidine for essential hypertension. *Chest*, 93:941–945.

Howie, H., Heimer, G. (1995): A multicenter randomized parallel group study comparing a new estradiol matrix patch and a registered reservoir patch. *Menopause*, 2:43–48.

Ito, M. K., O'Conner, D. T. (1991): Skin pretreatment and the use of transdermal clonidine. *The American Journal of Medicine*, 91:42S–49S.

Jordan, W. (1997): Allergy and topical irritation associated with transdermal testosterone administration: A comparison of scrotal and nonscrotal transdermal systems. *American Journal of Contact Dermatitis*, 108–113.

Jordan, W. P. (1992): Clinical evaluation of the contact sensitization potential of a transdermal nicotine system (Nicoderm). *J. Family Practice*, 34:709–712.

Kalish, R. S., Wood, J. A., Kydonieus, A., Wille, J. J. (1997): Prevention of contact hypersensitivity to topically applied drugs by ethacrynic acid: Potential application to transdermal drug delivery. *Journal of Controlled Release*, 48:79–87.

Kalish, R., Wood, J. A., Willw, J. J., Kydonieus, A. (1996): Sensitization of mice to topically applied drugs: Albuterol, chlorpheniramine, clonidine, and nadolol. *Contact Dermatitis*, 35:76–82.

Kanerva, L., Laine, R., Jolanki, R., Tarvaines, K., Estlander, T., Helander, I. (1991): Occupational allergic contact dermatitis caused by nitroglycerin. *Contact Dermatitis*, 24:356–362.

Kounis, N. G., Zavras, G. M., Papadaki, P. J., Soufras, G. D., Poulos, G. A., Gourlevenos, J., Alangoussis, A., Antonakopoulos, K., Cargies, C., Peristeropoulou, S. A., Koutsojoanis, C. (1996): Allergic reactions to local glyceryl trinitrate administration. *British Journal of Clinical Practice*, 50:437–439.

Landro, A. D., Valsecchi, R., and Cainelli, T. (1989): Contact dermatitis from Nitroderm. *Contact Dermatitis*, 21:115–116.

Machet, L., Martin, L., Toledano, C., Jan, V., Lorette, G., Valliant, L. (1999). Allergic contact dermatitis from nitroglycerin contained in 2 transdermal systems. *Dermatology*, 198:106–107.

Maibach, H. I. (1987): Oral substitution in patients sensitized by transdermal clonidine treatment. *Contact Dermatitis*, 16:1–8.

Matzek, K. A. (1988): Boehringer Ingelheim. Written communication, December 2.

McBurney, E. I., Noel, S. B., Collins, J. H. (1989): Contact dermatitis to transdermal estradiol system. *J. Am. Acad. Dermatol.*, 20:508–510.

McCarthy, T., Dramusic, V., Ratnam, S. H. (1992): Use of two types of estradiol-releasing skin patches for menopausal patients in a tropical climate. *Am. J. Obstet. Gynecol.*, 166:2005–2010.

McChesney, J. A. (1991): Preventing the contact dermatitis caused by a transdermal clonidine patch, (letter). *Western Journal of Medicine*, 154:736.

Physician's Desk Reference 2000. Medical Economics Co., Oradell, NJ.

Pecquet, C., Pradlier, A., Dry, J. (1992): Allergic contact dermatitis from ethanol in a transdermal estradiol patch. *Contact Dermatitis*, 27:275–276.

Polster, A. M., Warner, M. R., Camisa, C. (1999): Allergic contact dermatitis from transdermal clonidine in a patient with mycosis fungoides. *Cutis*, 63:154–155.

Quirce, S., Garde, A., Baz, G., Gonzalez, A., Durana, M. D. (1996): Allergic contact dermatitis from estradiol in a transdermal therapeutic system. *Allergy*, 51:62–63.

Robinson, M. K., Cruze, C. A. (1996): Preclinical skin sensitization testing of antihistamines: Guinea pig and local lymph node assay responses. *Food and Chemical Toxicology*, 34:495–506.

Robinson, M. K., Sozeri, T. J. (1990): Immunosuppressive effects of clonidine on the induction of contact sensitization in the Balb/c Mouse. *Journal of Investigative Dermatology*, 95:587–591.

Ross, D., Rees, M., Godfree, V., Cooper, A., Hart, D., Kingsland, C., Whitehead, M. (1997): Randomised crossover comparison of skin irritation with two transdermal oestradiol patches. *BMJ*, 315:288.

Schper, R. J., von Blomberg, B. M. E., de Groot, J., Goeptar, A. R., Lang, M., Oostendorp, R. A. J., Bruynzeel, D. P., van Tol, R. G. L. (1990): Low allerencity of clonidine impede\s studies of sensitization mechanisms in guinea pig models. *Contact Dermatitis*, 23:81–89.

Schwartz, B. K., Clendenning, W. E. (1988): Allergic contact dermatitis from hydroxypropyl cellulose in a transdermal estradiol patch. *Contact Dermatitis*, 18:106–107.

Shaw, J. E., Prevo, M. E., Amkraut, A. A. (1987): Testing of controlled-release transdermal dosage forms. *Arch. Dermatol.*, 122:1548–1556.

Silva, S. K., Berman, B. (1992): The effect of topical maalox on transdermal clonidine-induced contact dermatitis. *American Journal of Contact Dermatitis*, 3:79–82.

Smith, E. W., Smith, K. A., Maibach, H. I. (1992): The local side effects of transdermally absorbed nicotine. *Skin Pharmacology*, 5:69–76.

The Transdermal HRT Investigators Group. 1993. A randomized study to compare the effectiveness, tolerability and acceptability of two different transdermal estradiol replacement therapies. *Int. J. Fertil.*, 38:5–11.

Torres, V., Lopes, C., and Leite, L. (1992): Allergic contact dermatitis from nitroglycerin and estradiol transdermal therapeutic systems. *Contact Dermatitis*, 26:53–54.

Transdermal nicotine study group. (1991): Transdermal nicotine for smoking cessation: six month results from two multi-center controlled clinical trials. *JAMA*, 266:3133–3138.

Trozak, D. J. (1985): Delayed hypersensitivity to scopolamine delivered by a transdermal device. *J. Am. Acad. Dermatol*, 13:247–251.

Utiah, W. H. (1987): Transdermal estradiol overall safety profile. *Am. J. Obstet. Gynecol.*, 156:1335–1338.

Valliant, L., Bietter, S., Machet, L., Constans, T., Monpiere, C. (1990): Skin acceptance of transcutaneous nitroglycerin patches: A prospective study of 33 patients. *Contact Dermatitis*, 23:142–145.

Van der Willigen, A. H., Oranje, A. P., Stolz, E., Van Joost, T. (1988): Delayed hypersensitivity to scopolamine in transdermal therapeutic systems. *J. Am. Acad. Dermatol.*, 18:146–147

Vermeer, B. J. (1991): Skin irritation and sensitization. *Journal of Controlled Release*, 15:261–266.

Vincenzi, C., Tosti, A., Cirone, M., Guarrera, M., Cusano, F. (1993): Allergic contact dermatitis from transdermal nicotine systems. *Contact Dermatitis*, 29:104–105.

Von Bahr, B. and Wahlberg, J. E. (1997): Reactivity to nicotine patches wrongly blamed on contact allergy. *Contact Dermatitis*, 37:44–45.

Wainright, R. J., Foran, J. P. M., Padaria, S. F., Akhras, F., Jackson, G., Clark, A. R. (1993): The long-term safety and tolerability of transdermal glyceryl trinitrate, when used with a patch-free interval in patients with stable angina. *British Journal of Clinical Practice*, 47:178–182.

Weber, M. A., Drayer, J. I. M., McMahon, F. G., Hamburger, R., Shah, A. R., Kirk, L. N. (1984): Transdermal administration of clonidine for treatment of high blood pressure. *Arch. Int. Med.*, 144:1211–1213.

Wille, J. J., Kydonieus, A., Kalish, R. S. (1999): Several different ion channel modulators abrogate contact hypersensitivity in mice. *Skin Pharmacol Appl. Skin Physiol.*, 12:12–17.

SECTION IV
Contact Urticaria

Toxicology of Skin
Edited by Howard I. Maibach
Copyright © 2001 Taylor & Francis

26

Animal Models for Immunologic Contact Urticaria and Nonimmunologic Contact Urticaria

Antti I. Lauerma

Department of Dermatology, University of Helsinki, Helsinki, Finland

Howard I. Maibach

Department of Dermatology, University of California, San Francisco, California, USA

INTRODUCTION

Contact urticaria is not an uncommon problem. Its importance has increased recently, especially during natural rubber latex allergy epidemics (Turjanmaa et al., 1996). The symptoms experienced in contact urticaria range from mere tingling sensations on skin to generalized, life-threatening reactions.

Contact urticaria can be distinguished into two types: immunologic contact urticaria (ICU) and nonimmunologic contact urticaria (NICU). These two forms are separate in their mechanisms and etiology and somewhat also in their clinical profile. The most important factor, though, is whether immune memory plays a role in the reactions. ICU occurs only in patients earlier sensitized to the causative agent, whereas NICU can be produced on any person regardless of their immunologic profile. These factors are, of course, of paramount importance when diagnosis is established through testing.

As an increasing number of new chemicals and substances enter skin care and medication fields each year, possible contact urticarias caused by these products also occur. To avoid these possibilities, predictive testing would be needed. Also, as NICU and ICU often are skin diseases which, in the case of ICU, could be fatal, medications for these diseases are needed. Predictive tests would therefore be needed for this purpose, too. This chapter deals with animal assays. In the future, though, nonanimal alternatives for predictive testing of ICU and NICU would hopefully become available.

IMMUNOLOGIC CONTACT URTICARIA: MECHANISMS

Immunologic contact urticaria is a skin reaction mediated via IgE antibodies against molecules perceived foreign by the immune system. The responsible molecule enters the dermis through stratum corneum, viable epidermis, and basement membrane. The allergenic molecules usually are large-sized polypeptides or proteins. Alternatively, they are chemicals bound to proteins present in patient's skin, thus forming immunogenic complexes. Either way, the allergen binds to IgE bound to a dermal mast cell. This binding results in release of histamine and possibly other mediators (e.g., serotonine) that cause leakage of water from dermal vessels, as well as itch and inflammation. The release of water is seen as edema, which is the principal clinical finding in urticaria, whereas itching is the main subjective symptom. As rodent skin contains mast cells, such animals are possible as models for ICU (Lauerma and Maibach, 1997).

RESPIRATORY CHEMICAL ALLERGY AS AN ANIMAL MODEL FOR IMMUNOLOGIC CONTACT URTICARIA

In the 1970s it was observed that anhydrides, such as trimellitic anhydride (TMA), cause asthma-like symptoms in persons exposed to them (Zeiss et al., 1977). The immunologic reactions seen in lungs of patients and experimental animals feature anaphylactic (Coombs–Gell Type I), complement-mediated (Coombs–Gell type II), antibody-complex-mediated (Coombs–Gell Type III), and cell-mediated (Coombs–Gell Type IV) reactions. Nonimmunologic (irritant) reactions may participate (Hayes et al., 1992), possibly due to metabolization of trimellitic anhydride to trimellitic acid.

TMA causes skin reactions if sensitization is done through skin contact (Dearman et al., 1992b). The skin reactions consist of two phases, immediate and delayed, possibly implying both immediate- and delayed-type hypersensitivity (Dearman et al., 1992b; Lauerma et al., 1997).

Sensitization to TMA may be performed either through airways (Obata et al., 1992) or through skin. Cutaneous sensitization is done either intradermally (Hayes et al., 1992) or topically (Dearman et al., 1992b). Cutaneous sensitization seems to induce both immediate (1 hour) and delayed (24 hours) skin reactions when TMA is subsequently encountered.

Intradermal sensitization to TMA may be done with the free hapten. When guinea pigs have been sensitized, TMA has been suspended in corn oil at 30% and injected into skin at a 0.1 mL dose. Guinea pigs sensitized in this manner have been used for challenges 3–4 weeks after the injection.

Immediate skin reactions to TMA have been studied using topical skin sensitization (Dearman et al., 1992b; Lauerma et al., 1997). The study animal has been the BALB/C mouse, and the animals have been sensitized topically on shaven skin on the trunk that has been tape-stripped prior to application. The dose used has been 100 μL TMA at 500 mg/ml. To enhance development of anti-TMA-IgE-antibodies, a second sensitization has been performed with 50 uL 250 mg/ml TMA at the same site. The animals have been used for elicitation 1 week after the second dosing.

In mice sensitized to TMA, a two-phase reaction is seen after TMA application on ears; there is a first phase with swelling peaking at 1 hour and a second phase peaking at 24 hours (Dearman et al., 1992b); however, a dose-dependent early swelling after TMA application is observed also in nonsensitized mice (Lauerma et al., 1997). Nonimmunologic, possibly irritant reactions in lungs have been suggested to be caused by trimellitic acid, a hydrolization product of TMA (Patterson et al., 1982). The skin reactions in nonsensitized animals could possibly be considered a form of NICU.

The effect of topical treatment with three immunomodulating chemicals (i.e., an antihistamine, diphen-hydramine, a glucocorticosteroid, betamethasone dipropionate, and a nonsteroidal anti-inflammatory drug, indomethacin hydrochloride) was studied. Treatment with diphenhydramine or betamethasone dipropionate suppress the reaction whereas indomethacin slightly increases swelling (Lauerma et al., 1997).

There also are other haptens capable of inducing respiratory allergy. Diphenylmethane-4,4-diisocyanate (MDI) and phtalic anhydride are respiratory allergens that induce IgE antibody production in animals exposed by epicutaneous application (Dearman et al., 1992a). These haptens could be used to establish a similar animal model for ICU as with TMA.

CONTACT CHEMICAL ALLERGY AS AN ANIMAL MODEL FOR IMMUNOLOGIC CONTACT URTICARIA

In a recent study, BALB/C mice repeatedly sensitized for up to 48 days with the contact allergen 2,4,6-trinitro-1-chlorobenzene (TNCB) had the time-course of TNCB-specific hypersensitivity reactions shifted from delayed-type to an immediate-type response. The reaction kinetic shift coincided with an increase of mast cell number in the skin area used for sensitization and elicitation. These early reactions in mouse ears, also followed by a later phase, were dependent on the presence of skin mast cells on application site as well as antigen-specific IgE (Kitagati et al., 1995). Therefore, such reactions could resemble ICU.

PROTEIN ALLERGY AS AN ANIMAL MODEL FOR IMMUNOLOGIC CONTACT URTICARIA

Rabbits sensitized through airways or through the skin with natural rubber latex (NRL) have, in addition to respiratory symptoms, wheal-and-flare skin prick test reactions (Reijula et al., 1994). Therefore, it could be that such animals could possibly be used as models for ICU; however, it has to be determined whether percutaneous penetration of NRL proteins is sufficient to cause ICU reactions in this model. Additionally, mice exposed to NRL have elevated IgE levels and eosinophilia and IgG antibodies against NRL antigens (Kurup et al., 1994). Similar models of immediate-type hypersensitivity to other proteins may be worthwhile to investigate.

NONIMMUNOLOGIC CONTACT URTICARIA: MECHANISMS

Nonimmunologic immediate contact reactions range from erythema to urticaria and occur in individuals not sensitized to the compounds. Nonimmunologic contact reactions are probably more common than immunologic contact reactions. They are possibly due to the causative agent's ability to release inflammatory mediators, such as histamine, prostaglandins, and leukotrienes, from skin cells without the participation of IgE molecules. Agents capable of causing nonimmunologic contact reactions are numerous; the most potent and well-studied agents are benzoic acid, sorbic acid, cinnamic aldehyde, and nicotinic acid esters. Provocative skin tests for diagnosis of nonimmunologic contact urticaria include the rub test and open test. In nonimmunologic contact reactions, the reactions appear within 45–60 minutes after application, whereas in immunologic reactions they should appear in 15–20 minutes (Gollhausen and Kligman, 1984).

ANIMAL MODELS FOR NONIMMUNOLOGIC CONTACT URTICARIA: MOUSE, RAT, GUINEA PIG

Animal models have been searched in a hope to find a screening method to identify compounds able to cause NICU (Lahti and Maibach, 1984). Such research has resulted in methods where the thickness of animal ear pinnae have been the primary end-point. Such in vivo approach seems to be the only possibility at present stage, as NICU reactions differ from each other greatly. As the mechanism is more phramacological than immune, the different compounds elicit different reaction mechanisms (Lahti and Maibach, 1985).

Mice and rats seem to be less sensitive to NICU than guinea pigs. Therefore, guinea pig is the animal of choice when NICU is studied (Lahti and Maibach, 1985). Substances studied are applied openly on guinea pig earlobe, and edema is quantitated with micrometer, which is a rapid and reliable method. Dose-responses are seen when optimal concentration ranges are studied. The maximal response is 100% increase in thickness, and it is seen at approximately 50 minutes after application (Lahti and Maibach, 1984), which is later than that of the 30 minutes found in mouse model for ICU (Lauerma and Maibach, 1997). NICU model can also be used to study pharmacological agents to treat the disease (Lahti et al., 1986).

CONCLUSIONS

It seems that mice sensitized through skin contact to TMA or possibly to other respiratory hapten allergens could be used as animal models for ICU. Guinea pig ear lobe method would be the most useful model for NICU. It must be noted, though, that these models need constant refinement and standardization to enable more specific identification of compounds causing ICU or NICU or medications to treat these diseases. Therefore, further studies are needed. The development of nonanimal-based reliable screening methods are another future goal.

REFERENCES

Dearman, R. J., Basketter, D. A., and Kimber, I. (1992): Variable effects of chemical allergens on serum IgE concentration in mice: Preliminary evaluation of a novel approach to the identification of respiratory sensitizers. *J. Appl. Toxicol.*, 12:317–323.

Dearman, R. J., Mitchell, J. A., Basketter, D. A., and Kimber, I. (1992): Differential ability of occupational chemical contact and respiratory allergens to cause immediate and delayed dermal hypersensitivity reactions in mice. *Int. Arch. Allergy Appl. Immunol.*, 97:315–321.

Gollhausen, R., and Kligman, A. M. (1985): Human assay for identifying substances which induce non-allergic contact urticaria: The NICU-test. *Contact Dermatitis*, 13:98–106.

Hayes, J. P., Daniel, R., Tee, R. D., Barnes, P. J., Chung, K. F., and Newman Taylor, A. J. (1992): Specific immunological and bronchopulmonary responses following intradermal sensitization to free trimellitic anhydride in guinea pigs. *Clin. Exp. Allergy*, 22:694–700.

Kitagati, H., Fujisawa, S., Watanabe, K., Hayakawa, K., and Shiohara, T. (1995): Immediate-type hypersensitivity response followed by a late reaction is induced by repeated epicutaneous application of contact sensitizing agents in mice, *J. Invest. Dermatol.*, 105:749–755.

Kurup, V. P., Kumar, A., Choi, H., Murali, P. S., Resnick, A., Kelly, K. J., and Fink J. N. (1994): Latex antigens induce IgE and eosinophils in mice. *Int. Arch. Allergy Immunol.*, 103:370–377.

Lahti, A., and Maibach, H. I. (1984): An animal model for nonimmunologic contact urticaria. *Toxicol. Appl. Pharmacol.*, 76:219–224.

Lahti, A., and Maibach, H. I. (1985): Species specifity of nonimmunologic contact urticaria: Guinea pig, rat and mouse. *J. Am. Acad. Dermatol.*, 13:66–69.

Lahti, A., McDonald, D. M., Tammi, R., and Maibach, H. I. (1986): Pharmacological studies on nonimmunologic contact urticaria in guinea pigs. *Arch. Dermatol. Res.*, 279:44–49.

Lauerma, A. I., Fenn B., and Maibach, H. I. (1997): Trimellitic anhydride sensitivity as an animal model for contact urticaria. *J. Appl. Toxicol.*, 17:357–360.

Lauerma, A. I., and Maibach, H. I. (1997): Model for immunologic contact urticaria. In: *Contact Urticaria Syndrome*. Edited by Amin, S., et al., pp. 27–32. CRC Press, Boca Raton.

Obata, H., Tao, Y., Kido, M., Nagata, N., Tanaka, I., and Kurowa, A. (1992): Guinea pig model of immunologic asthma induced by inhalation of trimellitic anhydride. *Am. Rev. Respir. Dis.*, 146:1553–1558.

Patterson, R., Zeiss, C. R., and Pruzansky, J. J. (1982): Immunology and immunopathology of trimellitic anhydride pulmonary reactions. *J. Allergy Clin. Immunol.*, 70:19–23.

Reijula, K. E., Kelly, K. J., Kurup, V. P., Choi, H., Bongard, R. D., Dawson, C. A., and Fink, J. N. (1994): Latex-induced dermal and pulmonary hypersensitivity in rabbits. *J. Allergy Clin. Immunol.*, 94:891–902.

Turjanmaa. K., Alenius, H., Makinen-Kiljunen, S., Reunala, T., and Palosuo, T. (1996): Natural rubber latex allergy. *Allergy*, 51:593–602.

Zeiss, C. R., Patterson, R., Pruzansky, J. J., Miller, M. M., Rosenberg, M., and Levitz, D. (1977): Trimellitic anhydride-induced airway syndromes: Clinical and immunologic studies. *J. Allergy Clin. Immunol.*, 60:96–103.

27

Diagnostic Tests in Dermatology

Patch and Photopatch Testing and Contact Urticaria

Smita Amin and Howard I. Maibach

*Department of Dermatology , University of California, School of Medicine,
San Francisco, California, USA*

Antti I. Lauerma

Department of Dermatology, University of Helsinki, Helsinki, Finland

---

INTRODUCTION

Diagnostic in vivo skin tests are used in dermatology to detect and define the possible exogenous chemical agent that causes a skin disorder, and hence are critical in their scientific documentation. Such chemical agents often cause skin disorders by hypersensitivity mechanisms, which can thus be diagnosed by a provocative test (Lauerma and Maibach, 1995). The anatomical advantage of studying skin disorders is that the skin is the foremost frontier of the human body and therefore easily accessible for testing. Although it has been shown that differences in the reactivity of different skin sites exist, many causative agents may be tested locally on one skin site, thus exposing only limited areas of skin to the diagnostic procedures. Such procedures include patch, intradermal, prick, scratch, scratch-chamber, open, photo, photopatch, and provocative use tests. In cases of some generalized skin reactions, however, systemic exposure to the external agent may be necessary for diagnosis.

TABLE 27.1. *Chemically related skin disorders diagnosable through diagnostic testing*

Disorder	Mechanism*	Test method
Drug eruption	Type I	Prick test or Open Test
		Scratch test
		Intradermal test
		Systemic challenge
	Type II	Patch test
		Systemic challenge
	Type III	Intradermal test
		Systemic challenge
	Type IV	Patch test
		Systemic challenge
	Nonimmunological Systemic challenge	
Allergic contact dermatitis	Type IV	Patch test
		Intradermal test
		Open test or Provocative Use
		Test (Repeated Open Application Test)
Contact Urticaria Syndrome (Immediate contact reaction)	Type I	Open test (Single Application)
		Prick test
		Scratch test
		Scratch-chamber test
	Nonimmunological open test (Single Application)	
Subjective irritation	Unknown	Lactic acid test
		Open test (single application)

*Types I–IV: Coombs–Gell classification of immunological mechanisms.

The value of diagnostic tests is identification of the causative agent, which enables restarting of those chemicals and/or medications not responsible for the eruption. This chapter briefly describes the in vivo test methods used for making diagnoses of skin disorders. The skin disorders in which such tests are useful include drug eruptions, contact dermatitis and immediate contact reactions (contact urticaria), and possibly sensory (subjective) irritation (Table 27.1).

DRUG ERUPTIONS

Drug eruptions are a heterogeneous class of adverse skin reactions due to ingestion or injection of therapeutic drugs. The drug eruptions should ideally be diagnosed through systemic rechallenge because many factors (e.g., systemic drug metabolism) may contribute to the process, and skin tests therefore are not as reliable. Because systemic challenge is not always easy to perform, however, skin tests may precede such challenges, according to reaction type. If skin tests do not provide information about the causative agent and a medication needs to be restarted, the next step is a controlled drug rechallenge, preferably in a hospital environment (Kauppinen and Alanko, 1989).

The choice of provocation protocols depends on the type of reaction involved. Although much work has been directed toward classifying drug eruptions and elucidating their mechanisms, they are still not well understood. Many of them are presumably

TABLE 27.2. *Systemic challenge: Protocol**

The patient should be monitored under hospital conditions, and emergency resuscitation equipment should be available throughout the study.
Especially if the initial drug eruption was strong, challenge should be started at a low dose (i.e., no more than one tenth of the initial dose).
A dose of the suspected drug is given orally in the morning. The patient's skin, temperature, pulse, and other signs are followed at 1 hour intervals for 10 hours and recorded.
If no reactions appear during 24 hours, the challenge is repeated at a higher dose (e.g., one third of the dose) the next morning.
If no reactions appear on days 1 and 2, then on the third morning a full therapeutic dose is given as a third challenge. If necessary, different drug challenges may be repeated every 24 h.

*See Kauppinen and Alanko (Kauppinen and Alanko, 1989) for detailed instructions. The Kauppinen publication provides an equaled clinical experience and offers many valuable short cuts in making scientifically based diagnoses.

mediated by immunological mechanisms, but there are also *nonimmunological drug eruptions*, idiosyncrasies in genetically predisposed persons. In cases of nonimmunological drug eruptions, skin tests usually are negative and systemic provocations are also often negative (Tables 27.2 and 27.3). Immunological drug eruptions may be classified into the four reaction types according to Coombs and Gell (Bruynzeel and Ketel, 1989).

Anaphylactic (Coombs-Gell type I) reactions include *anaphylaxis, urticaria,* and *angioneurotic edema*. They usually are mediated by immunoglobulin E (IgE) antibodies. Penicillin is one well-known causative agent for type I reactions. Prick (Table 27.4) and scratch (Table 27.5) tests are used in diagnosis of type I reactions and are a relatively safe way of detecting the causative agent. Intradermal tests (Table 27.6) may also be used in such cases, although a much larger amount of the antigen is introduced into the body, which makes systemic reactions more likely. Also, in vitro

TABLE 27.3. *Systemic challenge: Precautions**

Challenge is not advisable if the patient has had
   Anaphylaxis
   Toxic epidermal necrolysis (TEN)
   Stevens–Johnson syndrome and/or erythema multiforme
   Systemic lupus erythematosus-like reaction
Extreme care should be exercised if the patient has had
   Urticaria
   Asthma
   Any other immediate-type reaction
   Fixed drug eruption or its most severe form: Generalized bullous fixed drug eruption (special variant of TEN)
Usually performed 1–2 months after the original eruption, except in severe reactions, a longer interval (6 months–1 year) is advisable.
Minimum provocative dose is generally less than one single therapeutic dose, except in cases of severe bullous fixed drug eruption, the initial test dose must be smaller: i.e., 1/10–1/4 of a single therapeutic dose.

*See Kauppinen and Alanko (Kauppinen and Alanko, 1989) for detailed instructions.

TABLE 27.4. *Prick test*

Materials:	(1) Allergens in vehicles.
	(2) Vehicle (negative control).
	(3) Histamine in 0.9% NaCl (positive control).
	(4) Prick lancets.
Method:	One drop of each test allergen, vehicle, and histamine control is applied to the volar aspects of forearms. The test site is pierced with a lancet to introduce the allergen into the skin.
Reading time:	15–30 minutes.
Interpretation:	An edematous reaction (wheal) of at least 3 mm in diameter and at least half the size of the histamine control is considered positive in the absence of such reaction in the vehicle control.
Precautions:	General anaphylaxis not very likely because of the small amount of allergen introduced, but a physician should always be available for such occurrences. The patient should not leave the premises during the first 30 minutes after the test.
Controls:	Required.

tests, such as the radioallergosorbent test (RAST), are used in diagnosis (Bruynzeel and Ketel, 1989). Because type I reactions are potentially life-threatening, systemic challenges (Tables 27.2 and 27.3) (Kauppinen and Alanko, 1989), if done, should be performed with extreme care, starting with very low doses, under hospital conditions. A physician should always be readily available, and the patient should be monitored frequently. *Cytotoxic (type II)* reactions are mediated by cytotoxic mechanisms; quinine and quinidine are examples of causative agents. Patch tests (Table 27.7) may be attempted (Calkin and Maibach, 1993) before systemic challenges (Tables 27.2 and 27.3) (Kauppinen and Alanko, 1989), for example, in the case of *thrombocytopenic purpura* caused by carbromal or bromisovalum (Bruynzeel and Ketel, 1989). *Immune complex-mediated (type III)* reactions include *Arthus* and *vasculitic* reactions. Type III reactions are mediated by immunoglobulins, complement, and the antigen itself, which form complexes. For example, sulfa preparations, pyrazolones, and hydantoin

TABLE 27.5. *Scratch test*

Materials:	(1) Allergens in vehicles.
	(2) Vehicle (negative control).
	(3) Histamine in 0.9% NaCl (positive control).
	(4) Needles.
Method:	A drop of each test allergen, vehicle, and histamine control is applied to the volar aspects of forearms or back, and needles are used to scratch the skin slightly at these sites.
Reading time:	Up to 30 minutes.
Interpretation:	Difficult because of unstandardized procedure. Edematous reaction at least as wide as the histamine control is considered positive in the absence of such reaction in the vehicle control.
Precautions:	As with prick test.
Controls:	Required.

TABLE 27.6. *Intradermal test*

Materials:	(1) Allergens in isotonic solution vehicles. (2) Solution vehicle (negative control). (3) Tuberculin (1 cc) syringes and needles.
Method:	0.05–0.1 ml of allergen solution and vehicle solution is applied intradermally to the skin of the volar aspects of the forearms.
Reading time:	30 minutes 24 hours, and 48 hours.
Interpretation:	Erythematous and edematous reaction at 30 minutes is suggestive of immediate type (type I) allergy in the absence of such a reaction in the vehicle control. Arthus reaction with polymorphonuclear leukocyte infiltration appearing in 2–4 hours, which may progress into necrosis in hours or days, suggests cytotoxic (type III) reaction. Erythema and edema of at least 5 mm in diameter at 48 hours indicates delayed-type hypersensitivity (type IV), for, example, contact allergy.
Precautions:	The risk of general anaphylaxis is higher than in prick or scratch tests because of larger amount of allergen introduced; therefore, a physician should always be available for such occurrences. The risk is greater in asthmatic patients. The patient should not leave the premises during the first 30 minutes after the test.
Controls:	Required.

derivatives have caused vascular purpura via type III mechanisms. Intradermal tests (Table 27.6) may be tried for diagnosis before systemic challenges (Tables 27.2 and 27.3) (Bruynzeel and Ketel, 1989).

Delayed hypersensitivity (type IV) reactions are cell-mediated immune reactions involving the antigen, antigen-presenting cells, and T lymphocytes. Drug reactions of

TABLE 27.7. *Patch test*

Materials:	(1) Allergens in vehicle (e.g., petrolatum, ethanol, water). (2) Vehicles. (3) Aluminum chambers (Finn Chamber), Scanpor tape, and filter papers (for solutions). OR: (1) Ready-made patch test series (TRUE test).
Method:	Patches on tape or ready-made patch test series are applied on intact skin of the back. Filter papers are used for solutions, and 17 μl of allergen in vehicle is used for each patch. Ready-made patch series are applied as is on similar skin sites. The patches are removed after 48 hours.
Reading time:	48 hours; and 96 hours.
Interpretation:	Erythema and edema or more is positive. Distinguishing between allergic and irritant reaction is important. If the reaction spreads across the boundaries of the patch site, the reaction is more likely to be allergic. If the reaction peaks at 48 hours and starts to fade rapidly after that, it may be irritant.
Precautions:	Intense skin reactions possible; these can be treated with topical glucocorticosteroids. Active sensitization possible.
Controls:	Required.

TABLE 27.8. *Photo test*

Materials:	Ultraviolet (UV) radiation source.
Method:	Minimal erythema dose (MED) of UVA or UVB is measured (1) while the subject is taking the suspected medication and (2) after discontinuing the same medication.
Interpretation:	If MED (UVA or UVB) is much lower while the subject is taking the medication, this suggests a photosensitive (phototoxic or photoallergic) reaction to the drug.

this type are often *maculopapular* or *eczematous*, although *photoallergic* reactions and *fixed drug eruptions* are also presumably mediated by type IV mechanisms. Other type IV reactions include some cases of *erythroderma, exfoliative dermatitis, lichenoid* and *vesicobullous eruptions, erythema exudavitum multiforme,* and *toxic epidermal necrolysis.* Type IV reactions may be detected by patch tests with the causative agent (Table 27.7) (Bruynzeel and Ketel, 1989). In the case of a fixed drug eruption, in which the reaction reoccurs in the same skin site every time the drug is ingested, the patch test should be done in that particular skin site for a positive result (Alanko et al., 1987). For photosensitivity reactions, photo (Table 27.8) or photopatch tests (Table 27.9) should be done (Rosen, 1989). A negative patch test does not rule out the possibility that the tested drug may be causative. This is because patch testing involves potential limitations, such as insufficient penetration. In this case a systemic provocation (Tables 27.2 and 27.3) should be considered (Kauppinen and Alanko, 1989).

CONTACT DERMATITIS

Contact dermatitis is commonly divided into irritant contact dermatitis and allergic contact dermatitis. *Irritant contact dermatitis*, the more common of the two, is initiated by *nonimmunological* toxic mechanisms and is not diagnosed by patch testing, whereas allergic contact dermatitis is. In *allergic contact dermatitis* the patient becomes topically sensitized to a low-molecular-weight hapten, and in subsequent topical contact develops an eczematous skin reaction, which is mediated by *delayed*

TABLE 27.9. *Photopatch test*

Materials:	(1) Ultraviolet (UV) radiation source. (2) Patch test materials (see Table 27.8).
Method:	Two sets of patch tests are applied for 48 hours. After removal, one set is irradiated with UVA at a dose below minimal erythema dose (MED) ($5–10$ J/cm^2 or 50% of MED, whichever is smaller), and the other set is protected from UV dose.
Reading time:	48 and 96 hours.
Interpretation:	Reaction only at irradiated site suggests photoallergy. Reaction at both sites suggests contact allergy. Reaction at both sites and a much stronger reaction at the irradiated site suggests both contact allergy and photoallergy.
Controls:	Required.

TABLE 27.10. *Provocative use test (open test or repeated open application test)*

Materials:	(1) Allergen in vehicle (petrolatum, ethanol, water).
	(2) Vehicle.
	(3) Cotton-tipped applicators or other devices to spread the preparations.
Method:	Patient applies allergen and vehicle on antecubital fossa (outpatient) or shoulder regions of upper back, two times a day for 14 days or until a positive reaction appears.
Reading time:	Patient reports whether positive reaction appears. At 7 and 14 days the patient returns for reading of the test site.
Interpretation:	Erythema and edema or more is positive.
Precautions:	Active sensitization possible but not yet documented,
Controls:	May be required.

hypersensitivity (type IV) mechanisms. Allergic contact dermatitis is diagnosed with patch tests (Table 27.7), intradermal tests (Table 27.6), or open tests (repeated application) (Table 27.10). Of these three methods, patch testing is the most common and standardized. The problems involved in patch testing are insufficient penetration of the allergenic compound, which may result in false-negative results, and irritation from the test compound, which may cause a false-positive result. Also, patch testing may cause a worsening of eczema in other skin sites (Excited Skin Syndrome) or active sensitization to patch compounds (Fischer and Maibach, 1990).

Two widely used methods for patch testing exist: the Finn Chamber and TRUE Test methods. Both have been shown to be reliable, especially when stronger reactions (contact allergies) are investigated (Ruhnek-Forsbeck et al., 1988). The TRUE Test is somewhat easier to handle, as it is ready to use; however, the Finn Chamber method provides more flexibility for the dermatologist and the allergist to test substances not in routine patch test use. Regardless of the test method, the most important factor in successful patch testing is the experience and skill of the interpretator. A standard patch test series is shown in Table 27.11. It is the standard series of the International Contact Dermatitis Research Group and the European Environmental and Contact Dermatitis Research Group (Andersen et al., 1991). A standard patch test series has been compiled to represent the most commonly encountered contact allergens, and it is meant to act as a screening tray. Its content is subject to change due to research findings about contact allergy (Andersen et al., 1991). A multitude of other patch test series are available when the causative agents of the individual patient's contact dermatitis are known better; these include, for example, patch test series for preservatives, rubber chemicals, topical drugs, and clothing chemicals. There also are special series to investigate occupational contact allergies in, for example, dental personnel or hairdressers.

Patch tests should be applied on the back for 48 hours and be read after removal. A second reading 24–48 hours after patch removal is necessary, as irritant reactions, which are often easily misinterpreted as allergic, often tend to fade during the third and fourth days, whereas allergic reactions tend to persist. Additionally, with some allergens, such as corticosteroids and neomycin, late reactions often occur, possibly

TABLE 27.11. *Standard patch test series**

Potassium dichromate	0.5% pet.
Neomycin sulfate	20.0% pet.
Thiuram mix	1.0% pet.
p-Phenylenediamine free base	1.0% pet.
Cobalt chloride	1.0% pet.
Benzocaine	5.0% pet.
Formaldehyde	1.0% aq.
Colophony	20.0% pet
Quinoline mix	6.0% pet.
Balsam of Peru	25.0% pet.
PPD-black rubber mix	0.6% pet.
Wool alcohols (lanolin)	30.0% pet.
Mercapto mix	2.0% pet.
Epoxy resin	1.0% pet.
Paraben mix	15.0% pet.
p-tert-Butylphenol-formaldehyde resin	1.0% pet.
Fragrance mix	8.0% pet.
Ethylenediamine dihydrochloride	1.0% pet.
Quaternium 15	1.0% pet.
Nickel sulfate	5.0% pet.
Kathon CG	0.01% aq.
Mercaptobenzothiazole	2.0% pet.
Primin	0.01% pet.

*Of the International Contact Dermatitis Research Group and the European Environmental and Contact Dermatitis Research Group.
pet. = petrolatum vehicle; aq. = aqueous vehicle.

because of low percutaneous penetration. Therefore, a third reading approximately 1 week after patch application may be advisable, although this may be difficult to do routinely in practice. Intradermal testing (Table 27.6) has recently been shown to be of value in diagnosing hydrocortisone contact allergy (Wilkinson et al., 1991). See Herbst for a review of intradermal testing for allergic contact dermatitis (Herbst et al., 1993). Open tests or repeated open-application tests (Table 27.10) are not as sensitive as patch or intradermal tests, possibly because of insufficient penetration of the compound under unoccluded conditions (Hannuksela and Salo, 1986; Hannuksela, 1991).

CONTACT URTICARIA SYNDROME: IMMEDIATE CONTACT REACTIONS

Contact urticaria syndrome includes a group of skin reactions (i.e., immediate contact reaction) that usually appear within 1 hour of skin contact with the causative agent. Immediate skin reactions are divided into immunological (IgE-mediated) and nonimmunological immediate contact reactions. The symptoms range from mere itching and tingling to local wheal and flare. In cases of intense sensitivity, a generalized urticaria, systemic symptoms, and even anaphylaxis (contact urticaria syndrome) may occur (Lahti and Maibach, 1992).

TABLE 27.12. *Scratch-chamber test*

Materials:	(1) Scratch test materials (see Table 27.5).
	(2) Chambers.
Method:	As with scratch test, but scratch sites are covered with aluminum chambers for 15 minutes.
Reading time:	30 minutes.
Interpretation:	See Table 27.5.
Precautions:	As with prick and scratch tests.
Controls:	Required.

Immunological immediate contact reactions usually are urticarial, although they may range from mere tingling in the skin to a generalized anaphylactic reaction in the whole body. Immunological immediate reactions are Coombs-Gell type I reactions mediated mainly via allergen-specific IgE bound to skin mast cells. Coupling of membrane-bound IgE by allergen causes mast cells to liberate histamine, which with other inflammatory mediators makes skin vessels permeable, and edema (urticaria) results. The sensitization in IgE-mediated contact urticaria may occur through skin or possibly the respiratory or gastrointestinal tract. Exposure through skin is the most likely route in occupational latex allergy in health personnel. The provocative in vivo methods usually performed first are a prick test (Table 27.4), scratch test (Table 27.5), and scratch-chamber test (Table 27.12); however, the test method simulating the clinical contact situation more realistically is the open application test (single application) (Table 27.13). A previously affected skin site is more sensitive to immunological skin reactions than a nonaffected site. In addition to in vivo methods, the diagnosis of immunological immediate contact reactions can be done with RAST, which detects antigen-specific IgE molecules from the patient's serum (Lahti and Maibach, 1992).

Nonimmunological immediate contact reactions range from erythema to urticaria and occur in persons not sensitized to the compounds (Lahti and Maibach, 1992). Nonimmunological contact reactions are probably more common than immunological contact reactions. They are possibly due to the causative agent's ability to release inflammatory mediators, such as histamine, prostaglandins, and leukotrienes, from skin cells without the participation of IgE molecules. Agents capable of causing

TABLE 27.13. *Open test (single application) for contact urticaria syndrome (immediate contact reactions)*

Materials:	(1) Allergen in vehicle (petrolatum, ethanol, water).
	(2) Vehicle.
	(3) Cotton-tipped applicators or other devices to spread the preparations.
Method:	Allergen and vehicle are applied to skin.
Reading time:	Up to 1 hour.
Interpretation:	Urticarial reaction is positive.
Precautions:	See Table 27.4.
Controls:	Required to aid in discriminating immunological (ICU) from nonimmunological contact urticaria (NICU); in NICU the reaction will be noted in most controls.

TABLE 27.14. *Lactic acid test: Model for sensory irritation*

Materials:	1. Facial sauna
	2. 5% lactic acid in water
	3. Vehicle (water)
	4. Soap, paper towels, cotton-tipped applicators
Method:	Facial area below eyes is cleansed with soap, paper towels, and water, rinsed with water, and patted dry. Face is exposed to sauna heat for 15 minutes. Moisture of face is blotted away. Lactic acid in water is rubbed on one side of face (cheek) and water on other. Face is exposed to sauna again.
Reading time:	2 and 5 minutes after second sauna exposure.
Interpretation:	Any subjective sensation is graded by patient: 0 = none; 1 = slight; 2 = moderate; 3 = severe. If cumulative score of two time points is 3 or more, patient is "stinger."
Precautions:	Irritation may occur.

nonimmunological contact reactions are numerous; the most potent and best-studied agents are benzoic acid, sorbic acid, cinnamic aldehyde, and nicotinic acid esters. The test for diagnosis of nonimmunological contact urticaria is the open application test (single application) (Table 27.13).

SUBJECTIVE IRRITATION

Although *sensory (subjective) irritation* is not fully characterized, there is evidence for a group of such persons, known as "stingers" (Maibach et al., 1989). The lactic acid test (Table 27.14) has been used experimentally to distinguish between "stingers," who more often have subjective irritation, and "nonstingers" (Lammintausta et al., 1988).

REFERENCES

Alanko, K., Stubb, S., and Reitamo, S. (1987): Topical provocation of fixed drug eruption. *Br. J. Dermatol.*, 116:561–567.

Andersen, K., Burrows, D., and White, I. R. (1991): Allergens from the standard series. In: *Textbook of Contact Dermatitis.* Edited by R. J. G. Rycroft, T. Menne, P. J. Frosch, and C. Benezra, pp. 416–456. Springer-Verlag, Berlin.

Bruynzeel, D. P., and Ketel, W. G. V. (1989): Patch testing in drug eruptions. *Semin. Dermatol.*, 8:196–203.

Calkin, J. M., and Maibach, H. I. (1993): Delayed hypersensitivity drug reactions diagnosed by patch testing. *Cont. Derm.*, 29:223–233.

Fischer, T., and Maibach, H. (1990): Improved but not perfect patch testing. *Am. J. Cont. Derm.*, 1:73–90.

Hannuksela, M., and Salo, H. (1986): The repeated open application test (ROAT). *Cont. Derm.*, 14:221–227.

Hannuksela, M. (1991): Sensitivity of various skin sites in the repeated open application test. *Am. J. Cont. Derm.*, 2:102–104.

Herbst, R. A., Lauerma, A. I., and Maibach, H. I. (1993): Intradermal testing in the diagnosis of allergic contact dermatitis: a reappraisal. *Cont. Derm.*, 29:1–5.

Kauppinen, K., and Alanko, K. (1989): Oral provocation: Uses. *Semin. Dermatol.*, 8:187–191.

Lahti, A., and Maibach, H. I. (1992): Contact urticaria syndrome. In: *Dermatology*. Edited by S. L. Moschella and H. J. Hurley, pp. 433–440. Philadelphia, W. B. Saunders.

Lammintausta, K., Maibach, H. I., and Wilson, D. (1988): Mechanisms of subjective (sensory) irritation: Propensity to non-immunologic contact urticaria and objective irritation in stingers. *Dermatosen*, 36: 45–49.

Lauerma, A. I., and Maibach, H. I. (1995): Provocative tests in dermatology. In: *Provocation in Clinical Practice*. Edited by S. L. Spector, pp. 749–760. Marcel Dekker, New York, 1995.

Maibach, H. I., Lammintausta, K., Berardesca, E., and Freeman, S. (1989): Tendency to irritation: Sensitive skin. *J. Am. Acad. Dermatol.*, 21:833–835.

Rosen, C. (1989): Photo-induced drug eruptions. *Semin. Dermatol.*, 8:149–157.

Ruhnek-Forsbeck, M., Fischer, T., Meding, B., Petterson, L., Stenberg, B., Strand, A., Sundberg, K., Svensson, L., Wahlberg, I. E., Widstrdm, L., Wrangsjd, K., and Billberg, K. (1988): Comparative multi-center study with TRUE Test and Finn Chamber patch test methods in eight Swedish hospitals. *Acta. Dermatol. Venereol. (Stockh)*, 68:123–128.

Wilkinson, S. M., Cartwright, P. H., and English, J. S. C. (1991): Hydrocortisone: An important cutaneous allergen. *Lancet*, 337:761–762.

SECTION V
MISCELLANEOUS

28

Wound Healing Products

David W. Hobson

DFB Pharmaceuticals, Inc., San Antonio, Texas, USA

One writes of scars healed, a loose parallel to the pathology of the skin, but there is no such thing in the life of an individual. There are open wounds, shrunk sometimes to the size of a pin-prick but wounds still. The marks of suffering are more comparable to the loss of a finger, or the sight of an eye. We may not miss them, either, for one minute in a year, but if we should there is nothing to be done about it.

—**F. Scott Fitzgerald,** *Tender is the Night,* bk. 2, chapter 11 (1934)

INTRODUCTION

Less metaphorically, wounds are breaks in the continuity of the soft parts of any body structure, and are commonly defined as "An injury, especially one in which the skin or other external surface is torn, pierced, cut, or otherwise broken" (*American Heritage Dictionary of the English Language*, 1992). Wounds to the skin result from a myriad of causes, involve damage to the structural integrity of the epidermis, and may penetrate into the dermis as well as underlying tissues. Wound severity can range from being life threatening to cosmetically undesirable. In some cultures, wounds may even be inflicted intentionally. In any case, wounds can become quite problematic if not properly cared for, particularly during the healing process. Chronic wound conditions can be quite debilitating and cause much suffering. Wound treatments have always been developed to promote healing or to make the chronic wound more bearable. The current state of the art in wound healing product development still shares these same objectives with the past; however, over the past decade our knowledge of the biological processes occurring in wound healing has been rapidly improving. As a result, there has been a marked increase in the development and marketing of topical wound care products of various types. Along with the increase in wound care product development, a related concern about the exposure of broken skin to toxicants and their effects on wound healing is emerging. Although this chapter's primary focus on dermatoxicity relates to the development of wound care products, this chapter, wherever relevant, will also identify significant toxicants that may enter the wound environment from sources other than products.

Wound treatments have been available since antiquity. The Ebers Papyrus dating from ancient Egypt describes many different wound healing prescriptions and dressings containing various combinations of vegetable, animal, and mineral ingredients (Ebbell, 1937). The treatment of a life-threatening wound is described in 2 Kings 20:7 of the Old Testament of the Bible: "And Isaiah said, 'Take a lump of figs.' And they took and layed it on the boil, and he recovered." No doubt, various formulated treatments for wounds have existed since before the written word, and many have been lost to antiquity.

Modern wound healing products are currently developed using various principles scientifically proven during the past century. These include the principles of antisepsis of Baron Joseph Lister and others at the end of the 19th century (Lister, 1867), and the principles of moist wound healing demonstrated in the mid 20th century (Winter, 1962). Most recently, during the past decades, biotechnology-derived products of different types have become routinely available for clinical application. Use of these treatments will no doubt increase in the future.

During the past decade and continuing well into the next millennium, scientists will be involved in finding solutions to a myriad of medical conditions confronting increasing proportions of worldwide populations age 65 or older (Kinsella and Gist, 1998). The proportions of the population that require some form of caregiving is also increasing worldwide, particularly in developing countries. By 2025 it is estimated that the worldwide population over 60 will be about 1.2 billion, with 839 million being in developing countries (Velkoff and Lawson, 1998). The proportions of these individuals that will live to the age of 85 is also increasing worlwide. The United States population is currently about 272 million (U.S. Census Bureau, 1999). Over 35 million of these are adults over the age of 65, and over 4 million are over the age of 85, and these numbers and proportions are increasing steadily. While skin care and aging present significant concerns for this portion of our population, the development of new and more effective products to promote wound healing in these individuals may be an equally, if not more significant, new involvement for toxicologists well into the future. Older adults have a greater likelihood of developing chronic, nonhealing wounds, and the dimension of the worldwide wound care problem for these persons is becoming epidemic. Our knowledge of the toxicity of various agents on the wound healing process is quite limited, particularly as it applies to aging patients and the more chronic nature of their wounds. The toxicology of wound healing products should, therefore, be considered an emerging discipline within the science when taken in this context, with many important opportunities for the advancement of our knowledge.

WOUND CLASSIFICATION

A description for a wound to the skin usually involves at least the following:

1. Anatomical location (right hand, left forearm, abdomen, sacrum, etc.)
2. Surface area
3. Depth of injury
4. Etiology (laceration, ulcer, burn, etc.)

Often, based on descriptive information, wounds are then classified in other ways. For example, "surgical" wounds—those caused intentionally—usually heal relatively quickly under proper care. These wounds generally heal by first intention, as the wound edges are held together by sutures or other means. In contrast, "nonsurgical" wounds can be caused in many different ways, including mechanical trauma (e.g., lacerations, abrasions, etc.), thermal burns (flame, scalds, frostbite), radiation burns, chemical burns, arterial insufficiency, venous insufficiency, pressure ulcer (capillary insufficiency), incontinence (stool burn, urine scald), toxic ulceration (insect bites, spider bites, bacterial toxins), and various combinations of the above. Nonsurgical wounds usually heal by second intention (formation of granulation tissue for acute, healing wounds), or these wounds may become chronic, nonhealing wounds. Due to many different factors, the probability for developing a chronic, nonhealing wound tends to increase with age.

Wounds can also be classified by the extent of tissue damage and involvement of underlying structures. For example, thermal burns can be classified as first, second, or third degree, with pressure ulcers often classified as either stage I, stage II, stage III, or stage IV (see Table 28.1). The progressive depth and extent of tissue involvement in the different stages are shown in Figures 28.1, 28.2, 28.3, 28.4, and 28.5. Other classification schemes describe appearance and treatment types as well (van Rijswijk, 1997).

Probably the most straightforward and often-used classification for dermatoxicology research is the depth of skin damage involved (Figure 28.6). Thus, wounds can be classified as (1) superficial, (2) partial thickness, (3) full thickness, or (4) complex:

1. Superficial wounds are most often caused by shearing friction, mild burns, and shallow cuts and scrapes to the stratum corneum. These wounds heal readily, in most cases, by regeneration of epithelial cells across the wound surface. There is a loss of contact inhibition, and keratinocytes and other epidermal cell types migrate across the surface of the wound. Keeping the wound environment moist facilitates this migration and can reduce the time to closure. Closure generally completes in 5–7 days. These wounds usually heal without leaving a scar, and the skin accessory structures are left intact. On occasion, a superficial wound

TABLE 28.1. *Examples of the classification of wound types by the extent of tissue damage or involvement of underlying structures*

Wound type	Classification	Description
Thermal burn	First Degree	Erythema and pain evident (e.g., sunburn).
	Second Degree	Erythema marked by blisters (e.g., scald by hot liquid).
	Third Degree	Destruction of the epidermis and dermis. Underlying tissue may also be damaged (charring of tissue by flame).
Pressure ulcer	Stage I	Erythema not resolving in 30 minutes of pressure relief. Epidermis intact. Reversible with intervention.
	Stage II	Partial thickness loss of skin layers involving epidermis and possibly dermis but not penetrating through the dermis.
	Stage III	Full-thickness tissue loss extending through the dermis to involve subcutaneous tissue.
	Stage IV	Deep tissue destruction extending through subcutaneous tissues to fascia and possibly involving muscle, joint, and/or bone tissues.

may have some underlying tissue trauma and there can be an iceberg-like effect. Careful assessment can identify the depth of trauma. Underlying trauma can result in a chronic wound condition and greater depth of involvement.

2. Partial thickness wounds exhibit damage into the dermis with accessory structures generally spared and follow a healing pattern of cellular migration similar to superficial wounds. When blood vessels are damaged, an eschar or scab may develop. These wounds also develop as the result of the same types of trauma that cause superficial wounds, except greater depth of insult occurs. Pressure can also produce these wounds, resulting in a pressure ulcer or bed sore. Partial thickness wounds can be acute healing or chronic, depending on the patient and the degree of vascular damage.

3. Full thickness wounds include those involving the entire epidermis and dermis, and also may involve subcutaneous structures. These wounds heal by "primary intent" when created surgically and kept clean and moist. Full thickness wounds

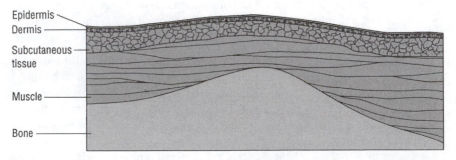

Epidermis
Dermis
Subcutaneous tissue
Muscle
Bone

FIG. 28.1. Intact, healthy skin tissue cross-section showing epidermis, dermis, subcutaneous tissue, muscle, and bone.

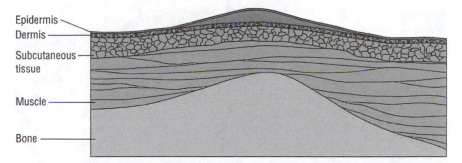

FIG. 28.2. Stage I pressure ulcer characterized by intact skin exhibiting erythema that is warm and tender to the touch. The ulcer is reversible at this stage with minimal intervention.

that heal by primary intent generally have smooth edges and there is little or no tissue loss during the healing process. Delayed primary intent wounds are full thickness wounds that are sutured or stapled together following some delay after the injury; they may be contaminated, have some tissue loss, and exhibit some risk of infection. As indicated previously, secondary intent wounds generally have some tissue loss, have irregular edges, may be necrotic, exhibit high microbial counts, and contain foreign-body debris. These wounds can be acute healing or chronic, nonhealing, depending on the level of tissue trauma and vascular damage. The health of the patient can also significantly affect whether this type of wound becomes chronic or not. Secondary intent wounds exhibit the classical phases of wound healing in essentially all cases, except when they are chronic and the wound may become arrested in the early phases of healing.

In addition to depth of involvement, location is also important. For example, it is well known that lower-extremity ulcerations are the most frequently occurring chronic wound type (Kane and Krasner, 1997).

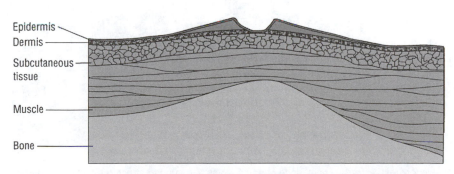

FIG. 28.3. Stage II pressure ulcer showing partial thickness skin lesion involving the epidermis and possibly the dermis. Clinical presentation of this stage is a distinct compromise of skin integrity resulting in a blister, abrasion, or shallow crater. There is more defined erythema and pain than stage I and no necrotic tissue.

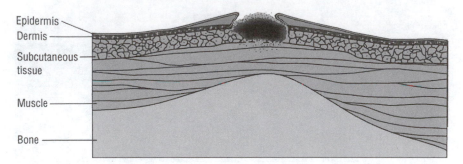

FIG. 28.4. Stage III pressure ulcer characterized by full thickness skin loss and damage to the subcutaneous tissue that may extend down to, but not through, underlying muscle. These lesions present clinically as deep craters, often with sinus tract formation, wound exudate, infection, and some necrosis. The base of the stage III ulcer is generally not painful.

Other factors to consider in the classification or characterization of a wound, particularly chronic wounds, include necrosis and infection. In the United States and probably worldwide, the number of palliative-care-status patients with chronic, nonhealing, infected wounds grows. Patients with leg ulcers are particularly increasing, and the pathology of these lesions is often quite complex (Browse and Burnand, 1982; Gogia, 1995; McCulloch, 1995). These patients often have compromised immune systems, decreased circulatory function, poor nutritional status, incontinence, etc., that act individually or in combination to produce wounds for which complete healing is essentially impossible. The rare occasions where improvement in healing can be obtained prove very costly to maintain. This is an area where new products are increasingly in demand to help improve the quality of life for these individuals.

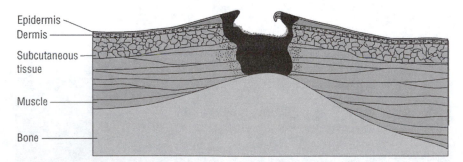

FIG. 28.5. Stage IV pressure ulcer is the most damaging stage characterized by full thickness skin loss with extensive damage to underlying tissues, support structures and considerable necrosis. The true extent of the underlying lesion may not be discernible from the surface appearance. Overlying tissue may fold in over the edges, giving the appearance of a smaller wound. Sepsis, infection, undermining, and tunneling are all potential risks at this stage.

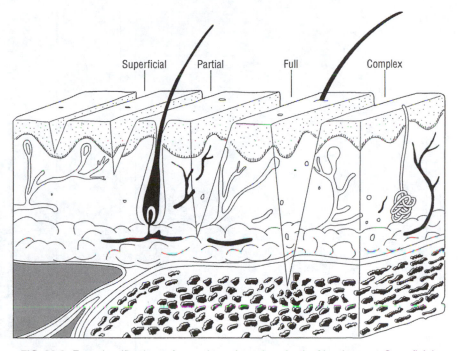

FIG. 28.6. Four classifications of wound type based on depth of involvement. Superficial: describes a wound that does not penetrate into the dermis; partial thickness: describes a wound that penetrates the dermis but not the underlying facia; full thickness: a wound that penetrates the dermis and through the underlying facia, complex thickness: wounds penetrate the dermis, underlying facia, and into the underlying bone.

Other classification schemes are used or proposed, including schemes based on tissue color, and as our knowledge of the treatment of various wound types continues to increase, no doubt new wound classification schemes will be developed.

THE WOUND HEALING PROCESS

As indicated above, wounds of various types that heal by normal healing processes are considered acute healing wounds. Wounds that do not heal readily and that demonstrate impaired healing are typically considered chronic or chronic, nonhealing wounds. The process of wound healing usually is described as it applies to acute healing wounds; research into the nature of various factors that impair this process, resulting in impaired healing or chronic nonhealing conditions, is relatively new. In fact, a complete and detailed understanding of the acute wound healing process is still the subject of much current research.

The wound healing process for acute-healing wounds includes three general phases: (1) inflammatory, (2) proliferative, and (3) remodeling. Figure 28.7 shows the general

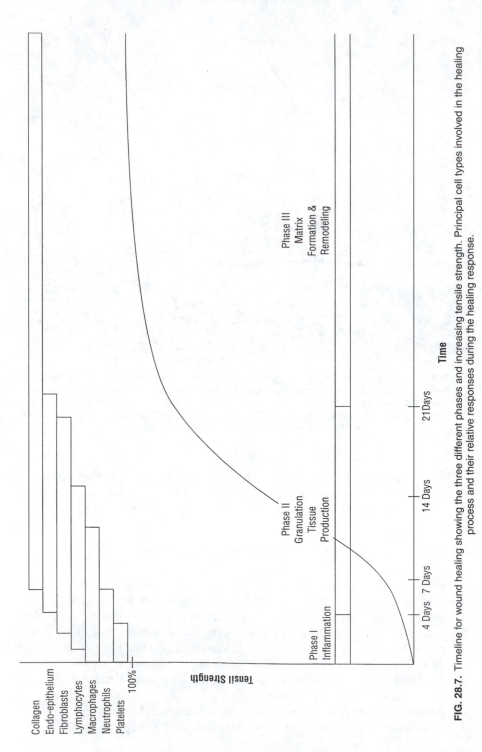

FIG. 28.7. Timeline for wound healing showing the three different phases and increasing tensile strength. Principal cell types involved in the healing process and their relative responses during the healing response.

relationships between these phases, the relative times involved, and some of the principal events and cell types involved. The principal events occurring in these phases are as follows:

Inflammatory Phase

This phase usually begins with an inflammatory response to injury due to vascularized tissue involvement. The cardinal signs of inflammation include heat, redness, swelling, and pain. This phase is known to involve the following events:

A. Activation of platelets. This results in platelet aggregation, initial vasoconstriction, growth factor release, hemostasis, fibrin formation.
B. Eschar (scab) formation. Fibrin combines with blood to form eschar, and epithelial cells migrate below the eschar.
C. Chemotaxis. Macrophages and neutrophils migrate into the wound site in response to chemical signaling.
D. Exudate production. Vasodilation, with vascular escape of cells and proteins, produces wound exudate. The inflammatory response is characterized by the production of exudate fluid.
E. Leukocyte migration. White cell activity content and activity changes the color and consistency of exudate. For example, pus is primarily fluid whereas slouch contains solubilized necrotic tissue produced by phagocytosis and enzymatic digestion of necrotic tissue by leukocytes (autolytic débridement). Neutrophils phagocytize and release proteases to remove necrotic debris. Monocytes also release proteolytic enzymes as well as chemotactic and growth factors that promote fibroplasia and angiogenesis.

The inflammatory phase requires all activities to occur with proper timing and to occur simultaneously to the greatest degree possible.

Chronic inflammation of a wound occurs when one or more aspects of the inflammatory phase is inhibited or arrested in some fashion. Chronic inflammation is often confused with infection because it is typically caused by infection, the presence of foreign matter, or repeated damage to the wound bed. In some cases, attempts to speed healing with particular products or product combinations may result in slower healing (e.g., particulate material from wound dressings, vigorous cleansing, adherent dressings, or cytotoxic agents used to prevent infection). Chronic inflammation can also be produced by malnutrition or decreased vascularity (e.g., venous insufficiency). Chronic inflammation is often characterized by a lack of cardinal signs—the presence of granulomas, adhesions, and the production of transudate with an absence of cells.

Proliferative Phase

This phase begins with the production of granulation to fill in dermal defects, and then continues with

angiogenesis to provide vascularity to the granulating tissue;
contraction to minimize the amount of granulation tissue required for closure;
(ends with) epithelialization to fill in epidermal defects.

In some cases, grafting of skin may be required to rapidly cover granulation tissue. Proper timing of the granulation and epithelialization processes is necessary for a smoothly healing surface.

A. Granulation tissue formation: Fibroblasts enter the wound site following chemo-tactic signaling and begin to proliferate and to produce growth factors. Fibroblasts secrete numerous substances to create matrix, including the fibers collagen, elastin, and reticulin:

Collagen forms a triple helix and provides strength.
Elastin forms wavy fibers and provides extensibility.
Reticulun, a ground substance, contains glycosaminoglycans and glycoproteins (laminin and fibronectin) that prevent water from moving freely, cement cells, and that impart a gel quality, respectively.

B. Angiogenesis begins in the new matrix produced by the fibroblasts. Angioblasts are recruited into the wound site and begin to form temporary blood vessels, causing redness. These new blood vessels leak a little and impart a shiny appearance to the wound bed. This process forms a reddish-pink, bud-like tissue mass termed "granulation tissue." Due to toxic or physical effects, cleansing and dressings can easily damage granulation tissue.

C. Wound contraction. The wound begins to shrink in size as it closes. Contractile elements in fibroblasts (myofibroblasts) provide the principal contractile force. This process reduces amount of granulation tissue needed and increases the tensile or breaking strength of the newly formed skin tissue. Wound contraction can be a problem on flexor surfaces, as contraction may damage newly formed tissue; it can also cause cosmetically unacceptable scars. In some cases, the use of compression garments can prevent excessive scarring.

D. Epithelialization. Epithelial cells migrate into and over the wound site and form new skin. Epidermal cells are the principal cell types. The formation of basement membrane and fibronectin deposition is necessary for attachment. Epithelial cells can be easily displaced deliberately or accidentally before attachment. Deliberate removal can prevent premature epithelialization of a defect or prevent repair of epiboly. Accidental removal can occur during cleansing and dressing and may be the result of mechanical or toxic effects. A loss of contact inhibition stimulates proliferation. Cells line skin appendages and wound edges as the process proceeds from the wound edge as well as from the appendages. Epithelialization creates a barrier to trauma and infection. The presence of eschar slows epithelialization, making débridement by surgical, mechanical, or chemical (pharmaceutical) means desirable in many cases.

Remodeling Phase

This phase completes the healing process and is directed toward consolidating and organizing the tissue structure to restore normal function to the area and for strength. In this phase, the initial random fiber orientation and weakness is "remodeled" to a more normal and organized orientation. The rates of fiber production and breakdown are more balanced in this phase. Remodeling of collagen along lines of stress is an important part of this process. Reduction in fiber density may occur where less stress occurs. Further proliferation of different cells and contraction fills in defects. The number of fibroblasts decreases, as does the number of capillaries. The wound becomes more white in appearance and avascular. Accessory structures may be lost and the repaired skin may exhibit dryness, little or no temperature regulation, no hair, and may also lack some sensation.

The wound healing process generally proceeds normally in most individuals, and it has been shown clearly by (Winter, 1962) that it can be improved over the use of dry, sterile dressings if the wound bed is maintained in a moist condition; however, many factors can complicate, slow, or even arrest the healing process. A few of these, of interest to toxicologists, are presented in succeeding paragraphs.

FACTORS THAT COMPLICATE WOUND HEALING

Wound healing rate and quality can be complicated by a number of factors. These factors can be placed into two broad categories: (1) factors related to the patient, and (2) factors related to wound management.

1. Patient-related factors can include the following;

    Various forms of vascular disease that reduce tissue perfusion, especially diabetes mellitis.

    Poor nutrition (particularly amino acids, protein, and vitamins A, C, copper and zinc, etc.).

    Infection.

    Dementia, debilitation, decreased sensation.

    Incontinence or vomiting due to bowel disease or cancer therapy.

    Skin aging (decreased moisture content makes skin more brittle, more slow to heal, with increased susceptibility to bruising and tearing; with less elasticity, the structure is less cohesive, and pressure, shear, and friction forces produce greater effects).

2. Wound management–related factors include the following:

    Improper dressing removal (particularly wet-to-dry dressings that remove eschar and granulating tissue).

    Too vigorous wound cleansing (use of too great a pressure stream or too frequent cleansing that removes granulating tissue).

Toxic effects of drugs, dressings, or cleansing agents: NSAIDs, corticosteroids, immunosuppressives, antineoplastic agents, anticoagulants, antimicrobials, chelators (penicillamine), iodine, hypochlorite, hydrogen peroxide, and silver compounds, etc.

The remainder of this chapter focuses primarily on the toxic effects of wound treatments, to include cleansing agents, different types of pharmaceutical products, wound cleansing agents, and other wound management issues related to dermatoxicity. From the previous information, it should be clear that optimal wound healing depends on coordination of the events in the three processes of (1) inflammation, (2) proliferation, and (3) remodeling.

When considering the adverse effects of different types of agents on the wound healing process, it will be important to keep in mind the following well-established principles:

1. A dry wound bed slows epithelialization (moist-, protected-, body-temperature-healing is optimal).
2. Healing rates can be enhanced following eschar removal (débridement).
3. Too forceful cleaning and too frequent dressing changes may damage granulation tissue.
4. Too forceful cleansing and adherent dressings may damage epithelialization.
5. Reduced rates of healing can be the result of patient- and therapy-related factors.
6. Aging skin is often more susceptible to damage.

DERMATOXICITY CONSIDERATIONS FOR WOUND HEALING PRODUCTS

There are several types of products available commercially to assist in different aspects of wound treatment. These include wound cleansers, débriding agents, antimicrobials, dressings of various types, wound healing accelerators or promoters, etc. Succeeding paragraphs describe several of the more prominent product types, along with their principal usage and with examples of some of the toxicological concerns that should be considered in their development and usage. A complete discussion of the toxicological and pharmacological considerations encountered for each product type is beyond the scope of this text; however, this chapter attempts to identify and describe some of the more prominent issues. The references at the end of this chapter are provided for more detail on specific aspects of each of these product types.

Wound Cleansers

Wound cleansing is the use of fluids to remove inflammatory agents from the wound surface. This usually is accomplished using various types of solutions to cleanse wounds prior to the application of a topical dressing. These agents are intended to wash away dead tissue and foreign particles from the wound bed, as the presence of dead tissue (black/gray necrotic tissue or yellow fibrin slough), as well as foreign particles, may interfere with the healing process and increase the risk of wound infection. Cleansing agents should not be confused with débriding agents that are generally applied and left in situ to facilitate removal of slough or necrotic tissue.

While isotonic saline is perhaps the most simple of the wound cleansing agents, because of the limited contact time between the wound and cleasing solution, it is not necessary that the solution be completely isotonic. To enhance the clinical efficacy of wound cleansing agents, commercial wound cleansing formulations typically contain surface-active agents or surfactants to improve the removal of wound contaminants.

The surfactants used in commercial wound cleansing formulations are categorized by the nature of their ionic charge in solution (i.e., cationic, anionic, or nonionic). Surfactants of different types are known to be cytotoxic to various cell types in vitro and to delay wound healing and inhibit host defenses to infection in vivo, depending on the concentration in solutions (Hellewell et al., 1993; Wright and Orr 1993). It is important when developing these solutions to carefully balance the capacity of the surfactant needed to cleanse wounds against the level at which toxicity occurs to inhibit wound healing processes. The concentrations of surfactants in these solutions vary but are on the order of generally 5% or less by weight of the product.

Wound cleansing products may contain other components as well, including antimicrobial agents such as chlorhexidine gluconate, povidone iodine, benzethonium chloride, benzalkonium chloride, imidurea, etc., and wound healing promotional agents, such as aloe vera, zinc, copper-peptide complex, etc. These agents, if present in too great a quantity, may also increase the toxicity of the solution to healing processes (Hellewell et al., 1993; Hellewell et al., 1997]. Ingredients such as humectants, moisturizers, emollients, fragrances, and preservatives may also be included in products used for cleansing wounded skin.

Skin cleansers, incontinence cleansers (those designed for cleaning the perineal area) and wound cleansers are differently formulated. Skin cleansers may contain detergents to remove surface dirt and debris, whereas incontinence cleansers will usually contain emollients and moisturizers to help prevent skin breakdown from repeated incontinent and cleansing episodes. Wound cleansers must be carefully formulated to be capable of removing necrotic debris and foreign matter while causing little or no harm to healing tissues and the cells involved in the healing process.

The knowledge base for assessing the toxicity of various types of wound cleansing solutions is currently limited to standardized in vivo irritation and sensitization tests, to include broken skin in some instances, and in vitro tests using various models and interpretive schemes. Although several studies have been conducted to compare and contrast the cytotoxicity of various wound cleansing formulations using different models, carefully conducted clinical studies or in vivo studies to validate these models have not been done. Recommendations for use of these in vitro results as a guide for the selection and use of a wound cleansing product must be taken quite cautiously, if at all. Many of these in vitro studies are conducted by scientists not properly trained in toxicology, and may be used to promote a particular product. Studies comparing the safety of various wound cleansing solutions are often muddled with opinion and with very limited in vivo clinical data to support their conclusions. One statement espoused popularly to wound care practitioners with respect to proper wound cleanser selection: "Don't do to a wound what you would not do to your own eye" (Rodeheaver, 1990) exemplifies a clear lack of understanding of the underlying toxicology and pathology involved in surfactant damage to tissues in the wound bed versus those of the eye. Although quite well intentioned, the concept could be misleading and is without much true toxicological or pathological merit. This is but one of many cases that exemplify that the current "state-of-the-art" and quality of toxicological knowledge in the wound care field is somewhat lacking.

At present, the U.S. Food and Drug Administration (FDA) has three avenues for the regulation and registration of wound cleansing products: (1) over the counter (OTC) monograph, (2) medical device or 510(k) registration, and (3) new drug approval (NDA). The toxicity testing required to demonstrate safety for each of these different avenues varies significantly and ranges from no testing at all, in some cases, to complete clinical safety assessment in the case of the NDA.

Toxicologists and consumers should insist, therefore, on seeing some form of in vivo toxicity test data, to include standardized irritation and sensitization tests as a minimum, and in vitro cytotoxicity testing in the form of a validated, FDA-approved test, such as the L929 mouse fibroblast contact test used commonly in medical device testing. When available, the results from clinical wound healing studies that had product safety as at least one objective should be reviewed. Manufacturers should be encouraged to obtain clinical safety data for their wound cleansing products, and caution is advised in the use of in vitro product comparison data not properly validated by clinical trial and/or in vivo studies.

A tremendous need exists for improving and developing toxicological models to assess the safety of wound cleansing agents; demand for these products is expected to continue to increase. Future work needs to focus on developing clinically valid in vitro models that consider the need to effectively remove dead tissue and foreign particles from the wound, without significantly adversely affecting underlying wound healing processes. Studies to either validate or eliminate current in vitro testing schemes for the selection of wound cleansing products should be conducted as expeditiously as possible to reduce confusion among clinicians with respect to the scientific merit of these schemes.

Débriding Agents

Débridement (French term that means to unbridle) is the process of removing necrotic tissue and debris, including foreign matter, from a wound until healthy tissue is exposed for the purpose of improving and promoting the wound healing process. In many cases, this necrotic tissue is infected and bacterial endotoxins may also be present that inhibit wound healing. There are four types of débridement: (1) surgical, (2) mechanical, (3) chemical, and (4) autolytic. The different types of débridement are described in detail in (Kennedy and Tritch, 1997). Only chemical débridement is of significant concern from a dermatoxicological perspective.

TABLE 28.2. *Comparison of different enzymatic agents used for wound débridement and their substrate specificities*

Enzyme	Typical source	Substrate specificity	Commercial availability
Bromelain	Plant–Pineapple	Denatured protein/nonspecific	No
Collagenase	Bacterial–*Clostridium histolyticum*	Collagen/specific	Yes
Deoxyribonuclease	Bovine–spleen, pancrease	DNA/specific	No
Fibrinolysin	Bovine–spleen	Fibrin/specific	No
Papain	Plant–Papaya	Denatured protein/nonspecific	Yes
Sutilain	Bacterial–*Bacillus subtilis*	Denatured protein/nonspecific	No
Trypsin	Bovine–pancreas	Denatured protein/nonspecific	Yes

Chemical débridement is accomplished using formulations that contain proteolytic enzymes. These enzymes can be derived from many different sources, including plants, animal tissues, and bacteria. These proteases may be specific to a specific substrate (e.g., collagenase), or they may be nonspecific, such as the cysteine protease papain or the serine protease, subtilisin. At present, products containing papain, derived from the tropical fruit papaya, and collagenase, obtained from the culture of clostridium bacteria, are the only enzymes commercially available in enzymatic débriding formulations. Other enzymes have been used in the past, or are available for use, in the development of these types of formulations. Table 28.2 summarizes some of these enzymes.

Chemical débriding formulations are considered effective débriding agents alone, or used as adjuncts to surgical débridement. These formulations have long histories of safe use and generally are of little toxicologic concern except for the occasional report of irritation or possible sensitization. Systemic absorption from the wound bed does not appear to occur or is not sufficient to the extent of producing observable effects. Patients treated with these formulations generally have burns or chronic wounds, with a significant amount of tissue damage or necrosis, and occasionally may report "burning" sensations upon treatment. The source of these sensations is not completely understood, but it is presumed that they are due to mechanical and/or chemical stimulation of nerve fibers and pain receptors in the damaged tissue.

Chemical débriding formulations must be developed, therefore, with concerns for the rapid removal of necrotic tissue while leaving healthy tissues of various types unharmed, and they must not produce significant discomfort to the patient while chemical débridement is occurring.

Some popular chemical débriding formulations contain urea. The urea in these products comes from chemical synthesis and meets United States Pharmacopeia (USP) purity specifications for pharmaceutical use. Urea reacts across sulfhydryl bonds between amino acids in the necrotic tissue to open up and hydrate the eschar for attack by proteolytic enzymes. Urea is also commonly used as a hydrating agent in dry skin conditions and to remove excess keratin in hyperkeratotic lesions. It can also enhance the penetration of other drugs, such as corticosteroids, into the skin. Urea in débriding formulations has a long history of safe use at concentrations of 10% or less.

These formulations are generally applied once a day to the necrotic areas of the wound. If eschar is present, cross-hatching of the eschar surface with a surgical blade is recommended prior to application of the enzymatic agent.

Wound-care practitioners understand that it is desirable for an enzymatic débriding agent to be selective in its action (i.e., removal of necrotic and damaged tissue while leaving healthy tissue intact). Selectivity of enzymatic action is accomplished in the formulation either by sparing use of the proteolytic enzyme, as in the case of collagenase, which does not distinguish necrotic tissue collagen from that of healthy tissues, or by the addition of another chemical agent that assists in directing the enzyme to the necrotic tissue, such as the use of urea in papain/urea débriding formulations.

Chemical débridement is used generally in wounds with large to moderate amounts of necrotic tissue, whereas mechanical débridement is more often used to Débride wounds with only moderate amounts of necrotic tissue.

Chemical débriding agents must be carefully formulated. The enzymes used in these formulations originate from bacterial, plant, and animal sources, and they may contain toxicants, toxins, or infectious agents that should not come into contact with healing tissue and enter the patient's circulation. Examples are plant latex, residuals from bacterial culture media, solvents used to purify the enzyme, bacterial endotoxins, animal viruses, and emerging agents, such as prions thought responsible for diseases such as bovine spongiform encephalopathy or mad cow disease. The enzyme supplier and débriding agent manufacturer must work very closely to ensure using only the highest quality raw materials, always free from such risk factors.

The base material used to deliver the enzyme must also be capable of releasing the enzyme to the necrotic tissue efficiently while not containing any agents that might be histotoxic to the underlying healing tissue or causing sensitization problems for the patient. Petrolatum is used as a vehicle for some older products, but this is giving way to more modern formulations that are more hydrophilic and more readily release the enzyme. It is recommended that an experienced toxicologist during the discovery and development of new chemical débriding formulations monitor the constituents of these more complex formulations. This will help ensure wise choices in selecting formulation components to be in repeated contact with compromised skin, wounds, and burns.

Mechanical débridement can be accomplished by applying a moist dressing, allowing the dressing to proceed from moist to wet, then manually removing the dressing. This action causes a form of nonselective débridement and is often painful even though this technique has been used for decades in wound care. Hydrotherapy is also a type of mechanical débridement. Its benefits versus risks are of issue to the wound-care practitioner and toxicologist. Hydrotherapy can cause tissue maceration, and waterborne pathogens and toxicants may cause contamination, infection, or toxic injury to healing tissues. In this regard, disinfecting additives to the hydrotherapy bath may be cytotoxic, irritating, and/or sensitizing. Many of these additives contain quaternary ammonium or chlorine compounds that must be carefully monitored so that the concentration, or time of exposure, does not produce dermatoxicity.

Antimicrobials

With any open wound, infection may occur. Many factors, such as age and general health status, may increase the likelihood of infection, but the size and depth of the wound prove critical factors in determining the chronicity of any wound. Infection greatly adds to the morbidity associated with open wounds. An infected wound not only heals more slowly, there is also the risk of systemic infection and even death. Infected wounds also scar more severely and are associated with more prolonged rehabilitation. Topical antimicrobial agents have been shown effective in the management of open skin wounds. They may assist less complicated healing and decrease the conversion of a partial-thickness injury to a full-thickness injury, thereby reducing wound-related morbidity (Bendy et al., 1964; Sapico et al., 1986; Ward and Saffle, 1995).

Wounds generally are not sterile and usually contain some level of microbial contamination. A wound is considered to be infected when it contains 10^5 or greater microorganisms per square inch of wound surface (Robson and Heggers, 1984). Products developed to control or reduce the microbiological contamination or bioburden in a wound include antimicrobial agents that can be roughly classified as being either antiseptic or antibiotic.

Antiseptics are substances that inhibit the growth and reproduction of disease-causing microorganisms. They are chemicals that produce their antimicrobial effects through toxicity to microorganisms either in a broad sense, with activity over a wide range of different types of microorganisms, or in a narrow manner by selectively killing only certain types of microorganisms. As these agents can be toxic to the tissues of the patient if used in excess, care must be exercised when applying these agents into a wound. Examples of antiseptic agents in common use include povidone iodine, chlorhexidine gluconate, hydrogen peroxide, triclosan, etc.

Antibiotics are biologically derived substances, such as penicillin or streptomycin, produced by or derived from certain fungi, bacteria, and other organisms that can destroy or inhibit the growth of other microorganisms. Antibiotics also can produce toxicity in a wound generally from their interaction and interference with the biological processes ongoing in various cells involved in the wound healing process.

Some antibiotics and antiseptics can produce skin sensitizaton and irritation. Examples are shown in Table 28.3.

The following provides an overview of some of the more popular antimicrobials that have been or are currently being used in wound healing.

Antiseptics

The Agency for Health Care Policy and Research (AHCPR) in the United States does not advocate the use of antiseptics in the treatment of pressure ulcers. Instead, they recommend a 2-week trial of topical antibiotics for clean ulcers that are not healing or that are continuing to produce exudate after 2–4 weeks of optimal care (AHCPR,1992,

TABLE 28.3. *Antibiotics and antimicrobial ingredients used commonly in wound care and their potentials to cause skin irritation and/or sensitizaton along with any other precautions*

Active ingredient	Typical formulation	Dosage regimen	Mode of action	Precautions
Bacitracin	400–500 units/g in a white petrolatum base	Four times a day	Interferes with cell wall synthesis	Allergy; Anaphylaxis occurs almost exclusively in topical applications of venous stasis dermatitis and ulcers (systemic absorption)
Bacitracin Zinc	400–500 units/g, white petrolatum	Four or five times per day	Cell wall synthesis inhibitor zinc increases potency	Generalized itching
Gentamycin Sulfate	0.1% in a cream base consisting of stearic acid, propylene glycol, stearate, isopropyl myristate, propylene glycol, polysorbate 40, sorbitol, butylparaben, methylparaben, purified water.	Four to six times a day	Inhibits protein synthesis.	Cross-resistance; renal or hepatic failure; allergy unusual
Metronidazole	0.75% gel consisting of purified water, propylene glycol, carborner 940, methylparaben, propylparabin, and edetate disodium.	Thin film twice a day	Mechanism unknown, may include an antibacterial and/or antiprotozoal effect.	Blood dyscrasia; interference with anticoagulant therapy
Mupirocin	2.0% in a water miscible, polyethylene glycol base.	Two or three times a day; 5–14 days	Inhibits bacterial protein and RNA synthesis	Burning, itching, and contact dermatitis has been reported
Neomycin Sulfat	3.5 mg/g, white petrolatum	Four to six times a day	Inhibits protein synthesis	Greater incidence of allergic sensitivity and cross-sensitivity to other aminoglyosides
Polymyxin B Sulfate	5,000–10,000 units/g, white petrolatum	Three or four times a day	Alters lipoprotein membranes and produces cellular lysis	Sensitization can occur after long-term use
Povidone-Iodine	0.5% in an oil and water emulsion	As needed	Disrupts cellular membranes and interferes with metabolism	Sensitization, irritatition
Silver Sulfadiazine	1.0% cream consisting of white petrolatum, stearyl alcohol, isopropyl myrislate, sorbitan monooleate, polyoxyl 40 stearate, polypropylene glycol, methylparaben.	Daily or twice a day	Acts on cell, walls and membranes; silver is slowly released and is toxic to bacteria.	Allergy to sulfa; renal or hepatic impairment

1994). The antibiotic should be effective against gram-negative, gram-positive, and anaerobic organisms (e.g., triple antibiotic or silver sulfadiazene), with monitoring for allergic responses; however, despite the AHCPR guidance, some clinicians still use antiseptics in wounds and claim them beneficial in some cases. Several studies, however, have documented the toxic effects of exposing wound-healing cells to antiseptics (Fleming, 1919; Lineaweaver et al., 1985; Teepe, 1993; Wright and Orr, 1993). Wound-care practitioners are often reminded of this to the extent that even more modern antiseptics may be thought to be unsafe, in general, without strong toxicologic evidence to support or refute the concern at the concentrations used, or in a particular formulation.

Iodine and Iodophors

Povidone-iodine, for example, has been used for preventing cross-contamination of wounds in different patients. A concentration of 0.001% is reported to be noncytotoxic even though, on occasion, it is used on gauze-packed wounds in much higher concentrations. Views on the benefits and risks of using iodine and iodophors in wounds are quite mixed and have been the subject of much debate.

(Rodeheaver et al., 1982) concluded from the results of in vitro and in vivo studies with different iodine-containing solutions that the bactericidal activity of iodophors was inferior to that of uncomplexed aqueous iodine. These authors went on to conclude that although povidone iodophors did not enhance the rate of wound or infection, they offer no therapeutic benefit when compared with control wounds treated with saline solution. Furthermore, they observed that the addition of detergents to the povidone iodophor was deleterious, with the wounds exposed to this combination displaying significantly higher infection rates than untreated control wounds. Based on these results, they recommended that surgical scrub solutions containing iodophors should not be used on broken skin.

From a review of the literature, (Hess, 1990) concluded that povidone iodine should not be used in open wounds due to its cytotoxicity and the lack of documented clinical evidence supporting its safety and efficacy for use in open wounds. In an even more recent review, povidone iodine was noted to cause toxicity to wounds, and the authors concluded that indicated antibiotics were more desirable (Cho and Lo, 1998). In another recent review, (Niedner, 1997) showed that although in vitro experiments with povidone iodine showed a certain cytotoxicity, it is not easy to transfer these toxicological data to in vivo situations. In vivo investigations in animals and in humans indicate that the cytotoxic effects of povidone-iodine, measured by the wound healing process, is essentially nondiscernible. Only when administered in combination with detergents was an obvious cytotoxicity seen in wounds, but not on the intact skin. In comparison to the frequently used antibiotic neomycin, it was concluded that the sensitization rate of povidone-iodine is very low. Thus, the toxicity of povidone-iodine to wounds may be dose dependent as well as attributable, in part, to the detergents in the formulation.

In contrast, studies by (Lamme et al., 1998) have shown that an absorbent wound dressing containing 0.9% iodine, as povidone iodine bound to a cadexomer matrix, can be quite beneficial to wound healing. Povidone iodine is probably not as bioavailable from this dressing as its use is intended by design to produce its antimicrobial action within the cadexomer matrix and not on the wound surface. Preclinical and clinical studies (Mertz et al., 1994; Hunt, 1995) with this particular type of povidone iodine containing dressing formulation have demonstrated clearly that povidone iodine can be useful in wounds if incorporated into an absorbent dressing.

Povidone iodine-containing liposomes are also reported to have beneficial effects on wound healing (Reimer et al., 1997). Povidone iodine-containing liposomes in deep dermal burn wounds indicated an outstanding quality of wound healing with smooth granulation tissue, less inflammation, less wound contraction, and no hyperkeratotic reactivity.

There is also recent evidence that free molecular iodine may be effective for topical use at iodine concentrations well below those found in antimicrobial cleansing products containing povidone iodine (Gottardi, 1995). Development of wound cleansing product formulations that provide controlled release of free molecular iodine at safe and effective concentrations is in progress, and products based on this new technology may soon be in use (Hickey, 1997).

Chlorine and Chlorine-containing Solutions

Chlorine-sodium hypochlorite (household bleach) is also sometimes used to prevent cross-contamination of wound patients. It is reported to be less cytotoxic than chlorazene, also used routinely to prevent cross-contamination.

Dakin's solution (sodium hypochlorite and boric acid) is reported to be effective against *Staphylococcus spp.* and *Streptococcus* spp. Dakin's solution is reported to be cytotoxic unless diluted.

Acetic Acid

Acetic acid (0.25% solution) is reported to be effective against *Pseudomonas* spp. and is used effectively to control outbreaks of antibiotic-resistant strains of *Staphylococcus aureus*. At high concentrations it can be a caustic agent that damages healthy tissue.

Hydrogen Peroxide

Hydrogen peroxide is not extremely effective as an antiseptic agent in open wounds. It has a mechanical effect of effervescence. It is not toxic in low concentrations, as the enzyme, catalase, in the eukaryotic cells of the wound tissue and blood converts hydrogen peroxide to water and oxygen. Hydrogen peroxide is considered to have little antibacterial effectiveness and is of mostly entertainment value. Contact lens disinfecting solutions containing hydrogen peroxide are known to be quite effective against a broad spectrum of microorganisms, however. It is possible that these stabilized hydrogen peroxide solutions may find use in wound antisepsis in the future.

Silver

Silver has been long known to have antimicrobial value and has been used in different forms since antiquity. The FDA considers silver a pre-1938 active ingredient. Historically, in the pre-1938 era, colloidal silver was used as a topical antiseptic to treat cuts, abrasions, and burns; as an oral antiseptic to treat gingivitis, and for oral hygiene; as an oral (and often-injected) antibiotic to treat a variety of infectious diseases; and as a tonic, or elixir, remedy for a variety of afflictions of the mucous membranes.

In recent years, colloidal silver has been rediscovered and promoted in the alternative health industry as an alternative to antibiotics and as a "wonder substance" capable of treating virtually every infectious disease imaginable. Some of the claims for silver use are marketing hype and others may have some scientific basis, but many of these claims lack the rigorous safety and clinical testing that the FDA would consider proof of safety and effectiveness for a new drug.

Silver nitrate solutions have been shown to be effective against gram-negative bacteria, but are not often used in wound healing. Apparently they are more useful as hemostatic agents.

Silver-coated or impregnated dressings are a very recent development in the wound care industry, having just been introduced over the last 2 years. Presumably, these recently introduced products have undergone some toxicity testing and have gained FDA approval. Their use history is very short at present, and any of these newly approved silver impregnated dressings should be monitored carefully for any sign of adverse effect until such time that a reasonable safe-use history can be established.

The best-studied consequence of overconsumption of silver is with argyria. With argyria, silver is taken internally in excess and the excess is deposited in the skin, organs, and other tissues. This causes the skin to turn a gray or bluish-gray color, due to the formation of silver sulfides that form insoluable complexes in elastic fibers. Upon exposure to strong sunlight, the affected individuals can turn a dark brown or black color. This coloration is irreversible. Silver accumulates in the human body over the individual's entire lifetime. The excretion of silver is extremely slow. Silver complexes with albumin and accumulates in the skin, liver, spleen, bone marrow, lungs, and muscle. Small doses of silver appear harmless for most people, while large doses taken in great excess can be toxic, or even lethal (Goyer, 1991).

The new silver-impregnated wound dressings claim to solve the problem of potential toxicity by administering small doses of silver for up to 7 days. This steady release is controlled by sodium phosphate and calcium phosphate in one commercial dressing. When the dressing comes in contact with moisture from the skin, it begins to dissolve, releasing silver ions.

Topical antimicrobial products containing silver sulfadiazine have long been used safely, so these new silver dressings may become more popular in the near future. As some of the silver-impregnated or silver-coated wound dressings are used more widely, the safety of these items will become more defined.

When considering the development or use of a silver-containing antimicrobial, it is necessary to remember that some individuals are sensitive to specific metals. Nickel, copper, silver, and other metals have been known to cause contact allergic reactions in the skin. It is a good idea to be certain that the patients you intend to treat are not allergic to silver before using this type of topical antimicrobial agent.

Mercurachrome

Mercurachrome is a mercury compound that dates back to antiquity for the prevention of sepsis in open wounds. Its modern use dates back to experiments by Koch in 1881, where mercuric chloride solutions were observed to be very effective in killing spores. Mercurochrome is bacteriostatic, and its antimicrobial action may be useful on small, partially healed, superficial wounds. Due to reports of toxicity, anaphylaxis, and aplastic anemia, wound-care professionals generally find this agent to be unacceptable (Rund, 1997). The FDA has recently moved to eliminate the use of this agent in wound healing products in the United States.

Antibiotics

In contrast to antiseptic agents, topical antibiotics generally are thought to have low toxicity to wounds, and some are reported to be very effective when applied to control-sensitive organisms. For example, silver sulfadiazine cream and metronidazole gel have been used effectively to reduce bacterial contamination in infected wounds and to control odors (Kucan, 1981; Witkowski and Parish, 1991). Tests to determine the sensitivity of wound microorganisms to topical antibiotics are available (Nathan et al., 1978; Rodeheaver et al., 1980). To reduce the risk of adverse reaction, it is recommended that topical antibiotic use should be limited to no more than 2 weeks, and patients must be monitored for any signs of reaction. There is a need to study the absorption of antibiotics from wound sites and to examine the localized histopathological effects following their use. Toxicity studies for new antibiotics and antibiotic preparations intended for use in wound sites need to be designed such that systemic effects, as well as localized adverse effects, on wound healing and closure are examined.

The following paragraphs present a few of the more popular antibiotics.

Neomycin

Neomycin is an aminoglycoside antibiotic that can be used alone or in combination with other antibiotics in topical antibiotic preparations and wound irrigation solutions.

Hypersensitivity reactions, generally skin rashes, are reported in 6%–8% of patients when neomycin is applied topically; sensitive individuals may develop cross-reactions to other aminoglycosides (Sande and Mandell, 1990).

Neomycin irrigations are sometimes regarded as benign because it is thought that little of the drug is absorbed; however, it has been demonstrated that toxic amounts of neomycin can be attained by instilling the drug into surgical wounds (Weinstein, et al.,1977). Unlike other aminoglycosides, routine serum concentration monitoring is not possible with neomycin, but research has shown that toxic serum concentrations of neomycin are approximately 5–10 mg/L (Kucers et al., 1987). Serum concentrations do not correlate with the dose, and absorption from wounds is variable and unpredictable (Weinstein, et al., 1977).

There are multiple reports in the literature of toxicity from neomycin irrigation (Trimble, 1969; Davia et al., 1970; Anderson, 1978). Toxicity has been reported with intraoperative instillation, wound irrigation, and bowel irrigation, and topically, and as an aerosolized spray to burn wounds.

In 1972, sterile neomycin for intramuscular injection was decertified by the FDA because of substantial toxicity and the availability of safer, more effective agents. In 1988, the FDA amended regulations regarding nonsterile neomycin for prescription compounding so that the label must state that the powder is for oral products. The FDA changed this regulation because the powder was being used for indications, including irrigations of infected ulcers and wounds, for which proof of efficacy is lacking and for which there is substantial risk of toxicity.

Because of the relative risk versus the benefit of using neomycin irrigating solution, there is significant liability associated with its use. Several reports of litigation against physicians, hospitals, and pharmacists are published in the literature. In one report (Watson, 1988), the plaintiff was awarded $1.5 million because of neomycin-induced cytotoxicity.

Neomycin is available over the counter in the form of triple antibiotic ointment preparations containing neomycin, polymixin B, and gramicidin. Triple antibiotic preparations are routinely used to treat minor burns and superficial wounds and have a history of safe use.

Polymixin B

The polymixins have been available for nearly 50 years and are derived from various strains of *Bacillus polymyxa*. These are simple polypeptides with molecular weights of approximately 1000. The antimicrobial activity is restricted to gram-negative bacteria. They are surface-active agents that interact with phospholipids and disrupt cell membranes. Systemically, these antimicrobials can produce nephrotoxicity, but when applied topically, they are not well absorbed and there is little or no risk of systemic toxicity. They are quite useful for treating infections of the skin and mucous membranes with little risk of toxicity.

Polymixin B is available over the counter in ointment and other topical forms. It is commonly found in triple antibiotic formulations with neomycin and gramicidin.

Mupirocin

Mupirocin is produced by *Pseudomonas fluorescens* and acts by blocking bacterial protein synthesis. The mechanism by which this occurs is unique to mupirocin such that there is essentially no cross-resistance to other antimicrobials. Mupirocin is particularly effective against most all gram-positive bacteria. In the form of its soluable calcium salt, its use is increasing in the treatment of infected wounds. Mupirocin safety data for use in wounds is not extensive at this time.

Metronidazole

Topical metronidazole is being used ever more increasingly in the management of chronic wounds. It is quite useful in reducing infection with anaerobic bacteria that may cause reduced rates of healing and malodor. Its use in wound healing is fairly recent and it is gaining popularity currently, particularly in palliative care applications, where the severe odor of chronic, nonhealing wounds can cause a detraction for the patient, caregivers, and the patient's family. At present, the most frequently used concentrations are 0.75% and 1.0% in the United States, Canada, and Europe.

Topical metronidazole has a good use history at present for topical indications such as acne. Currently it is not approved by the FDA for use in wound therapy, even though it is enjoying a fair degree of off-label use for this purpose. Topically, it is expected to be relatively safe, particularly for use in aging patients with chronic, nonhealing infected wounds. There are reports that topical metronidazole may cause skin dryness and irritation in some individuals; however, its effect on the wound healing process and absorption from wound sites has not been studied.

Mafenide

Mafenide is a sulfonamide marketed as mafenide acetate cream containing 85 mg/g. It has been used effectively to control a variety of gram-negative and gram-positive bacteria in burns, but is contraindicated for use in deep infections. Nevertheless, its use is mentioned in wound-care literature, and there appears to be some use of this agent in wound healing in addition to burns. It is rapidly absorbed systemically and then metabolized to carboxybenzenesulfonamide. Burn-surface absorption studies indicate that peak plasma concentrations occur 2–4 hours after application. Absorption from an open wound would likely be somewhat faster. Among the adverse effects are intense pain at the application site and allergic reactions (Mandell and Sande,1990).

Nitrofurazone

This agent is available in a 0.2% topical cream formulation intended for use on second- or third-degree burns when bacterial resistance, *Pseudomonas* spp. in particular, to other agents occurs. The manufacturer indicates no evidence of effectiveness on minor burns or chronic wounds, even though it has been mentioned in recent wound-healing texts as used to control bacteria in chronic wounds (Roberts Pharmaceutical Inc., 1997; Rodeheaver, 1997). Clinical skin reactions, including contact dermatitis, have been reported with use of nitrofurazone.

Penicillin

Although once placed in surgical sites to reduce the risk of infection, penicillins are not used topically in wounds to any significant degree at present. The risk of hypersensitivity reactions, skin rashes, contact dermatitis, etc., and the development of resistance to this antibiotic are concerns for topical use. Similar concerns for use of cephalosporins, which are also beta-lactam class antimicrobials, have kept them from widespread use in wound healing.

Silver Sulfadiazine

Silver sulfadiazine is a broad-spectrum antibiotic especially useful for burns. It is used as a cream or ointment to cover burns and other wounds and is effective in reducing the microbial colonization by both bacteria and fungi. Silver sulfadiazine preparations are reported to have a soothing effect and to prevent gauze bandages from adhering to the wound. Adverse effects are not a great concern based on a long history of safe use; however, silver is released slowly from the preparation at concentrations toxic to microorganisms. Care must be exercised when treating large wounds or burns, as the systemic absorption of silver increases with increasing surface area. Although rare, reported adverse reactions include burning sensation, rash, and itching (Mandell and Sande,1990a).

In its final rule on first aid antiseptic products approved for OTC use, the FDA recently listed these antibiotic active ingredients as safe and effective: bacitracin, bacitracin zinc, chlortetracycline hydrochloride, tetracycline hydrochloride, neomycin sulfate, oxytetracycline hydrochloride, and polymyxin B sulfate (the latter two being only for combination products because of their limited effectiveness against certain microorganisms when used alone).

At the present time the following antibiotics and combinations are approved by the FDA for use without prescriptions. They are ointments unless otherwise noted:

Single-ingredient products: bacitracin, bacitracin zinc, chlortetracycline hydrochloride, neomycin sulfate, tetracycline hydrochloride.

Combination products: bacitracin-neomycin, bacitracin-polymyxin B aerosol, bacitracin-neomycin-polymyxin B, bacitracin zinc-neomycin, bacitracin zinc-polymyxin B ointment, aerosol or powder, bacitracin zinc-neomycin-polymyxin B, neomycin-polymyxin B ointment or cream, oxytetracycline-polymyxin B ointment or powder.

Antifungals

There are several antifungal agents used in or around skin wounds. Antifungals present a different problem toxicologically than antimicrobials intended to control bacterial sepsis in a wound. Bacteria are prokaryotes that have a cell wall, and their key membrane lipids are hydroxylated glycerols. In contrast, fungi are eukaryotes that have a cell wall, and their key membrane lipid is ergosterol. These differences become important toxicologically when contrasted to the fact that vertebrate cells are also eukaryote but do not have a cell wall, and their key membrane lipid is cholesterol. Therefore, the strategy for developing a safe and effective antibacterial agent is necessarily different than that for an antifungal agent. It should be expected that antiseptics claiming to kill both bacterial and fungal organisms should have different margins of safety relative to vertebrate toxicity.

There are few antifungal agents used in or around wounds at present. Wound sepsis is generally more concerned with bacterial colonization rather than fungal colonization at the present time; however, fungal infections can be a problem for incontinent patients and those receiving radiation or chemotherapy. Skin breakdown in these patients can rapidly become an infected wound colonized with different types of microbes, including fungi. Cutaneous candidiasis and candidal infections of mucous membranes and mucocutaneous junctions is a particular problem.

At present, the agent of choice is miconazole nitrate. Other agents are emerging, and it is likely that other antifungal agents will increasingly enter the wound care arena. Particular care should be exercised in selection and use of these products in open wounds unless proper safety studies have been conducted.

Miconazole

This is an azole antifungal agent that has been in topical use for a long time and is currently available in OTC cream and powder formulations containing up to 2%. These formulations are used in skin care of the incontinence patient and can find their way into wounds in this fashion. Miconazole acts by interfering with ergosterol synthesis. It does not apparently exhibit the synthesis of steroids (i.e., testosterone) in vertebrates observed with other azole antifungal agents, particularly ketoconazole. Hypersensitivity can occur with miconazole nitrate in the form of a rash or itching. The systemic toxicity is low.

It should be mentioned that, in some instances, the vehicle used to deliver an antifungal may be a toxicologic concern. This is particularly true for powders and other ingredients that cannot be readily biodegraded or absorbed by the wound tissue.

This is the case with the use of talc as a vehicle for antifungal agent delivery around wound sites. There are no studies at present showing the biodegradation and removal of talc entering a wound site; however, corn starch is currently being substituted for talc in products intended to be used near wound sites and on the compromised skin of incontinent patients.

Wound Healing Promoters

There are many different substances and chemical compounds that have been claimed to stimulate the rate of healing of wounds. Only a few, however, have been studied clinically enough to demonstrate their value. Others are currently in development by biotechnology companies for use as active ingredients in wound healing products. Some have been around a long time, and only recently have the molecular biological mechanisms by which these agents stimulate wound healing become understood. Wound healing promotional agents can be placed into two groups based on the nature of their biological action: wound stimulators and growth factors.

Wound Stimulators

This group of agents consists of agents that act in some biochemical manner to stimulate proliferation and maturation of cell types involved in the healing process. They may also act to improve the structural integrity of the newly forming tissue and increase its breaking strength. Examples in this class would include the following.

Chlorophyllin

Chlorophyllin is short for sodium copper chlorophyllin complex. This water-soluble form of chlorophyll obtained from green plants has been used since antiquity to help heal wounds, but has only been studied clinically since the mid portion of this century (Chernomorsky and Segelman, 1988). Its mode of action at the molecular level is still being unraveled, but it appears that it may aid in wound healing through more than one action. It appears to exhibit antagonistic properties on bacterial flora that colonize wounds and protagonist properties on tissue regeneration (Smith, 1955). Its current use is in the form of ointments containing 0.5% chlorophyllin by weight. The toxicity of chlorophyllin is low and it does not appear to have any significant potential to cause irritation or skin sensitization; in fact, it is known to produce a soothing sensation when applied to burn patients with painful injuries (Gruskin, 1940).

Others

Other wound-stimulating agents have been proposed, including copper-peptides, zinc, aloe vera, honey, peptide fragments, etc. To date, however, none of these agents has proven to be effective in controlled clinical trials with human subjects having chronic wounds. Their effectiveness remains to be proven clinically.

Growth Factors

Growth factors are agents recognized by the host as being part of the natural wound healing process that exert their action on growth-factor receptors.

Platelet-Derived Growth Factor-Beta (PGDF-β)

Recombinant human PGDF-β (rhPDGF-B or becaplermin) is currently the only growth factor developed and marketed successfully as a new drug. It is produced using recombinant DNA technology by insertion of the gene for the B chain of PDGF into the yeast, *Saccharomyces cerevisiae*. It is intended for neuropathic ulcers with good circulation that extend into the subcutaneous tissue, or beyond, and have an adequate blood supply. The product is available in a gel base that contains 100 μg of becaplermin per gram. The studies reported below provide an overview of the safety studies conducted with this product and reported in the package insert (McNeil Pharmaceutical, 1998).

The systemic bioavailability of becaplermin was studied in a rat model and was less than 3% in rats with full thickness wounds receiving single or multiple (5 days) topical applications of 127 μg/kg (20.1 μg/cm^2 of wound area) of becaplermin gel (McNeil Pharmaceutical, 1998).

The improvement in wound healing observed in early clinical trials and clinical safety assessment appears less than dramatic in the results from initial trials. For example, one study was a multicenter-controlled trial of 172 patients to assess the safety of vehicle gel (placebo; n = 70), as compared to good ulcer care alone (n = 68). The study included a small (n = 34) becaplermin gel 0.01% arm. Incidences of complete ulcer closure were 44% for becaplermin gel, 36% for placebo gel, and 22% for good ulcer care alone. In another multicenter, evaluator-blind, controlled trial of 250 patients, the incidences of complete ulcer closure in the becaplermin gel 0.01% arm (n = 128) (36%) and good ulcer care alone (n = 122) (32%) were not statistically different (McNeil Pharmaceutical, 1998).

Safety labeling indicates that preservative sensitivity to parabens or m-cresol in the gel base may be of concern for the becaplermin-containing product.

Becaplermin was not genotoxic in a battery of in vitro assays (including those for bacterial and mammalian cell point mutation, chromosomal aberration, and DNA damage/repair). Becaplermin was also not mutagenic in an in vivo assay for the induction of micronuclei in mouse bone marrow cells.

Carcinogenesis, reproductive toxicity, and pediatric safety studies have not been conducted. It is not known whether becaplermin is excreted in human milk.

Patients receiving becaplermin gel, placebo gel, and good ulcer care, alone, had a similar incidence of ulcer-related adverse events such as infection, cellulitis, or osteomyelitis; however, erythematous rashes occurred in 2% of patients treated with becaplermin gel and placebo, and none in patients receiving good ulcer care alone. The incidence of cardiovascular, respiratory, musculoskeletal, and central and peripheral nervous system disorders was not different across all treatment groups. Mortality rates were also similar across all treatment groups. Patients treated with becaplermin gel did not develop neutralizing antibodies against becaplermin.

Keratinocyte Growth Factor-2

Keratinocyte growth factor-2 (KGF-2) is a newly discovered human growth factor that has not yet been introduced into the wound-care market (Human Genome Sciences, 1998). It is believed that KGF-2 is always present in humans and is activated only when there is an epithelial tissue wound. It may work by attracting fibroblasts, collagen, and connective tissues to the site of the wound to heal it through the creation of new tissue. For example, when an injury to the skin occurs, thick pink skin forms around the edge of the wound, and a scab forms. When healing is complete, the skin returns to normal thickness.

Several different models were conducted successfully in preparation for the two phase I human clinical trials. In topical wound-healing models, KGF-2 healed full-thickness wounds through the creation of new skin in a short period of time with little scarring.

KGF-2 was identified through a systematic review of more than 300 full-length genes corresponding to what scientists believed to be potential therapeutic protein candidates. Many of these have been expressed in recombinant form and have been tested.

Toxicology testing requirements for this product would be assumed to be similar to those required for becaplermin. It remains to be seen if this new growth factor is safe and effective enough to be approved by the FDA and what safety testing the agency will require.

Dressings

Ever since (Winter, 1962) first observed that a moist wound bed improved wound healing, many clinicians and scientists in the wound care industry have been improving wound dressings to promote moist wound healing and attempting to convert healthcare practitioners to their use. Today, many different kinds of dressings are manufactured from a variety of materials, and many companies are striving to produce the "ideal" wound dressing.

Wounds come in a wide array of sizes, depths, etiologies, and healing conditions, so it is very difficult to produce a single dressing to promote moist healing optimal for all types of wounds. This has led to a situation today where most companies offer a number of different options to cover a wound. These options range from completely absorbent dressings that remove heavy exudate from the wound bed to dressings that actually will donate fluid to a dry wound bed. This situation has resulted in the use of a range of materials—some natural and others synthetic—that come into contact with the wound bed. These materials may be solid bandages made of woven cellulose or cellulosic material, cast or formed as gel sheets or foams, cut from transparent polymer films, or formulated as amorphous gels delivered from a tube or other suitable semisolid dispenser.

Wound dressings are regulated as medical devices in the United States and most other countries. This requires a level of toxicity testing commensurate with the level of exposure to the systemic vasculature and internal organs. Because wound dressings are typically intended to be applied topically, not used for extended periods, and not assimilated into the body, the degree of toxicologic evaluation in the safety review and approval process is usually not as extensive as required for many topical pharmaceutical agents. At present, most new wound dressings require at minimum the following tests to demonstrate their safety:

Primary dermal irritation (usually in the rabbit).
Dermal sensitization (usually in the guinea pig).
In vitro cytotoxicity (usually using mouse lymphoma cells).

Other tests may be required, depending on the specific use claims; these may include in vitro hemolysis, acute dermal toxicity, and, in some rare instances, subchronic dermal toxicity.

Even though it is not specifically required and requires judgment as to the specific need for some types of data, it is a good idea to perform some of these tests using abraded or broken skin in order to maximize the potential for causing effects in the open wound. It is also useful to perform some irritation and sensitization testing on human subjects, using cosmetic protocols for sensitization and irritation in human subjects.

During formulation and production, attention must be given to the raw materials used in the compounding and manufacturing of the dressing, particularly the concentrations of potentially harmful ingredients, such as some preservatives, pH adjusters, monomer residuals, lanolin, sterilization byproducts, wound healing promoters, wound healing promoters, aromatherapy ingredients, fragrances, etc. Relatively recently there has been concern over the presence of natural latex in some raw materials that might be introduced during the manufacturing process. This concern has been evidenced to the extent that in some cases, manufacturers may be requested to provide letters of assurance to some healthcare facilities, indicating that a product contains no latex. This subject is discussed in more detail in the section on "skin protectants."

It is beyond the scope of this chapter to discuss the many different forms of wound coverings, including calcium alginate dressings, hydrocolloids, hydrogels, polymeric films, gel sheets, foams, etc. There are many excellent reviews that discuss each type of product in detail, and the reader is referred to these for more information (Krasner and Kane, 1997; Wiseman et al., 1992; Varghese et al., 1986; Davies, 1983).

Preservatives

It is appropriate, however, to provide a brief discussion regarding the use of preservative agents in wound dressings. This has become a recent concern to some and may amount to marketing hype or a problem for which there is a lack of extensive research. Many wound dressings are sterilized and contain no preservatives. Others, however, contain very small amounts of preservative agents, such as imidurea, parabens, benzyl alcohol, DMDM hydantoin, etc., to inhibit microbial growth during the shelf life and use of the product. In some cases, a product may be sterilized and also contain some preservatives either by design or because a preservative was used to provide increased shelf life to one of the raw materials used in manufacturing the dressing.

Parabens (methylparaben and propylparaben) are used in many hydrogel wound dressings. Paraben esters are typically used at concentrations ranging from 0.1%–0.8% by weight in cosmetics, and their use in combination is generally more effective than when used as single esters (Haag and Loncrini, 1984). In wound hydrogels, the total paraben concentration appears to be on the order of between 0.15%–0.28% by weight (White and Anderson, 1995).

Paraben sensitivity has been extensively studied in the cosmetic ingredient safety literature and been shown to have little or no significant risk to the public applying their use in cosmetics and for having very low to negligible potential for causing contact dermatitis (Elder, 1984). The results of a small patch testing study performed on leg-ulcer patients in the United Kingdom showed parabens to produce an incidence of sensitization (13%) in these patients approximately 10 times greater than the frequency observed in any previous cosmetic literature (Cameron et al., 1992). Despite the small number of patients in the test (81 subjects), the authors concluded that leg-ulcer patients would benefit from patch testing to identify contact allergens such as lanolin, neomycin, and framycetin. While not specifically concluding that parabens in products for leg-ulcer patients were a particular problem, this article was used as evidence in product marketing to show that the use of parabens as a preservative system in wound care products might cause a sensitization problem. Thus, if a product did not contain parabens, it was somehow superior. Two years later, the results of paraben sensitization testing on 2295 consecutive outpatients by the Swiss Contact Dermatitis Research Group were published (Perrenoud et al., 1994) and showed the incidence of paraben sensitivity to be 1.7% in this group of patients. Interestingly, (Fisher, 1979) observed that paraben-sensitive patients reacting with allergic contact dermatitis when paraben-containing pharmaceuticals were applied to eczematous or ulcerated skin could tolerate paraben-containing cosmetics applied to normal, unbroken skin. No sensitization was induced, even when the cosmetics were applied to the very thin skin of the eyelid. Fisher called this the "paraben paradox." It is interesting that this paradox did not appear to evidence itself in the results of numerous animal tests, including many where the parabens were administered intradermally (Elder, 1984). Even so, it may be that preservative agents and other ingredients applied to broken skin, such as parabens, may exhibit differential contact sensitization responses than when they are applied to normal, unbroken skin. Further research is needed to develop testing protocols that can more effectively identify these agents.

The Swiss Contact Dermatitis Research Group study (Perrenoud et al., 1994) also identified formaldehyde as the chemical with the highest incidence of positive test results (5.7%). This is an interesting finding in that DMDM hydantoin, a preservative also commonly used in wound dressings, exhibited the same positive incidence as observed for the parabens (1.7%), and DMDM hydantoin is structurally related to formaldehyde (Rosen, 1984). There are no reports currently of DMDM hydantoin sensitization in wound patients.

The Cosmetic Toiletry and Fragrance Association (CTFA) has published many other ingredient reviews for preservatives and other ingredients that may be used in wound dressings. For example, there is an excellent review published for imidazolidinyl urea, another common preservative used in hydrogels that should be examined before formulating a new product (Elder, 1980).

Skin Protectants

Skin protectant products are an important part of the treatment and care of some types of wounds, particularly those on areas of the body that need protection from the urinary and fecal excretions of incontinent patients. In addition to providing protection against contamination, skin protectants used in wound healing can also help to reduce shear forces acting to pull the healing tissue apart and prolong the healing process. There are many different types of products that are used as skin protectants. A few are presented below along with comments on any significant concerns of a toxicologic nature. Skin protectants are typically considered drugs regulated as either OTC products, medical devices, or legend drugs.

Petrolatum, Mineral Oil, and Other Natural Products

Natural oils and lipids of animal origin have been used since antiquity as skin protectants and lubricants. Petrolatum and mineral oil have since become popular replacements for these oils in many pharmaceutical products. Natural oils and extracts, such as castor oil, Peruvian balsam, and various animal lipids, are occasionally found in formulations intended for would care. Products intended for pharmaceutical use usually employ purified components of these natural products as either active ingredients or excipients. Due to its documented safe-use history, petrolatum is a skin protectant ingredient listed for use in OTC skin protectant products by the FDA.

White Petrolatum, USP, is a purified pharmaceutical ingredient used alone or in some parts of many skin-care and wound-care products to help prevent moisture loss and to form a protective barrier on the skin. Petrolatum is not well absorbed into the skin and does not penetrate into the dermis, but does enter the stratum corneum and can help prevent moisture loss detrimental to the repair of damaged skin. Despite some consumer advocacy groups' concerns about clogging of pores, skin respiratory impairment, photosensitivity, and skin drying, there is little or no toxicologic evidence to support these concerns for use in skin products used in or around wound healing sites. A general lack of water solubility is a concern for some applications, as well as a greasy after-feel, so manufacturers have developed forms more easily emulsified, with less drag to decrease the greasy after-feel. The purified white petrolatums generally will meet the OTC monograph requirements for skin-care protectants. Their ability to create a prolonged, effective barrier against the excretory products from incontinence is generally not as effective as the dimethicone-containing products and polymeric films, but it can be effective if applied frequently. They can also work to reduce shear force. Their low cost and availability makes them still popular for some applications today.

Petrolatum and mineral oil have been shown to break down natural latex; this makes them a concern for use around wound sites where appliances containing latex might be worn or on sites over which surgical gloves may be worn.

Products containing high percentages of petrolatum are also a concern for use on patients receiving hyperbaric oxygen therapy (HBO), due to their potential flammability.

Zinc Oxide

Zinc is essential for healthy skin. As a drying agent and astringent, zinc oxide has been used for generations to soothe diaper rash and relieve itching, and zinc is a natural sun screen, protecting chapped lips and skin from the sun's harmful rays.

Zinc also improves healing of wounds, like surgical incisions, burns, and other skin irritations. Used as an anti-inflammatory, zinc soothes skin and skin tissue. Poison ivy, sunburn, blisters, and gum disease are all improved when treated with zinc. It is even a natural insect repellent. Zinc also stimulates the transport of vitamin A from the liver to the skin, helping to protect body tissue from damage.

Zinc occurs naturally in the earth, in the air, and in the foods we eat. It is the second most common trace metal, after iron, naturally found in the body.

Zinc oxide is used in a variety of wound healing formulations in the form of ointments, pastes, creams, etc. Its healing and barrier properties make it useful as a protective treatment for many different types of wounds, including skin damage from exposure to incontinent episodes and diaper rash. Sensitive individuals generally do well using zinc-oxide–containing products, as they are generally fragrance free and formulated for use on broken skin as well as post-laser-treatment skin care.

Dimethicone-Containing Products

Dimethicone is often used in skin protectant products used on broken skin under an OTC monograph published by the FDA. The safety record for dimethicone and related skin protectants, such as cyclomethicone, is excellent; however, other ingredients may be added to these skin protectant formulations to impart better cosmetic feel (lanolin or lanolin derivatives, aloe vera, etc.) or a bioadhesive quality (i.e. gantrez). It is important to review any additional excipient ingredients prior to using them in a formulation that may be applied in or around a wound site and to test the full formulation for toxicity prior to marketing, even if it meets OTC monograph requirements.

Film-Forming Polymers

Few in number, these products are the newest of the skin protectants and currently include products that contain some form of acrylate copolymer. These can be painted or sprayed on and represent the state of the art in convenience. No-sting, quick-drying agents are employed to rapidly deliver the polymer film without the stinging of alcohols. The formulations are usually not complex and do not contain preservatives. These products are marketed as medical devices such that the toxicity resulting from subchronic and chronic exposure to the mixtures is not of great issue, but testing for irritation and sensitization should be conducted as a minimum. It is possible that monomers or related chemicals known to be irritating or to cause contact sensitization may find their way into the product (e.g., methacrylic acid).

Latex

Natural latex, obtained from the sap of the *Hevea brasiliensis* and other rubber trees, is used in the manufacture of many medical products, including diaphragms, bandages, stethoscopes, tourniquets, catheters, disposable syringes, surgical gloves, and ostomy products. Latex is also used in some older formulations for skin care and burn treatment. Despite the current trend to eliminate the use of latex in wound healing formulations, concern about latex allergy continues.

The term *latex allergy* or *natural rubber latex allergy* is the term used to describe contact sensitization to the proteins contained in natural latex. Some irritant or contact dermatitis occurring upon exposure to some latex products may actually be due to chemicals used in the manufacture of latex products, such as accelerators and polymerizing agents. After extensive research and testing, the FDA created a rule in Chapter 21, Code of Federal Regulations, Part 801, "User Labeling of Natural Rubber-Containing Medical Devices Requiring Warning Labels On All Medical Devices and Medical Device Packaging Containing Natural Rubber That May Contact Humans." The rule calls for warning labels on all medical products containing natural latex rubber, including the powder inside latex surgical gloves.

Warning labels follow almost a decade of increasing research and concern by the FDA. From 1989 through 1990, patients began dying from anaphylactic shock related to the use of barium enema kits with latex cuffs. As the number of deaths increased during that period, the manufacturer voluntarily recalled the devices and started using silicone cuffs. From 1990 through 1991, anaphylactic reactions to latex anesthesia equipment were reported among children. The Centers for Disease Control and Prevention discovered that all the children who had reactions had spina bifida or other conditions involving the genitourinary tract.

Establishing procedures for latex-allergic and high-risk patients in wound care facilities can provide a formal mechanism to minimize allergic reactions. Procedures should include the following:

Identifying at-risk patients.
Providing written guidelines informing staff how to prepare a latex-free examination room.
Establishing clinical protocols for examinations and treatments requiring the use of nonlatex medical devices.
Cleaning areas exposed to latex-containing dust.
Changing ventilation and vacuum filter bags frequently.
Requiring all personnel to wash their hands immediately following the use of latex equipment, especially powdered latex gloves, to minimize the transportation of latex particles.
Converting to nonpowdered latex gloves where possible.
Ensuring that everyone involved in the care of the latex sensitive is aware of the need for nonlatex medical devices and a latex-free environment.

Surgical preparations are complex, and many hospitals have assembled nonlatex surgical carts and protocols that include a full array of substitute items.

The wound care industry is becoming more aware of the need and is responding by eliminating the use of latex in devices, formulations, and manufacturing processes wherever possible.

Hyperbaric Oxygen

Oxygen applied at pressure greater than one atmosphere (greater than 760 mm Hg or 14.7 psi) is considered to be hyperbaric and is well known to help improve topical wound healing (Van Meter, 1997). Hyperbaric oxygen (HBO) can be applied systemically or topically. An example of a typical protocol for the application of topical hyperbaric oxygen (THBO) to promote wound healing is shown in Table 28.5.

Oxygen is required for all new cell growth. Tissue at the base of chronic or non-healing wounds tends to be ischemic. Application of hyperbaric oxygen induces the growth of new blood vessels at the wound base. The new blood vessels allow an increased flow of oxygenated blood to the wound, which begins the healing process.

TABLE 28.4. *Topical hyperbaric oxygen treatment protocol*

WEEK ONE:	
Days 1 through 4	One 90-minute THBO treatment per day
Days 5 through 7	Discontinue THBO. Note: This period of relative hypoxia is an important step in the healing process.
WEEK TWO:	
Days 1 through 7	Repeat same treatments as in week one. Continue using preferred conventional would healing modalities between THBO sessions

As healing progresses, new granulation tissue exposed to hyperbaric oxygen is better vascularized. This, in turn, leads to higher tensile strength collagen being formed during wound healing, which reduces scarring and the risk of recidivism.

Another important benefit of hyperbaric oxygen is that it is bactericidal for anaerobic bacteria (e.g., *S. aureus* and *Escherichia coli*).

Systemic Hyperbaric Oxygen (SHBO)

Hyperbaric oxygen is administered in full-body chambers. The patient breathes 100% oxygen intermittently while the pressure of the treatment chamber is increased to 2–3 atmospheres, equivalent to 1500–2500 mmHg, or 30–45 psi. The systemic increase in oxygen concentration improves tissue oxygenation, thereby aiding in control of infection and improved wound healing.

Topical Hyperbaric Oxygen (THBO)

Oxygen applied directly to the base of an open wound at a pressure slightly above atmospheric (e.g., 1.03 atmospheres: 22 mm Hg, or 0.4 psi) is also thought by some to have a beneficial effect on wound healing. THBO is often applied using disposable devices designed to be used one time and discarded. There are also multiple-use extremity chamber devices that are heavy, awkward to handle, and difficult to clean and disinfect. These are becoming less desirable, particularly for home use.

The difference between SHBO and THBO in therapeutic approach is that SHBO increases blood oxygen levels; however blood oxygen levels are normally adequate for wound healing. The problem is that oxygen delivery to the wound site can be limited by poor wound tissue vascularization.

THBO, on the other hand, delivers oxygen directly to the wound. Transcutaneous oxygen levels are increased despite the lack of well-vascularized wound tissue. In addition, because this therapy is topical and relatively low pressure, there is no systemic absorption of oxygen, and therefore no risk of pulmonary or central nervous system toxicity that can result from breathing high-pressure (30–45 psi) oxygen in full-body chambers.

WOUND HEALING PRODUCT TOXICITY EVALUATION: CURRENT STATE OF THE SCIENCE

Toxicity testing requirements for wound care products depend on their drug classifications. The different types of wound care products can fall under different drug classifications, even within their own wound care product group. For example, wound cleansing agents can be new drugs, OTCs. Even medical devices and topical antimicrobials can be new drugs or OTCs. Wound healing promoters can be divided into new drugs and old drugs, with each having different regulatory requirements for safety evaluation. There are even products that probably fall under the rather dubious term "cosmeceutical" with respect to their claims related to use in wound healing.

Regulatory classifications of different wound care products are ever changing at present, as the number and type of products increases rapidly in the marketplace. Regulatory agencies do all that they can to evaluate and approve new products and to streamline regulatory requirements, but they must struggle with the classification issue as well as the manufacturer's claims that can often cross product type and regulatory classification guidelines. For example, manufacturers may want to have a wound cleanser with an additional antimicrobial claim, or a skin protectant with an antimicrobial and a promotion of wound healing claim. This creates some rather interesting regulatory dialogue, to say the least.

Another problem is the lack of standardized models for assessing the safety of wound care products. Wound cleansers, for example, are evaluated using cytotoxicity assays that dilute the products from the start and do not take into consideration that the surfactancy needed for good wound cleansing alone will produce cytotoxicity in the models (Hellewell et al., 1997). The results of these models are then used to construct a "toxicity index" delivered to healthcare providers with the message that products with lower index values are better for use than those with greater values. These models lack proper validation and may, in fact, lead the industry toward producing substandard products. These products might be nothing more than buffered saline solutions with perhaps a touch of surfactant for wound cleansing, with prices competitive with cleansers designed to rapidly and efficiently mobilize necrotic Débris from the wound site such that proliferation of wound healing cells is not inhibited. Clearly there is a need for standardized and validated models.

The only standardized models for product safety evaluation available for the evaluation of new products for use in wound care are those designed to evaluate new drugs, medical devices, or cosmetic ingredients. There are no toxicity models specifically designed for wound care product safety assessment. Manufacturers and regulatory agencies typically rely on a battery of tests using standardized models to gain some insight as to the safety of a new product to be placed into a wound. Whenever necessary, specialized models may be developed to evaluate a particular aspect of concern, such as the bioassimilation of fibers or other product residues in the wound bed, or the inhalation of aerosols from a spray-on product.

Manufacturers of wound care products range from very large corporations to very small companies. Their resources and knowledge concerning the need and desirability to obtain good safety data prior to marketing a new product are equally as wide ranging. For example, in the case of OTC products, this can lead to the use of a wide variety of excipients in products, such as wound cleansers and antimicrobial creams that may not be examined for irritation, sensitization, or cytotoxicity prior to their entry into the market. Of course, user experience and preference, as well as product liability claims and adverse event reports to the FDA, will sort out some of the worst products, but not before they have caused some problem that might have been avoided with appropriate safety testing.

Defining what appropriate safety testing is for a particular product is a problem for many of the newer or less-experienced companies wanting to market safe wound-care products. Larger corporations usually either have a toxicologist on staff, use experienced consultants, or both. Smaller wound care companies may not have the resources to have a toxicologist on staff or to afford the services of an experienced toxicology consultant; in fact, they may not even realize that such assessment is necessary until confronted by the FDA or they receive an adverse event complaint. Regulatory agencies, such as the FDA and the Occupational Safety and Health Administration (OSHA), have recently opened internet web sites that provide information and guidance documents to industry for product-safety-related matters. In the future, consideration should be given to adding minimum-product, full-formulation safety testing requirements for OTC wound care products into the monographs for each product type. This would help ensure that new OTC formulations did not contain problematic excipients and that manufacturers had on file a minimum of useful safety information to support material safety data sheet (MSDS) preparation and for reference in responding to adverse event reporting.

(Bronaugh and Maibach, 1982, 1984) have provided excellent reviews and descriptions of different methods used for the evaluation of the safety of cosmetics and cosmetic preservatives. These reviews can provide a basis for the development of approaches for the testing of product formulations intended for use in wounds. The reader is referred to these documents for more detail on the mechanics of irritation, sensitization, phototoxicity, photosensitization, human patch testing, and other assays (e.g., mutagenicity, in vitro cytotoxicity, etc.) conducted for cosmetics and cosmetic ingredients. Additional information on the range of available tests for assessment of the dermatoxicity of products are provided by (Hobson, 1991; Patrick and Maibach, 1994; Marzulli and Maibach, 1996). The testing of products intended for use in open wounds should include similar considerations to those for cosmetics and other topical products with the addition of use on broken or scarified skin in in vivo irritation and sensitization testing, and with perhaps less concern about phototoxicity and photosensitization.

In developing guidelines for the proper evaluation of a particular type of product, first consideration should be given to the intended use of the product and the phase in the wound healing process for which it is to be used. Secondly, collection and review of the available safety data on the formulation and its ingredients will identify any missing information or concerns for further testing of the formulation. Safety testing of wound care products generally involves the use of both in vitro and in vivo models. *In vitro* models may include mutagenicity tests and cytotoxicity tests on cell types that may be active during the particular phase of healing during which the product is to be applied. Cytotoxicity tests using cell types such as fibroblasts, keratinocytes, and leukocytes are readily available in the literature. In vivo models should examine and address primary dermal irritation, skin sensitization, effects on wound closure rate, the kinetics of systemic absorption of new active ingredients, histopathological effects on the wound healing process, etc. Suitable in vivo models are available in the literature and could be easily developed in standardized formats.

What then is to motivate the industry and regulatory agencies to develop, perform, and require such safety testing? At present, the wound care industry is growing to the pace of an increasing population of individuals over the age of 65 that will demand the best consumer protection possible from their healthcare providers and regulatory agencies. They will not likely tolerate the regulation of wound care products based on the frequency of adverse effects reports received when there is little safety testing performed in the premarketing phase. They will also not tolerate poor quality science to help position products in the marketplace or to promote a particular feature without respect to proper safety assessment.

CONCLUSIONS

Wound care is an ever-increasing worldwide concern. The process of wound healing is complex but being better understood at a rapid rate. Wound care products of various types are available and are constantly changing to suit market demands for better products. The promotion of wound healing is a very active area of research, and the first commercial growth-factor-containing product has entered the market. This situation has led to a need for the development and requirement for standardized models designed to properly assess the safety of new wound healing products. The demand for these standardized studies is likely to come from the consumer base and industry and regulatory agencies.

Wound care providers are also becoming more experienced and more knowledgeable about the need for good quality safety data on the wound care products they use and recommend. They can also demand the development and validation of suitable tests for product safety evaluation. As clinicians become experienced with the merits of good toxicological data collection and interpretation, statements made today, such as, "do not put anything in a wound that you would not put into your eye," will be considered carefully and will, hopefully, become less acceptable to wound care practitioners and disappear into the past.

REFERENCES

AHCPR (1992): Pressure Ulcers in Adults: Prediction and Prevention, Clinical Practice Guideline Number 3, Agency for Health Care Policy and Research (AHCPR) Pub. No. 92-0047, AHCPR Publications Clearinghouse, Silver Spring, MD 20907.

AHCPR (1994): Treatment of Pressure Ulcers, Clinical Practice Guideline Number 15, Agency for Health Care Policy and Research, AHCPR Publication No. 95-0652, AHCPR Publications Clearinghouse, Silver Spring, MD 20907.

American Heritage Dictionary of the English Language, 3rd edition. Houghton Mifflin Co., 1992.

Anderson, M. G. (1978): Neomycin ototoxicity associated with wound irrigation in the local treatment of osteomyelitis. *J. Fl. Med. Assoc.*, 65:20–21.

Bendy, R. H. Jr., Nuccio, P. A., Wolfe, E., Collins, B., Tamburro, C., Glass, W., and Martin, C. M. (1964): Relationship of quantitative wound bacterial counts to healing of decubiti: Effect of topical gentamicin. *Antimicrob Agents Chemother*, 4:147–55.

Bronaugh, R. L., and Maibach, H. I. (1982): Adverse reactions to cosmetic ingredients. In: *Principles of Cosmetics for the Dermatologist*. Edited by Frost, P., Horwitz, S., C. V. Mosby, St Louis, M. O.

Bronaugh, R. L., and Maibach, H. I. (1984): Safety evaluation of cosmetic preservatives. In: *Cosmetic and Drug Preservation: Principles and Practice*. Edited by Kabara, J. J., Marcel Dekker, Inc., New York.

Browse, W. L., and Burnand, K. G. (1982): The cause of venous ulceration. *Lancet*, 2:243–245.

Chernomorsky, S. A., and Segelman, A. B. (1988): Review article: Biological activities of chlorophyll derivatives. *New Jersey Medicine*, 85:669–73.

Cho, C. Y., Lo, J. S. (1998): Dressing the part. *Dermatol. Clin.*, 16:25–47.

Davia, J. E., Siemsen, A. W., Anderson, R. W. (1970): Uremia, deafness, and paralysis due to irrigating antibiotic solutions. *Arch. Intern. Med.*, 125:135–9.

Davies, J. W. L. (1983): Synthetic materials for covering burn wounds: Progress towards perfection. *Burns*, 10:94–108.

Ebbell, B. (1937): *The Papyrus Ebers: The Greatest Egyptian Medical Document*. London, Oxford University Press.

Elder, R. L. (1980): Cosmetic ingredients: Their safety assessment: Final report of the safety assessment for imidazolidinyl urea. *J. Environ. Path. Toxicol.*, 4:133–146.

Elder, R. L. (1984): Safety assessment of cosmetic ingredients: Final report on the safety assessment of methylparaben, ethylparaben, propylparaben, and butylparaben. *J. Am. Coll. Toxicol.*, 3:147–209.

F. Scott Fitzgerald (1896–1940), *Tender is the Night*, (1934). New York, Scribner.

Fisher, A. A. (1979): Paraben dermatitis due to a new medicated bandage: The "paraben paradox." *Contact Dermatitis*, 5:273–274.

Fleming, A. (1919): The action of chemical and physiological antiseptics in a septic wound. *Br. J. Surg.*, 7:99–129.

Gogia, P. (1995): *Clinical Wound Management*. Thorofare, NJ, Slack Inc.

Gottardi, W. (1995): The uptake and release of molecular iodine by the skin: Chemical and bactericidal evidence of residual effects caused by povidone-iodine preparations. *J. Hospital Infect.*, 29:9–18.

Goyer, R. A. (1991): Toxic effects of metals. Chap.19 In. Amdur, M. O., Doull, J., Klaassen, C. D., Eds. Cassarett and Doull's Toxicology the basic science of poisons, fourth edition., Pergamon Press, New York, NY.

Gruskin, B. (1940): Chlorophyll: Its therapeutic place in acute and suppurative disease. *Am. J. Surgery*, 49:49–53.

Haag, T. E., and Loncrini, D. F. (1984): Esters of para-hydroxybenzoic acid. In: *Cosmetic and Drug Preservation: Principles and Practice*. Edited by Kabara, J. J., Marcel Dekker, Inc., New York.

Hellewell, T. B., Major, D. A., Forseman, P. A., and Rodeheaver, G. T. (1993): A cytotoxicity evaluation of an antimicrobial and non-antimicrobial wound cleanser. *Wounds*, 5:226–231.

Hellewell, T. B., Major, D. A., Forseman, P. A., and Rodeheaver, G. T. (1997): A cytotoxicity evaluation of antimicrobial and non-antimicrobial wound cleansers. *Wounds*, 9:15–20.

Hess, C. T. (1990): Alert: Wound healing halted with the use of povidone-iodine. In: *Chronic Wound Care: A Clinical Source Book for Healthcare Professionals*. Edited by Krasner, D. Health Management Publications, Inc., King of Prussia, PA.

Hickey, J., Panicucci, R., Duan, Y., Dinehart, K., Murphy, J., Kessler, J., and Gottardi, W. (1997): Control of the amount of free molecular iodine in iodine germicides. *J. Pharm. Pharmacol.*, 49:1195–1199.

Hobson, D. W. (1991): *Dermal and Ocular Toxicology: Fundamentals and Methods*. CRC Press, Boca Raton, FL.

Human Genome Sciences. (1998): Initiates Phase I Studies of KGF-2, A New Wound-Healing Therapy, February 10, 1998, Information release in the public domain.

Hunt, T. (1995): Iodine and wound physiology. Symposium proceedings, Hunt, T., Chairman. Padua, Italy, September 1, 1995, Information Transfer Ltd., Cambridge, UK.

Kane, D. P., and Krasner, D. (1997): Wound healing and wound management. In: *Chronic Wound Care: A Clinical Source Book for Healthcare Professionals*. Edited by Krasner, D., Kane, D., second edition. Health Management Publications, Inc., Wayne, PA.

Kennedy, K. L., and Tritch, D. L. (1997): Debridement. In: *Chronic Wound Care: A Clinical Source Book for Healthcare Professionals*, second edition. Edited by Krasner, D., Kane, D. Health Management Publications, Inc., Wayne, PA.

Kinsella, K., and Gist, Y. J. (1998): International brief: Gender and aging, mortality and health. U.S. Dept. of Commerce, Economics and Statistics Administration, Bureau of the Census, report no. IB/98–2.

Krasner, D., and Kane, D. (1997): *Chronic Wound Care: A Clinical Source Book for Healthcare Professionals*, second edition. Health Management Publications, Inc., Wayne, PA.

Kucan, J. O., Robson, M. C., Heggers, J. P., and Ko, F. (1981): Comparison of silver sulfadiazine, povidone-iodine and physiologic saline in the treatment of chronic pressure ulcers. *J. Am. Geriatr. Soc.*, 29:232–5.

Kucers, A., and Bennett, M. (1987): Neomycin, framycetin, and paromycin. In: *The Use of Antibiotics*, 4th edition. Lippincott, Philadelphia. pp. 739–50.

Lamme, E. N., Gustafsson, T. O., and Middelkoop, E. (1998): Cadexomer-iodine ointment shows stimulation of epidermal regeneration in experimental full-thickness wounds. *Arch. Dermatol. Res.*, 290: 18–24.

Lineaweaver, W., Howard, R., Soucy, D., McMorris, S., Freeman, J., Crain, C., Robertson, J., and Rumley, T. (1985): Topical antimicrobial toxicity. *Arch. Surg.*, 120:267–70.

Mandell, G. L., and Sande, M. A. (1990): Antimicrobial agents: Sulfonamides, trimethoprim-sulfamethoxazole, quinolones and agents for urinary tract infections. In: *The Pharmacological Basis of Therapeutics*, 8th Edition. Edited by Gilman, A. G., Rall, T. W., Nies, A. S., and Taylor, P. Pergamon Press, New York.

Marzulli, F. N., and Maibach, H. I. (1996): *Dermatotoxicology*, 5th edition. Taylor and Francis, Washington, DC.

McCulloch, J. M., Kloth, L. C., and Feedar, J. A. (1995): *Contemporary Perspectives in Rehabilitation: Wound Healing*, 2nd edition. Philadelphia, F. A. Davis Co.

McNeil Pharmaceutical. (1998): Regranex package insert. 635-10-240-1, Raritan, NJ.

Nathan, P., Law, E. J., Murphy, D. F., and MacMillan, B. G. (1978): A laboratory method for selection of topical antimicrobial agents to treat infected burn wounds. *Burns*, 4:177–187.

Mertz, P., Davis, S., Brewer, L., and Franzen, L. (1994): Can antimicrobials be effective without impairing wound healing? The evaluation of a cadexomer iodine ointment. *Wounds: A Compendium of Clinical Research and Practice*, 6:23–8.

Niedner, R. (1997): Cytotoxicity and sensitization of povidone-iodine and other frequently used anti-infective agents. *Dermatology*, 195 (Suppl 2):89–92.

Patrick, E., and Maibach, H. (1994): Dermatotoxicology, In: *Principles and Methods of Toxicology*, 3rd edition. Edited by Hayes, A. W., Raven Press, New York.

Perrenoud, D., Bircher, A., Hunziker, T., Suter, H., Bruckner-Tuderman, L., Stager, J., Thurlimann, W., Schmid, P., Suard, A., and Hunziker, N. (1994): Frequency of sensitization to 13 common preservatives in Switzerland. *Contact Dermatitis*, 30:276–279.

Reimer, K., Fleischer, W., Brogmann, B., Schreier, H., Burkhard, P., Lanzendorfer, A., Gumbel, H., Hoekstra, H., and Behrens-Baumann, W. (1997): Povidone-iodine liposomes: An overview, Dermatology, 195 (Suppl 2):93–9.

Roberts Pharmaceutical Inc. (1997): Furacin topical cream. In: *Physician's Desk Reference*, 51st edition. Medical Economics Company, Inc., Montvale, NJ.

Robson, M. C., and Heggers, J. P. (1984): Quantitative bacteriology and inflammatory mediators in soft tissue. In: *Soft and Hard Tissue Repair*. Edited by Hunt, T. K., Heppenstall, R. B., Pines, E., Rovee, D. pp. 483–507. Praeger Publications, New York.

Rodeheaver, G., Bellamy, W., Kody, M., Spatafora, G., Fitton, L., and Leyden, K. (1982): Bactericidal activity and toxicity of iodine-containing solutions in wounds. *Archives of Surgery*, 117:181–6.

Rodeheaver, G. T. (1990): Controversies in topical wound management: Wound cleansing and wound disinfection. In: *Chronic Wound Care: A Clinical Source Book for Healthcare Professionals*. Edited by Krasner, D. Health Management Publications, Inc., King of Prussia, PA.

Rodeheaver, G. T. (1997): Wound cleansing, wound irrigation, wound disinfection. In: *Chronic Wound Care: A Clinical Source Book for Healthcare Professionals*, second edition. Edited by Krasner, D., Kane, D. Health Management Publications, Inc., Wayne, PA.

Rodeheaver, G. T., Gentry, S., Safferm, I., and Edlich, R. F. (1980): Topical antimicrobial cream sensitivity testing. *Surg. Gynec. Obstet.*, 151:747–752.

Rosen, M. (1984): Glydant and MDMH as cosmetic preservatives. In: Kabara, J. J., Ed. Cosmetic and drug preservation: principles and practice. Marcel Dekker, Inc., New York.

Rund, C. R. (1997): Alternative topical therapies for wound care. In: *Chronic Wound Care: A Clinical Source Book for Healthcare Professionals*, second edition. Edited by Krasner, D., Kane, D. Health Management Publications, Inc., Wayne, PA.

Sande, M. A., and Mandell, G. L. (1990): Antimicrobial agents: The aminoglycosides. In: *The Pharmacological Basis of Therapeutics*, 8th Edition. Edited by Gilman, A. G., Rall, T. W., Nies, A. S., Taylor, P. Pergamon Press, New York.

Sapico, F. L., Ginunas, V. J., Thornhill-Joynes, M., Canawati, H. N., Capen, D. A., Klein, N. E., Khawam, S., and Montgomerie, J. Z. (1986): Quantitative microbiology of pressure sores in different stages of healing. *Diagn. Microbiol. Infect. Dis.*, 5:31–8.

Smith, L. W. (1955): The present status of topical chlorophyll therapy. *NY State J. of Med.*, 55:2041–50.

Teepe, R. G., Koebrugge, E. J., Lowik, C. W., Petit, P. L., Bosboom, R. W., Twiss, I. M., Boxma, H., Vermeer, B. J., and Ponec, M. (1993): Cytotoxic effects of topical antimicrobial and antiseptic agents on human keratinocytes in vitro. *J. Trauma*, 35:8–19.

Trimble, G. X. (1969): Neomycin ototoxicity dossier and doses. *N. Engl. J. Med.*, 281:219.

Van Meter, K. (1997): Systemic hyperbaric oxygen therapy as an aid in resolution of selected chronic problem wounds. In: *Chronic Wound Care: A Clinical Source Book for Healthcare Professionals*, second edition. Edited by Krasner, D., Kane, D. Health Management Publications, Inc., Wayne, PA.

U. S. Census Bureau (1999): On line database estimates obtained January 31, 1999.

Van Rijswijk, L. (1997): Wound assessment and documentation. In: *Chronic Wound Care: A Clinical Source Book for Healthcare Professionals*, second edition. Edited by Krasner, D., Kane, D. Health Management Publications, Inc., Wayne, PA.

Varghese, M. C., Balin, A. K., and Carter, M. (1986): Local environment of chronic wounds under synthetic dressings. *Arch. Dermatol.*, 122:52–57.

Velkoff, V. A, and Lawson, V. A. (1998): International brief: Gender and aging, caregiving. U. S. Dept. of Commerce, Economics and Statistics Administration, Bureau of the Census, report no. IB/98–3.

Ward, R. S., and Saffle, J. R. (1995): Topical agents in burn and wound care. *Phys. Ther.*, 75:526–538.

Watson, D. (1988): Pharmacist liability for injury resulting from use of neomycin irrigation. *Am. J. Hosp. Pharm.*, 45:73–4.

Weinstein, A. J., McHenry, M. C., and Gavan, T. C. (1977): Systemic absorption of neomycin irrigation solution. *JAMA*, 238:152–3.

White, E., and Anderson, L. (1995): Characterization of wound gel products. *Private Communication*.

Winter, G. (1962): The formation of the scab and rate of epithelisation of superficial wounds in the skin of the domestic pig. *Nature*, 193:293–94.

Wiseman, D. M., Rovee, D. T., and Alvarez, O. M. (1992): Wound dressings: Design and use. In: *Wound Healing: Biochemical and Clinical Aspects*. Edited by Cohen, I. K., Diegelmann, R. F., Lindblad, W. J., W. B. Saunders, Philadelphia, PA.

Witkowski, J. A. and Parish, L. C. (1991): Topical metronidazole gel. The bacteriology of decubitus ulcers. *Int J. Derm.*, 30:L660–661.

Wright, R., and Orr, R. (1993): Fibroblast cytotoxicity and blood cell integrity following exposure to dermal wound cleanser. *Ostomy/Wound Management*, 39:33–40.

Toxicology of Skin
Edited by Howard I. Maibach
Copyright © 2001 Taylor & Francis

29

Issues of Toxicity with Dermatologic Drugs During Pregnancy

Barbara R. Reed

University of Colorado Health Sciences Center, Denver, Colorado, USA

INTRODUCTION

Prior to the discovery that thalidomide was capable of causing severe developmental toxicity, embryology experts believed that the fetus and mother were completely separate beings. In 1958, Patten's text noted, "It should be emphasized that at no time during pregnancy is there any mingling of fetal and maternal blood streams. The fetal circulation is, from its first establishment, a closed circuit (Patten, 1958)." Subsequent to the discovery of the severe malformations attributed to thalidomide, the perceived placental separation was reworked. It was clear that placental and maternal circulatory systems commingled in some fashion and that transfer of drug between mother and child had occurred.

Dermatologists' awareness of teratogenic risk was heightened after 1981, when isotretinoin (Accutane) was released. Isotretinoin was a known teratogen, and despite wide publicity and labeled precautions, some women became pregnant while on the drug. Some women who carried the pregnancies to term had infants with congenital anomalies. Cautions were strengthened amidst threats of removal of the drug from use by the Food and Drug Administration (FDA).

Many events, such as the discovery of thalidomide and retinoid teratogenicity, co-incided in such a way that awareness of potential toxicity of drugs was a threat known to all physicians and to most patients. As drug companies began advertising in the public media, patients became better informed. Physicians had wider communication through use of the internet. The National Library of Medicine, Food, and Drug Administration and various teratology services went online. Physicians and patients are now more aware of the problems with use of drugs in pregnancy. In the future the medical profession will be under greater pressure to sift through data, gather epidemiologic information, and identify and quantify risk for a particular stage of pregnancy or for a particular genetic profile of mother and/or infant.

CAUSE OF BIRTH DEFECTS

In the United States, 65% of birth defects are of unknown origin. When the cause is known, birth defects are accounted for by genetic transmission in 15%–20% of cases, by chromosomal abnormalities in 5%, by maternal infections in 3%, by maternal conditions, including diabetes and various addictions, in 4%, and by maternal or infant deformities in 1%–2%. Drugs, including chemicals, radiation, and hyperthermia, are responsible for less than 1% of all birth defects (Sever and Mortensen, 1996).

Wilson's six principles of teratogenesis aid in the understanding of how teratogenesis occurs. These principles are as follows:

1. Genotype. Susceptibility of the organism depends on the manner in which the genotype of the conceptus interacts with environmental factors.
2. Developmental Stage. Gestational age at the time of exposure is crucial in determination of the type of injury to which the fetus is susceptible. Most structural defects occur when exposure to the agent has occurred between week 2 and week 8 of gestation. Exposure at the stage of germ cell development may lead to germ-line

mutation and/or infertility. Exposure during the stages of fertilization, ovum transport, implantation, and placenta formation may lead to embryotoxicity. Exposure during organogenesis, fetal maturation, and fetal survival (days 18–55) may lead to teratogenesis or fetotoxicity.

3. Drug Mechanism of Action. True teratogenic agents always act in specific ways. Thalidomide, for example, repeatedly produces limb reduction defects. Cleft lip, on the other hand, is reported only very rarely.

4. Final Manifestation vs. Exposure Time. The manifestations of teratogenesis include death, malformation, growth retardation, and functional disorder. Early exposure is more likely to result in death or malformation; later exposure to the same drug may produce growth retardation or functional disorders.

5. Agent Access. Access of environmental influences to the conceptus depends on the nature of the agent. Agents must have direct access to the fetus to produce an effect.

6. Dose-Effect. As drug dose, duration, and type of exposure (continual vs. intermittent) increase, developmental toxicity responses appear and may approach lethality. High doses of drug may produce death of the embryo or fetus, whereas low doses may produce an observable congenital anomaly, possibly confusing the interpretation of the drug-dose response. At low levels there is a no-effect dose. It is not until a concentration threshold is exceeded that injury to the fetus occurs. Factors contributing to exceeding the concentration threshold include degree of drug absorption by the mother, ease of placental crossing by the drug, and ability of drug to deposit in fetal tissue (Selevan and Lemasters, 1987).

DRUGS IN PREGNANCY: MATERNAL CONSIDERATIONS

Drug Distribution

Body water increases substantially, with mean plasma volume increasing by 20% during the second trimester and 50% by term. This results in an increased apparent volume of distribution and can prolong the half-life of drugs. Body fat also increases and may function as a harbor for fat-soluble drugs. Plasma albumin concentrations decrease during pregnancy. As a result, some drugs, especially anticoagulants, anticonvulsants, and benzodiazepines, pass more easily into tissues (Murray and Seger, 1994).

Drug Absorption

Although there is some decrease in gastric emptying time and intestinal transit time, the effects on absorption kinetics are minimal (Cruikshank et al., 1996).

Drug Metabolism: Hepatic and Renal

Although it is possible that some hepatic enzyme systems may be induced, there are no significant changes in blood flow to the liver (Cruikshank et al., 1996). Blood flow to the kidney increases by almost 50%, as does glomerular filtration rate (Cruikshank et al., 1996). The net result is a corresponding increased rate of clearance for drugs metabolized by the kidney (Morgan, 1997).

Drugs Crossing the Placenta

Most drugs cross the placenta by simple diffusion, although there may be a poorly understood mechanism by which the placenta metabolizes drugs, especially those related to endogenous compounds (Schatz and Pettiti, 1997). Most drugs at steady state in the mother result in drug level of 50%–100% in infant. Lipophilic drugs pass the placental membrane more easily than hydrophilic drugs (Morgan, 1997).

DRUGS IN PREGNANCY: FETAL CONSIDERATIONS

Drug Metabolism

Fetal and neonatal hepatic enzymes are much less active than that of adults, and protein binding may be impaired, especially if the infant is hyperbilirubinemic. This results in slowed drug excretion, which may prolong fetal exposure to a drug (Murray and Seger, 1994). Also in the fetus and neonate, glomerular filtration rate and renal secretory processes may be immature, possibly prolonging exposure to the infant. Drugs excreted through the kidneys enter the amniotic fluid and are re-ingested by the fetus through swallowing (Morgan, 1997).

In summary, fetal response to a drug is a result of drug concentration in the fetus. This is a net effect of the amount of drug the mother ingests, protein binding and ionization in maternal serum, placental transport, protein binding and ionization in fetal serum, and fetal drug clearance (Garland, 1998).

SOURCES OF INFORMATION ON DRUGS

1. *Physician's Desk Reference* and FDA Pregnancy Categories.
   Through the (*Physicians' Desk Reference*, 1999), most physicians are aware of the FDA Pregnancy Categories (Code of Federal Regulations, 1999); FDA Pregnancy Categories are listed in Table 29.1. Labeling for all pregnancy categories is also required to contain a description of available data on the effect of the drug on the later growth, development, and functional maturation of the child. Revisions of FDA Pregnancy Categories may be forthcoming, as the categories are found to be confusing and lacking in helpful data.
2. Other printed sources of information on drug use during pregnancy.
   These are listed in Table 29.2.
3. Online Sources of Information.
   There are a number of national and international online sources of information on developmental toxicity. These are listed in Tables 29.3a and 29.3b.

TABLE 29.1. *FDA pregnancy categories*

X	Positive evidence of human fetal risk in studies on animals or humans, and/or positive human fetal risk from investigational or post-marketing reports, and risk of use clearly outweighs any benefit. Label contains information under "Contraindications" section.
D	Positive human fetal risk from investigational or post-marketing reports. Potential benefit to mother may be acceptable despite risk to fetus. Label contains information under "Warnings" section.

C	Positive animal fetal risk and no human studies OR no studies in animals or humans.
	Benefits acceptable despite possible risk.
	Label contains information on studies, if any.
	Label describes information on effect of drug on
	    later growth
	    development
	    functional maturation of child.

B	Negative animal fetal risk and no studies in humans OR positive animal risk and negative human studies in all trimesters.
	Label describes animal and human studies.
	Label describes information on effect of drug on
	    later growth
	    development
	    functional maturation of child.

A	Negative human studies in all trimesters.
	Label describes human studies and animal studies, if any.
	Label describes information on effect of drug on
	    later growth
	    development
	    functional maturation of child.

4. States with Teratology Information Services.
A number of states have information on developmental toxicity. These are listed in Table 29.4.

RISK RELATED TO TIME OF DRUG CONSUMPTION

Risks from drug intake during pregnancy differ from those during the prenatal period or during lactation. Drug-related risks during pregnancy may vary with the trimester

TABLE 29.2. *Printed sources on drug use during pregnancy and lactation*

United States Pharmacopeial Dispensing Information, Volume 1.	This volume, updated yearly, contains detailed information on mutagenicity, carcinogenicity, effects on fertility, teratogenicity, and lactation. Information is available through USPC, 12601 Twinbrook Parkway, Rockville, MD 02780.
Briggs et al., Drugs in Pregnancy and Lactation	A regularly updated text reviewing risks of drugs used during pregnancy and lactation.

TABLE 29.3a. *National on-line sources of information on developmental toxicity*

NIH Birth Defects and Teratology Interests
Wesbite: *http://www.nih.gov/sigs/birth-def/index.html*

Organization of Teratology Information Services (OTIS) (801)/328-2229
Wesbite: *http://orpheus.ucsd.edu./otis*

of pregnancy as well. Drugs labeled as teratogenic may put a fetus at risk for only a few weeks of pregnancy. Nevertheless, it has been recommended that these drugs be avoided for the entire pregnancy.

Drugs that place a fetus or mother at risk during lactation may be different from those that affect an infant during pregnancy. During pregnancy there is a circulatory continuity between mother and infant that is interrupted at birth. After birth, gastrointestinal absorption by the infant is a factor. Relatively, alkaline drugs, such as penicillin, may be preferentially absorbed, whereas drugs bound to milk, such as tetracycline, may be absorbed minimally if at all. Nevertheless, drugs for which use is prohibited during pregnancy are routinely contraindicated during lactation. Drugs that have minimal risk to mother and fetus or infant have been reported elsewhere (Reed, 1997a; 1997b; 1998).

TABLE 29.3b. *International on-line sources of information on developmental toxicity*

Japan
Japanese Teratology Society
c/o Department of Anatomy and Developmental Biology,
Kyoto University Faculty of Medicine
Konoe-cho, Yoshida, Sakyo-ku, Kyoto 606-01, Japan
E-mail: *miyabi@med.kyoto-u.ac.jp*

Europe
ENTIS European Network of Teratology Information
    Services (ENTIS).
ENTIS liaison: Elisabeth Robert, M.D., Ph.D.
Institut Europeen des Genomutations,
86 rue Docteur Edmond Locard
69005 Lyon, France
(33) 78 25 82 10
e-mail robieg@univ.-lyon1.fr.

European Teratology Society
Wesbite: *http://www.etsoc.com*

Britain
British Toxicology Society
Wesbite: *http:www/bts/org/toxicology*

Canada
Motherisk Program
(416) 813-6780
Geographic Region: Ontario (and across Canada)
Wesbite: *http://www.motherisk.org*

TABLE 29.4. *States with information on developmental toxicity*

ALABAMA	University South Alabama Department of Medical Genetics (800) 423-8324 or (334) 460-7500 Geographic Areas Served: Alabama
ARIZONA	Arizona Teratogen Information Program University of Arizona, Tucson, Arizona (888) 285-3410 or (520) 626-3410 Geographic Area Served: Arizona/National
ARKANSAS	Arkansas Genetics Program Arkansas Teratogen Information Service University of Arkansas for Medical Sciences Department of Obstetrics and Gynecology (800) 358-7229 or (501) 296-1700 Geographic Area Served: Arkansas
CALIFORNIA	California Teratogen Information Service & Clinical Research Program UCSD Medical Center Department of Pediatrics San Diego, California (800) 532-3749 (CA Only) or (619) 543-2131 Geographic Area Served: California
CONNECTICUT	Connecticut Pregnancy Exposure Information Service University of Connecticut Health Center Division of Human Genetics MC6310 (800) 325-5391 (CT Only) or (860) 679-8850 Geographic Region Served: Connecticut
DISTRICT OF COLUMBIA	Reproductive Toxicology Center Columbia Hospital for Women Medical Center (202) 293-5137 Geographic Region Served: Unrestricted
FLORIDA	Florida Teratogen Information Service University of Florida Health Science Center Gainesville, Florida (800) 392-3050 (FL Only) or (352) 392-3050 Geographic Region Served: FL, National
	Florida Teratogen Information Service University of Miami Miami, Florida (305) 243-6464 or (305) 243-3919 Geographic Region Served: Florida
	Teratogen Information Service University of South Florida Birth Defects Center Department of Pediatrics Tampa, Florida (813) 233-2627 Geographic area served: Central Florida
ILLINOIS	Illinois Teratogen Information Service Northwestern Memorial Hospital Division of Reproductive Genetics Chicago, Illinois (800) 252-4847 (IL Only) or (312) 926-7441 Geographic Region Served: Illinois

(Continued)

TABLE 29.4. (Continued)

INDIANA KENTUCKY OHIO	Indiana Teratogen Information Service Indiana University Medical Center Department of Medical and Molecular Genetics 1B130 Indianapolis, Indiana (317) 274-1071 Geographic Region Served: Indiana, Kentucky, Ohio
MASSACHUSETTS MAINE NEW HAMPSHIRE RHODE ISLAND	Massachusetts Teratogen Information Service (MTIS) Pregnancy Environmental Hotline Waltham, Massachusetts (800) 322-5014 (MA Only) or (781) 466-8474 Geographic Region Served: Massachusetts, New Hampshire, Maine, Rhode Island
	Genetics & Teratology Unit, Pediatric Service Massachusetts General Hospital Boston, Massachusetts (617) 726-1742 Geographic Region Served: Northeast U.S.
MICHIGAN	Michigan Teratogen Information Service Detroit, Michigan (877) 52-MITIS [526-4748] or (313) 966-9368 Geographic Region Served: Michigan and surrounding states
MINNESOTA	MN Perinatal Physicians Abbott-Northwest Hospital Minneapolis, Minnesota (612) 863-4502 Geographic Region Served: Minnesota
MISSOURI	Missouri Teratogen Information Service University of Missouri Hospital and Clinics Columbia, Missouri (800) 645-6164 or (573) 884-1345 Geographic Region Served: Missouri
NEBRASKA	Nebraska Teratogen Project University of Nebraska Medical Center 600 S. 42nd St. Omaha, NE 68198-5440 (402) 559-5071 Geographic Region Served: Nebraska
NEW JERSEY DELAWARE PENNSYLVANIA	Pregnancy Healthline Southern New Jersey Perinatal Cooperative Pennsauken, New Jersey (888) 722-2903 or (856) 665-6000 (NJ) Geographic Region Served: New Jersey, Delaware, Pennsylvania
NEW YORK	Pregnancy Risk Network Buffalo, New York (800) 724-2454 or (716) 882-6791 Geographic Region: New York Hours: 8:30 AM - 4:00 PM
	PEDECS University of Rochester Medical Center Department of Obstetrics and Gynecology Rochester, New York (716) 275-3638 Geographic Region Served: New York

(Continued)

TABLE 29.4. *(Continued)*

NORTH DAKOTA	North Dakota Teratogen Information Services UND School of Medicine and Health Sciences Department of Pediatrics, Division of Medical Genetics Grand Forks, North Dakota (701) 777-4277; Fax: (701) 777-3220 Geographic Region Served: North Dakota
TEXAS	Texas Teratogen Information Service UNT Department of Biology Denton, Texas (800) 733-4727 or (940) 565-3892 Geographic Region Served: Texas
UTAH MONTANA	Pregnancy Riskline Salt Lake City, Utah (801) 328-2229 Geographic Region Served: Utah, Montana
VERMONT	Pregnancy Risk Information Service Vermont Regional Genetics Center (802) 658-4310 (Upstate NY and VT) Geographic Region Served: Vermont and Upstate New York
WASHINGTON ALASKA IDAHO	CARE Northwest Box 357920 University of Washington Seattle, WA 98195-7920 (900) 225-2273 ($8/call) Geographical Region: Alaska, Idaho, Oregon, Washington
WASHINGTON	Department of Pediatrics University of Washington Seattle, Washington (900) 225-2273 Geographic Region Served: Pacific Northwest, National
WISCONSIN	Wisconsin Teratogen Project Wisconsin Clinical Genetics Center Madison, Wisconsin (800) 442-6692 Geographic Region Served: Wisconsin

Source: Organization of Teratology Information Services (http://www.orpheus.ucsd.edu/otis)

Prior to Conception

Physicians prescribing a drug for women of childbearing age should have two concerns: (1) contraceptive failure due to medication interactions and (2) potential risk to mother and/or fetus caused by the drug should pregnancy occur.

Some medications have been associated with a possible risk of contraceptive failure. These are listed in Table 29.5.

Azathioprine (Zerner et al., 1981) and nonsteroidal anti-inflammatory drugs (NSAIDs) (Papiernik et al., 1989) have been associated with increased risk of contraceptive failure of the intrauterine device. Oral contraceptive failure may occur

TABLE 29.5. *Medications associated with contraceptive failure*

Drug	Contraceptive device	Purported mechanism
Azathioprine	IUD	Unknown
NSAID	IUD	Unknown
Griseofulvin	Oral contraceptive	Increased estrogen metabolism by hepatic microsomal enzyme induction
Rifampin	Oral contraceptive	Increased estrogen metabolism by hepatic microsomal enzyme induction or reduced enterohepatic circulation of estrogen
Tetracycline	Oral contraceptive	Reduced enterohepatic circulation of estrogen
Sulfonamides	Oral contraceptive	Reduced enterohepatic circulation of estrogen
Itraconazole	Oral contraceptive	Unknown

when the oral contraceptive is taken along with hepatic enzyme inducers, such as griseofulvin (Griseofulvin, 1998), which may increase metabolism of estrogen due to induction of hepatic microsomal enzymes (USP-DI, 1995). Other oral medications that may reduce oral contraceptive effectiveness include (Rifampin, 1998), which may stimulate estrogen metabolism or reduce enterohepatic circulation of estrogens, and (Penicillins, 1998), which reduce enterohepatic circulation of estrogens. (Tetracyclines, 1998) and (Sulfonamides, 1998) may increase breakthrough bleeding and reduce contraceptive effectiveness. Pregnancy has been reported when itraconazole was used with an oral contraceptive (Hingston, 1993; Pillans and Sparrow, 1993), suggesting that itraconazole should be considered as a possible cause of oral contraceptive failure.

Males and females should avoid three medications if pregnancy is anticipated: (1) griseofulvin (Vallance, 1995), (2) (Methotrexate, 1997), and (3) (Thalidomide, 2000).

First Trimester

During very early pregnancy, encompassing the first 22.5 weeks, cells are undifferentiated. Drugs administered during this period affect all cells equally, either killing the organism or having no effect. The net effect tends to be spontaneous abortion rather than any congenital anomalies. Animal studies reflect this finding as toxicity and are considered as a potentially adverse finding in the current FDA Pregnancy Categories. Human toxicity studies have not been done.

Organogenesis lasts from week 28 of the pregnancy. Differentiating cells may be affected by particular drugs and will result in congenital anomalies. During this period, it is important to avoid all medications known to have possible teratogenic effects, including medications listed as FDA Pregnancy Category X or D, as well as some drugs that are unrated. Exceptions on an individual basis may be considered when use is essential. A list of medications included in FDA Pregnancy Categories X, D, and Unclassified is in Table 29.6. This is not intended to be an inclusive list of drugs with a potential to produce developmental toxicity.

TABLE 29.6. *FDA pregnancy category X, D, unclassified*

Androgens	(X)	Meclorethamine	(D)
Azathioprine	(D)	Methotrexate	(X)*
Bleomycin	(D)	Penicillamine	(D)
Colchicine	(D)	Potassium iodide	(D)
Cyclophosphamide	(D)	Retinoids	(X)
Finasteride	(X)	Spironolactone	(U)
Fluorouracil	(X)	Stanozolol	(X)
Griseofulvin	(U)	Tetracycline	(D)
		Thalidomide	(X)

*Both males and females should avoid if pregnancy is anticipated.

Second Trimester

During mid-pregnancy, maturation of various organ systems occurs. Drug metabolism by the fetus may be slower than that of the mother, with a potential effect for allowing prolonged fetal exposure to medications.

Drug excretion by the fetus into amniotic fluid also may promote prolonged exposure by virtue of skin contact with amniotic fluid by ingestion of drug-containing amniotic fluid by the infant (Livezey and Rayburn, 1992).

Third Trimester and Preterm

Late in pregnancy, especially near the time of delivery, nonteratogenic conditions, such as persistent fetal circulation, kernicterus, fetal hemorrhage, and fetal excitability, may occur.

INTERPRETING AVAILABLE INFORMATION

Types of Reports of Developmental Toxicity Possibly Caused by Drugs

There are several types of reports that may appear in the literature, including case reports, case-control studies, cohort studies, meta-analyses, and randomized clinical trials. Case reports offer the least validity and provide the weakest link between agent and disease. Validity of cause-effect relationship increases with case-control studies, cohort studies, and meta-analyses, and are probably strongest with randomized clinical trials.

Case Reports

Case reports are individual reports of isolated cases. They are numerators without denominators; no comparison group is identified. No causal association is proven by such studies, and although they may be valuable, they may also serve to generate undue anxiety on the part of physicians and patients, as well as unnecessary liability claims.

Case-Control Studies

In case-control studies, groups of individuals with a particular disease are compared to controls with regard to exposure history. This type of study may be used when an outcome is rare or is not apparent within a short time frame. Patients who have developed the outcome being studied are chosen and compared with an otherwise similar group of patients who have not been exposed to the drug. Although inexpensive to conduct, they have the disadvantage of relying on patient's recollection of exposure and in selecting controls (Koren et al., 1998; Sever and Mortensen, 1996). Case-control studies are best used to answer questions about prognosis or harm (Bigby, 1998).

Cohort Studies

Cohort studies identify a group of people at risk by virtue of exposure to a potential toxicant and follow them over a period of time. Cohort studies are used when it is not possible (or ethical) to assign patients randomly to an agent that may cause harm. There are two types of cohort studies. In concurrent cohort studies, a group is identified and followed prospectively. In noncurrent cohort studies, a cohort from the past is identified and followed to the present time. Post-marketing cohort studies may be performed by drug companies. These surveillance programs identify a population at risk and report outcomes. Two programs are available in the United States: the Birth Defects Monitoring Program and the Metropolitan Atlanta Congenital Defects Program; these are operated by the Centers for Disease Control and Prevention in Atlanta, Georgia (Edmonds et al., 1981). There are also a number of states that conduct the birth defect surveillance systems listed above. Also, there is a program for self-reporting of problems by exposed women to manufacturers of some drugs.

Drug companies complain that regulations on inclusion of post-market surveillance information on effects of drugs during pregnancy (Code, 1994) are so restrictive that much available information is excluded from drug labeling.

Meta-Analyses

Studies of similar design may be compared. Such studies consist of an analysis of smaller studies of developmental toxicity with similar design. Meta-analyses may be analyzed statistically and are regarded as having more validity than single case reports. Results of meta-analyses may differ from those of large clinical trials because of differences in study populations, treatment protocols, or other changes (Bigby, 1998).

Randomized Controlled Trials

In randomized controlled trials, patients are assigned to one of several groups of patients. One group will receive the agent to be tested; the other groups have some sort of alternative experience. Ideally, the two groups differ only in that one is exposed to the agent in question and the other is not. Patients are followed prospectively and outcomes are analyzed. Relationships between agents and adverse events may be established with confidence. Randomized controlled trials are best used to answer questions about therapy and prevention (Bigby, 1998). The Cochrane Collaboration Skin Group is a complete database of randomized controlled clinical trials that provides information on treatment of diseases. Results are available by subscription (Williams et al., 1998).

Animal Studies and Developmental Toxicity

Animal studies may serve as prototypes for development of human toxicity with exposure to a drug during pregnancy. Some drugs do not produce problems in animals but do in humans. Thalidomide, for example, produces no congenital anomalies in rats (Ashby et al., 1997), although anomalies in humans are well known. Salicylates produce cardiac malformations in animals but not humans (Slone et al., 1976; Werler et al., 1989).

In animals, testing is often done at high doses. The same drug may be used in comparatively low doses in humans. Comparison of degree of teratogenicity between animals and humans may not be valid when high animal versus low human doses are given.

Human Studies and Developmental Toxicity: Confounding Factors

Self-Reporting

The recall of the woman who took the drug may not be exact; women who self-report are much more likely to report adverse events than normal events (Murray and Seger, 1994). This produces a bias in favor of adverse events unbalanced by a denominator of women who took the drug without problems.

Maternal and fetal genetic predisposition may also be a factor in production of fetal injury. Drugs and chemicals other than the one being studied may have been ingested by the mother during the period of study.

Inconsistency of Type of Toxicity Reported

Malformations reported should have a pattern or consistency. When case reports are depended upon to delineate teratogenicity, inconsistency of malformation is frequent.

Many drugs reported as teratogenic have only case reports to substantiate the claim (Murray and Seger, 1994).

Delay in Appearance of Developmental Toxicity

Structural malformations often are identified early after birth. Functional problems are more difficult to identify and often appear later in childhood. Had there been in fact a direct association between drug use and functional deficits, identification of this association becomes nearly impossible.

Frequency of Toxicity

When large numbers of women have taken a drug and case reports are small in number, it may reflect a spontaneous occurrence of malformations in the population. Contrarily, if a small number of women have taken a drug and a rare type of developmental toxicity is consistently noted, a strong association between malformation and drug is likely.

Disease vs. Drug Risk

Risk assessment for medications given to pregnant women must be balanced against the risk of the disease for which the drug is being given. Some diseases have an inherent risk. For example, pregnant women with hypertension or cancer are more likely to have infants with intrauterine growth retardation; those with epilepsy or diabetes mellitus are more likely to have infants with malformations (Gonen et al., 1994).

Unplanned Exposure

Women today take a large number of drugs, both prescription and over-the-counter. Whether this is due to lack of information on the effect of drugs during pregnancy or the patient's concern for personal comfort is not certain. Many women do not report for medical care until the 6th week of pregnancy. By this time, organogenesis is well underway and drug exposure may already have occurred. A World Health Organization (WHO) study of over 14,000 women in 22 countries revealed that 79% of women received an average of 3.3 prescription drugs during pregnancy (Collaborative Group on Drug Use in Pregnancy, 1991).

DERMATOLOGIC DRUGS WITH PROVEN DEVELOPMENTAL TOXICITY (TERATOGENICITY)

(Koren et al., 1998) listed a small number of drugs with proven developmental toxicity. Of these, only a few are used routinely by dermatologists; these are listed in Table 29.7 and are discussed in detail below.

TABLE 29.7. *Dermatologic drugs with proven developmental toxicity*

Drug	Pregnancy category	Type of toxicity	At-risk trimester
Androgenic Drugs (Stanozolol)	X	Clitoromegaly Masculinizes Female fetus	1-2-3; late 1
Methotrexate	X	Abnormal skull Large fontanelle Craniosynostosis Ocular hypertelorism	Mid 1 (weeks 6–8)
NSAIDs	B, C	Oligohydramnios Pulmonary hypertension Persistent fetal circulation	3
Retinoids	X	Vitamin A embryopathy	1 (weeks 2–5)
Tetracycline	D	Dental staining	2–3
Thalidomide	X	Limb reduction Autism Mental retardation unknown	

Androgenic Drugs

Stanozolol is an anabolic androgen used in the treatment of some leg ulcers and hereditary angioedema. *Stanozolol* is FDA Pregnancy Category X.

There are no human studies on use of stanozolol during pregnancy; however, testosterone, to which stanozolol is related, has been implicated in production of masculinizing features in the female fetus. The time of most risk for labioscrotal fusion appears to be days 80–90 of gestation, although clitoromegaly can occur with administration of testosterone at any time during gestation (Grumbach, 1959). In rats, masculinization of the female fetus was found with use of stanozolol (Kawashima et al., 1977).

Methotrexate

Methotrexate is FDA Pregnancy Category X. Usage during early pregnancy has resulted in miscarriage (Donnenfeld et al., 1994; Kozlowski et al., 1990). Its use as an abortifacient is widely known. Several case reports indicate that methotrexate also produces characteristic congenital anomalies consisting of abnormal skull, large fontanelles, craniosynostosis, ocular hypertelorism, and bony defects (Milunsky et al., 1968; Powell and Ekert, 1971; Warkany, 1978). The malformations appear to be quite time specific, related to weeks 6–8 of gestation, and also dose related, with infants of women who took low-dose methotrexate (10 mg or less per week) appearing normal at birth (Altschuler and Kenney, 1986). A small study showed no increase in teratogenicity when methotrexate was taken between week 2 and 20 of pregnancy (mean 7.5 weeks), although there was an increased number of abortions (Kozlowski et al., 1990). There are multiple case reports of apparently normal infants born to women who took methotrexate during pregnancy (Briggs et al., 1998).

Pregnancy should be avoided for both males and females while methotrexate is received. For male patients, pregnancy should be avoided for a minimum of 3 months. Females who have discontinued methotrexate should have one normal ovulatory cycle before attempting to conceive (Methotrexate Product Information, 1999).

In males there are reports of defective spermatogenesis, oligospermia, and infertility (Shamberger et al., 1981; Sherins and DeVita, 1973; Sussman and Leonard, 1980), although others fail to indicate a relationship between problems with semen and methotrexate (El-Behelry et al., 1979; Grunnet et al., 1977).

There are rare reports of congenital defects or adverse events in infants of fathers who had been taking methotrexate prior to or during conception. In one case, a woman whose husband had been receiving an unknown dose of methotrexate had a spontaneous abortion. In another case, deformities of the hands and feet were observed in a child whose father had taken methotrexate at the time of conception. Another case reported an infant who had respiratory problems since birth. The father had taken an unknown dose of methotrexate for 3–4 years prior to conception. Finally, an infant with atrial septal defect and left-to-right shunt, as well as a ventricular defect, was born to a man who had been taking 15 mg/wk for over 2 years (Personal Communication, 1998).

Nonsteroidal Anti-Inflammatory Drugs

Nonsteroidal anti-inflammatory drugs, such as *ibuprofen, ketoprofen, naproxen,* and *indomethacin*, have been associated with oligohydramnios, persistence of the fetal circulation (Anti-inflamatory drugs, 1998) and pulmonary hypertension in the infant (Niebyl, 1992; Turner and Levin, 1984); they are FDA Pregnancy Categories B and C. Congenital anomalies were no more frequent than expected in a large surveillance study (Aselton et al., 1985). A small case-control study revealed an association of first-trimester use of ibuprofen with gastroschisis (Torfs et al., 1996); a larger case-control study failed to confirm the association (Werler et al., 1992).

Animal studies have shown that use of NSAIDs causes premature closure of the ductus arteriosus in rats (Momma and Takao, 1990; Momma and Takeuchi, 1983). In humans there are several case reports of oligohydramnios and neonatal renal failure with ibuprofen (Hendricks et al., 1990; Kaplan et al., 1994; Wiggins and Elliott, 1990), ketoprofen (Simeoni et al., 1989), and naproxen (Wilkinson et al., 1979). The greatest risk appears to be after 32 weeks' gestation (Eronen, 1993; Moise, 1993).

Indomethacin is also associated with an increased risk of necrotizing enterocolitis in low-birth-weight infants whose mothers used the drug (Major et al., 1994; Norton et al., 1993). Oligohydramnios has also been reported in infants of mothers who used indomethacin late in pregnancy (Norton, 1997). This is believed to be due to a decrease in fetal urinary output and amniotic fluid (Kirshon et al., 1991) and is associated with prostaglandin synthetase inhibitors in preterm labor (Levin et al., 1978).

Retinoids

Acitretin, etretinate, and *isotretinoin* are retinoids used in dermatology. All are FDA Pregnancy Category X. Acitretin, the active metabolite of etretinate, has supplanted use of etretinate because the half-life of acitretin in the human is 50–60 hours, as compared with etretinate, whose half-life is 100–120 days (Geiger et al., 1994; Pilkington and Brogden, 1992). Use of acitretin within 2–3 weeks before or during pregnancy results in abortion and miscarriage (Geiger et al., 1994) as well as vitamin A embryopathy, consisting of central nervous system (CNS) malformations, microtia, anotia, micrognathia, eye anomalies, cleft palate, thymic anomalies, and cardiac and great vessel anomalies (Rosa et al., 1986; Thomson and Cordero, 1989).

Isotretinoin has a half-life of about 50 hours (Teratology, 1991); it is also associated with miscarriages (Dai et al., 1989). Characteristic vitamin A embryopathy also occurs (Dai et al., 1989), and Subnormal intelligence has also been reported (Adams, 1990; Adams and Lammer, 1993; Adams and Lammer, 1992; Adams et al., 1991). The period of most risk for development of vitamin A embryopathy occurs during the first trimester, 2–5 weeks after conception (Rosa et al., 1986).

Tetracycline

Tetracycline is FDA Pregnancy Category D. Tetracycline administered to pregnant women results in dental staining of the infant's deciduous teeth. The time of most susceptibility is after the 4th month of gestation (Baden, 1970). Similar staining has been found in the bones and lenses of fetuses whose mothers were treated with tetracycline during gestation (Cohlan et al., 1963; Glorieux et al., 1991; Totterman and Saxen, 1969).

Thalidomide

Thalidomide is used in dermatology as an immunosuppressive agent. Thalidomide is FDA Pregnancy Category X. Congenital anomalies include limb reduction defects (Newman, 1985; Smithells, 1973). Infants born to mothers who took thalidomide have also been reported to have cardiovascular anomalies (Smithells, 1973) and eye and ear anomalies (Smithells, 1973; Stromland and Miller, 1993). Human clinical series have also shown growth retardation (Brook et al., 1977), malformations of the urogenital, gastrointestinal systems, or spine (Smithells, 1973), mental retardation (McFie and Robertson, 1973), and autism (Smithells, 1973). The period of most risk for limb reduction anomalies is from days 34–50 from the mother's last menstrual period (Mellin and Katzenstein, 1962; Mellin and Katzenstein, 1962; Newman, 1986; Stromland and Miller, 1993). Ear anomalies appear with earliest exposure (days 21–27 of gestation); arm anomalies appear with exposure from days 27–30 of gestation, and leg anomalies occur with exposure around day 30 of gestation (Iffy et al., 1967; Pembrey et al., 1970; Schumacher, 1975).

Although there are no epidemiological studies, there was a direct correlation between number of sales of thalidomide in Britain and Germany and the frequency of thalidomide embryopathy (Eskes, 1984).

OTHER DERMATOLOGIC DRUGS AND
DEVELOPMENTAL TOXICITY

With other drugs, possible problems have been identified, although not necessarily confirmed by exhaustive studies. Knowledge of available information assists the clinician in determining when it is best to minimize use of drugs.

Antibiotics

In a large surveillance study of over 200,000 patients, *cephalosporins*, including cephalexin, cefaclor, and cephradine, have been reported to have a possible increase in risk of congenital defects with use during the first trimester of pregnancy; however, this is controversial (Briggs et al., 1998). Cefadroxil use has not been associated with risk (Briggs et al., 1998).

In a large surveillance study of over 200,000 patients and in a smaller surveillance study of over 50,000 mother-child pairs, there was no increase in the risk of congenital malformations with maternal use of *erythromycin* (Briggs et al., 1998). *Erythromycin estolate* use has been associated with a subclinical maternal hepatotoxicity when used for longer than 3 weeks and should be avoided (Landers et al., 1983; Sullivan et al., 1980). *Clarithromycin* has been associated with cardiovascular defects and fetotoxicity in animals at high doses (Clarithromycin, 1998) and with several diverse birth defects in humans, including spina bifida, cleft lip, pulmonary hyperplasia, and *C*oloboma, *H*eart malformations, *A*tresia choanae, *R*etarded growth and development and/or CNS abnormalities, *G*enital hypoplasia, *E*ar abnormalities and/or deafness (CHARGE) syndrome, which includes cardiac and CNS anomalies (Briggs et al., 1998). Small case-control studies (Briggs et al., 1998; Schick et al., 1996) and a larger prospective multicenter study of 122 women (Einarson et al., 1998) who used clarithromycin during the first and early second trimester showed outcomes of pregnancy similar in the exposed and nonexposed groups. The manufacturer advises that clarithromycin not be used in pregnant women except in clinical circumstances where no alternative therapy is appropriate, and that if pregnancy occurs while a patient is using the drug, the patient should be warned of the potential hazard to the fetus (Biaxin Product Information, 1997). *Dirithromycin* has been associated with fetal growth retardation and incomplete ossification in animals (Briggs et al., 1998); however, there are no reports of use in humans during pregnancy. There are no studies of the use of *azithromycin* in humans. Azithromycin is derived from and chemically related to erythromycin.

Animal studies have shown that *fluoroquinolones* have a potential risk of arthropathy (Fluoroquinolones, 1998; NegGram Product Information, 1998). A small prospective study of 38 pregnant women who used quinolones during pregnancy has failed to confirm an increased risk of malformations or of musculoskeletal problems (Berkovitch et al., 1994). Another prospective study of 549 pregnancies in women who used fluoroquinolones at varying times during pregnancies had a malformation rate that was comparable to the background rate (Schaefer et al., 1996).

Penicillins are FDA Pregnancy Category B. Two case-control studies (Carter and Wilson, 1965; Kullander and Kallen, 1976), the Collaborative Perinatal Project (Heinonen et al., 1977), in which over 3500 women were given penicillin during the first trimester, and two cohort studies (Aselton et al., 1985; Jick et al., 1981), failed to confirm a relationship between use of the drug and congenital malformations. Briggs et al. (1998) reports a large series of over 8500 infants born to mothers with first-trimester exposure to *amoxicillin* (Briggs et al., 1998) and over 10,000 infants of mothers who used *ampicillin* (Briggs et al., 1998). No consistent anomalies were reported, and the observed rate of malformations was comparable to that expected. Two case series of 309 (Briggs et al., 1998) and 409 (Aselton et al., 1985) women given ampicillin during the first trimester likewise showed a rate of congenital anomalies similar to controls. A case-control study showing congenital heart malformations in 390 infants exposed to ampicillin early in pregnancy (Rothman et al., 1979) has not been confirmed in later studies (Bracken, 1986; Zierler and Rothman, 1985).

Rifampin has been associated with neonatal hemorrhage in a small case series. Only three infants were affected, and the condition is preventable with vitamin K (Eggermont et al., 1976). A study of 442 women who used rifampin with or without other drugs did not associate use of the drug with fetal malformations (Snider et al., 1980).

Sulfonamides are FDA Pregnancy Category C. The third-term risk of use of systemic sulfonamides includes jaundice, hemolytic anemia, and kernicterus, which is theorized to occur by displacement of bilirubin from plasma-binding sites by the drug (Briggs et al., 1998). Several small epidemiologic studies showed no increased risk of congenital anomalies (Briggs et al., 1998; Heinonen et al., 1977; Williams et al., 1969). The Collaborative Perinatal Project studied over 1400 women who had taken sulfonamides during the first trimester and found no increase in congenital malformations over baseline (Heinonen et al., 1977). In addition, a study of the wives of 300 men who worked in a factory that produced sulfonamides revealed that there was an increased incidence of abortions and a decrease in pregnancies that were carried to term (Prasad et al., 1996). Topical application of *silver sulfadiazine* during the late third trimester is believed to pose a potential risk for kernicterus and hemorrhage in the premature infant or infant with G6PD deficiency; manufacturers have contraindicated use during the third trimester (Silvadene Product Information, 1991; SSD Product Information, 1993).

Trimethoprim is FDA Pregnancy Category C. Usually it is used in combination with sulfonamides. Three case series (Bailey, 1984; Colley et al., 1982; SSD Product Information, 1993) and a case-control series (Czeizel, 1990) showed a rate of congenital anomalies that was no greater than expected.

Miscellaneous Anti-Acne Topical Products

Adapalene and *tretinoin* are topical retinoids for treatment of acne; both are FDA Pregnancy Category C. Oral retinoids are known to be associated with congenital anomalies. With topical application, only 5%–7% of that which is applied is absorbed (Van Hoogdalem, 1998). In two prospective cohort studies, the incidence of congenital anomalies was not increased among women who had received a prescription for or had used tretinoin during the first trimester (Jick et al., 1993; Shapiro et al., 1998). Several case reports have associated use of tretinoin cream with dysplastic ears (Camera and Pregliasco, 1992), small ear canals with coarctation of the aorta (Navarre-Belhassen et al., 1998), and multiple congenital anomalies (Lipson et al., 1993), but no definitive causal relationship has been established. (Koren et al., 1998) has labeled use of topical tretinoin as "safe." The manufacturers of topical tretinoin note that five of the thirty case reports during two decades of one formulation of tretinoin (Retin A) were associated with incomplete midline development of the forebrain (Avita Product Information, 1999; Renova Product Information, 1999; Retin A Product Information, 1999). The manufacturers of Renova recommend that the product be avoided during pregnancy. The manufacturers of Retin A recommend that the product should be used only if the potential benefit justifies the potential risk to the patient.

There is a case report of ophthalmic congenital anomalies associated with use of *adapalene* (Autret et al., 1997). The manufacturer of adapalene advises that the pregnant patient should use adapalene only if the potential benefit justifies the potential risk to the patient (Differin Product Information, 1999).

Clindamycin is FDA Pregnancy Category B. A large surveillance study has failed to indicate an increased risk over baseline for use of the oral drug during the first trimester (Briggs et al., 1998). No epidemiologic studies have been done.

Metronidazole is FDA Pregnancy Category B. It has been associated with congenital anomalies in animals (Finegold, 1980). Two large meta-analyses of infants with congenital anomalies failed to find an association between first-trimester use of metronidazole and anomalies (Burtin et al., 1995; Caro-Paton et al., 1997). A case series of over 4200 women had similar findings (Rosa et al., 1987), as have several cohort studies (Aselton et al., 1985; Briggs et al., 1998; Heinonen et al., 1977; Morgan, 1978; Piper et al., 1993). Nevertheless, (Briggs et al., 1998) notes that "The use of metronidazole in pregnancy is controversial."

Antifungal Agents

Amphotericin B is FDA Pregnancy Category B. No association between congenital malformations and use of the drug during any trimester of pregnancy has been noted in various case reports (Briggs et al., 1998).

Fluconazole is FDA Pregnancy Category C. Congenital anomalies have been reported in infants born to mothers who used high-dose (400-800 mg/d) fluconazole during pregnancy (Aleck and Bartley, 1997; Briggs et al., 1998; Lee et al., 1992;

Pursley et al., 1996). A retrospective review (Inman et al., 1994) and a prospective cohort study (Mastroiacovo et al., 1996) with fluconazole in doses of 150 mg/day or less has failed to establish a relationship between use of the drug and congenital anomalies. Use of single-dose fluconazole has not been associated with congenital malformations in humans (Inman et al., 1994; King et al., 1998).

Griseofulvin is FDA Pregnancy Category Unclassified. It has been associated with conjoined twins (Knudsen, 1987; Rosa et al., 1987b); however, in two studies involving 86 cases of conjoined twins, none of the mothers had used griseofulvin (Metneki and Czeizel, 1987). Although this phenomenon, if proven, would contraindicate use only early during the first trimester, manufacturers advise against griseofulvin use during pregnancy (Fulvicin, 1999; Grifulvin, 1999; Gris-Peg, 1999). One manufacturer (Grifulvin) advises contraceptive use for a month before treatment, during treatment, and 1 month after treatment.

Itraconazole is FDA Pregnancy Category C. Information on use of itraconazole in humans is limited. Briggs reports 14 cases of malformations, including digital anomalies, but notes that there is too little information to conclude whether or not a true risk is present and recommends that the drug be avoided during organogenesis (Briggs et al., 1998). The manufacturer now advises contraception during and for two months following cessation of treatment in women of child-bearing age (Sporanox, Titusville, NJ; Janssen Pharmaceutica, 1999).

Ketoconazole is FDA Pregnancy Category C. It has been associated with animal teratogenicity with use during the first trimester (Buttar et al., 1989; Nishikawa et al., 1984). Ketoconazole may also inhibit ovarian progesterone synthesis early in pregnancy, with possible adverse effects on the male fetus (Cummings et al., 1997). Briggs reports six cases of limb defects (Briggs et al., 1998). Ketoconazole use during the third trimester for treatment of hypercortisolism of Cushing's syndrome may be used if the fetus is female (Amado et al., 1990).

Oral nystatin is FDA Pregnancy Category C. A large surveillance study failed to show an increase in congenital malformations over background with use of nystatin (Heinonen et al., 1977). Two cohort studies (Aselton et al., 1985; Briggs et al., 1998) and a case-control study (Rosa et al., 1987) also failed to establish an association between drug use and malformations.

Potassium iodide is FDA Pregnancy Category D. It is a nonmetallic halogen element. Use of iodine during pregnancy produces fetal goiter and hypothyroidism (Vicens-Calvet et al., 1998; Wolff, 1969) or goiter with cardiomegaly (Mehta et al., 1983). Use of povidone-iodine douches raises serum iodine levels in the pregnant woman and her fetus (l'Allemand et al., 1983).

Terbinafine is FDA Pregnancy Category B. Studies have shown rat tumorigenicity but not animal toxicity (Terbinafine-Systemic, 1998), studies in humans have not been done. Although it is rated FDA Pregnancy Category B, the manufacturer advises against use for onychomycosis, as treatment during pregnancy is elective (Lamisil, 1999).

Most topical antifungal and anticandidal agents have no contraindication for use during pregnancy other than by the manufacturer. *Clotrimazole* and *Miconazole* use have been associated with an increased risk of spontaneous abortion when used early in pregnancy (Rosa et al., 1987a), but that study had confounding factors.

According to the manufacturer (Selsun, 1999), *selenium sulfide* should not be used by pregnant women for tinea versicolor because no animal or human studies have been performed and the risk to the fetus is not known.

Antihistamines

Antihistamines are FDA Pregnancy Categories B and C. A meta-analysis of antihistamine use during pregnancy has failed to establish an association between drug use and congenital malformations (Seto et al., 1997).

Use of antihistamines during the last 2 weeks of pregnancy has been associated with an increased risk of retrolental fibroplasia in a large collaborative study (Purohit et al., 1985; Zerler and Purohit, 1986). In a retrospective cohort study of over 3000 neonates with retrolental fibroplasia and a birth weight of less than 1750 g, use of antihistamines during the last 2 weeks of pregnancy in premature infants was associated with an increased incidence of retrolental fibroplasia, most marked in infants with birth weights between 500 and 749 g (Purohit et al., 1985).

Astemizole is FDA Pregnancy Category B. A cohort study of 104 infants whose mothers had taken astemizole during the first trimester of pregnancy did not show an increase in congenital anomalies (Pastuszak et al., 1996). Metabolites are present for up to 4 months after use of the medication (Antihistamines, 1998).

Brompheniramine is FDA Pregnancy Category B. The Collaborative Perinatal Project, a prospective study of over 50,000 mother-child pairs, studied 65 pairs with a history of exposure to brompheniramine during the first trimester. There was a small but statistically significant association between brompheniramine and congenital malformations (Heinonen et al., 1977). Another cohort study of 34 women who took brompheniramine during the first trimester failed to confirm the observation (Seto et al., 1993).

Chlorpheniramine is FDA Pregnancy Category B. First-trimester exposure of over 1000 mother-child pairs to chlorpheniramine did not show a relationship between use of the drug and congenital malformations (Heinonen et al., 1977). Another surveillance study of 61 newborns exposed to the drug during the first trimester did not support an association between use of chlorpheniramine and congenital malformations (Hocking, 1968).

Diphenhydramine is FDA Pregnancy Category B. In the Collaborative Perinatal Project, 595 mother-child pairs were exposed to diphenhydramine during the first trimester with no increase in congenital malformations (Heinonen et al., 1977). Two later cohort studies involving 631 women confirmed this finding (Aselton et al., 1985; Briggs et al., 1998). In a retrospective case-control study of 590 infants with oral clefts, maternal use of diphenhydramine was slightly higher than expected (Saxen, 1974), but no later studies have confirmed this finding.

Cimetidine is FDA Pregnancy Category C. Use of cimetidine is controversial. Cimetidine is a human antiandrogen. Use during pregnancies in animals has resulted in decreased weight of androgenic tissues (Vigersky et al., 1980) and has been questionably associated with transient liver impairment in a single case report (Sawyer et al., 1981). No studies have confirmed an increase in risk of congenital malformations with use of cimetidine during any trimester of pregnancy (Hocking, 1968). The potential for feminization has resulted in the recommendation that cimetidine not be used during pregnancy (Smallwood et al., 1995). No studies in humans have been done.

Systemic doxepin is FDA Pregnancy Category C. It has been associated with fetal ileus (Falterman and Richardson, 1980), as well as cardiac problems, fetal irritability, respiratory distress muscle spasm, and seizures in infants. Maternal use of doxepin at birth may place infants at risk for respiratory depression (Antidepressants, 1998).

Hydroxyzine is FDA Pregnancy Category Unrated. It has been possibly associated with fetal abnormalities in the Collaborative Perinatal Project, which studied 50 infants of mothers who had first-trimester exposure to the drug (Heinonen et al., 1977). In another large surveillance study of 828 infants born to mothers who used hydroxyzine during the first trimester, there was a higher-than-expected incidence of oral clefts (Briggs et al., 1998). A smaller study, but one that was double blind and controlled, showed no increase in congenital malformations in 74 infants whose mothers had used hydroxyzine during the first 2 months of pregnancy (Erez et al., 1971). The manufacturer (Atarax Product Information, 1999) contraindicates its usage during early pregnancy. *Cetirizine* is FDA Pregnancy Category B. It is available by prescription in the United States and over-the-counter in Canada. Cetirizine is a metabolite of hydroxyzine. A double-blind study of 33 infants whose mothers took cetirizine and 35 infants whose mothers took hydroxyzine during early pregnancy did not reveal an increase in the number of congenital malformations (Einarson et al., 1997). The manufacturer states that cetirizine should be used only if clearly needed (Zyrtec Product Information, 1999), but others say that cetirizine should not be used during the first trimester of pregnancy (Antihistamines, 1998).

Loratidine is FDA Pregnancy Category B. Although no studies in humans have been done, there have been several reports of adverse outcomes, including cleft palate (Briggs et al., 1998).

Terfenadine is FDA Pregnancy Category C. In two studies of over 1000 infants whose mothers had received prescriptions for terfenadine during the first trimester (Schatz and Pettiti, 1997) and over 100 women who had taken terfenadine during the first or early second trimester of pregnancy (Schick et al., 1994), the incidence of congenital malformations was no more frequent than expected.

Antimalarials

Chloroquine is FDA Pregnancy Category Unrated. Cohort studies of infants born to women using chloroquine for weekly prophylaxis have not shown an increased risk of congenital malformations (Phillips-Howard and Wood, 1996; Wolfe and Cordero, 1985). There are several small case series of infants born to women using chloroquine for systemic lupus erythematosus (Levy et al., 1991; Martinez-Rueda et al., 1996; Parke, 1988). The risk of congenital anomalies was not increased. In one study

(Armenti et al., 1994), several women had miscarriages or stillbirths, which are recognized complications of lupus. Chloroquine in therapeutic doses has been associated with ototoxicity, congenital deafness, retinal hemorrhages, and abnormal pigmentation of the retina (Chloroquine, 1999).

Hydroxychloroquine is FDA Pregnancy Category Unrated. A small case series failed to reveal any increase in risk of congenital malformations in women who took hydroxychloroquine during the first trimester (Buchanan et al., 1996). Two other case series of women who took hydroxychloroquine during all trimesters of pregnancy likewise showed no increase in risk of congenital malformations (Parke and West, 1996; Parke, 1988). Given the risk of maternal lupus to the fetus, several authors have concluded that use of hydroxychloroquine for treatment of lupus during pregnancy was a reasonable alternative (Buchanan et al., 1996; Parke and West, 1996; Parke, 1988). Others recommend against use of both chloroquine and hydroxychloroquine during pregnancy, except for prophylaxis of malaria (Chloroquine, 1999; Hydroxychloroquine, 1999).

Chloroquine and hydroxychloroquine are both accumulated in the fetal eye and retained in ocular tissue for at least 5 months (Aralen Product Information, 1999; Lindquist, 1970; Physicians' Desk Reference, 1999). Both drugs carry official indications only for treatment and suppression of malaria or hepatic amebiasis (Aralen Product Information, 1999; Plaquenil Product Information, 1999).

Quinacrine has been associated with renal agenesis and hydrocephalus in an infant whose mother was exposed (Quinacrine, 1999). Pregnancies without adverse effects in the infant have also been reported (Humphreys and Marks, 1988). United States Pharmacopaeia Drug Information (USPDI) (Quinacrine, 1999) and the manufacturer (Atabrine Product Information, 1994) advise against use during pregnancy, "except when in the judgment of the physician the benefit outweighs the possible hazard."

Antineoplastic Drugs

Azathioprine is FDA Pregnancy Category D. In case series, anomalies occurred at a higher rate in infants born to women taking azathioprine during pregnancy than in those born to unmedicated women, although nearly all women studied were transplant patients and were on multiple medications (Penn et al., 1980; Registration, 1980). There was no specific pattern to the anomalies. In another small case series, women using azathioprine for inflammatory bowel disease during the first trimester had no congenital anomalies (Alstead et al., 1990). Several authors have stated that there is no increased risk of congenital anomalies with use of azathioprine during pregnancy (Alstead et al., 1990; Ostensen, 1992; Ramsey-Goldman and Schilling, 1997; Tanis, 1998). In several case reports there an increased risk of serious and sometimes fatal neonatal anemia, thrombocytopenia, and lymphopenia in infants born to women using azathioprine and prednisone during pregnancy has been noted (Alter, 1985; Atabrine Product Information, 1994; Davison et al., 1985; De Witte et al., 1984; Lower et al., 1971; Price et al., 1976; Rudolph et al., 1979).

Bleomycin is FDA Pregnancy Category D. It is an antibiotic with antineoplastic activity. There have been no large studies indicating use of bleomycin with congenital defects, although chromosomal changes (Bornstein et al., 1971) have been reported. In a case report, use during the second and third trimester has resulted in transient neonatal leukopenia and neutropenia in a premature infant; the mother had also received other medications (Raffles et al., 1989). In other case reports, use of bleomycin with other antineoplastic agents during the second and third trimesters resulted in the birth of normal infants (Falkson et al., 1980; Lowenthal et al., 1982; Ortega, 1977). Use during the first trimester has resulted in the birth of normal children (Aviles et al., 1991).

Cyclophosphamide is FDA Pregnancy Category D. It is an alkylating antineoplastic agent. Use of cyclophosphamide appears to be related to an increased risk of spontaneous abortion. Several cases of congenital anomalies involving digits and the nose have been reported (Kirshon et al., 1988; Scott, 1977). The drug had been given from day 77 to day 82 of gestation in the Scott case and from days 15 to days 45 of gestation in the Kirshon case. Low birth weight is reported in many infants who receive cyclophosphamide and other drugs during pregnancy (Nicholson, 1968). Pancytopenia has been reported in infants born to women given cyclophosphamide with other antineoplastic agents during the third trimester (Pizzuto et al., 1980).

Normal infants have also been born to women who used cyclophosphamide with other antineoplastic agents (Kim and Park, 1989; Kirshon et al., 1988; Malfetano and Goldkrand, 1990; Rustin et al., 1984; Weed et al., 1979).

Men who used cyclophosphamide have had azoospermia (Hinkes and Plotkin, 1973; Lendon et al., 1978; Qureshi et al., 1972; Schilsky et al., 1980; Stewart, 1975).

Pregnancies following high-dose cyclophosphamide with or without high-dose busulfan or total-body irradiation and bone marrow transplantation. Several cases of congenital anomalies involving digits and the nose have been reported (Kirshon et al., 1988; Nicholson, 1968; Sanders et al., 1996; Toledo et al., 1971). The drug had been given from day 77 to day 82 of gestation in one case and from days 15 to days 45 of gestation in another. Pancytopenia has been reported in infants born to women given cyclophosphamide with other antineoplastic agents during the third trimester (Pizzuto et al., 1980).

Normal infants have also been born to women who used cyclophosphamide with other antineoplastic agents (Dayoan et al., 1998; Kim and Park, 1989; Weed et al., 1979).

Men who used cyclophosphamide have had azoospermia (Qureshi et al., 1972; Stewart, 1975). In one case of azoospermia, DNA testing confirmed that a man fathered a child (Pakkala et al., 1994). Chromosomal anomalies have also been reported (Schleuning and Climm, 1987; Tolchin et al., 1974).

Use of cyclophosphamide appears to be most risky during the first trimester (Ostensen, 1992).

Fluorouracil is FDA Pregnancy Category X. In dermatology it is used topically. There are several reports of chromosomal abnormalities in infants of women who used topical vaginal fluorouracil in the periconceptional period (Otano et al., 1992; Van Le et al., 1991). Normal infants have also been born to mothers who used topical vaginal fluorouracil (Kopelman and Miyakaza, 1990; Turchi and Villasis, 1988; Van Le et al., 1991).

Hydroxyurea is FDA Pregnancy Category Unrated. It is an antineoplastic drug that inhibits DNA synthesis. No epidemiologic studies in humans have been reported. Use of the drug during pregnancies has resulted in birth of a stillborn infant (Tertian et al., 1992). Normal infants have also been born to women who used hydroxyurea during pregnancy in case reports (Cinkotai et al., 1994; Delmer et al., 1992; Doney et al., 1979; Fitzgerald and McCann, 1993; Jackson et al., 1993; Patel et al., 1991). In animal studies, multiple species have shown development toxicity, including fetal death, growth retardation, and various malformations (Aliverti et al., 1980; Chaube et al., 1963; Ferm, 1965; Murphy and Chaube, 1964; Wilson et al., 1975). Changes in behavior have also been observed in rats treated with high doses of hydroxyurea (Adlar and Dobbing, 1975; Vorhees et al., 1979).

Mechlorethamine is FDA Pregnancy Category D. It is an alkylating agent, and there are no epidemiologic studies in humans. Isolated anomalies have been reported (Mennuti et al., 1975; Thomas and Andes; Zemlickis et al., 1992) that were believed to be unrelated to drug exposure. In various clinical series, normal infants have been born to mothers who took mechlorethamine during pregnancy (Andrieu and Ochoa-Molina, 1983; Aviles et al., 1991; Green et al., 1991; Schilsky et al., 1981; Whitehead et al., 1983; Zuaza et al., 1991).

Antiparasitic Drugs

Albendazole is FDA Pregnancy Category C. It is teratogenic and embryotoxic in animals (Cristofol et al., 1997; Macri et al., 1993; Mantovani et al., 1995; Whittaker et al., 1990). There are no epidemiologic studies in humans. Pregnancy testing is recommended before use in a woman of childbearing age, and contraception should be used during the course of treatment and for 1 month thereafter (Albendazole, 1999).

Pentamidine is FDA Pregnancy Category C. It has shown embryotoxicity and delayed ossification in rats (Pentamidine, 1998). In humans, a case series of six infants born to HIV-infected mothers who used pentamidine as well as other drugs during all trimesters of pregnancy had no increase in congenital anomalies related to the use of pentamidine (Sperling et al., 1991). No epidemiologic studies of use of pentamidine in humans are available, and use of pentamidine during pregnancy is controversial (Briggs et al., 1998). The manufacturer recommends use only if clearly needed (Pentam Product Information, 1992).

Thiabendazole is FDA Pregnancy Category C. It has been associated with cleft palate and vertebral fusion in rats given high doses in olive oil, but not in the aqueous preparation (Ogata et al., 1984; Thiabendazole, 1998); no studies in humans are available. Topical thiabendazole is probably absorbed. The manufacturer recommends use of oral thiabendazole only if risk to the fetus justifies the potential benefit to the mother (Mintezol Product Information, 1999).

Antiscabetic Drugs

Crotamiton is FDA Pregnancy Category C. No epidemiologic studies in humans are available. The manufacturer advises that animal studies have not been done so risk is not fully understood, and that crotamiton should be used "only if clearly needed" (Eurax Product Information, 1999).

Ivermectin is FDA Pregnancy Category C. Several case series in humans have failed to demonstrate an increased risk of congenital anomalies (Chippaux et al., 1993; Duombo et al., 1992; Pacque et al., 1990). In animals, developmental toxicity is not seen until doses of ivermectin are maternotoxic (Stromectol Product Information, 1999). The manufacturer recommends that the drug not be used during pregnancy until safety has been established (Stromectol Product Information, 1999).

Lindane is FDA Pregnancy Category B. It has been both discouraged (Hurwitz, 1973) and recommended for limited use for treatment of scabies in pregnant women (Cristofol et al., 1997; Macri et al., 1993; Mantovani et al., 1995; Whittaker et al., 1990). In high doses in animal studies, it produced testicular degeneration (Chowdhury et al., 1990), and in a large surveillance study there was a possible association with hypospadias (Briggs et al., 1998). Misuse of the product or use on traumatized or otherwise abnormal skin is believed to be responsible for adverse occurrences (Lee and Groth, 1977; Rasmussen, 1981). Briggs notes that pyrethrins are recommended for the treatment of lice infestations during pregnancy (Briggs et al., 1998).

Careful use of lindane is not contraindicated by the manufacturer; lindane should be used for no longer than the label advises and no more than twice during a pregnancy (Lindane, 1998). Briggs recommends use of pyrethrins for treatment of scabies or lice during pregnancy (Briggs et al., 1998).

Precipitated sulfur has been associated with fatalities following use in animals, as reported by Rasmussen (Rasmussen, 1981).

Permethrin is FDA Pregnancy Category B; it is a pyrethrin. Animal studies have not shown teratogenicity (Miyamoto, 1976). No epidemiologic studies in humans are available. The manufacturer recommends that permethrin be used "only if clearly needed" (Elimite Product Information, 1999). Several authors have stated that pyrethrins are the preferred treatment for scabies and lice during pregnancy (Altschuler and Kenney, 1986; Haustein and Hlawa, 1989; Briggs et al., 1998).

Antiviral Drugs

Acyclovir is FDA Pregnancy Category C. Acyclovir is not teragotenic in animals but causes fetal death, growth retardation, and malformations in rats at maternotoxic doses (Chahoud et al., 1988; Stahlmann et al., 1988). Large post-market surveillance studies by the manufacturers have revealed that the risk of congenital anomalies in human pregnancies is not increased (Andrews et al., 1988; Andrews et al., 1992; Reiff-Eldridge et al., 2000). The manufacturer states that acyclovir should not be used in pregnancy unless the potential benefit justifies the potential risk to infants (Zovirax Product Information, 1999). Others have stated that acyclovir use for disseminated herpesvirus (HSV) infections has been helpful in reducing mortality for these conditions but that use for recurrent HSV infections is not as convincing an indication (Briggs et al., 1998).

Famciclovir is FDA Pregnancy Category B. Its metabolite, penciclovir, is the active compound. Famciclovir has not shown teratogenicity in animals but is a tumorigen in rats (Famciclovir, 1998); there are no epidemiologic studies in humans. The manufacturer states it is to be used only if the potential benefit to the mother justifies the potential risk to the fetus (Famvir, 2000) (Famvir Product Information, 1999).

Valacyclovir is FDA Pregnancy Category B. Valacyclovir is nearly completely converted to acyclovir. Uneventful use during human pregnancy has been reported by the manufacturer, who recommends use only if potential benefit to the mother justifies the potential risk to the infant (Valtrex Product Information, 1999). Guidelines for use of acyclovir should be followed.

Corticosteroids

Corticosteroids are FDA Pregnancy Category C. A study of 43 women who used prednisone during the first and early second trimesters of pregnancy showed no increased risk of congenital malformations (Heinonen et al., 1977). A case-control series of pregnancies in mothers with systemic lupus erythematosus showed no association between use of prednisone and stillbirth, congenital malformations, or abortion (Martinez-Rueda et al., 1996). Another case-control study showed no association between use of corticosteroids during the first trimester and congenital malformations (Czeizel and Rockenbauer, 1997). There has been a single case report of intrauterine growth retardation in the infant of a mother who used 40 mg of topical triamcinolone per day (Katz et al., 1990); however, risk of use of both oral and topical corticosteroids is regarded as minimal (Czeizel and Rockenbauer, 1997; Fraser and Sajoo, 1995).

Miscellaneous

Allopurinol is FDA Pregnancy Category C. It is used for treatment of leishmaniasis. Animal studies using high doses have shown an increase in cleft palate, growth retardation, and death (Fujii and Nishimura, 1972). With topical use, fetal toxicity was not seen until maternal toxicity was approached (Zyloprim, 1998). There are no epidemiologic studies in humans. In a small, randomized controlled trial in which allopurinol was used during the third trimester of pregnancy, there were no adverse effects noted in the infants (Gulmezoglu et al., 1997).

Colchicine is FDA Pregnancy Category D. It is an alkaloid antimitotic agent. No controlled studies of infants born to women who used colchicine during pregnancy have been performed. Azoospermia has been reported in males who have taken colchicine (Bremer and Paulsen, 1976; Ehrenfeld et al., 1986; Haimov-Kochman and Ben-Chetrit, 1998; Merlin, 1972). Use during early pregnancy has resulted in Down's syndrome in several series (Ferreira and Buoniconti, 1968; Rabinovitch et al., 1992;

Zemer et al., 1993). There is a case report of atypical Down's syndrome in a child born to a father who used colchicine prior to conception (Cestari et al., 1965). Normal children have also been born to women who were using colchicine during pregnancy (Cohen et al., 1977; Ehrenfeld et al., 1987; Levy et al., 1991). Normal infants have been born to males as well (Cohen et al., 1977).

Cyclosporine is FDA Pregnancy Category C. It is embryotoxic, fetotoxic, and carcinogenic in some animal studies (Cyclosporine, 1999; Sangalli et al., 1990) but not in others (Ryffel et al., 1983), and is not considered an animal teratogen (Cyclosporine, 1999). In one case series, use in humans has been associated with congenital anomalies in 2 of 154 infants born to mothers who used cyclosporine (Armenti et al., 1994). An increased risk of premature births has been reported in other case series, but this may have been related to the underlying disease (Armenti et al., 1995; Radomski et al., 1995). The manufacturer reports an increased risk of premature birth and low birth weight, with 47% of women experiencing pre-term delivery and 27% of infants being small for gestational age (Neoral Product Information, 1999).

A number of successful pregnancies with cyclosporine have been noted (Cristofol et al., 1997; Macri et al., 1993; Mantovani et al., 1995; Whittaker et al., 1990). Briggs states that use of cyclosporine during pregnancy poses no major risk to the fetus (Briggs et al., 1998).

Dapsone is FDA Pregnancy Category C. Use during pregnancy for leprosy or dermatitis herpetiformis is supported by case series in the literature (Dayoan et al., 1998; Pakkala et al., 1994; Sanders et al., 1996; Schleuning and Climm, 1987; Tolchin et al., 1974; Van Le et al., 1991). There is a case report of hemolytic anemia in a mother and her newborn (Hocking, 1968). A theoretic risk of neonatal kernicterus may be minimized by stopping therapy during the last month of pregnancy (Thornton and Bowe, 1989). Briggs notes that use of dapsone during pregnancy poses no apparent major risk to the fetus (Briggs et al., 1998).

Finasteride is FDA Pregnancy Category X. It is a 5-alpha reductase inhibitor. There are no epidemiologic studies in humans. Use in animals during pregnancy has resulted in anogenital anomalies, including hypospadias in the rat (Anderson and Clark, 1990; Clark et al., 1990; Prahalada et al., 1997). Female animals were not affected. Finasteride is not approved for use in women by the FDA.

Glycolic acid topical has been associated with congenital anomalies in the offspring of animals who were given maternotoxic doses (Munley and Hurtt, 1996). No studies have been done in humans.

Gold is FDA Pregnancy Category C. Gold administered to animals has been associated with congenital abnormalities and gastroschisis (Szabo et al., 1978a; 1978b). In one series, 2 out of 119 infants had congenital anomalies, one with a dislocated hip and one with a flattened acetabulum (Miyamoto et al., 1972). The relationship to drug use is not clear. A surveillance study in humans has not shown significant effects (Freyberg et al., 1979). Normal infants have been born to mothers who used gold (Needs and Brooks, 1985; Tarp and Graudal, 1985).

Interferons are FDA Pregnancy Category C. At doses used in hematology and oncology, interferons have been associated with fetotoxic effects in humans (Faulds and Benfield, 1994; Frieden et al., 1996). Normal pregnancies have been reported, however (Baer, 1991; Baer et al., 1992; Haggstrom et al., 1996; Lipton et al., 1996). No epidemiologic studies have been done in humans. Although doses used for treatment of warts in dermatology may be below the level to produce adverse effects, use for elective treatment of warts during pregnancy should probably be avoided.

Methoxsalen is FDA Pregnancy Category C. Several case series reported no problems in infants whose mother and/or father had been treated with oral methoxsalen and ultraviolet A (UVA) in the periconceptional period or early pregnancy (Garbis et al., 1995; Gunnarskog et al., 1993; Stern and Lange, 1991). One study included pregnancies that occurred 1 or more years after treatment and did not identify any increased risk of harm to the fetus (Narron et al., 1992); however, these studies were small. It has been suggested that women who become pregnant during or following psoralen with ultraviolet A (PUVA) treatment should be offered prenatal karyotyping to rule out the possibility of chromosomal abnormalities (Narron et al., 1992).

Minoxidil topical is FDA Pregnancy Category C. Use of the oral drug by mothers has been associated in case reports with transient hypertrichosis in infants (Kaler et al., 1987; Rosa et al., 1987c). The manufacturer recommends that the medication not be used in pregnant and lactating women (Rogaine, 1994). Others have recommended that pregnancy be delayed until 1 month after discontinuation of topical minoxidil (Carlson and Feenstra, 1977).

Penicillamine is FDA Pregnancy Category Unclassified. It is a chelating agent. Case reports of penicillamine use during pregnancy describe similar and unusual connective anomalies in infants (Harpey et al., 1983; Linares et al., 1979; Mjolnerod et al., 1971; Rosa, 1986; Solomon et al., 1977). Although reports are isolated, the association of use of a drug with an unusual anomaly, which is similar to one seen as a complication of the drug in adults (Narron et al., 1992), supports a causal relationship between drug use and the anomaly. Nevertheless, a case series of 93 infants whose mothers used penicillamine during pregnancy revealed only four with congenital anomalies, all involving connective tissue (Roubenoff et al., 1988). There is a case report of oral clefts with use of penicillamine as well (Martinez-Frias et al., 1998). Normal infants born to mothers who used penicillamine during pregnancy have been reported (Berghella et al., 1997; Hartard and Kunze, 1994; Roubenoff et al., 1988; Mustafa and Shamina, 1998). Zinc may offer some protection from cutis laxa in infants whose mothers use penicillamine during pregnancy (Anderson et al., 1998; Brewer, 1995).

Pentoxifylline is FDA Pregnancy Category C. It has been associated with a possible increased risk of cardiovascular and other defects (Briggs et al., 1998). In animals, pentoxifylline interfered with the efficacy of the IUD (Ramey et al., 1994). In humans, pentoxifylline has been used to enhance sperm motility (Tasdemir et al., 1998). No epidemiologic studies have been done in humans. The manufacturer recommends use only if the potential benefit to the mother outweighs the potential risk to the fetus (Trental, 1999).

Podophyllum topical is FDA Pregnancy Category C. It is cytotoxic and teratogenic. There are case reports of minor anomalies (Cullis et al., 1962; Karol et al., 1980) and miscarriage (Chamberlain et al., 1972) after use early in pregnancy. Normal infants have been born to mothers who were given podophyllum during pregnancy, described in a case report (Balucani and Zellers, 1964) and in the Collaborative Perinatal Project (Heinonen et al., 1977). Its use during any trimester of pregnancy is not recommended because of the risk to the mother of severe myelotoxicity and neurotoxicity (Podophyllum, 1998). The manufacturer notes that the drug should be used only if the potential benefit justifies the potential risk to the fetus (Condylox Product Information, 1998).

Spironolactone is FDA Pregnancy Category Unrated. It is a diuretic that competitively inhibits aldosterone and has antiandrogenic activity in humans. There are no epidemiologic studies in humans. Its use during pregnancy has been contraindicated because of its known antiandrogen effects in humans and because of feminization of male fetuses in rats (Hecker et al., 1990; Messina et al., 1979). Several pregnancies resulting in the birth of normal males have been reported in humans (Groves and Corenblum, 1995; Rigo et al., 1996).

Tacrolimus is FDA Pregnancy Category C. Tacrolimus has been associated with intrauterine growth retardation, premature birth, hyperkalemia, and reversible renal function impairment in humans. Postnatal growth was normal (Tacrolimus, 1998). No epidemiologic studies have been done in humans.

Tranexamic acid is FDA Pregnancy Category B. No epidemiologic studies have been done in humans. Healthy infants have been born to women who used tranexamic acid during pregnancy (Tranexamic acid, 1998).

CONCLUSION

A few drugs used by dermatologists have been proven to be teratogenic. Many drugs have isolated case reports, with or without animal studies, indicating possible problems with drug use. Often there are very limited data of support these associations, and Koch's postulates are far from being demonstrated. Specifically, the mechanism of toxicity, if any, is unknown.

As data become more plentiful and sophisticated, more valid epidemiologic studies may prove or disprove an actual association between drug use and potential problems during pregnancy. Meanwhile, a cautious approach, which includes identifying and quantifying what is known, sharing information with the pregnant woman, her family, and her obstetrician, and avoiding potentially harmful drugs as well as unnecessary drug use, is recommended.

REFERENCES

Adams, J. (1990): High incidence of intellectual deficits in 5-year-old children exposed to isotretinoin "in utero." *Teratology*, 41:614.
Adams, J., and Lammer, E. J. (1993): Neurobehavioral teratology of isotretinoin. *Reprod. Toxicol.*, 7:175–177.

Adams, J., and Lammer, E. J. (1992): Relationship between dysmorphology and neuro-psychological function in children exposed to isotretinoin "in utero." In: *Functional Neuroteratology of Short Term Exposure to Drugs.* Edited by T. Fujii, G. J. Boer, Tokyo, Teikyo University Press, pp 159–170.

Adams, J., Lammer, E. J., and Holmes, L. B. (1991): A syndrome of cognitive dysfunctions following human embryonic exposure to isotretinoin. [abstract]. *Teratology,* 43:497.

Adlard, B., and Dobbing, J. (1975): Maze learning by adult rats after inhibition of neuronal multiplication in utero. *Pediatr. Res.,* 9:139–42.

Albendazole (systemic). (1999): In: *USP-DI,* monograph.

Aleck, K. A., and Bartley, D. L. (1997): Multiple malformation syndrome following fluconazole use in pregnancy: Report of an additional patient. *Am. J. Med. Genet.,* 72:253–256.

Aliverti, V., Bonanomi, L., and Giavini, E. (1980): Hydroxyurea as a reference standard in teratological screening: Comparison of the embryotoxic and teratogenic effects following single intraperitoneal or repeated oral administrations to pregnant rats. *Arc. Toxicol. Suppl.,* 4:239–47.

Alstead, E. M., Ritchie, J. K., Lennard-Jones, J. E., Farthing, M. J., and Clark, M. L. (1990): Safety of azathioprine in pregnancy in inflammatory bowel disease. *Gastroenterology,* 99:443–6.

Alter, B. P. (1985): Neonatal pancytopenia after maternal azathioprine therapy [letter]. *J. Pediatr.,* 106:691.

Altschuler, D. Z., and Kenney, L. R. (1986): Pediculocide performance, profit, and the public health. *Arch. Dermatol.,* 122:259–61.

Amado, J. A., Pesquera, C., Gonzalez, E. M., Otero, M., Freijanes, J., and Alvarez, A. (1990): Successful treatment with ketoconazole of Cushing's syndrome in pregnancy. *Postgrad. Med. J.,* 66:221–3.

Anderson, C., and Clark, R. (1990): External genitalia of the rat: Normal development and the histogenesis of 5-alpha reductase inhibitor-induced abnormalities. *Teratology,* 42:483–96.

Anderson, L., Hakojarvi, S., and Boudreaux, S. (1998): Zinc acetate treatment in Wilson's disease. *Ann. Pharmacother.,* 32:78–87.

Andrews, E. B., Tilson, H. H., Hurn, B. A., and Cordero, J. F. (1988): Acyclovir in pregnancy registry: An observational epidemiologic approach. *Am. J. Med.,* 85(suppl. 2A): 123–8.

Andrews, E. B., Yankaskas, B. C., and Cordero, J. F., et al. (1992): Acyclovir in pregnancy registry: Six years' experience. *Obstet. Gynecol.,* 79:7–13.

Andrieu, J., and Ochoa-Molina, M. (1983): Menstrual cycle, pregnancies and offspring before and after MOPP therapy for Hodgkin's disease. *Cancer,* 52:435–8.

Anti-inflammatory drugs, nonsteroidal (Systemic). In: *USP-DI:Drug Information for the Health Care Professional,* Vol. 1. Taunton, MA, Rand McNally, 1998, pp. 376–422.

Antidepressants, tricyclic (Systemic). In: *USP-DI:Drug Information for the Health Care Professional,* Vol. 1. Taunton, MA, Rand McNally, 1998, pp. 260–72.

Antihistamines, Systemic. In: *USP-DI: Drug Information for the Health Care Professional,* Vol. 1. Taunton, MA, Rand McNally, 1998, pp. 315–333.

Antihistamines, Systemic. In: *USP-DI: Drug Information for the Health Care Professional,* Vol. 1. Taunton, MA, Rand McNally, 1998, pp. 315–333.

Aralen Product Information, New York, Sanofi Pharmaceuticals, 1999.

Armenti, V. T., Ahlswede, K. M., and Alhswede, B. A. (1994): National transplantation pregnancy registry: outcomes of 154 pregnancies in cyclosporine-treated female kidney transplant recipients. *Transplantation,* 57:502–6.

Armenti, V. T., Ahlswede, K. M., and Ahlswede, B. A., et al. (1995):Variables affecting birthweight and graft survival in 197 pregnancies in cyclosporine-treated kidney transplant recipients. *Transplantation,* 59:476–9.

Aselton, P., Jick, H., Milunsky, A., Hunter, J. R., and Stergachis (1985): First-trimester drug use and congenital disorders. *Obstet. Gynecol.,* 65:451–5.

Ashby, J., Tinwell, H., Callander, R. D., Kimber, I., Clay, P., Galloway, S. M., Hill, R. B., Greenwood, S. K., Gaulden, M. E., Ferguson, M. J., Vogel, E., Nivard, M., Parry, J. M., and Williamson, J. (1997): Thalidomide: Lack of mutagenic activity across phyla and genetic endpoints. *Mutat. Res.,* 396:45–64.

Atabrine Product Information, New York, Sanofi Winthrop Pharmaceuticals, Inc., 1994.

Atarax Product Information, New York, Pfizer, Inc., 1999.

Autret, E., Berjot, M., Jonville-Bera, A., and Moraine, C. (1997): Anophthalmia and agenesis of optic chiasma associated with use of adapalene gel in early pregnancy. *Lancet,* 350(9074):339.

Aviles, A., Diaz-Maqueo, J. C., Talavera, A., Guzman, R., and Garcia, E. L. (1991): Growth and development of children of mothers treated with chemotherapy during pregnancy: Current status of 43 children. *Am. J. Hematol.,* 36:243–8.

Avita Product Information. Foster City, CA, Penederm Incorporated, 1999.

Baden, E. (1970): Environmental pathology of the teeth. In: *Thomas Oral Pathology*, 6th edition. Edited by R. J. Gorlin, and H. M. Goldman, St. Louis, C. V. Mosby Co.; pp. 189–191.

Baer, M. R. (1991): Normal full-term pregnancy in a patient with chronic myelogenous leukemia treated with alpha-interferon [letter]. *Am. J. Hematol.*, 37:66.

Baer, M., Ozer, H., and Foon, K. (1992): Interferon alpha therapy during pregnancy in chronic myelogenous leukaemia and hairy cell leukaemia. *Br. J. Haematol.*, 81:167–9.

Bailey, R. (1984): Single-dose antibacterial treatment for bacteriuria in pregnancy. *Drugs*, 27:183–6.

Balucani, M., and Zellers, D. (1964): Podophyllum resin poisoning with complete recovery. *JAMA*, 189: 639–40.

Berghella, V., Steele, D., Spector, T., Cambi, F., and Johnson, A. (1997): Successful pregnancy in a neurologically impaired woman with Wilson's disease. *Am. J. Obstet. Gynecol.*, 176:712–4.

Berkovitch, M., Pastuszak, A., Gazarian, M., Lewis, M., and Koren, G. (1994): Safety of the new quinolones in pregnancy. *Obstet. Gynecol.*, 84:535–8.

Berkowitz, R. L., Capriotti, E. A., and Cote, J. R., et al. (1986): *Handbook for Prescribing Medications during Pregnancy.* 2nd edition. Boston/Toronto: Little, Brown, and Co, p. 54.

Biaxin Product Information, North Chicago, IL, Abbott Laboratories, 1997.

Bigby, M. (1998): Evidence-based medicine in a nutshell. *Arch. Dermatol.*, 134:1609–18.

Bornstein, R. S., Hungerford, D. A., Haller, G., Engstrom, P. F., and Yarbro, J. W. (1971): Cytogenetic effects of bleomycin therapy in man. *Cancer. Res.*, 31:2004–7.

Bracken, M. (1986): Drug use in pregnancy and congenital heart disease in offspring [letter]. *N. Engl. J. Med.*, 314:1120.

Bremer, W., and Paulsen, C. (1976): Colchicine and testicular function in man. *N. Engl. J. Med.*, 294:1384–5.

Brewer, G. J. (1995): Practical recommendations and new therapies for Wilson's disease. *Drugs*, 50:240–9.

Briggs, G. G., Freeman, R. K., and Yaffe, S. L. (1998): *Drugs in Pregnancy and Lactation,* 5th ed., Baltimore, Williams and Wilkins, pp. 702–6.

Brook, C. G. D., Jarvis, S, N., and Newman, C. G. H. (1977): Linear growth of children with limb deformities following exposure to thalidomide in utero. *Acta. Paediatr. Scand.*, 66:673–5.

Buchanan, N., Toubi, E., Khamashta. M., Lima, F., Kerslake, S., and Hughes, G. R. (1996): Hydroxy-chloroquine and lupus pregnancy: Review of a series of 36 cases. *Ann. Rheum. Dis.*, 55:486–8.

Burrows, D. A., O'Neil, T. J., and Sorrells, T. L. (1988): Successful twin pregnancy after renal transplant maintained on cyclosporine A immunosuppression. *Obstet-Gynecol*, 72(3 Pt 2):459–61.

Burtin, P., Taddio, A., Ariburnu. O., Einarson, T. R., and Koren, G. (1995): Safety of metronidazole in pregnancy: A meta-analysis. *Am. J. Obstet. Gynecol.*, 172:525–9.

Buttar, H. S., Mofatt, J. H., and Bura, C. (1989): Pregnancy outcome in ketoconazole-treated rats and mice [abstract]. *Teratology*, 39:444.

Camera, G., and Pregliasco, P. (1992): Ear malformation in baby born to mother using tretinoin cream. *Lancet*, 339:687.

Carlson, T. G., and Feenstra, E. S. (1977): Toxicologic studies with the hypotensive agent minoxidil. *Toxicol. Appl. Pharmacol.*, 39:1–11.

Caro-Paton, T., Carbajal, A., de Diego, I., Martin-Arias, L. H., Alvarez Requejo, A., and Rodriguez Pinilla, E. (1997): Is metronidazole teratogenic? A meta-analysis. *Br. J. Clin. Pharmacol.*, 44:179–82.

Carter, M., and Wilson, F. (1965): Antibiotics in early pregnancy and congenital malformations. *Dev. Med. Child. Neurol.*, 7:353–9.

Cestari, A., Vieira Filho, J., and Yonenaga, Y. (1965): A case of human reproductive abnormalities possibly induced by colchicine treatment. *Rev. Bras. Biol.*, 25:253–6.

Chahoud, I., Stahlmann, R., Bochert, G., Dillmann, I., and Neubert, D. (1988): Gross structural defects in rats after acyclovir application on day 10 of gestation. *Arch. Toxicol.*, 62:8–14.

Chamberlain, M., Reynolds, A., and Yeoman, W. (1972): Toxic effect of podophyllum in pregnancy. *Br. Med. J.*, 3:391–2.

Chaube, S., Simmel, E., and Lacon, C. (1963): Hydroxyurea, a teratogenic chemical. *Proc. Am. Assoc. Cancer. Res.*, 4:10.

Chippaux, J., Gardon-Wendel, N., Gardon, J., and Ernould, J. (1993): Absence of any adverse effect of inadvertent ivermectin treatment during pregnancy. *Trans. R. Soc. Trop. Med. Hyg.*, 87:318.

Chloroquine (Systemic). In: *USP-DI: Drug Information for the Health Care Professional*, Vol. 1. Taunton, MA, Rand McNally, pp. 849–852.

Chowdhury, A. R., Gautam, A. K., and Bhatnagar, V. K. (1990): Lindane-induced changes in morphology and lipids profile of testes in rats. *Biomed. Biochim. Acta.*, 49:1059–65.

Cinkotai, K. I., Wood, P., Donnai, P., and Kendra, J. (1994): Pregnancy after treatment with hydroxyurea in a patient with primary thrombocythaemia and a history of recurrent abortion. *J. Clin. Pathol.*, 47:769–70.

Clarithromycin, systemic. In: *USP-DI: Drug Information for the Health Care Professional*, Vol. 1. Taunton, MA, Rand McNally, 1998, pp. 840–3.

Clark, R., Antonello, J., Grossman, S., Wise, L. D., Anderson, C., Bagdon, W. J., Prahalada, S., MacDonald, J. S., and Robertson, R. T. (1990): External genitalia abnormalities in male rats exposed in utero to finasteride, a 5 alpha-reductase inhibitor. *Teratology*, 42:91–100.

Code of Federal Regulations. (1987): 21 (Part 201–57):23.

Code of Federal Regulations. Title 21, volume 4, parts 200–299, pp. 24–26, Revised April 1999.

Code of Federal Regulations. Title 21, part 201, Section 57, page 30 (m) [44 FR 37462], June 26, 1979, as amended at 55 FR 11576, Mar. 29, 1990; 59 FR 64249, Dec. 13.

Cohen, M. M., Levy, M., and Eliakim, M. (1977): A cytogenetic evaluation of long-term colchicine therapy in the treatment of Familial Mediterranean Fever (FMF). *Am. J. Med. Sci.*, 274:147–52.

Cohlan, S. Q., Bevelander, G., and Tiamsic. T. (1963): Growth inhibition of prematures receiving tetracycline. *Am. J. Dis. Child.*, 105:453–461.

Collaborative Group on Drug Use in Pregnancy: An International Survey on Drug Utilization during Pregnancy. *International Journal of Risk and Safety in Medicine* 1:1, 1991.

Colley, D., Kay, J., and Gibson, G. (1982): A study of the use in pregnancy of co-trimoxazole and sulfamethizole. *Aust. J. Pharm.*, 63:570–5.

Collier, P., Kelly, S., and Wojnarowska, F. (1994): Linear IgA disease and pregnancy. *J. Am. Acad. Dermatol.*, 30:407–11.

Condylox Product Information, North Chicago, IL, Oclassen Pharmaceuticals, Inc., 1998.

Cristofol, C., Navarro, M., Franquelo, C., Valladares, J. E., Carretero, A., Ruberte, J., and Arboix, M. (1997): Disposition of netobimin, albendazole, and its metabolites in the pregnant rat: developmental toxicity. *Toxicol. Appl. Pharmacol.*, 144:56–61.

Cruikshank, D. P., Wigton, T. R., and Hays, P. M. (1996): Maternal physiology in pregnancy. In: *Obstetrics*, 3rd edition. Edited by S. G. Gabbe, J. R. Niebyl, and J. L. Simpson. New York: Churchill Livingstone, pp. 91–109.

Cullis, J. (1962): Congenital deformities and herbal "slimming tablets" [letter]. *Lancet*, 2:511–12.

Cummings, A. M., Hedge, A. L., and Laskey, J. (1997): Ketoconazole impairs early pregnancy and the decidual response via alterations in ovarian function. *Fundam. Appl. Toxicol.*, 40:238–46.

Cyclosporine (Systemic). In: *USP-DI: Drug Information for the Health Care Professional*, Vol. 1. Taunton, MA, Rand McNally, 1999, pp. 1136–43.

Czeizel, A. (1990): A case-control analysis of the teratogenic effects of co-trimoxazole. *Reprod. Toxicol.*, 4:305–13.

Czeizel, A., and Rockenbauer, M. (1997): Population-based case-control study of teratogenic potential of corticosteroids. *Teratology*, 56:335–40.

Dai, W. S., Hsu, M. A., and Itri, L. M. (1989): Safety of pregnancy after discontinuation of isotretinoin. *Arch. Dermatol.*, 125:362–365.

Dapsone (systemic). In: *USP-DI: Drug Information for the Health Care Professional*, Vol. 1. Taunton, MA, Rand McNally, pp. 1126–8.

Davison, J. M., Dellagrammatikas, H., and Parkin, J. M. (1985): Maternal azathioprine therapy and depressed haemopoiesis in the babies of renal allograft patients. *Br. J. Obstet. Gynaecol.*, 92:233–9.

Dayoan, E. S., Dimen, L. L, and Boylen C. T. (1998): Successful treatment of Wegener's granulomatosis during pregnancy: A case report and review of the medical literature. *Chest.*, 113:836–8.

De Witte, D. B., Buick, M. K., Cyran, S. E, and Maisels, M. J. (1984): Neonatal pancytopenia and severe combined immunodeficiency associated with antenatal administration of azathioprine and prednisone. *J. Pediatr.*, 105:625–8.

Delmer, A., Rio, B., Bauduer, F. Ajchenbaum, F., Marie, J. P., and Zittoun, R. (1992): Pregnancy during myelosuppressive treatment for chronic myelogenous leukaemia [letter]. *Br. J. Haematol.*, 82: 783–4.

Differin Product Information, Fort Worth, TX, Galderma Laboratories Inc., 1999.

Doney, K., Kraemer, K., and Shepard, T. (1979): Combination chemotherapy for acute myelocytic leukemia during pregnancy: three case reports. *Cancer. Treat. Rep.*, 63:369–71.

Donnenfeld, A. E., Pastuszak, A., Noah, J. S., Schick, B., Rose, N. C., and Koren, G. (1994): Methotrexate exposure prior to and during pregnancy [letter]. *Teratology*, 49:79–81.

Doria, A., DiLenardo, L., Vario, S., Calligaro, A., Vaccaro, E., and Gambari, P. F. (1992): Cyclosporin A in a pregnant patient affected with systemic lupus erythematosus. *Rheumatol. Int.*, 12:77–81.

Duombo, O., Soula, G., Kodio, B., and Perrenoud, M. (1992): Ivermectin and pregnancy in mass treatment in Mali. *Bull. Soc. Pathol. Exot.*, 85:247–51.

Edmonds, L. D., Layde, P. M., James, L. M., Flynt, J. W., Erickson, J. D., and Oakley, G. P. Jr. (1981): Congenital malformation surveillance: Two American systems. *Int. J. Epidemiol.*, 10:247–52.

Eggermont, E., Logghe, N., Van De Casseye, W., Casteels-Van Daele, M., Jaeken, J., Cosemans, J., Verstraete, M., and Renaer, M. (1976): Haemorrhagic disease of the newborn in the offspring of rifampicin and isoniazid treated mothers. *Acta. Paediatr. Belg.*, 29:87–90.

Ehrenfeld, M., Levy, M., Margalioth, E. J., and Eliakim, M. (1986): The effects of long-term colchicine therapy on male fertility in patients with familial Mediterranean fever. *Andrologia*, 18:420–6.

Ehrenfeld, M., Brzezinski, A., Levy, M., and Eliakim, M. (1987): Fertility and obstetric history in patients with familial Mediterranean Fever on long-term colchicine therapy. *Br. J. Obstet. Gynaecol.*, 94:1186–91.

Einarson, A., Phillips, E., Mawji, F., D'Alimonte, D., Schick, B., Addis, A., Mastroiacova, P., Mazzone, T., Matsui, D., and Koren, G. (1998): A prospective controlled multicentre study of clarithromycin in pregnancy. *Am. J. Perinatol.* 15(9):523–5.

Einarson, A., Bailey, B., Jung, G., Spizzirri, D., Baillie, M., and Koren, G. (1997): Prospective controlled study of hydroxyzine and cetirizine in pregnancy. *Ann. Allergy. Asthma. Immunol.*, 78:183–6.

El-Behelry, A., El-Mansy, E., Kamel, N., and Salama, N. (1979): Methotrexate and fertility in men. *Arch. Androl.*, 3:177–9.

Elimite Product Information, Irvine, CA, Allergan, 1999.

Erez, S., Schifrin, B., and Dirim, O. (1971): Double-blind evaluation of hydroxyzine as an antiemetic in pregnancy. *J. Reprod. Med.*, 7:57–9.

Eronen, M. (1993): The hemodynamic effects of antenatal indomethacin and a beta-sympathomimetic agent on the fetus and the newborn: A randomized study. *Pediatr. Res.*, 33:615–619.

Eskes, T. (1984): Classic illustration. *Eur. J. Obstet. Gynecol. Reprod. Biol.*, 16:365.

Eurax Product Information, Buffalo, NY, Westwood-Squibb Pharmaceuticals, Inc., 1999.

Falkson, H. C., Simson, I. W., and Falkson, G. (1980): Non-Hodgkin's lymphoma in pregnancy. *Cancer*, 45:1679–82.

Falterman, C. G., and Richardson, C. J. (1980): Small left colon syndrome associated with maternal ingestion of psychoactive drugs. *J. Pediatr.*, 97:308–10.

Famciclovir (Systemic). In: *USP-DI: Drug Information for the Health Care Professional*, Vol. 1. Taunton, MA, Rand McNally, 1998, pp. 1434–6.

Famvir Product Information. Philadelphia, PA, SmithKline Beecham Pharmaceuticals, 2000.

Faulds, D., and Benfield, P. (1994): Interferon beta-1b in multiple sclerosis: An initial review of its rationale for use and therapeutic potential. *Clin. Immunother.*, 11:79–87.

Feldkamp, M., and Carey, J. C. (1994): Clinical teratology counseling and consultation case report: Low dose methotrexate exposure in the early weeks of pregnancy [letter]. *Teratology*, 49:79–81.

Ferm, V. (1965): Teratogenic activity of hydroxyurea. *Lancet*, 11:1138–9.

Ferreira, N. R., (1968): Buoniconti A. Trisomy after colchicine therapy [letter]. *Lancet*, 2:1304.

Finegold, S. (1980): Metronidazole. *Ann. Intern. Med.*, 93:585–7.

Fitzgerald J, and McCann S. (1993): The combination of hydroxyurea and leucapheresis in the treatment of chronic myeloid leukaemia in pregnancy. *Clin. Lab. Haematol.*, 15:63–5.

Fluoroquinolones (Systemic). In: *USP-DI: Drug Information for the Health Care Professional*, Volume 1, Taunton, MA, The US Pharmacopeial Convention, 1998, pp. 1458–65.

Fraser FC, and Sajoo A. Teratogenic potential of corticosteroids in humans. *Teratology*, 51:45–6, 1995.

Freyberg, R., Ziff, M., and Baum, J. (1979): Gold terapy for rheumatoid arthritis. In: *Arthritis and Allied Conditions*. 8th edition. Edited by J. Hollander, D. McCarty, Jr., Philadelphia, PA, Lea and Febiger, p. 479.

Frieden, I. J., Reese, V., and Cohen, D. (1996): PHACE syndrome. The association of posterior fossa brain malformations, hemangiomas, arterial anomalies, coarctation of the aorta and cardiac defects, and eye abnormalities. *Arch. Dermatol.*, 132:307–11.

Fujii, T., and Nishimura, H. (1972): Comparison of teratogenic action of substances related to purine metabolism in mouse embryos. *Jpn. J. Pharmacol.*, 22:201–6.

Fulvicin P/G Product Information, Kenilworth, NJ, Schering Corporation, 1999.

Garbis, H., Elefant, E., Bertolotti, E., Robert, E., Serafini, M. A., and Prapas, N. (1995): Pregnancy outcome after periconceptional and first-trimester exposure to methoxsalen photochemotherapy [letter]. *Arch. Dermatol.*, 131:492–3.

Garland, M. (1998): Pharmacology of drug transfer across the placenta. *Obstet. Gynecol. Clin. North. Am.*, 25:21–42.

Geiger, J. M., Baudin, M., and Saurat, J. H. (1994): Teratogenic risk with etretinate and acitretin treatment. *Dermatology*, 189:109–116.

Glorieux, F. H., Salle, B. L., Travers, R., and Audra, P. H. (1991): Dynamic histomorphometric evaluation of human fetal bone formation. *Bone*, 12:377–381.

Gonen, R., Shilalukey, K., Magee, L., Koren, G., and Shime, J. (1994): Maternal disorders leading to increased reproductive risk. In: *Maternal-Fetal toxicology: A Clinician's Guide*. 2nd edition. Edited by G. Koren, New York, Marcel Dekker, pp. 641–82.

Green, D., Zevon, M., Lowrie, G., Seigelstein, N., and Hall, B. (1991): Congenital anomalies in children of patients who received chemotherapy for cancer in childhood and adolescence. *N. Engl. J. Med.*, 325:141–6.

Grifulvin V Product Information, Skillman, NJ, Ortho Dermatological, 1999.

Gris-Peg Product Information, Irvine, CA, Allergan Inc., 1999.

Griseofulvin, Systemic. In:*USP-DI:Drug Information for the Health Care Professional*, Vol. 1. Taunton, MA, Rand McNally, 1998, pp. 1527–9.

Groves, T., and Corenblum, B. (1995): Spironolactone therapy during human pregnancy [letter]. *Am. J. Obstet. Gynecol.*, 172:1655–6.

Grumbach, M., Ducharme, J., and Moloshok, R. (1959): On the fetal masculinizing action of certain oral progestins. *J. Clin. Endocrinol. Metab.*, 19:1369–1380.

Grunnet, E., Nyfors, A., and Hansen, K. B. (1977): Studies on human semen in topical corticosteroid-treated and in methotrexate treated psoriatics. *Dermatologica*, 154:78–84.

Gulmezoglu, A., Hofmeyer, G., and Costhuisen, M. (1997): Antioxidants in the treatment of severe pre-eclampsia: An explanatory randomized controlled trial. *Br. J. Obstet. Gynaecol.*, 104:689–96.

Gunnarskog, J. G., Kallen, A. J. B., Lindelop, B. G., and Sigurgeirsson, B. (1993): Psoralen photochemotherapy and pregnancy. *Arch. Dermatol.*, 129:320–323.

Haggstrom, J., Adriansson, M., Hybbinette, T., Harnby, E., and Thorbert, G. (1996): Two cases of CML treated with alpha-interferon during second and third trimester of pregnancy with analysis of the drug in the newborn immediately postpartum. *Eur. J. Haematol.*, 57:101–2.

Haimov-Kochman, R., and Ben-Chetrit, E. (1998): The effect of colchicine treatment on sperm production and function: A review. *Hum. Reprod.*, 13:360–2.

Harpey, J. P., Jaudon M. C., Clavel J. P., Galli, A., and Darbois, Y. (1983): Cutis laxa and low serum zinc after antenatal exposure to penicillamine [letter]. *Lancet*, 2:858.

Hartard, C., and Kunze, K. (1994): Pregnancy in a patient with Wilson's disease treated with D-penicillamine and zinc sulfate: A case report and review of the literature. *Eur. Neurol.*, 34:337–40.

Haustein, U., and Hlawa, B. (1989): Treatment of scabies with permethrin versus lindane and benzyl benzoate. *Acta. Derm. Venereol.*, 69:348–351.

Hecker, A., Hasan, S., and Neumann, F. (1990): Disturbances in sexual differentiation of rat foetuses following spironolactone treatment. *Acta. Endocrinol.*, (Copenh) 95:540–5.

Heinonen, O. P., Slone, D., and Shapiro, S. (1977): *Birth Defects and Pregnancy*. Littleton, MA, Publishing Sciences Group, pp. 297–313.

Hendricks, S., Smith, J., Moore, D., and Brown, Z. (1990): Oligohydramnios associated with prostaglandin synthetase inhibitors in preterm labour. *Br. J. Obstet. Gynaecol.*, 97:312–6.

Hingston, G. R. (1993): Pregnancy associated with a combined oral contraceptive and itraconazole. *New Z. Med. J.*, 106:528.

Hinkes, E., and Plotkin, D. (1973): Reversible drug-induced sterility in a patient with acute leukemia. *JAMA*, 223:1490–1.

Hocking, D. (1968): Neonatal haemolytic disease due to dapsone. *Med. J. Aust.*, 1:1130–1.

Humphreys, F., and Marks, J. M. (1988): Mepacrine and pregnancy [letter]. *Br. J. Dermatol.*, 118:452.

Hurwitz, S. (1973): Scabies in babies. *Am. J. Dis. Child.*, 126:226–8.

Hussein, M. M., Mooij, J. M., and Roujoule, H. (1993): Cyclosporine in the treatment of lupus nephritis including two patients treated during pregnancy. *Clin. Nephrol.*, 40:160–3.

Hydroxychloroquine (Systemic). In: *USP-DI: Drug Information for the Health Care Professional*, Vol. 1. Taunton, MA, Rand McNally, 1999, pp. 1663–6.

Iffy, L., Shepard, T. H., Jakobovits, A., Lemire, R. J., and Kerner, P. (1967): The rate of growth in young human embryos of Streeters horizons XIII to XXIII. *Acta. Anatomica.*, 66:178–86.

Inman, W., Pearce, G., and Wilton, L. (1994): Safety of fluconazole in the treatment of vaginal candidiasis: A prescription-event monitoring study, with special reference to the outcome of pregnancy. *Eur. J. Clin. Pharmacol.*, 46:115–8.

Jackson, N., Shukri, A., and Ali, K. (1993:) Hydroxyurea treatment for chronic myeloid leukaemia during pregnancy. *Br. J. Haematol.*, 85:203–4.

Jick, H., Holmes, L., Hunter, J., Madsen, S., and Stergachis, A. (1981): First trimester drug use and congenital disorders. *JAMA*, 246:343–6.

Jick, S., Terris, B., and Jick, H. (1993): First trimester topical tretinoin and congenital disorders. *Lancet*, 341:1181–2.

Kahn, G. (1985): Dapsone is safe during pregnancy. *J. Am. Acad. Dermatol.*, 13:838–9.

Kaler, S. G., Patrinos, M. E., Lambert, G. H., Myers, T. F., Karlman, R., and Anderson, C. L. (1987): Hypertrichosis and congenital anomalies associated with maternal use of minoxidil. *Pediatrics*, 79: 434–6.

Kaplan, B., Restaino, I., Raval, D., Gottlieb, R. P., and Bernstein, J. (1994): Renal failure in the neonate associated with in utero exposure to non-steroidal anti-inflammatory agents. *Pediatr. Nephrol.*, 8: 700–4.

Karol, M. D., Conner, C. S., Watanabe, A. S., and Murphrey, K. J. (1980): Podophyllum: Suspected teratogenicity from topical application. *Clin. Toxicol.*, 16:283–6.

Katz, F. H., Thorp, J. M. Jr., and Bowes, W. A. Jr. (1990): Severe symmetric intrauterine growth retardation associated with the topical use of triamcinolone. *Am. J. Obstet. Gynecol.*, 162:396–7.

Kawashima, K., Nakaura, S., Nagao, S., Tanaka, S., and Kuwamura, T. (1977): Virilizing activities of various steroids in female rat fetuses. *Endocrinol. Japan.*, 24:77–81.

Kim, D., and Park, M. (1989): Maternal and fetal survival following surgery and chemotherapy of endodermal sinus tumor of the ovary during pregnancy: A case report. *Obstet. Gynecol.*, 73:503–7.

King, C., Rogers, P., Cleary, J., and Chapman, S. (1998): Antifungal therapy during pregnancy. *Clin. Infect. Dis.*, 27:1151–60.

Kirshon, B., Wasserstrum, N., Willis, R., Herman, G. E., and McCabe, E. R. (1988): Teratogenic effects of first trimester cyclophosphamide therapy. *Obstet. Gynecol.*, 72:462–4.

Kirshon, B., Moise, K. J. Jr., Mari, G., and Willis, R. (1991): Long-term indomethacin therapy decreases fetal urine output and results in oligohydramnios. *Am. J. Perinatol.*, 8:86–88.

Knudsen, L. (1987): No association between griseofulvin and conjoined twinning [letter]. *Lancet*, 2:1097.

Kopelman, J., and Miyakaza, K. (1990): Inadvertent 5-fluorouracil treatment in early pregnancy: A report of three cases. *Reprod. Toxicol.*, 4:233–5.

Koren, G., Patuszak, A., and Ito, S. (1998): Drugs in pregnancy. *N. Eng. J. Med.*, 338:1128–37.

Kozlowski, R. D., Steinbrunner, J. V., Mackenzie, A. H., Clough, J. D., Wilke, W. S., and Segal, A. M. (1990): Outcome of first trimester exposure to low-dose methotrexate in eight patients with rheumatic disease. *Am. J. Med.*, 88:589–92.

Kullander, S., and Kallen. B. (1976): A prospective study of drugs and pregnancy 4: Miscellaneous drugs. *Acta. Obstet. Gynecol.*, Scand. 55:287–95.

l'Allemand, D., Gruters, A., Heidemann, P., and Schumbrand, P. (1983): Iodine-induced alteration of thyroid function in newborn infants after prenatal and perinatal exposure to povidone iodine. *J. Pediatr.*, 102:935–8.

Lamisil Product Information, Novartis Pharmaceuticals Corporation, East Hanover, NJ, 1999.

Landers, D. V., Green, J. R., and Sweet, R. L. (1983): Antibiotic use during pregnancy and the postpartum period. *Clin. Obstet. Gynecol.*, 26:391–406.

Lee, B. E., Feinberg, M., Abraham, J. J., and Murthy, A. R. (1992): Congenital malformations in an infant born to a woman treated with fluconazole. *Pediatr. Infect. Dis. J.*, 11:1062–3.

Lee, B., and Groth, P. (1977): Scabies: Transcutaneous poisoning during treatment [letter]. *Pediatrics*, 59:643.

Lendon, M., Hann, I. M., Palmer, M. K., Shalet, S. M., and Jones, P. H. (1978): Testicular histology after combination chemotherapy in childhood for acute lymphoblastic leukaemia. *Lancet*, 2:439–41.

Levin, D. L., Fixler, D. E., Morriss, F. C., and Tyson, J. (1978): Morphologic analysis of the pulmonary vascular bed in infants exposed in utero to prostaglandin synthetase inhibitors. *H. Pediatr.*, 92: 478–83.

Levy, M., Buskila, D., and Gladman, D., et al. (1991): Pregnancy outcome following first trimester exposure to chloroquine. *Am. J. Perinatol.*, 8:174–8.

Levy, M., Spino, M., and Read, S. (1991): Colchicine: A state-of-the-art review. *Pharmacotherapy*, 11: 196–211.

Linares, A., Zarranz, J., Rodriguez-Alarcon, J., and Diaz-Perez, J. (1979): Reversible cutis laxa due to maternal D-penicillamine treatment [letter]. *Lancet*, 2:43.

Lindane (topical). In: *USP-DI: Drug Information for the Health Care Professional*, Volume 1. Taunton, MA, Rand McNally, 1998, pp. 1876–9.

Lindquist, N. G., Sjostrand, S. E., and Ullberg, S. (1970): Accumulation of chorioretinotoxic drugs in the foetal eye. *Acta. Pharmaceutica. Toxicol. Suppl.*, (Copenhagen) 28:64.

Lipson, A., Collins, F., and Webster, W. (1993): Multiple congenital defects associated with maternal use of topical tretinoin. *Lancet*, 341:1352–3.

Lipton, J., Derzko, C., and Curtis, J. (1996): Alpha-interferon and pregnancy in a patient with CML. *Hematol. Oncol.*, 14:119–22.

Livezey, G. T., and Rayburn, D. R. (1992): Principles of perinatal pharmacology. In: *Drug Therapy in Obstetrics and Gynecology*, 3rd. edition. Edited by W. F. Rayburn, and F. P. Zuspan. Mosby-Year Book, St. Louis.

Loebstein, R., Lalkin, A., Addis, A., Costa, A., Lalkin, I., Bonati, M., and Koren, G. (1994): Terfenadine (Seldane) exposure in early pregnancy [abstract]. *Teratology*, 49:417.

Lowenthal, R., Funnell, C. F., Hope, D. M., Stewart, I. G., and Humphrey, D. C. (1982): Normal infants after combination chemotherapy including tenoposide for Burkitt's lymphoma in pregnancy. *Med. Pediatr. Oncol.*, 10:165–9.

Lower, G. D., Stevens, L. E., Najarian, J.S., and Reetsma, K. (1971): Problems from immunosuppressives during pregnancy. *Am. J. Obstet. Gynecol.*, 111:1120–1.

Lyde, C. (1997): Pregnancy in patients with Hansen disease. *Arch. Dermatol.*, 133:623–7.

Macri, C., Ricciardi, C., Stazi, A., and Mantovani, A. (1993): Histologic lesions in rat embryos exposed in utero to albendazole. *Teratology*, 48:29A.

Major, C. A., Lewis, D. F., Harding, J. A., Porto, M. A., and Garite, T. J. (1994): Tocolysis with indomethacin increases the incidence of necrotizing enterocolitis in the low-birth-weight neonate. *Am. J. Obstet. Gynecol.*, 170:102–106.

Malfetano, J., and Goldkrand, J. (1990): Cis-platinum combination chemotherapy during pregnancy for advanced epithelial ovarian carcinoma. *Obstet. Gynecol.*, 75:545–7.

Mantovani, A., Ricciardi, C., Stazi, A., and Macri, C. (1995): Effects observed on gestational day 13 in rat embryos exposed to albendazole. *Reprod. Toxicol.*, 9:265–73.

Martinez-Frias, M., Rodriguez-Pinella, E., Bermejo, E., and Blanco, M. (1998): Prenatal exposure to penicillamine and oral clefts: Case report. *Am. J. Med. Genet.*, 76:274–5.

Martinez-Rueda, J., Arce-Salinas, C., Kraus, A., Alcocer-Varela, J., and Alarcon-Segovia, D. (1996): Factors associated with fetal losses in severe systemic lupus erythematosus. *Lupus*, 5:113–9.

Mastroiacovo, P., Mazzone, T., Botto, L. D., Serafini, M. A., Finardi, A., Caramelli, L., and Fusco, D. (1996): Prospective assessment of pregnancy outcomes after first-trimester exposure to fluconazole. *Am. J. Obstet. Gynecol.*, 175:1645–50.

Maurus, J. N. (1978): Hansen's disease in pregnancy. *Obstet. Gynecol.*, 52:22–25.

McFie, J., and Robertson, J. (1973): Psychological test results of children with thalidomide deformities. *Dev. Med. Child. Neurol.*, 15:719–27.

Mehta, P., Mehta, S., and Vorherr, H. (1983): Congenital iodide goiter and hypothyroidism: A review. *Obstet. Gynecol.*, Surv., 38:237–47.

Mellin, G. W., and Katzenstein, M. (1962): The saga of thalidomide: Neuropathy to embryopathy, with case reports of congenital anomalies. *N. Engl. J. Med.*, 267:1184–93.

Mellin, G. W., and Katzenstein, M. (1962): The saga of thalidomide (concluded): Neuropathy to embryopathy, with case reports of congenital anomalies. *N. Engl. J. Med.*, 267:1238–44.

Mennuti, M., Shepard, T., and Mellman, W. (1975): Fetal renal malformation following treatment of Hodgkin's disease during pregnancy. *Obstet Gynecol.*, 46:194–6.

Merlin, H. (1972): Azoospermia caused by colchicine: A case report. *Fertil. Steril.*, 23:180–1.

Messina, M., Biffignandi, P., and Ghiga, E., et al. (1979): Possible contraindication of spironolactone during pregnancy. *J. Endocrinol. Invest.* 2:222.

Methotrexate Product Information, Seattle, WA, Immunex Corporation, 1997.

Methotrexate Product Information, Seattle, WA, Immunex Corporation, 1999.

Metneki, J., and Czeizel, A. (1987): Griseofulvin teratology including two thoracopagus conjoined twins. *Lancet*, 1:1042.

Milunsky, A., Graef, J. W., and Gaynor, M. F., Jr. (1968): Methotrexate-induced congenital malformations. *J. Pediatr.*, 72:790–5.

Mintezol Product Information, West Point, PA, Merck and Co., 1999.

Miyamoto, J. (1976): Degradation, metabolism and toxicity of synthetic pyrethroids. *Environ. Health. Presp.*, 14:15–28.

Miyamoto, T., Mijaya, S., Horiuchi, Y., Hara, M., and Ishihara, K. (1972): Gold therapy in bronchial asthma-special emphasis upon blood levels of gold and its teratogenicity. *Nippon. Naika. Gakai. Zasshi.*, 63:1190–7.

Mjolnerod, O., Rasmussen, K., Dommerud, S., and Gjeruldsen, S. (1971): Congenital connective-tissue defect probably due to D-penicillamine treatment in pregnancy. *Lancet*, 1:673–5.

Moise, K. J., Jr. (1993): Effect of advancing gestational age on the frequency of fetal ductal constriction in association with maternal indomethacin use. *Am. J. Obstet. Gynecol.*, 168:1350–1353.

Momma, K., and Takao, A. (1990): Transplacental cardiovascular effects of four popular analgesics in rats. *Am. J. Obstet. Gynecol.*, 162:1304–10.

Momma, K., and Takeuchi, H. (1983): Constriction of fetal ductus arteriosus by non-steroidal anti-inflammatory drugs. *Prostaglandins*, 26:631–43.

Morgan, D. J. (1997): Drug disposition in mother and foetus. *Clin. Exp. Pharmacol. Physiol.*, 24:869–73.

Morgan, I. (1978): Metronidazole treatment in pregnancy. *Int. J. Gynaecol. Obstet.*, 15:501–2.

Munley, S., and Hurtt, M. (1996): Developmental toxicity study of glycolic acid in rats [abstract]. *Teratology*, 53:117.

Murphy, M., and Chaube, S. (1964): Preliminary survey of hydroxyurea (NSC-32065) as a teratogen. *Cancer. Chemother. Rep.*, 40:1–7.

Murray, L., and Seger, D. (1994): Drug therapy during pregnancy and lactation. *Emergency Medicine Clinics of North America*, 12: 129–49.

Mustafa, M., and Shamina, A. (1998): Five successful deliveries following 9 consecutive spontaneous abortions in a patient with Wilson's disease. *Aust. N. Z. J. Obstet. Gynaecol.*, 38:312–4.

Narron, G., Zec, M., Neves, R., Manders, E. K., and Sexton, F. M. Jr. (1992): Penicillamine-induced pseudoxanthoma elasticum-like changes requiring rhytidectomy. *Ann. Plast. Surg.*, 29:367–70.

Navarre-Belhassen, C., Blanchet, P., Hillaire-Buys, D., Sarda, P., and Blayac, J. P. (1998): Multiple congenital malformations associated with topical tretinoin [letter]. *Ann. Pharmacother.*, 32:505–6.

Needs, C. J., and Brooks, P. M. (1985): Antirheumatic medication in pregnancy. *Br. J. Rheumatol.*, 24:282–90.

NegGram Product Information. New York, Sanofi Winthrop Pharmaceuticals, 1998.

Neoral Product Information, East Hanover, Novartis Pharmaceuticals, 1999.

Newman, C. G. H. (1985): Teratogen update: Clinical aspects of thalidomide embryopathy: A continuing preoccupation. *Teratology*, 32:133–44.

Newman, C. G. H. (1986): The thalidomide syndrome: Risks of exposure and spectrum of malformations. *Clin. Perinatol.*, 13:555–73.

Nicholson, H. (1968): Cytotoxic drugs in pregnancy: Review of reported cases. *J. Obstet. Gynaecol. Br. Commonw.*, 75:307–12.

Niebyl, J. R. (1992): Drug therapy during pregnancy. *Curr. Opin. Obstet. Gynecol.*, 4:43–7.

Nishikawa, S., Hara, T., Miyazaki, H., and Ohguro, Y. (1984): Reproduction studies of KW-1414 in rats and rabbits. *Clin. Rep.*, 18:1433–88.

Norton, M. E., Merrill, J., Cooper, B. A. B., Kuller, J. A., and Clyman, R. I. (1993): Neonatal complications after the administration of indomethacin for preterm labor. *N. Engl. J. Med.*, 329:1602–1607.

Norton, M. E. (1997): Teratogen update: Fetal effects of indomethacin administration during pregnancy. *Teratology*, 56:282–292.

Ogata, A., Ando, H., Kubo, Y., and Hiraga, K. (1984): Teratogenicity of thiabendazole in ICR mice. *Food. Chem. Toxicol.*, 22:509–20.

Ortega, J. (1977): Multiple agent chemotherapy including bleomycin of non-Hodgkin's lymphoma during pregnancy. *Cancer*, 40:2829–35.

Ostensen, M. (1992): Treatment with immunosuppressive and disease modifying drugs during pregnancy and lactation. *Am. J. Reprod. Immunol.*, 28: 148–52.

Otano, L., Amestoy, G., Paz, J., and Gadow, E. (1992): Periconceptional exposure to topical 5-fluorouracil [letter]. *Am. J. Obstet. Gynecol.*, 166:263–4.

Pacque, M., Munoz, B., Poetsche, G., Foose, J., Greene, B. M., and Taylor, H. R. Pregnancy outcome after inadvertent ivermectin treatment during community-based distribution. *Lancet*, 2:1486–9, 1990.

Pakkala, S., Lukka, M., Helminen, P., Koskimies, S., and Ruutu, T. (1994): Paternity after bone marrow transplantation following conditioning with total body irradiation. *Bone Marrow Transplant*, 13: 489–90.

Papiernik, E., Rozenbaum, H., Amblard, P., Dephot, N., and de Mouzon, J. (1989): Intra-uterine device failure: Relation with drug use. *Eur. J. Obstet. Gynecol. Reprod. Biol.*, 32:205–12.

Parke, A., and West, B. (1996): Hydroxychloroquine in pregnant patients with systemic lupus erythematosus. *J. Rheumatol.*, 23:1715–8.

Parke, A. (1988): Antimalarial drugs and pregnancy. *Am. J. Med.*, 85 (Suppl 4A): 30–3.

Pastuszak, A., Schick, A., D'Alimonte, D., Donnenfeld, A., and Koren, G. (1996): The safety of astemizole in pregnancy. *J. Allergy. Clin. Immunol.*, 98:748–50.

Patel, M., Dukes, I., and Jull, J. (1991): Use of hydroxyurea in chronic myeloid leukemia during pregnancy: A case report. *Am. J. Obstet. Gynecol.*, 165:565–6.

Patten, B. M. (1958): *Foundations of Embryology*. New York, McGraw-Hill Publications.

Pembrey, M. E., Clarke, C. A., and Frais, M. M. (1970): Normal child after maternal thalidomide ingestion in critical period of pregnancy. *Lancet*, 1:275–277.

Penicillins, Systemic. In: *USP-DI: Drug Information for the Health Care Professional*, Vol. 1. Taunton, MA, Rand McNally, 1998, pp. 2253–75.

Penn, I., Makowski, E. L., and Harris, P. (1980): Parenthood following renal transplantation. *Kidney. Int.*, 18:221–233.

Pentamidine (systemic). In: *USP-DI: Drug Information for the Health Care Professional*, Vol. 1. Taunton, M. A., Rand McNally, 1998, pp. 2286–9.

Pentam Product Information. Deerfield, IL, Fujisawa, USA, Inc., 1992.

Personal Communication, (1998): Wyeth-Ayerst Pharmaceuticals, September 30.

Phillips-Howard, P., and Wood, D. (1996): The safety of antimalarial drugs in pregnancy. *Drug Saf.*, 14:131–45.

Physicians' Desk Reference. Medical Economics Data Production Company, Montvale, NJ, 1999.

Pilkington, T., and Brogden, R. N. (1992): Acitretin. A review of its pharmacology and therapeutic use. *Drugs*, 43:597–627.

Pillans, P. I., and Sparrow, M. J. (1993): Pregnancy associated with a combined oral contraceptive and itraconazole. *New. Z. Med. J.*, 106:436.

Piper, J., Mitchel, E., and Ray, W. (1993): Prenatal use of metronidazole and birth defects: No association. *Obstet. Gynecol.*, 82:348–52.

Pizzuto, J., Aviles, A., Noriega, L., Niz, J., Morales, M., and Romero, F. (1980): Treatment of acute leukemia during pregnancy: Presentation of nine cases. *Cancer. Treat. Rep.*, 64:679–83.

Plaquenil Product Information, New York, Sanofi Pharmaceuticals, 1999.

Podophyllum (topical) In: *USP DI: Drug Information for the Health Care Professional*, Vol 1. Taunton, MA, Rand McNally, 1998, pp. 2371–3.

Powell, H. R., and Ekert, H. (1971): Methotrexate-induced congenital malformations. *Med. J. Aust.*, 2:1076–7.

Prahalada, S., Tarantal, A., Harris, G., Ellsworth, K. P., Clarke, A. P., Skiles, G. L., MacKenzie, K. I., Kruk, L. F., Ablin, D. S., Cukierski, M. A., Peter, C. P., vanZwieten, M. J., and Hendrickx, A. G. (1997): Effects of finasteride, a type 2 5-alpha reductase inhibitor, on fetal development in the Rhesus monkey (Macaca mulatta). *Teratology*, 55:119–131.

Pramanik, A. K., and Hansen, R. C. (1979): Transcutaneous gamma benzene hexachloride absorption and toxicity in infants and children. *Arch. Dermatol.*, 115:1224–5.

Prasad, M. H., Pushpavathi, K., Devi, G. S., and Reddy, P. P. (1996): Reproductive epidemiology in sulfonamide factory workers. *J. Toxicol. Environ. Health.*, 47:109–14.

Price, H. V., Salaman, J. R., Laurence, K. M., and Langmaid, H. (1976): Immunosuppressive drugs and the foetus. *Transplantation*, 21:294–8.

Purohit, D. M., Ellison, R. C., Zierler, S., Miettinen, O. S., and Nadas, A. S. (1985): Risk factors for retrolental fibroplasia: Experience with 3,025 premature infants. National Collaborative Study on Patent Ductus Arteriosus in Premature Infants. *Pediatrics*, 76:339–44.

Pursley, T. J., Blomquist, I. K., Abraham, J., Andersen, H. F., and Bartley, J. A. (1996): Fluconazole-induced congenital anomalies in three infants. *Clin. Infect. Dis.*, 22:336–40.

Quinacrine, (Systemic). In: *USP-DI monograph*, 1999.

Qureshi, M., Pennington, J., Goldsmith, H., and Cox, P. (1972): Cyclophosphamide therapy and sterility. *Lancet*, 2:1290–1.

Rabinovitch, O., Zemer, D., Kukia, E., Sohar, E., and Mashiach, S. (1992): Colchicine treatment in conception and pregnancy: Two hundred thirty-one pregnancies in patients with Familial Mediterranean Fever. *Am. J. Reprod. Immunol.*, 28:245–6.

Radomski, J. S, Ahswede, B. A., Jarrell, B. E., Mannion, J., Cater, J., Moritz, M. J., and Armenti, V. T. (1995): Outcomes of 500 pregnancies in 335 female kidney, liver, and heart transplant recipients. *Transplant. Proc.*, 27:1089–90.

Raffles, A., Williams, J., Costeloe, K., and Clark, P. (1989): Transplacental effects of maternal cancer chemotherapy: Case Report. *Br. J. Obstet. Gynaecol.*, 96:1099–1100.

Ramey, J., Starke, M., Gibbons, W., and Archer, D. (1994): The influence of pentoxifylline (Trental) on the antifertility effect of intrauterine devices in rats. *Fertil. Steril.*, 62:181–5.

Ramsey-Goldman, R., and Schilling, E. (1997): Immunosuppressive drug use during pregnancy. *Rheum. Dis. Clin. North. Am.*, 23: 149–67.

Rasmussen, J. E. (1981): The problem of lindane. *J. Am. Acad. Dermatol.*, 5:507–16.

Rasmussen, J. E. (1987): Lindane, a prudent approach. *Arch. Dermatol.*, 123:1008–9.

Reed, B. R. (1997): Dermatologic drugs, pregnancy, and lactation. *Arch. Dermatol.*, 133:894–898.

Reed, B. R. (1997): Dermatologic drug use during pregnancy and lactation. *Dermatology Clinics*, 15: 197–202.

Reed, B. R. (1998): Dermatologic drug therapy. In: *Comprehensive Dermatologist drug therapy*, ed. S. Wolverton. In press.

Registration Committee of the European Dialysis and Transplant Association: Successful pregnancies in women treated by dialysis and kidney transplantation. *Br. J. Obstet. Gynaecol.*, 87:839–45, 1980.

Reiff-Eldridge, R., Heffner, C. R., Ephross, S. A., Tennis, P. S., White, A. D., and Andrews, E. B. (2000): Monitoring pregnancy outcomes after prenatal drug exposure through prospective pregnancy registries: A pharmaceutical company commitment. *Am J Obstet Gynecol.* 182:159–63.

Renova Product Information. Skillman, NJ, Ortho Dermatological, 1999.

Retin A Product Information. Skillman, NJ, Ortho Dermatological, 1999.

Rifampin, Systemic. In: *USP-DI: Drug Information for the Health Care Professional*, Vol. 1. Taunton, M. A., Rand McNally, 1998, pp. 2509–16.

Rigo, J. Jr., Glaz, E., and Papp, Z. (1996): Low or high doses of spironolactone for treatment of maternal Bartter's syndrome. *Am. J. Obstet. Gynecol.*, 174:297.

Rogaine Topical Solution Product Information, Kalamazoo, MI, The Upjohn Company, 1994.

Rosa, F., Baum, C., and Shaw, M. (1987): Pregnancy outcomes after first-trimester vaginitis drug therapy. *Obstet. Gynecol.*, 69:751–5.

Rosa, F. W., Hernandez, C., and Carlo, W. A. (1987): Griseofulvin teratology [letter]. *Lancet*, 1:171.

Rosa, F. W., Indanpaan-Heikkila, J., and Asanti, R. (1987): Fetal minoxidil exposure [letter]. *Pediatrics*, 80:120.

Rosa, F. (1986): Teratogen update: Penicillamine. *Teratology*, 33:127–31.

Rosa, F. W., Wilk, A. L., and Kelsey, F. O. (1986): Teratogen update: Vitamin A congeners. *Teratology*, 33:355–364.

Rothman, K., Fyler, D., Goldblatt, A., and Kreidberg, M. (1979): Exogenous hormones and other drug exposures of children with congenital heart disease. *Am. J. Epidemiol.*, 109:433–9.

Roubenoff, R., Hoyt, J., Petri, M., Hochberg, M. C., and Hellmann, D. B. (1988): Effects of antiinflammatory and immunosuppressive drugs on pregnancy and fertility. *Semin. Arthritis. Rheum.*, 18:88–110.

Rudolph, J. E., Schweizer, R. T., and Bartus, S. A. (1979): Pregnancy in renal transplant patients. *Transplantation*, 27:26–9.

Rustin, G., Booth, M., Dent, J., Salt, S., Rustin, F., and Bagshawe, K. D. (1984): Pregnancy after cytotoxic chemotherapy for gestational trophoblastic tumors. *Br. Med. J.*, 288:103–6.

Ryffel, B., Donatsch, P., Madorin, M., Matter, B. E., Ruttimann, G., Schon, H., Stoll, R., and Wilson, J. (1983): Toxicological evaluation of cyclosporin A. *Arch. Toxicol.*, 53:107–41.

Sanders, J., Hawley, J., Levy, W., Gooley, T., Buckner, C. D., Deeg, H. J., Doney, K., Storb, R., Sullivan, K., Witherspoon, R., and Appelbaum, F. R. (1996): Pregnancies following high-dose cyclophosphamide with or without highdose busulfan or total body irradiation and bone marrow transplantation. *Blood*, 87:3045–52.

Sangalli, L., Bortolotti, A., Passerini, F., and Bonati, M. (1990): Placental transfer, tissue distribution, and pharmacokinetics of cyclosporine in the pregnant rabbit. *Drug. Metab. Dispos.*, 18:102–6.

Sawyer, D., Conner, C. S., and Scalley, R. (1981): Cimetidine: Adverse reactions and acute toxicity. *Am. J. Hosp. Pharm.*, 38:188–97.

Saxen, I. (1974): Cleft palate and maternal diphenhydramine intake. *Lancet*, 1:407–8.

Shacter, B. (1981): Treatment of scabies and pediculosis with lindane preparations: An evaluation. *J. Am. Acad. Dermatol.*, 5:517–27.

Schaefer, C., Amoura-Elefant, E., Vial, T., Ornoy, A., Garbis, H., Robert, E., Rodriguez-Pinilla, E., Pexieder, T., Prapas, N., and Merlob, P. (1996): Pregnancy outcome after prenatal quinolone exposure: Evaluation of a case registry of the European Network of *Teratology*, Information Services (ENTIS). *Eur. J. Obstet. Gynecol. Reprod. Bio.*, 69:83–9.

Schatz, M., and Pettiti, D. (1997): Antihistamines and pregnancy. *Ann. Allergy Asthma Immunol.*, 78:157–9.

Schick, B., Hom, M., Librizzi, R., and Donnenfeld, A. (1996): Pregnancy outcome following exposure to clarithromycin. *Reprod. Toxicol.*, 10:162.

Schilsky, R., Lewis, B., Sherins, R., and Young, R. (1980): Gonadal dysfunction in patients receiving chemotherapy for cancer. *Ann. Intern. Med.*, 93:109–14.

Schilsky, R., Sherins, R., Hubbard, S., Wesley, M. N., Young, R. C., and DeVita VT SOURCE: Am J Med. (1981): Long-term follow-up of ovarian function in women treated with MOPP chemotherapy for Hodgkin's disease. *Am. J. Med.*, 71:552–6.

Schleuning, M., and Climm, C. (1987): Chromosomal aberrations in a newborn whose mother received cytotoxic treatment during pregnancy [letter]. *N. Engl. J. Med.*, 317:1666–7.

Schumacher, H. J. (1975): Chemical structure and teratogenic properties. In: *Methods for Detection of Environmental Agents that Produce Congenital Defects.* Edited by T. H. Shepard, J. R. Miller, M. Marois. Amsterdam: North Holland-American Elsevier, pp. 65–77.

Scott, J. (1977): Fetal growth retardation associated with maternal administration of immunosuppressive drugs. *Am. J. Obstet. Gynecol.*, 128:668–76.

Selevan, S. G., and Lemasters, G. K. (1987): The dose-response fallacy in human reproductive studies of toxic exposures. *J. Occup. Med.*, 29:451–454.

Selsun Product Information, Columbus, OH, Ross Laboratories, 1999.

Seto, A., Einarson, T., and Koren, G. (1997): Pregnancy outcome following first trimester exposure to antihistamines: Meta-analysis. *Am. J. Perinatol.*, 14:119–24.

Seto, A., Einarson, T., and Koren, G. (1993): Evaluation of brompheniramine safety in pregnancy. *Reprod. Toxicol.*, 7:393–5.

Sever, L. E., and Mortensen, M. E. (1996): Teratology and the epidemiology of birth defects: Occupational and environmental perspectives. In: *Obstetrics*, 3rd edition. Edited by S. G., Gabbe, J. R., Niebyl, and J. L., Simpson. NY, Churchill Livingstone, pp. 185–213.

Shamberger, R. C., Rosenberg, S. A., Seipp, C. A., and Sherins, R. J. (1981): Effects of high-dose methotrexate and vincristine on ovarian and testicular functions in patients undergoing post-operative adjuvant treatment of osteosarcoma. *Cancer. Treat. Rep.*, 65:739–45.

Shapiro, L., Pastuszak, A., Curto, G., and Koren, G. (1998): Is topical tretinoin safe during the first trimester? *Can. Fam. Physician*, 44:495–8.

Sherins, R. J., and DeVita, D. T. (1973): Effect of drug treatment for lymphoma on male reproductive capacity. *Ann. Intern. Med.*, 79:216–220.

Silvadene Product Information, Kansas City, MO, Marion Merrell Dow, 1991.

Simeoni, U., Messer, J., Weisburd, P., Haddad, J., and Willard, D. (1989): Neonatal renal dysfunction and intrauterine exposure to prostaglandin synthesis inhibitors. *Eur. J. Pediatr.*, 148:371–3.

Slone, D., Siskind, V., Heinonen, O. P., Monson, R. R., Kaufman, D. W., and Shapiro, S. (1976): Aspirin and congenital malformations. *Lancet*, 1:1373–5.

Smallwood, R., Berlin, R., Castagnoli, N., Festen, H. P., Hawkey, C. J., Lam, S. K., Langman, M. J., Lundborg, P., and Parkinson, A. (1995): Safety of acid-suppressing drugs. *Dig. Dis. Sci.*, 40 (Suppl): 63S–80S.

Smithells, R. W. (1973): Defects and disabilities of thalidomide children. *Br. Med. J.*, 1:269–72.

Snider, D. E., Jr., Layde, P. M., Johnson, M. W., and Lyle, M. A. (1980): Treatment of tuberculosis during pregnancy. *Am. Rev. Respir. Dis.*, 122:65–79.

Solomon, L., Abrams, G., Dinner, M., and Berman, L. (1977): Neonatal abnormalities associated with D-penicillamine treatment during pregnancy [letter]. *N. Engl. J. Med.*, 296:54–5.

Sperling, R., Stratton, P., O'Sullivan, M., Boyer, P., Watts, D. H., Lambert, J. S., Hammill, H., Livingston, E. G., Gloeb, D. J., and Minkoff, H. (1991): A survey of zidovudine use in pregnant women with human immunodeficiency virus infection. *N. Engl. J. Med.*, 326:857–61.

SSD Product Information, Lincolnshire, IL, Boots Pharmaceutical, 1993.

Stahlmann, R., Klug, S., Lwendowski, C., Bochert, G., Chahoud, I., Rahm, U., Merker, H. J., and Neubert, D. (1988): Prenatal toxicity of acyclovir in rats. *Arch. Toxicol.*, 61:468–79.

Stern, R. S., and Lange, R. (1991): Outcomes of pregnancies among women and partners of men with a history of exposure to methoxsalen photochemotherapy (PUVA) for the treatment of psoriasis. *Arch. Dermatol.*, 127:347–50.

Stewart, B. (1975): Drugs that cause and cure male infertility. *Drug. Ther.*, 5:42–8.

Stromectol Product Information, West Point, PA, Merck & Co., Inc., 1999.

Stromland, K., and Miller, M. T. (1993): Thalidomide embryopathy: Revisited 27 years later. *Acta. Ophthalmol.*, 71: 238–245.

Sulfonamides, Systemic. In: *USP-DI: Drug Information for the Health Care Professional*, Vol. 1. Taunton, MA, Rand McNally, 1998, pp. 2681–6.

Sullivan, D., Csuka, M. E., and Blanchard, B. (1980): Erythromycin ethinylsuccinate hepatotoxicity. *JAMA*, 243:1074.

Sussman, A., and Leonard, J. M. (1980): Psoriasis, methotrexate, and oligospermia. *Arch. Dermatol.*, 116:215–7.

Szabo, K. T., Guenriero, J., and Kang, Y. J. (1978): The effect of gold-containing compounds on pregnant rats and their fetuses. *Vet. Pathol.*, 15 (suppl 5):89.

Szabo, K. T., DiFebbo, M. E., and Phelan, D. G. (1978): The effects of gold-containing compounds on pregnant rabbits and their fetuses. *Vet. Pathol. Suppl.*, 15:97–102.

Tacrolimus (systemic). In: *USP-DI: Drug Information for the Health Care Professional*, Vol. 1. Taunton, MA, Rand McNally, pp. 2683–8.

Tanis, A. A. (1998): Azathioprine in inflammatory bowel disease, a safe alternative? *Mediators Inflamm.*, 7: 141–4.

Tarp, U., and Graudal, H. A. (1985): A follow-up study of children exposed to gold compounds in utero. *Arthritis Rheum*, 28:235–6.

Tasdemir, I., Tasdemir, M., and Tavukcuoglu, S. (1998): Effect of pentoxifylline on immotile testicular spermatozoa. *J. Assist. Reprod. Genet.*, 15:90–2.

Teratology, Society. (1991): Recommendations for isotretinoin use in women of childbearing potential. *Teratology*, 44:1–6.

Terbinafine-Systemic. In: *USP-DI: Drug Information for the Health Care Professional*, Vol. 1. Taunton, MA, Rand McNally, 1998, pp. 2787–90.

Tertian, G., Tchernia, G., Papiernik, E., and Elefant, E. (1992): Hydroxyurea and pregnancy [letter]. *Am. J. Obstet. Gynecol.*, 166:1868.

Tetracyclines, Systemic. In: *USP-DI: Drug Information for the Health Care Professional*, Vol. 1. Taunton, MA, Rand McNally, 1998, pp. 2801–10.

Thalomid Product Information, celgene Corp., Warren Grove, NJ, 2000.

Thiabendazole, (systemic) (1998): In: *USP-DI: Drug Information for the Health Care Professional*, Vol. 1. Taunton, MA, Rand McNally, 1998, pp. 2820–3.

Thomas, L., and Andes, W. (1982): Fetal anomaly associated with successful chemotherapy for Hodgkin's disease during the first trimester of pregnancy. *Clin. Res.*, 30:424A.

Thomson, E. J., and Cordero, J. F. (1989): The new teratogens: Accutane and other vitamin-A analogs. *Amer. J. Matern. Child. Nurs.*, 14:244–248.

Thornton, Y. S., and Bowe, E. T. (1989): Neonatal hyperbilirubinemia after treatment of maternal leprosy. *S. Med. J.*, 82:668.

Toledo, T., Harper, R., and Moser, R. (1971): Fetal effects during cyclophosphamide and irradiation therapy. *Ann. Intern. Med.*, 74:87–91.

Tolchin, S., Winkelstein, A., Rodnan, G., Pan, S. F., and Nankin, H. R. (1974): Chromosome abnormalities from cyclophosphamide therapy in rheumatoid arthritis and progressive systemic sclerosis (scleroderma). *Arthritis Rheum.*, 17:375–82.

Torfs, C. P., Katz, E. A., Bateson, T. K., Lam, P. K., and Curry, C. J. (1996): Maternal medications and environmental exposures as risk factors for gastroschisis. *Teratology*, 54:84–92.

Totterman, L. E., and Saxen, L. (1969): Incorporation of tetraycline into human foetal bones after maternal drug administration. *Acta. Obstet. Gynecol. Scand.*, 48:542–549.

Tranexamic acid (systemic) In: *USP-DI: Drug Information for the Health Care Professional*, Vol. 1. Taunton, MA, Rand McNally, 1998, pp. 2862–5.

Trental Product Information, Kansas City, MO, Hoechst Marion Roussell Pharmaceuticals Inc., 1999.

Tuffanelli, D. L. (1982): Successful pregnancy in a patient with dermatitis herpetiformis treated with low-dose dapsone. *Arch. Dermatol.*, 118:876.

Turchi, J., and Villasis, C. (1988): Anthracyclines in the treatment of malignancy in pregnancy. *Cancer*, 61:435–40.

Turner, G. R., and Levin, D. L. (1984). Prostaglandin synthesis inhibition in persistent pulmonary hypertension of the newborn. *Clin. Perinatol.*, 11:581–9.

USP-DI: Drug Information for the Health Care Professional, Vol. 1. Taunton, MA, Rand McNally, 1995, pp. 1258–66.

Vallance, P. (1995): Drugs and the fetus, BMJ 312:1053–4.

Valtrex Product Information, Research Triangle Park, NC, Burroughs Wellcome Co., 1999.

van der Aa, E., Perreboom-Stegeman, J. H., Noordhoek, J., Gribnau, F. W., and Russel, F. G. (1998): Mechanisms of drug transfer across the human placenta. *Pharm. World. Sci.*, 20:139–48.

Van Hoogdalem, E. J. (1998): Transdermal absorption of topical anti-acne agents in man: Review of clinical pharmacokinetic data. *J. Eur. Acad. Dermatol. Venereol.*, 11(Suppl 1):s13–9.

Van Le, L., Pizzuti, D., Greenberg, M., and Reid, R. (1991): Accidental use of low-dose 5-fluorouracil in pregnancy. *J. Reprod. Med.*, 36:872–4.

Vicens-Calvet, E., Potau, N., Carreras, E., Bellart, J., Albisu, M. A., and Carrascosa, A. (1998): Diagnosis and treatment in utero of goiter with hypothyroidism caused by iodide overload. *J. Pediatr.*, 133: 147–8.

Vigersky, R. A., Mehlman, I., Glass, C. R., and Smith, C. E. (1980): Treatment of hirsute women with cimetidine: A preliminary report. *N. Engl. J. Med.*, 303:1042.

Vorhees, C., Butcher, R., Brunner, R., and Sobotka, T. (1979): A developmental test battery for neurobehavioral toxicity in rats: A preliminary analysis using monosodium blutamate calcium carrageenan, and hydroxyurea. *Toxicol. Appl. Pharmacol.*, 50:267–82.

Warkany, J. (1978): Aminopterin and methotrexate: Folic acid deficiency. *Teratology* 17:353–7.

Weed, J., Jr., Roh, R., and Mendenhall, H. (1979): Recurrent endodermal sinus tumor during pregnancy. *Obstet. Gynecol.*, 54:653–6.

Werler, M. M., Mitchell, A. A., and Shapiro, S. (1989): The relation of aspirin use during the first trimester of pregnancy to congenital cardiac defects. *N. Engl. J. Med.*, 321:1639–42.

Werler, M. M., Mitchell, A., and Sjhapiro, S. (1992): First-trimester maternal use in relation to gastroschisis. *Teratology*, 45:361–7.

Whitehead, E., Shalet, S., Blackledge, G., Todd, I., Crowther, D., and Beardwell, C. G. (1983): The effect of combination chemotherapy on ovarian function in women treated for Hodgkin's disease. *Cancer*, 52:988–93.

Whittaker, S., Seeley, M., and Faustman, E. (1990): The effects of albendazole and albendazole sulfoxide on cultures of differentiating rat embryo limb bud cells [abstract]. *Teratology*, 41:598–9.

Wiggins, D., and Elliott, J. (1990): Oligohydramnios in each sac of a triplet gestation caused by Motrin—fulfilling Koch's postulates. *Am. J. Obstet. Gynecol.*, 162:460–1.

Wilkinson, A., Aynsley-Green, A., and Mitchell, M. (1979): Persitstent pulmonary hypertension and abnormal prostaglandin E levels in preterm infants after maternal treatment with naproxen. *Arch. Dis. Childhood*, 54:942–5.

Williams, J. D., Condie, A. P., Brumfitt, W., and Reeves, D. S. (1969): The treatment of bacteriuria in pregnant women with sulphamethoxsazole and trimethoprim: A microbiological, clinical, and toxicological study. *Postgrad. Med. J.*, 45 (Suppl 6): 71–6.

Williams, H., Adetugho, K., Po, A., Naldi, L., Diepgen, T., and Murrell, D. (1998): The Cochrane skin group: Preparing, maintaining and disseminating systematic reviews of clinical interventions in dermatology. *Arch. Dermatol.*, 134:1620–6.

Wilson, J., Scott, W., Ritter, R., and Fradkin, R. (1975): Comparative distribution and embryotoxicity of hydroxyurea in pregnant rats and rhesus monkeys. *Teratology*, 11:169–78.

Wolfe, M., and Cordero, J. (1985): Safety of chloroquine in chemosuppression of malaria during pregnancy. *Br. Med. J.*, 290:1466–7.

Wolff, J. (1969): Iodide goiter and the pharmacologic effects of excess iodide. *Am J. Med.*, 47:101–24.

Zemer, D., Livneh, A., Pras, M., and Shoar, E. (1993): Familial Mediterranean Fever in the colchicine era: The fate of one family. *Am. J. Med. Genet.*, 45:340–4.

Zemlickis, D., Lishner, M., Degendorfer, P., Panzarella, T., Sutcliffe, S. B., and Koren, G. (1992): Fetal outcome after in utero exposure to cancer chemotherapy. *Arch. Intern. Med.*, 152:573–6.

Zerler, S., and Purohit, D. (1986): Prenatal antihistamine exposure and retrolental fibroplasia. *Am. J. Epidemiol.*, 123:192–6.

Zerner, J., Doil, K. L., Drewry, J., and Leeber, D. A. (1981): Intrauterine contraceptive device failures in renal transplant patients. *J. Reprod. Med.*, 26:99–102.

Zierler, S., and Rothman, K. (1985): Congenital heart disease in relation to maternal use of Bendectin and other drugs in early pregnancy. *N. Engl. J. Med.*, 313:347–352.

Zovirax Product Information, Research Triangle Park, NC, Burroughs Wellcome Co., 1999.

Zuaza, J., Julia, A., Sierra, J., Valentin, M. G., Coma, A., Sanz, M. A., Batlle, J., and Flores, A. (1991): Pregnancy outcome in hematologic malignancies. *Cancer*, 67:703–9.

Zyloprim Product Information, Reasearch Triangle Park, NC, Burroughs Wellcome, 1998.

Zyrtec Product Information, New York, Pfizer Inc., 1999.

Toxicology of Skin
Edited by Howard I. Maibach
Copyright © 2001 Taylor & Francis

30

Roles of Calcium Ions in Skin Barrier

Hanafi Tanojo and Howard I. Maibach

Department of Dermatology, University of California, San Francisco, California, USA

INTRODUCTION

Recently, many publications have shown the relationship between calcium and the barrier regulation of skin. It would therefore be useful to look further into the mechanism in which calcium may contribute. Apart from the skin, this ion plays a crucial role in various processes in the body, including the growth, death, differentiation, and function of immune cells. Although the data are not yet abundant, there are clear indications that calcium in skin may play a significant role.

THE PRESENCE OF CALCIUM IN SKIN

Skin contains calcium ions in a balanced amount, depending on the skin site in the body of the organism or the skin layers. Whereas calcium is the most abundant metal ion and fifth (after hydrogen [H], oxygen [O], carbon [C], and nitrogen [N]) most abundant element in the whole body, over 90% resides in bones and tooth enamel. The rest is found throughout body fluids and membranes and take parts in various processes, including muscle contraction, blood clotting, nerve excitability, intercellular communication, membrane transport of molecules, hormonal responses, exocytosis and cell fusion, adhesion, and growth (Sigel, 1984). The calcium concentration in the skin is even lower than that of sodium (Na^+), potassium (K^+) and chloride (Cl^-). Analysis of human skin grafts showed that per 100 g skin graft, there is 6.1–9.3 mg calcium, compared to 3.6–5.0 mEq ($= 140.4$–195 mg) potassium, whereas Ca/N ratio $=$ 1.3–3.2 mg/g (N or nitrogen atom being an indicator of protein amount) (Isaksson et al., 1967).

In connection to the skin appendages, it has been known that sweat contains calcium (Mitchell and Hamilton, 1949). The 24-hour dermal losses of calcium in human varies interindividually and are reported to be between 70 and 230 mg at comfortable environmental conditions (Mitchell and Hamilton, 1949; Consolazio et al., 1962). It can be as high as 800–2000 mg/day for men at hard labor in the desert (Consolazio et al., 1962) or as low as 15 ± 10 mg for elderly people (women) (Gitelman and Lutwak, 1963). Generally there is no correlation between Ca intake and Ca dermal loss, as well as between Ca dermal loss and the body surface observed (Isaksson et al., 1967).

CALCIUM GRADIENT

Composition of calcium in the skin follows a certain pattern. Calcium concentrations in the extra- or intracellular domains of keratinocytes are significantly different and continuously maintained by a tight regulation. The free intracellular calcium concentration is about 1000 to 10,000 times lower than the extracellular calcium concentration. In keratinocytes, the free intracellular calcium concentration is around 100 nmol/l, whereas serum calcium is approximately 1 mmol/l (Fairley, 1991b). Human keratinocytes showed heterogeneity in their Ca_i basal levels, which varied between 60–150 nM, both within a cell and between individual cells (Pillai et al., 1993). Moreover, a calcium gradient is present within the epidermis with higher quantities of Ca^{2+} in the upper than in the lower epidermis (as cell moves from the basal layer to the stratum granulosum) (Menon et al., 1985). Ca^{2+} concentration increases steadily from the basal region to stratum corneum, although this is not the case with other ions (Forslind et al., 1995). Figure 30.1 illustrates the calcium gradient in human skin in comparison with an actual literature data (Malmqvist et al., 1987). Such a gradient is not observed in skin abnormalities related to the formation of abnormal barrier function (e.g., psoriasis) (Menon and Elias, 1991). Studies in mice and rats showed that this gradient exists at the same time as the formation of a maturing skin barrier at the end of gestation. The gradient is then maintained from the newborn throughout the adult life (Elias et al., 1998).

It is not yet clear whether the calcium gradient leads to the formation of a mature barrier or the barrier caused the gradient. It may even be both if the regulation uses a feedback mechanism, (i.e., as the differentiation will eventually form a barrier leading to the accumulation of calcium ions in the upper epidermis). This high level of calcium will, in turn, guarantee the ongoing process of differentiation toward the formation of corneocytes (horny cells in the stratum corneum). The mechanism is thus almost completely autonomous, perpetual, and, if runs smoothly, requires little correction from the body.

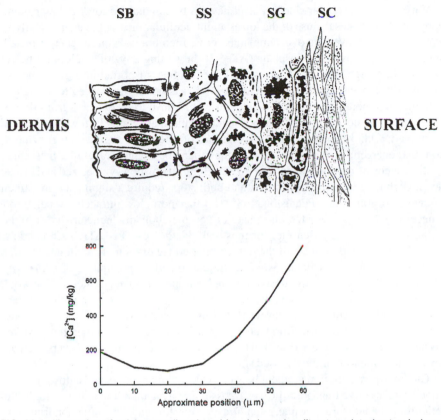

FIG. 30.1. Illustration of calcium gradient in epidermis based on literature data (proton-induced x-ray emission analysis of calcium in sectioned human skin) (Malmqvist et al., 1987). SB = stratum basale/basal layer; SS = stratum spinosum; SG = stratum granulosum; SC = stratum corneum.

REGULATION OF CALCIUM

The regulation of calcium in skin shows an ingenious adaptation of living organisms to the presence of ions. As Ca^{2+} cannot be metabolized like other second-messenger molecules, cells tightly regulate intracellular levels through numerous binding and specialized extrusion proteins (Clapham, 1995). The concentration of calcium in the cytosol ($\sim$0.1 mM) is kept at four orders of magnitude lower than in extracellular spaces (generally $\sim$1.5 μM). In excitable cells (e.g., muscle cells), the extracellular concentration of calcium must be closely regulated to keep it at its normal level of $\sim$1.5 mM so that it cannot accidentally trigger the muscle contraction, the transmission of nerve impulses, and blood clotting (Voet and Voet, 1990a). In other cells, including keratinocytes, the extracellular level is maintained in a specific equilibrium with the intracellular concentration.

Why does intracellular calcium level have to be maintained low? A low calcium concentration makes the use of the ion as an intracellular messenger energetically inexpensive. The movement of calcium ions across membranes requires energy, usually supplied by adenosine triphosphate (ATP). If the resting level of calcium in the cell were high, a large number of ions would need to be transported into the cytoplasm to raise the concentration by the factor of 10 that is ordinarily needed to activate an enzyme, and moreover, the excess calcium would have to be expelled from the cell afterward. The normally low calcium level means that relatively few ions need to be moved, with a relatively small expenditure of energy, to regulate an enzyme. In contrast, energetic cost of regulation by the other important intracellular messenger, cyclic adenosine monophosphate (cAMP), is high; it must by synthesized and broken down each time it carries a message, and both steps require a significant investment of energy (Carafoli and Penniston, 1985). Furthermore, low intracellular calcium is a necessary condition for the phosphate-driven metabolism characteristic of higher organisms. The energy-rich fuel for most cellular processes is ATP. Its breakdown releases inorganic phosphate. If the intracellular concentration of calcium were high, the phosphate and the calcium would combine to form a precipitate of *hydroxyapatite* crystals, the same stony substance found in bone. Ultimately, calcification would doom the cell (Carafoli and Penniston, 1985). This is likely the case with long-term occupational exposure to high levels of dissolved calcium (e.g., in miners (Sneddon and Archibald, 1958), agricultural laborers (Christensen, 1978), and oil field workers (Wheeland and Roundtree, 1985)), which can result in *calcinosis cutis*, a benign and reversible hardening of the exposed skin.

The large concentration gradient between extracellular spaces and cytosol is maintained by the active transport of Ca^{2+} across the plasma membrane, the endoplasmic reticulum (or the sarcoplasmic reticulum in muscle), and the mitochondrial inner membrane. Generally, plasma membrane and endoplasmic reticulum each contain a Ca^{2+}-ATPase that actively pumps Ca^{2+} out of the cytosol at the expense of ATP hydrolysis (Voet and Voet, 1990b). Mitochondria act as a "buffer" for cytosolic Ca^{2+}: If cytosolic concentration of calcium rises, the rate of mitochondrial Ca^{2+} influx increases whereas that of Ca^{2+} efflux remains constant, causing the mitochondrial concentration of calcium to increase, whereas the cytosolic concentration of calcium decreases to its original level (its set point). Conversely, a decrease in cytosolic concentration of calcium reduces the influx rate, causing net efflux of concentration of calcium and an increase of cytosolic concentration of calcium back to the set point (Voet and Voet, 1990c).

Besides the above-mentioned Ca^{2+}-ATPase, the transport of Ca^{2+} is regulated by a series of calcium pumps, transport systems, and ion channels. The availability of certain regulatory systems is dependent upon the activity of the cells. In excitable cells, such as cardiac muscle, the influx of Ca^{2+} to cytosol is regulated by voltage (or potential) dependent channels, whereas the efflux (out of cytosol) is regulated by cation exchanger, such as Na^+–Ca^{2+} exchanger (Fairley, 1991a). Undifferentiated keratinocytes in the basal layer have different sets of Ca^{2+} tranport system than differentiated cells in the upper layers. In basal layer, the system consists of 14 pS nonspecific cation channels (NSCC) (Mauro et al., 1993) and does not posses functional voltage-sensitive Ca^{2+} channels (Reiss et al., 1992). Differentiated keratinocytes are likely to possess at least 2 and possibly 3 pathways of Ca^{2+} influx: (1) nicotinic channel (nAChR); (2) voltage-sensitive Ca^{2+} channels (VSCC, which can be blocked by nifedipine or verapamil); (3) NSCC, which is not activated by nicotine (Grando et al., 1996).

CALCIUM SIGNALING

Calcium is a universal messenger, even in simple organisms and plants. The combination of its ionic radius and double charge may allow it tighter binding to receptors to the exclusion of other ions, such as magnesium, leading to strong, specific binding (Carafoli and Penniston, 1985). The specificity enables cells to form special receptors to assess signals from calcium. For many parts of the body, Ca^{2+} often acts as a second messenger in a manner similar to cAMP. Transient increases in cytosolic Ca^{2+} concentration trigger numerous cellular responses, including muscle contraction, release of neurotransmitters, and glycogen breakdown (glycogenolysis), also as an important activator of oxidative metabolism (Voet and Voet, 1990b). Ca^{2+} does not need to be synthesized and degraded with each message transmission, so it is an energy-efficient signal for the cell (Fairley, 1991a).

In skin, calcium can provide signals for the cells being either extracellular or intracellular (in the cytosol). The extra- and intracellular signaling may be connected to each other but may also act separately. In cultured keratinocytes, extracellular calcium levels influence growth and differentiation (Hennings et al., 1980; Pillai et al., 1988). Low extracellular calcium levels (<0.1 mM) induce the growth of keratinocytes as a monolayer with a high proliferation rate, rapidly becoming confluent. In this condition, keratinocytes never stratify or differentiate but possess many of the characteristics of basal cells; the cells synthesize keratin proteins and are connected by occasional gap junctions but not by desmosomes. High extracellular calcium levels (≥ 1 mM) induce differentiation of keratinocytes. Keratinocytes rapidly flatten, form desmosomes, and differentiate with stratification. Moreover, cornified envelopes form in cells of the uppermost layers (Hennings et al., 1980; Pillai et al., 1988).

Keratinocytes are very sensitive to the extracellular calcium, compared to other types of cells, such as fibroblasts (Kruszewski et al., 1991). The response to signaling also changes in a progressive way. Increased extracellular Ca^{2+} inhibits proliferation while it induces differentiation (Hennings et al., 1983). On the other hand, the differentiation of keratinocytes caused a decrease in the responsiveness to extracellular calcium. This actually may facilitate the maintenance of the high level of intracellular calcium required for further differentiation (Bikle et al., 1996).

Intracellular Ca^{2+} increases with raised extracellular Ca^{2+} (Pillai and Bikle, 1989; Kruszewski et al., 1991; Reiss et al., 1992). A change from 0.10 to 0.12 mM can already trigger a sustained increase in intracellular Ca^{2+} of murine keratinocytes

(Kruszewski et al., 1991). In mouse, it is also shown that the extracellular Ca^{2+} concentration that elicit intracellular Ca response corresponds to the concentration that control the optimal expression of specific differentiation-related mouse keratinocyte genes (Kruszewski et al., 1991), whereas blocking the transmembrane calcium inward flux prevents the rise of intracellular Ca and the calcium-induced differentiation (Reiss et al., 1992). This implies that increased intracellular Ca^{2+} is the actual signal to trigger keratinocyte differentiation.

CALCIUM AND BARRIER REPAIR MECHANISM

Disruption of the barrier with acetone treatment or tape stripping depletes Ca^{2+} from the upper epidermis, resulting in the loss of the Ca^{2+} gradient (Menon et al., 1992; Mauro et al., 1996; Man et al., 1997). This is due to accelerated water transit that leads to the increased passive loss of Ca^{2+} into and through the stratum corneum (Menon et al., 1992; Man et al., 1997). One in vitro study showed that the permeability of human stratum corneum to Ca^{2+} dramatically increased after the stratum corneum was pretreated with acetone or sodium lauryl sulfate solution (Tanojo et al., 1998). The decrease in Ca^{2+} levels in the outer epidermis is associated with enhanced lamellar body secretion and lipid synthesis (important components in repair responses) (Lee et al., 1992; Menon et al., 1992); however, if Ca^{2+} gradient is preserved by the addition of Ca^{2+} into the media, lamellar body secretion, lipid synthesis, and barrier recovery are inhibited (Lee et al., 1992). The inhibition raised by high extracellular concentration of calcium is potentiated by high extracellular K^+ (Lee et al., 1994). Another study confirmed that barrier recovery is accelerated by the low concentrations of calcium and also potassium during an increased water loss, as water loss may induce a decrease in the concentration of Ca^{2+} in the upper epidermis that, in turn, may stimulate lamellar body secretion and barrier repair (Grubauer et al., 1987). Furthermore, the inhibition raised by high extracellular concentration of calcium is reversed by nifedipine or verapamil, specific calcium channel blockers (Lee et al., 1994). In another study, administration of Ca^{2+} free solutions by sonophoresis results in a marked decrease in Ca^{2+} content in the upper epidermis, and subsequently the loss of the Ca^{2+} gradient is accompanied by accelerated lamellar body secretion (a sign of skin barrier repair) (Menon et al., 1994). The process of barrier repair in connection with transepidermal water loss and calcium gradient is illustrated in Figure 30.2. Other aspects of the skin barrier homeostasis has been extensively reviewed in (Feingold, 1997).

MECHANISM OF CELL SIGNALING

Both intracellular release and transmembrane flux contribute to the rise in intracellular Ca^{2+} (Kruszewski et al., 1991; Reiss et al., 1992). The rise in keratinocyte intracellular Ca^{2+} in response to raised extracellular Ca^{2+} has two phases: (i) an initial peak,

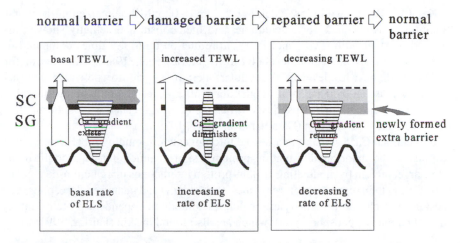

FIG. 30.2. Illustration of skin barrier repair in epidermis. SC = stratum corneum; SG = stratum granulosum; TEWL = transepidermal water loss; ELS = epidermal lipid synthesis.

not dependent on extracellular Ca^{2+}, and (ii) a later phase, which requires extracellular Ca^{2+} (Kruszewski et al., 1991). An early response of human keratinocytes to increases in extracellular Ca^{2+} is an acute increase in intracellular Ca^{2+}. Stepwise addition of extracellular Ca^{2+} to neonatal human keratinocytes is followed by a progressive increase in intracellular Ca^{2+}, where the initial spike of increased intracellular Ca^{2+} is followed by a prolonged plateau of higher intracellular Ca^{2+} (Pillai and Bikle, 1991). The response of intracellular Ca^{2+} to increased extracellular Ca^{2+} in keratinocytes is saturated at 2.0 mM extracellular Ca^{2+} (Sharpe et al., 1989; Pillai and Bikle, 1991). The response of intracellular Ca^{2+} to increased extracellular Ca^{2+} in keratinocytes resembles the response in parathyroid cells in that a rapid and transient increase in intracellular Ca^{2+} is followed by a sustained increase in intracellular Ca^{2+} above basal level. This multiphasic response is attributed to an initial release of Ca^{2+} from intracellular stores followed by an increased influx of Ca^{2+} through voltage-independent cation channels. The keratinocyte and parathyroid cell contains a similar cell membrane calcium receptor believed to mediate this response to extracellular Ca^{2+}. This receptor can activate the phospholipase-C pathway, leading to an increase in the levels of inositol 1,4,5-triphosphate (IP_3) and *sn*-1,2-diacylglycerol (DAG)—both are important messengers—as well as stimulating Ca^{2+} influx and chloride currents (Shoback et al., 1988; Brown et al., 1990). IP_3 causes release of Ca^{2+} from internal stores, such as endoplasmic reticulum, further increasing intracellular level to precede a number of calcium-stimulated cellular events (Berridge and Irvine, 1984). DAG forms a quarterary complex with phosphatidylserine, calcium, and protein kinase C to activate the kinase; this accelerates terminal differentiation (Hennings et al., 1983). The signal tranduction mediated through calmodulin induces other proteins (e.g., desmocalmin), which is associated with the formation of desmosomes (Tsukita and Tsukita, 1985).

Keratinocytes grown in low-calcium medium (0.02 mM) maintained intracellular calcium levels adequate for arachidonic acid metabolism and actually showed increased prostaglandin (PGE_2 and $PGF_{2\alpha}$) production up to 4.5 times compared to cells grown in nomal-calcium level (1.2 mM) (Fairley et al., 1988). If this is true for in vivo condition, a low level of extracellular calcium, (i.e., due to a defective skin barrier) may cause an increase in prostaglandin synthesis, leading to hyperproliferative epidermal disorders, such as psoriasis, which are often associated with abnormalities in prostaglandin production (Hammarström et al., 1979).

Intracellular Ca^{2+} signals are assessed through calcium-binding proteins to induce responses. The major calcium-binding protein in skin is calmodulin. Calmodulin regulates target protein by modulating protein-protein interactions in a calcium-dependent way. Calmodulin regulates many enzymes, (e.g., adenyl and guanyl cyclase, phosphodiesterase, ornithine decarboxylase, calcium-calmodulin-dependent protein kinase, transglutaminase, phospholipase), which are also found in skin (Fairley, 1991a).

TRANSPORT OF CALCIUM IN THE SKIN

The permeability of skin to Ca^{2+} ions has been known from some dermatoses, such as the calcinosis cutis (Sneddon and Archibald, 1958; Christensen, 1978; Wheeland and Roundtree, 1985) and perforating verruciform collagenoma (Moulin et al., 1995). In a shorter term, calcinosis cutis developed after a (at least) 24-hour topical application of an electrode paste containing saturated calcium chloride solution, bentonite and glycerin, used for examination by electroencephalography or electromyography (Mancuso et al., 1990; Johnson et al., 1993). The permeability of human skin to Ca^{2+} ions in vitro shows a marked dependence upon anatomic site. In agreement with the data observed for nonelectrolytes, permeation decreased in the following order: foreskin > mammary > scalp > thigh. Mouse and guinea pig skin show comparable permeability to that of human scalp. Ca^{2+} transport from dermis across epidermis is higher than that from epidermis to dermis (Stüttgen and Betzler, 1956, 1957).

Recently, a novel technique was developed to continuously monitor the low level of Ca^{2+} flux across human stratum corneum in vitro (Tanojo et al., 1998). The technique utilizes an ion-selective glass microelectrode containing a highly selective Ca^{2+} ionophore (type 1 cocktail A, Fluka). The microelectrode was positioned 50 μm above the surface of the SC and scanned over a specific area to measure the Ca^{2+} current. Using the technique, Ca^{2+} diffusion was quantified as a function of time for up to 3 hours, while measurements could be made every second, the spatial resolution was 2-4 μm, and the minimum concentration reliably detectable was 10^{-5} M calcium ion. The study showed that the flux through untreated human stratum corneum was sigmoidal. The steady state flux had an average of 7×10^{-12} mol/cm^2/s. After the stratum corneum was pre-treated with acetone or sodium laurylsulfate, the shape of the curve was similar but the Ca^{2+} flux was significantly higher (Figure 30.3). This method may be useful in examining the changes in Ca^{2+} flux that take place during barrier formation and repair (Tanojo et al., 1998).

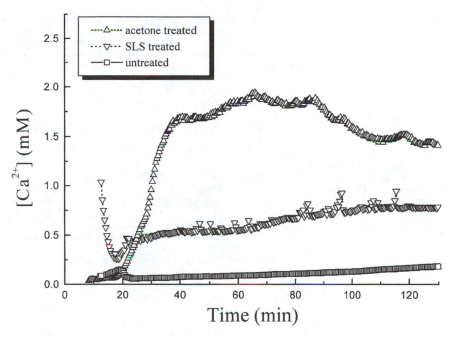

FIG. 30.3. In vitro calcium flux across human stratum corneum, untreated and pretreated with acetone or 10% sodium lauryl sulfate.

CONCLUSION

Calcium ions play an important role in the homeostasis of skin barrier. A change in the barrier will change the calcium ion gradient in skin and lead to barrier repair process. A severe change might lead to a high degree of calcium signaling, which may induce the activation of various processes from increased synthesis of skin components or messengers to the inflammatory reactions. Hence, the homeostasis of calcium in the skin should be taken into consideration while administering drug into the skin.

REFERENCES

Berridge, M. J., and Irvine, R. F. (1984): Inositol triphosphate, a novel second messenger in cellular signal transduction. *Nature (London)*, 312:315–321.

Bikle, D. D., Ratnam, A., Mauro, T. M., Harris, J., and Pillai, S. (1996): Changes in calcium responsiveness and handling during keratinocyte differentiation. *J. Clin. Invest.*, 97:1085–1093.

Brown, E. M., Chen, C. J., Kifor, O., Leboff, M. S., El-Hajj, G., Fajtova, V., and Rubin, L. T. (1990): Ca^{2+}-sensing, second messengers, and the control of parathyroid hormone secretion. *Cell Calcium*, 11:333–337.

Carafoli, E., and Penniston, J. T. (1985): The calcium signal. *Sci. Am.*, 253:70–78.

Christensen, O. B. (1978): An exogenous variety of pseudoxanthoma elasticum in old farmers. *Acta. Derm. Venereol., (Stockh.)*, 58:319–321.

Clapham, D. E. (1995): Calcium signaling. *Cell*, 80:259–268.

Consolazio, C. F., Matoush, Le R. O., Nelson, R. A., Hackler, L. R., and Preston, E. E. (1962): Relationship between calcium in sweat, calcium balance, and calcium requirements. *Journal of Nutrition*, 78:78–88.

Elias, P. M., Nau, P., Hanley, K., Cullander, C., Crumrine, D., Bench, G., Sideras-Haddad, E., Mauro, T. M., Williams, M. L., and Feingold, K. R. (1998): Formation of the epidermal calcium gradient coincides with key milestones of barrier ontogenesis in the rodent. *J. Invest. Dermatol.*, 110:399–404.

Fairley, J. A. (1991a): Calcium: A second messenger: In: *Physiology, Biochemistry, and Molecular Biology of the Skin*, 2nd ed. Edited by Goldsmith, L. A., pp. 314–328. Oxford University Press, New York.

Fairley, J. A. (1991b): Calcium metabolism and the pathogenesis of dermatologic disease. *Semin. Dermatol.*, 10:225–231.

Fairley, J. A., Weiss, J., and Marcelo, C. L. (1988): Increased prostaglandin synthesis by low calcium-regulated keratinocytes. *J. Invest. Dermatol.*, 86:173–176.

Feingold, K. R. (1997): Permeability barrier homeostasis: Its biochemical basis and regulation. *Cosmetics & Toiletries*, 112:49–59.

Forslind, B., Lindberg, M., Malmqvist, K. G., Pallon, J., Roomans, G. M., and Werner-Linde, Y. (1995): Human skin physiology studied by particle probe microanalysis. *Scanning Microscopy*, 9:1011–1026.

Gitelman, H. J., and Lutwak, L. (1963): Dermal losses of minerals in elderly women under nonsweating conditions. *Clin. Res.*, 11:42.

Grando, S. A., Horton, R. M., Mauro, T. M., Kist, D. A., Lee, T. X., and Dahl, M. V. (1996): Activation of keratinocyte nicotinic cholinergic receptors stimulates calcium influx and enhances cell differentiation. *J. Invest. Dermatol.*, 107:412–418.

Grubauer, G., Feingold, K. R., and Elias, P. M. (1987): Relationship of epidermal lipogenesis to cutaneous barrier function. *J. Lipid Res.*, 28:746–752.

Hammarström, S., Lindgren, J. A., Marcelo, C. L., Duell, E. A., Anderson, T. F., and Voorhees, J. J. (1979): Arachidonic acid transformations in normal and psoriatic skin. *J. Invest. Dermatol.*, 73:180–183.

Hennings, H., Michael, D., Cheng, C., Steinert, P., Holbrook, K. A., and Yuspa, S. H. (1980): Calcium regulation of growth and differentiation of mouse epidermal cells in culture. *Cell*, 19:245–254.

Hennings, H., Holbrook, K. A., and Yuspa, S. H. (1983): Factors influencing calcium-induced terminal differentiation in cultured mouse epidermal cells. *J. Cell. Physiol.*, 116:265–281.

Isaksson, B., Lindholm, B., and Sjörgen, B. (1967): A critical evaluation of the calcium balance technic. II: Dermal calcium losses. *Metabolism*, 16:303–313.

Johnson, R. C., Fitzpatrick, J. E., and Hahn, D. E. (1993): Calcinosis cutis following electromyographic examination. *Cutis*, 52:161–164.

Kruszewski, F. H., Hennings, H., Yuspa, S. H., and Tucker, R. W. (1991): Regulation of intracellular free calcium in normal murine keratinocytes. *Am. J. Physiol.*, 261:C767–C773.

Lee, S. H., Elias, P. M., Proksch, E., Menon, G. K., Man, M.-Q., and Feingold, K. R. (1992): Calcium and potassium are important regulators of barrier homeostatis in murine epidermis. *J. Clin. Invest.*, 89:530–538.

Lee, S. H., Elias, P. M., Feingold, K. R., and Mauro, T. M. (1994): A role for ions in barrier recovery after acute perturbation. *J. Invest. Dermatol.*, 102:976–979.

Malmqvist, K. G., Forslind, B., Themner, K., Hyltén, G., Grundin, T., and Roomans, G. M. (1987): The use of PIXE in experimental studies of the physiology of human skin epidermis. *Biological Trace Element Research*, 12:297–308.

Man, M.-Q., Mauro, T. M., Bench, G., Warren, R., Elias, P. M., and Feingold, K. R. (1997): Calcium and potassium inhibit barrier recovery after disruption, independent of the type of insult in hairless mice. *Exp. Dermatol.*, 6:36–40.

Mancuso, G., Tosti, A., Fanti, P. A., Berdondini, R. M., Mongiorgi, R., and Morandi, A. (1990): Cutaneous necrosis and calcinosis following electroencephalography. *Dermatologica*, 181:324–326.

Mauro, T. M., Isseroff, R. R., Lasarow, R., and Pappone, P. A. (1993): Ion channels are linked to differentiation in keratinocytes. *J. Membrane Biol.*, 132:201–209.

Mauro, T. M., Rassner, U., Bench, G., Feingold, K. R., Elias, P. M., and Cullander, C. (1996): Acute barrier disruption causes quantitative changes in the calcium gradient. *J. Invest. Dermatol.*, 106:919.

Menon, G. K., and Elias, P. M. (1991): Ultrastructural localization of calcium in psoriatic and normal human epidermis. *Arch. Dermatol.*, 127:57–63.

Menon, G. K., Grayson, S., and Elias, P. M. (1985): Ionic calcium reservoirs in mammalian epidermis: Ultrastructural localization by ion-capture cytochemistry. *J. Invest. Dermatol.*, 84:508–512.

Menon, G. K., Elias, P. M., Lee, S. H., and Feingold, K. R. (1992): Localization of calcium in murine epidermis following disruption and repair of the permeability barrier. *Cell. Tissue. Res.*, 270:503–512.

Menon, G. K., Price, L. F., Bommannan, B., Elias, P. M., and Feingold, K. R. (1994): Selective obliteration of the epidermal calcium gradient leads to enhanced lamellar body secretion. *J. Invest. Dermatol.*, 102:789–795.

Mitchell, H. H., and Hamilton, T. S. (1949): The dermal excretion under controlled environmental conditions of nitrogen and minerals in human subjects, with particular reference to calcium and iron. *J. Biol. Chem.*, 178:345.

Moulin, G., Balme, B., Musso, M., and Thomas, L. (1995): Perforating verruciform collagenoma, an exogenous inclusion-linked dermatosis? Report of one case induced by calcium chloride. *Ann. Dermatol. Venereol.*, 122:591–594.

Pillai, S., and Bikle, D. D. (1989): A differentiation-dependent, calcium-sensing mechanism in normal human keratinocytes. *J. Invest. Dermatol.*, 92:500.

Pillai, S., and Bikle, D. D. (1991): Role of intracellular-free calcium in the cornified envelope formation of keratinocytes: Differences in the mode of action of extracellular calcium and 1,25-dihydroxyvitamin D. *J. Cell. Physiol.*, 146:94–100.

Pillai, S., Bikle, D. D., Hincenbergs, M., and Elias, P. M. (1988): Biochemical and morphological characterization of growth and differentiation of normal human neonatal keratinocytes in a serum-free medium. *J. Cell. Physiol.*, 134:229–237.

Pillai, S., Menon, G. K., Bikle, D. D., and Elias, P. M. (1993). Localization and quantitation of calcium pools and calcium binding sites in cultured human keratinocytes. *J. Cell. Physiol.*, 154:101–112.

Reiss, M., Lipsey, L. R., and Zhou, Z. L. (1992): Extracellular calcium-dependent regulation of transmembrane calcium fluxes in murine keratinocytes. *J. Cell. Physiol.*, 147:281–291.

Sharpe, G. R., Gillespie, J. I., and Greenwell, J. R. (1989): An increase in intracellular free calcium is an early event during differentiation of cultured keratinocytes. *FEBS Lett.*, 254:25–28.

Shoback, D. M., Membreno, L. A., and McGhee, J. G. (1988): High calcium and other divalent cations increase inositol trisphosphate in bovine parathyroid cells. *Endocrinology*, 123:382–389.

Sigel, H. (1984): *Metal Ions in Biological Systems: Calcium and Its Role in Biology*, vol. 17. Marcel Dekker, New York.

Sneddon, I. B., and Archibald, R. M. (1958): Traumatic calcinosis of the skin. *Br. J. Dermatol.*, 70:211–214.

Stüttgen, G., and Betzler, H. (1956): Zur Frage der Permeation von Elektrolyten durch die Haut. I. Mitteilung: Vitroversuche mit radioaktivmarkierten Ca^{++}, SO_4^{--}, und PO_4^{3-}-Ionen an Meerschweinchen- und Mäusehaut. *Arch. klin. exp. Derm.*, 203:472–482.

Stüttgen, G., and Betzler, H. (1957): Zur Frage der Permeation von Elektrolyten durch die Haut. II. Mitteilung: In vitro- und vivo-Versuche an menschlicher Haut mit $^{45}Ca^{++}$. *Arch. klin. exp. Derm.*, 204:165–174.

Tanojo, H., Cullander, C., and Maibach, H. I.: Monitoring the permeation of calcium ion across human stratum corneum using an ion-selective microelectrode with high spatial resolution. In: *Perspective of Percutaneous Penetration*, vol 6a. Edited by K. R. Brain and K. A. Walters. STS Publishing, Cardiff. (1998)

Tsukita, S., and Tsukita, S. (1985): Desmocalmin: a calmodulin-binding high molecular weight protein isolated from desmosomes. *J. Cell Biol.*, 101:2070–2080.

Voet, D., and Voet, J. G. (1990a): *Biochemistry*, p. 1144. John Wiley & Sons, New York.

Voet, D., and Voet, J. G. (1990b): *Biochemistry*, pp. 496–498. John Wiley & Sons, New York.

Voet, D., and Voet, J. G. (1990c): *Biochemistry*, p. 531. John Wiley & Sons, New York.

Wheeland, R. G., and Roundtree, J. M. (1985): Calcinosis cutis resulting from percutaneous penetration and deposition of calcium. *J. Am. Acad. Dermatol.*, 12:172–175.

Toxicology of Skin
Edited by Howard I. Maibach
Copyright © 2001 Taylor & Francis

31

Nuclear Receptors for Psoralen in Cultured Human Skin Fibroblasts

K. Milioni, E. Bloom, and H. I. Maibach

Department of Dermatology, University of California, San Francisco, California, USA

- · **Introduction**
- · **Materials and Methods**
  - · Reagents · Cell Culture · Time Course and Binding Assay/Scatchard Plot
  - · Subcellular Localization of the Receptor · Cold Competition Studies · DNA Binding Studies
- · **Results**
  - · Structures · Kinetics of the [³H]8-MOP Binding · Determination of the Binding Constants of the Receptor · Subcellular Location of the Specifically Bound [³H]8-MOP · Specificity of Psoralen-Binding Sites: Cold Competition Study · Examination of Purified DNA vs. Intact Nuclei
- · **Discussion**

INTRODUCTION

Psoralens are structurally like furocoumarins and they occur naturally in many plants. They have been used for centuries by Egyptians and Indians for the treatment of a variety of skin disorders, such as vitiligo (Hearst, 1981; Pathak et al., 1977).

The term *photochemotherapy* was introduced in 1974, and the psoralens have been extensively used in combination with ultraviolet-light irradiation (Parrish et al., 1974; Walter and Voorhees, 1973). Three psoralens are currently in routine topical or oral use: 8-methoxypsoralen (methoxsalen, 8-MOP), 4,5′,8 trimethylpsoralen (trioxsalen, TMP), and 5-methoxypsoralen (bergaptene 5-MOP). The structures of these compounds are shown in Figure 31.1. The most potent oral photosensitizing psoralen is 8-MOP, followed by TMP. The difference in potency between these two compounds is especially pronounced when used in bath therapy. This may be due to differences in lipophilicity or to differences in the metabolism of TMP and 8-MOP

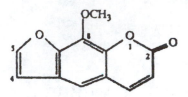

8-Methoxypsoralen

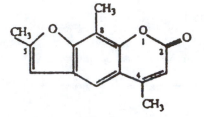

4,5',8-Trimethylpsoralen

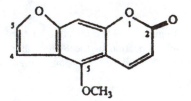

5-Methoxypsoralen

FIG. 31.1. Structures. The chemical structures of the three psoralens studied are shown.

(Fisher and Alsins, 1976; Hannuksela and Karnoven, 1978; Vaatainen, 1980). 5-MOP has not been extensively studied after either oral or topical administration.

In the skin, psoralen plus ultraviolet light A (UVA) (i.e., PUVA) treatment produces two clinically important, yet opposing effects on different epidermal cell populations (Gage and Parrish, 1982; Pathak, 1984). Photochemotherapy with PUVA results in inhibition of basal cell division, which causes a decreased rate of keratinocyte growth. Interestingly, the biological effects caused by PUVA are the opposite for melanocytes, which are stimulated to proliferate.

These observations are intriguing and could imply that psoralens have specific therapeutic biological interactions; however, the mode of therapeutic action of psoralens has been proposed to be a nonspecific covalent reaction with the pyrimidine bases of DNA after irradiation with UVA. Nonetheless, (Laskin et al., 1985) reported that specific receptors for psoralens in HeLa cells may be responsible for the PUVA effect. Also, (Jeong et al., 1995) have reported that 8-MOP may compete specifically for the aryl hydrocarbon (AH) – specific nuclear receptor in cultured mouse hepatoma cells.

In the present paper, we confirm the presence of specific binding sites for psoralens in human skin fibroblasts and that the majority of the ligand-bound receptors are located in the nucleus.

MATERIALS AND METHODS

Reagents

8-MOP, 5-MOP, TMP, and anthralin were from Sigma (St. Louis, MO); 8-[methoxy-^{3}H]psoralen ([^{3}H]8-MOP, >99% purity; 85Ci/mmol) was purchased from Amersham (Arlington Heights, IL). Purified calf thymus DNA was obtained from Sigma. Phosphate-buffered saline (PBS) contained 0.1 M NaCl and 0.025 M potassium phosphate, pH 7.6. Homogenization buffer contained 50 mM Hepes [(2-Hydroxy-ethyl) piperazine-N-(2-ethanesulfonic acid)] and 6 mM MgSO4, pH 7.4.

Cell Culture

Normal human fibroblasts were cultured from foreskin specimens using standard methods. Briefly, the culture medium was Dulbecco's Eagle's medium (DME) with 10% fetal calf serum and antimicrobials added. The cells were grown in a 37°C incubator in an 8% CO_2 atmosphere. All experiments were performed on third to fifth passage cells grown to confluency.

Time Course and Binding Assay/Scatchard Plot

The cell harvesting, incubation, and all subsequent steps were performed at 0–4°C. Confluent cells in the T-Flasks were washed twice with PBS-CMF (calcium, magnesium-free PBS), and then 1 mM EDTA and PBS were added in order to harvest them. The resulting suspension was centrifuged, the supernatant medium was discarded, and the cells taken up in 15 ml of PBS.

For the on-rate/off-rate receptor binding determination, the cells were distributed into 12 × 75 mm polypropylene tubes and incubated, at 0–4°C, with 10^{-8} M of [^{3}H]8-MOP. Some tubes had, in addition, 10^{-6} M nonradioactive 8-MOP added for the determination of nonspecific binding. The incubation was terminated after 2, 3, 5, 10, 15, 30, 45, or 70 minutes by aspirating the supernatant after centrifugation at 1500 × g for 5 minutes. (The supernatant medium was assayed for radioactivity, from which the free methoxypsoralen concentration was verified.) The cell pellet was immediately washed twice with 2.5 ml of ice-cold PBS and the pellet assayed for radioactivity using a filter paper collection method followed by scintillation counting. The off-rate of binding was determined by adding a 10^{-6} M "cold chase" of nonradioactive 8-MOP at approximately 30 minutes.

For determination of the binding site dissociation constant and the saturability of the binding site, the incubation was ended after 20 minutes, at which time point equilibrium is reached. A similar procedure to on-rate/off-rate described above was followed except that varying concentrations of [³H]8-MOP were added and the radioactive counts were done on the nuclear fraction, which was isolated after the incubation with [³H]8-MOP. The supernatant medium was collected and assayed for radioactivity, from which the free methoxypsoralen concentration was determined. The nuclei were isolated by freezing the cell pellets in liquid nitrogen, followed by thawing in homogenization buffer at 4°C for 30 minutes. The suspension was homogenized, vortexed, and then centrifuged at 30,000 × g for 20 minutes. The pellet, representing the nuclear fraction, was washed and assayed for radioactivity. The kinetic binding constants were calculated from the Scatchard plot data using the Lundon software package (Lundon Software Inc., Chagrin Falls, OH).

Subcellular Localization of the Receptor

In contrast to the Scatchard plot/binding curve study described above, the nuclei and cytosol were isolated before the incubation with [³H]8-MOP. The cells were harvested and centrifuged at 1500 g for 5 minutes, as described above. The supernatant cell harvest medium was discarded and the pellet frozen in liquid nitrogen. Afterward, the homogenization buffer was added and the cells were thawed at 4°C for 30 minutes. The resulting suspension was then homogenized, vortexed, and then recentrifuged at 30,000 × g for 20 minutes. The pellet (nuclei) and the supernatant (cytosol) were then separated. The nuclear fraction was examined by light microscopy after Azure C staining (Baxter et al., 1972). It contained some nuclei contaminated with membrane pieces and cytoplasmic "tags." The cytosol was also examined and was confirmed to be cell and nuclei free. The whole cells, nuclei, and cytosol fractions were then incubated with different concentrations of [³H]8-MOP with or without 10^{-6} M nonradioactive 8-MOP for determination of nonspecific binding. The whole cell and nuclear fractions were assayed for radioactivity directly as described above. The cytosol specimens were filtered over Sephadex G-25 columns in order to separate free from protein-bound ligand. The columns were eluted with homogenization buffer. The G-25 exclusion volume (i.e., the protein-bound fraction) was then assayed for radioactivity. The amount of [³H]8-MOP bound in the presence of excess (i.e., 10^{-6} M nonradioactive 8-MOP) was also determined for each data point; this represented the nonspecific binding. The specifically bound radioactivity was determined by subtracting the nonspecific bound data from the total bound. Each data point presented in Figure 31.4 represents the average of triplicate determinations.

Cold Competition Studies

The cold competition studies with TMP, 5-MOP, and anthralin are described in the text and the legend for Figure 31.5.

DNA Binding Studies

Purified thymus DNA, which had been pretreated with 400 mcg/ml molecular grade proteinase K (to remove any contaminating DNA associated protein), was used for these studies. The DNA binding studies were performed exactly as for the binding assay (described above) with the following exception: After the incubation step, the DNA (100 mcg/tube) and [^{3}H]8- MOP mixture was filtered through a 2 cm^2 Whatman GF/B filter followed by a 50 ml wash with ice-cold washing buffer. As a control for the purified DNA studies, intact nuclei were treated in an identical way. Total DNA in the purified DNA and in the isolated nuclei specimens was determined using standard methods (Giles and Myers, 1965).

RESULTS

Structures

The structures of three psoralen analogs are shown in Figure 31.1.

Kinetics of the [^{3}H]8-MOP Binding

We characterized the on-rate and off-rate of binding of radiolabelled 8-MOP to cultured human fibroblasts. As shown in Figure 31.2, the specific binding of [^{3}H]8-MOP to the cells was rapid, with a half-time of approximately 3 minutes at 0–4°C. Binding reached maximal values (i.e., equilibrium) within 20 minutes. The off-rate of the binding was examined by adding a large excess of unlabeled 8-MOP (10^{-6} M) to an equilibrated mixture of [^{3}H]8-MOP and human fibroblasts at 30 minutes. As shown in the figure, binding to the cells was readily reversible.

Determination of the Binding Constants of the Receptor

The dissociation constant (K_d) and maximal binding constant (B_{max}) of the nuclear binding sites for 8-MOP were determined by incubating the cells with various concentrations of [^{3}H]8-MOP. As shown in Figure 31.3(A), binding was saturable; Scatchard analysis of the data is shown in Figure 31.3(B). The K_d for binding was found to be 15.2 ± 4.9 nM, and the B_{max} was 16,000 ± 2,020 binding sites per nucleus.

Subcellular Location of the Specifically Bound [^{3}H]8-MOP

Figure 31.4 shows the subcellular locations of the 8-MOP binding; 98% of these high affinity binding sites are located in the nucleus. Thus, the receptor appears to be intranuclear before and after exposure to ligand.

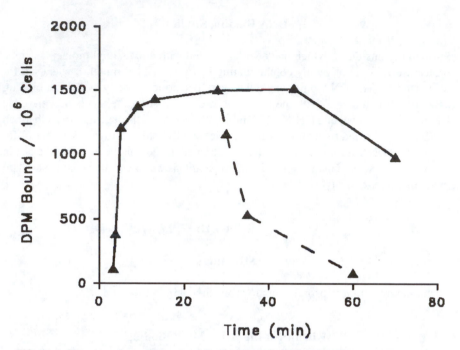

FIG. 31.2. The on-rate and off-rate of the binding of [³H]8-MOP to human fibroblasts. 10^{-8} M [³H]8-MOP was incubated with the fibroblasts. The solid line represents the on-rate of specific binding of [³H]8-MOP after initiating binding at time = 0 minutes. To measure the off-rate of binding, 10^{-6} M of unlabeled 8-MOP (i.e., a cold chase) was added at the 32-minute time point. The dashed line represents binding in those tubes to which the cold chase was added. This line represents the off-rate of [³H]8-MOP binding.

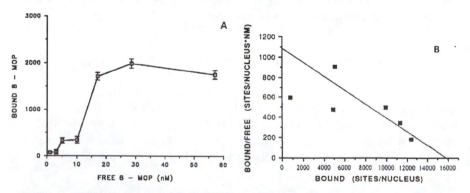

FIG. 31.3. (A) Binding curve of [³H]8-MOP in human skin fibroblast nuclei. The data are expressed as bound [³H]8-MOP (sites/nucleus) versus free ligand (i.e., free [³H]8-MOP). **(B) Scatchard plot of the data shown in Figure 31.3(A)**. The computer analysis of the data indicate that the K_d is 15.2 ± 4.9 nM and the B_{max} is $16,200 \pm 2020$ sites/nucleus.

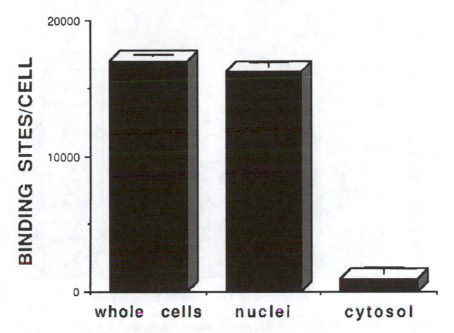

FIG. 31.4. Localization of binding sites in cultured human fibroblasts. The cell fractions were incubated with 3×10^{-8} M [^{3}H]8-MOP (see methods); 98% of the binding sites are located in the nuclei.

Specificity of Psoralen-Binding Sites: Cold Competition Study

To determine whether binding of psoralens to human fibroblasts is correlated with their relative clinical efficacies, we compared the relative abilities of several psoralen analogs to compete for [^{3}H]8-MOP binding sites; the data are shown in Figure 31.5. Cold competitors were added at a concentration of 10^{-8} M. 5-MOP was the most potent inhibitor, followed by TMP. To test whether these binding sites are specific for psoralens, we also examined anthralin, a medication that is used in psoriasis but is unrelated structurally to the psoralens. As shown in the figure, no significant inhibition by anthralin was observed.

Examination of Purified DNA vs. Intact Nuclei

To determine whether or not the 8-MOP is simply intercalating nonspecifically between DNA base pairs or is, in some way, reacting specifically with a receptor site in the nucleus, we incubated 10 nM ^{3}H8-MOP with either 100 mcg of protein-free calf thymus DNA or intact fibroblast nuclei. The latter specimens contained 20.6 mcg DNA/tube as determined by DNA assay. Bound and free 8-MOP were separated as described in the methods section. The ^{3}H8-MOP bound to the purified DNA at a

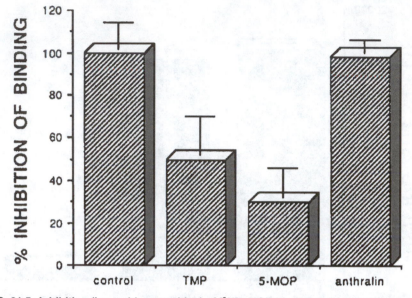

FIG. 31.5. Inhibition (i.e., cold competition) of [³H]8-MOP binding to human fibroblasts.
Nonradioactive TMP, 5-MOP, and anthralin were added at a concentration of 10^{-8} M to the
incubation tubes at the beginning of the binding assay. Ten minutes after adding the
nonradioactive ligands, 10^{-8} M [³H]8-MOP was added to the tubes. Anthralin, a nonpsoralen,
did not compete for the receptor. Values represent the average of triplicate determinations.

level of 2 ± 6 fm/mg DNA, a statistical result equivalent to no detectable binding.
In contrast, the ³H8-MOP bound avidly to the intact nuclei, which contained the
DNA-associated proteins, at a level of 790 ± 3 fm/mg DNA. This binding was highly
statistically significant.

DISCUSSION

The purpose of our studies was to determine if a receptor site for psoralens exists in skin
cells. We identified high-affinity specific receptors for psoralens in human fibroblasts;
98% of the binding was found in the nucleus and only 2% was in the cytosol. The
³H8-MOP binding we found by Scatchard analysis, as noted in the results, had a K_d
of approximately 15 nM with a B_{max} of approximately 16,000 sites/cell. We note that
these numbers are similar to those of the steroid receptor family.

To identify the specificity of this receptor, two psoralen analogs with different biological efficacy were used. We found that 5-MOP, a structural isomer of 8-MOP,
was the most potent inhibitor of [³H]8-MOP binding followed by TMP. Anthralin,
which is structurally unrelated to psoralen but also is used in the treatment of psoriasis, did not cause significant inhibition of [³H]8-MOP binding at 10^{-8} M or even
at higher concentrations (data not shown). This observation supports the notion of

a specific receptor. These results correlate approximately with the biological activities of these compounds in vivo in man. Thus, these receptors are of high affinity, saturable, and displayed approximate biologically appropriate structure-activity relationships.

The chemical nature of the receptor remains to be determined. Clearly the high-affinity binding does not occur with mammalian DNA stripped of protein, so simple intercalation between base pairs is not responsible for the observed binding. We speculate that a protein associated with the DNA is probably responsible for at least part of the observed high-affinity binding. This requires confirmation in future studies.

One in vitro effect of the psoralens that has been extensively studied is intercalation into cellular DNA, followed by UV light exposure (Averbeck, 1989; Hanawalt et al., 1981). After irradiation, the psoralens form mono- and bifunctional adducts with the adjacent pyrimidine bases in the DNA. These structural modifications of DNA may interfere with DNA replication and ultimately decrease proliferation of cells; however, we note that most of the in vitro studies of this type of mechanism have been done at concentrations far in excess of the in vivo free circulating concentration of 8-MOP.

In contrast to the relatively high free concentrations of 8-MOP used in the above studies, our receptor appears to function at the relevant biological free concentration. The K_d for 8-MOP is similar to the therapeutic circulating free concentration of 8-MOP following oral administration, as used in PUVA. Therapeutic plasma levels of 8-MOP are approximately 5×10^{-7} M 18 (with free levels of hormone much less than 5×10^{-8} M). Our observed Kd of approximately 1.5×10^{-8} M, therefore, is similar to but less than, the free plasma concentration of 8-MOP (i.e., the free levels of 8-MOP are appropriate to fill a majority of the receptor sites). Furthermore, the receptors are not oversaturated at the doses used in oral PUVA (i.e., there is no "excess" psoralen present at the recommended empirically determined oral dose). Thus, there is a possibility that they could be involved in the principal PUVA effect, side effects, or both.

Our study confirms what Laskin and his coauthors (Laskin et al., 1985; Laskin et al., 1986; Yurkow and Laskin, 1987) first reported about specific binding sites for psoralens. These authors, in contrast to our own findings for fibroblasts, demonstrated that the location of the specific binding sites for psoralens in HeLa cells is the cytosol; however, (Sasaki et al., 1988) used oral mucosa cells and studied the appearance of fluorescence after exposure to psoralen but without UV light exposure. He found that for these epithelial cells the uptake of the psoralen derivatives was by the nuclei, not by the cytoplasm, consistent with our findings.

Further evidence in support of a nuclear receptor for psoralen was found by (Jeong et al., 1995), who reported that 8-MOP appears to compete or interfere with the nuclear binding of dioxin to the AH-specific nuclear receptor in cultured mouse Hepa-1C1C7 cells. (Ledirac et al., 1997) also reported that 8-MOP interferes with AH receptor nuclear binding.

Interestingly, the mechanism of action of the AH receptor is similar to that of members of the steroid hormone receptor gene family, and it may be involved in neurodevelopment (Poellinger, 1996). This receptor is considered by some authors to be an "orphan" receptor in that there is no known endogenous ligand for the receptor (Poellinger, 1996).

The receptor binding studied here occurs without exposure to UVA light, and one might expect, if the receptor is biologically active, to see a biological effect without UVA exposure. In fact "dark" biological effects have been observed; 8-MOP has been observed to diminish lymphocyte surface markers without UV exposure (Schaller et al., 1987) and to regulate the cystic fibrosis associated chloride channel in T84 monolayers (Devor et al., 1997).

These findings, of course, do not explain the necessity for UVA light *in addition to* psoralens to achieve therapeutic effect in PUVA therapy; however, we speculate that this does not rule out that they have at least partial clinical relevance. In the future, the proposed clinical relevance of these receptors would be further supported if further quantitative nuclear binding studies with psoralen analogs demonstrated good rank-order correlation with observed clinical efficacy. Hopefully, these and other future studies of the biology of the receptor for psoralens will contribute to our increased understanding of the molecular mechanism(s) of this important group of medications.

REFERENCES

Averbeck, D. (1989): Recent advances in psoralen phototoxicity mechanisms. *Photochem. Photobiol.*, 50:859–882.

Baxter, J. D., Rousseau, C. G., Benson, M. C., Garcea, R. L., Ito, J., and Tomkins, G. M. (1972): Role of DNA and specific cytoplasmic receptors in glucocorticoid action. *Proc. Natl. Acad. Sci. USA*, 69:1892–1896.

Devor, D. C., Singh, A. K., Bridges, R. J., Frizzell, R. A. (1997): Psoralens: Novel modulators of Cl-secretion. *Am. J. Physiol.*, 272:C976–88.

Fisher, T., and Alsins, J. (1976): Treatment of psoriasis with trioxsalen baths and dysprosium lamps. *Acta. Derm. Veneorol* (Stockh), 56:383–390.

Gage, W., and Parrish, J. A. (1982): Cutaneous phytotoxicity due to psoralens. *Natl. Cancer Inst. Monogr.*, 66:117–124.

Giles, K. W., and Myers, A. (1965): An improved Diphenylamine method for the estimation of Deoxyri-bonucleic acid. *Nature*, 20:693.

Hanawalt, P. C., Su-chin L., and Parson, C. S. (1981): DNA repair responses in human skin cells. *J. Invest. Dermatol.*, 77:86–90.

Hannuksela, M., and Karnoven, J. (1978): Trioxsalen bath plus UVA effective and safe in the treatment of psoriasis. *Br. J. Dermatol.*, 99:703–707.

Hearst, J. E. (1981): Psoralen photochemistry and nucleic acid structure. *J. Invest. Dermatol.*, 77:39–44.

Jeong, H. G., Yun, C. H., Jeon, Y. J., Lee, S. S., and Yang, K. H. (1995): Suppression of cytochrome P450 (Cyp1a-1) induction in mouse hepatoma Hepa-1C1C7 cells by methoxsalen. *Biochem. Biophys. Res. Commun.*, 28:1124–30.

Laskin, J. D., Lee, E., Yurkow, E. J., Laskin, D. L., and Gallo, M. A. (1985): A possible mechanism of psoralen phototoxicity not involving direct interaction with DNA. *Proc. Natl. Acad. Sci. USA*, 82:6158–6162.

Laskin, J. D., Lee, E., Laskin, D. L., and Gallo, M. A. (1986): Psoralens potentiate ultraviolet light-induced inhibition of epidermal growth factor binding. *Proc. Natl. Acad. Sci. USA*, 83:8211–8215.

Ledirac, N., Delescluse, C., de Sousa, G., Pralavorio, M., Lesca, P., Amichot, M., Berge, J. B., and Rahmani, R. (1997): Carbaryl induces CYP1A1 gene expression in HepG2 and HaCaT cells but is not a ligand of the human hepatic Ah receptor. *Toxicol. Appl. Pharmacol.*, 144:177–82.

Mantelet, B., Pathak, M. A., and Dude, G. (1976): Identification of a metabolite of 4, 5′, 8-methyl psoralen. *Science*, 193:1131–1134.

Parrish, J. A., Fitzpatrick, T. B., Tanenbaum, L., and Pathak, M. A. (1974): Photochemotherapy of psoriasis with oral methoxsalen and long-wave ultraviolet light. *N. Engl. J. Med.*, 291:1207–1211.

Pathak, M. A. (1984): Mechanisms of psoralen photosensitization reactions. *Natl. Cancer Inst. Monogr.*, 66:41–66.

Pathak, M. A., Fitzpatrick, T. B., Parrish, J. A., and Biswas, R. (1977): Phototherapeutic, photobiological and photochemical properties of psoralens. *In: Proceedings of the Seventh International Congress on Photobiology*. Edited by A. Castellani, pp. 267–281, Plenum Press, New York.

Pibouin, M., Zin, R., Nguyen, P., Renouard, A., and Tillement, J. P. (1987): Binding of 8-methoxypsoralen to human serum proteins and red blood cells. *Br. J. Dermatol.*, 117:207–215.

Poellinger, L. (1996): The Anders Jahres Prize: Intracellular signal transmission via dioxin receptors. *Nord. Med.*, 111:45–8.

Sasaki, M., Meguro, F., Kumazawa, E., Fujita, H., Kakishima, H., and Sakata, T. (1988): Evidence for uptake of 8-methoxypsoralen and 5-methoxypsoralen by cellular nuclei. *Mutation Research*, 197:51–58.

Schaller, J., Gast, W., Walther, T., Haustein, U. E., Hoffman, C., and Rytter, M. (1987): Dark effect of 8-methoxypsoralen on lymphocyte surface markers. *Photodermatology*, 4:310–311.

Vaatainen, N. (1980): Phytotoxicity of topical psoralen. *Acta. Derm. Venereol* (Stockh), 60:327–331.

Walter, J. F., and Voorhees, J. J. (1973): Psoriasis improved by psoralen plus black light. *Acta. Derm. Veneorol* (Stockh), 53:469–472.

Yurkow, E. J., and Laskin, J. D. (1987): Characterization of a photoalkylated psoralen receptor in HeLa cells. *J. Biol. Chem.*, 282:8439–8432.

Toxicology of Skin
Edited by Howard I. Maibach
Copyright © 2001 Taylor & Francis

32

In Vitro Viability Assays

Mikyung Kwah

Department of Dermatology, University of California, San Francisco, California, USA

- · **Background**
- · **Choice of In Vitro Viability Assay**
- · **Viability Assay Types**
  - · Proliferation Studies · Histomorphology · Metabolic Assays · Mechanical Function · In Vivo Function
- · **Conclusion**

BACKGROUND

The term, *viability* has so many meanings that it is rarely defined in the literature of organ and tissue preservation. Although many methods exist to attempt quantification of true cellular or organ "death," mastery of the umbrella term viability continues to be a tremendous undertaking. Assays to determine organ viability often differ from cellular viability in endpoint. The choice of viability assay necessitates careful thought into the meaning of functional life or death for each undertaking or study goal. Perhaps the best way to address viability is to use several complementary measures of cellular or organ function in concert and not attempt a single definitive methodology to quantify "living" tissue.

To the organ cryobiologist, "The ultimate test of the viability of a preserved organ is the organ's ability to sustain life after either autograft or allografty transplantation" (Southard, 1989). This is indeed true in the clinical arena. In the research laboratory, however, viability assays should hold the dual role of (i) providing a reliable prediction of how an organ would function in vivo, and (ii) providing a reliable quantification of how time and study variables effect an organ.

In vitro, study tissue may involve anything from cell culture to biopsies of intact organ. The choice of viability assay is influenced by the needs of a single cell in culture against the characteristics of a community of cells in a three-dimensional organ system. For instance, an assay targeting melanocyte viability would not reflect the viability of an organ culture of whole skin. In vitro bioassays can often be adapted from cell culture to successful use in whole organ models. Bioassays of enzyme activity commonly used for single hepatocytes are also commonly used in whole organ samples.

In vivo bioassays can additionally be assays of physiological function specific to an organ. Kidneys maintained in artificial perfusion systems can be assessed by measuring reabsorptive and secretive functions. Such artificial perfusion systems can provide controlled quantification of metabolic substrates, electrolytes, and pharmacologic agents. Well-established criteria for kidney function currently involve sodium and glucose uptake (Fuller et al., 1977; Green and Pegg, 1979), oxygen consumption (Cohen and Folkman, 1968; Kearney et al., 1990), and the uptake and secretion of urine protein (Ploeg et al., 1988), and para-aminohypurate (Southard, 1989). Similarly, artificial perfusion of the liver can provide the means to assess viability by measuring metabolic activities and bile production/composition (Jamieson et al., 1988; Jamieson et al., 1988; Sundberg et al., 1988). In the working heart model, myocardium is similarly studied (Foreman et al., 1985), and artificially perfused pancreas is assessed for hormone production and enzymatic activity (Sanfey et al., 1985).

CHOICE OF IN VITRO VIABILITY ASSAY

Attempting to organize the multidisiciplinary universe of in vitro viability testing assays has been the subject of numerous reviews and symposia (Abouna, 1974; Bank and Schmehl, 1989; Fahy, 1981; Malinin and Perry, 1967; Southard, 1989; Pegg, 1989). Choosing the appropriate viability assay is essential in evaluating the cellular or organ system of choice. Optimization of any given assay relies greatly on the specific tissue's properties, as well as on the character of the assay. In choosing the appropriate viability assay, all characteristics of the tissue need to be considered.

Viability assessment has shown to be critical to transplantation medicine involving the liver, heart, kidney, pancreas, and skin. A review notes: "A clinically reliable viability assay could (i) reduce the need for retransplantation, (ii) reduce the induction of immunological sensitivity in retransplanted patients, (iii) reduce the need for extensive and expensive postoperative care, (iv) possibly reduce the number and severity of rejection episodes, and (v) improve long-term graft survival (Southard, 1989). To this end, several biochemical and physiological techniques have been employed. They can be classified grossly as (i) proliferation ability, (ii) histomorphology, (iii) metabolic assays, (iv) mechanical function, (v) in vivo function (Table 32.1).

Viability has been defined as "the ability of a living system to maintain itself in a steady state" (Fahy, 1981). Viability has also been defined as a transplanted organ's capability for self-repair (Bank and Schmehl, 1989). That is, a truly viable transplanted organ would be capable of the self-repair necessary to avoid gradual weakening, which occurs after transplantation of chemically crosslinked nonviable tissues, such as cardiac valves, fascia, or venous tissue (Bank and Schmehl, 1989).

There is a broad range to the definition of "viability"; however, a viability assay is very specific to the organ or cellular system under study. In the adaptation of assays

TABLE 32.1. *Classification of viability assays*

1 Proliferation ability
    Clonal potential
    DNA/RNA synthesis
    Protein synthesis
2 Histomorphology
    Light microscopy
    Gross pathology
3 Metabolic Assays
    Enzymatic activity
    Respiration
    Transport functions
4 Mechanical Function
    Motility
    Migration
    Contractility
5 In Vivo Function
    Organ-specific physiology

to individual organ systems, there exists a broad range of variables which will impact upon reliability, specificity, and sensitivity of a bioassay. Additionally, the choice of assay may differ between cell culture model to actual organ or tissue assessment.

Bioassays will address viability of groups of cells or of individual cells. For instance, populations of cells may be evaluated with such methods as immunologic reactions, oxygen uptake, enzyme activity, amino acid uptake, or staining techniques for cellular membrane integrity, but single cells may be assessed with such tests as flow cytometry or simple light microscopy.

Bioassays may also address, less directly, the quantitative responses of cell populations, such as urine output, or length of survival of an explant. Such assays lack the drawback of assuming an assay which may be applicable to a specific cell type that represents an entire population of cells; however, the activity of a community of cells may be the appropriate choice for evaluating the viability of whole tissue. For instance, the measure of hormone secretion or chemotaxis may be the definition of viability for one researcher but may not be the appropriate measure for the analysis of whole-organ structure and function. The measure of, say, fibroblast activity in skin would be too specific as a measure of the viability of the cryopreservation of whole skin.

Proliferation assays, or measures of reproductive potential, are an indirect measure of estimating the number of viable cells per unit volume of solution. These are tests of cells in suspended culture to attach and undergo clonal culture as a measure of functional viability. These tests of a cell's adhering and dividing capabilities provide little information on the abilities of cells in situ, however. The cell in situ has many external stimuli vital for its natural behavior, and the simple presence of an extracellular matrix often has great influence on viability. It is documented in cryobiology literature, for instance, that isolated chondrocytes can easily be cryopreserved but will not survive freezing within cartilage (Bank and Schmehl, 1989).

Biological variation is an inherent source of error that cannot be eliminated in biological assay systems (Bank and Schmehl, 1989). Differences between samples can be estimated by statistical methods which estimate variability; however, the errors created by variation can quickly become cumulative and affect an assay's final results. Methods which can help reduce the effects of variation include the use of appropriate standards when possible and the use of multiple experimental samples and controls.

VIABILITY ASSAY TYPES

Proliferation Studies

An indirect measure of cell viability is the limiting dilution assay. This assay measures the number of cells in suspended culture capable of adhering to a substrate and undergoing cell division. (Serial dilutions are used to obtain both positive and then negative replicate cultures.) These assays of cellular clonal potential are not always predictive of the integrity of the full tissue or 3-dimensional organ model, however. Also, attatchment ability to glass or plastic is a limiting factor, and although this ability has been the sole criteria of viability in some studies (Taylor et al., 1974), this is a limited quality which needs careful measurement (Bohmer et al., 1973).

Similarly, a simple count of the number of colonies yielded (the "plating" assay) can be counted on the assumption that each colony came from a single viable, surviving cell. This more direct assay was developed in 1959 by Puck for tissue culture (Puck, 1959), and has since been heavily utilized in cryobiology. A drawback to the plating assay is the variability of the plating efficiency, which is commonly seen in the growth of control populations. That is, for certain cell populations (e.g., bone marrow—where up to 2000 healthy cells can be needed to plate out a single culture colony), the normally low plating efficiency of untreated control cells can cause wide variation in the viability assay result. The in vivo adaptation of the plating assay is the spleen colonization assay of Till and McCulloch, which has been widely used for analysis of stem cell viability following cryopreservation (Leibo et al., 1969; Till and McCulloch, 1961).

Mitosis itself can be easily quantitated for a stained cell preparation with a light microscope. Mitotic indices are often measured by adding mitotic spindle poisons (e.g., colchicine) to a cell culture a few hours prior to harvesting and then visualizing the number of cells arrested in metaphase using dyes such as orcein (Moorhead et al., 1960). Cell cultures which are not actively mitotic require mitogenic factors; however, to yield a mitotic index. For instance, the induction of lymphocytes is achieved using phytohemagglutinin in the first few days of cell culture (Pegg, 1965). An actively mitotic tissue, such as bone marrow, would not require such stimulation.

Another indirect measure of viability is the analysis of DNA or RNA synthesis as measured by radiolabeled nucleic acids. Tritiated thymidine is a common radioactive label used to determine viable cell concentration (Gillis et al., 1978; Knight and Farrant, 1978). The use of tritiated uridine has also been well described (Ford et al., 1989). Recently, the nonradioactive labeling assay, 5-bromo-2-deoxyuridine (BrdU) ELISA, has been determined to be a more sensitive DNA proliferation assay for

CD4+ T-lymphocyte growth (compared to tritiated thymidine) (Maghni et al., 1999). Another commonly used precursor is 14C-formate (Thomas and Lochte, 1957) in determining DNA synthesis. DNA synthesis is a measure of mitotic activity and proliferation, as is protein synthesis (measured by the incorporation of amino acids). In cultured hair follicles, an established measure of proliferation is the uptake of radiolabeled amino acids (Uzuka et al., 1977). Vitamin A has been shown to inhibit protein synthesis in chicken embryonic epidermis, as shown by inhibition of uptake of 3H-leucine and 35S-cystine (Beckingham Smith, 1973).

On a higher level, the work on the cryopreservation of fertilized ova and early embryos has refined successful preservation techniques. The growth and continued proliferation (segmentation) of fertilized ova is an important criteria of viability after cryopreservation, and this assay looks at the ability of embryos to develop to the blastocyst stage in culture relative to the proportion of unfrozen control embryos. Numerous studies have shown this to correlate closely with the ability to produce viable young after reimplantation in vivo (Pegg, 1989).

Histomorphology

Morphological or pathological examination are useful in providing a descriptive evaluation of viable tissue. These evaluations are often subjective, however, and the reproducibility of the assay may be reliant on the consistency and expertise of the person conducting the test. Degrees of hyperplasia, atrophy, or edema are quantitative but also highly subjective. The simple presence or absence of immunologic label, for example, is more objective and direct. This binary use of histology provides greater objectivity to the subjective question of cell life or death but still addresses single, specific functions in the multifunctional cell. Stains which measure membrane viability, such as acridine orange or propidium iodide, do not address the many metabolic functions within a cell.

The appropriate choice of histological bioassay can be important, as even within a single species and tissue type, different fixation and staining procedures must be used not only for light microscope and ultrastructural studies, but also for enzyme localization or immunocytochemical localization (Bank and Schmehl, 1989). Details concerning a specific system can be crucial to the success of a bioassay, even when the function being measured is similar. Within an organ, the cell choice for histologic examination can also be critical, as demonstrated in a study of rat liver, where the abnormal morphology of only the cells of the sinusoidal lining was found to correlate with post-transplant organ viability (Iu et al., 1987).

Dye exclusion tests are measures of the ability of live cells to prevent uptake of certain dyes. This exclusion of dyes, such as eosin, trypan blue, nigrosin, or propidium iodide, is then correlated with the integrity of the cellular membrane. The actual enumeration of stained, nonvital cells can then be performed with the light microscope. Dye exclusion tests are widely accepted to reflect only certain types of trauma, however. High temperature, anoxia, and mechanical trauma are accurately measured by this technique, but dye exclusion has been misleading after cryopreservation (Greene et al., 1964; Pegg, 1964).

Lysosomal integrity has also been a measure of cellular viability. The amount of released versus intact lysosomal enzymes can be correlated with cellular viability. Measurement of lysosomal integrity include the acridine orange or neutral red assays—fluorescent vital dyes. Acridine orange is a particularly well-studied dye, and its uptake has been correlated with viability as measured by other methods, such as actual transplantation (bone marrow (Pegg, 1964) and cornea (Smith et al., 1963)). Neutral red values have been shown to be more sensitive than lactate dehydrogenase leakage testing in primary cultures of hepatocytes (Zhang et al., 1990). Acridine orange is popularly used in concert with the vital stain, ethidium bromide, to fluoresce viable cells green and nonviable cells orange (Detta and Hitchcock, 1990; Mischell and Shiigi, 1980).

The growing technology of fluorescent dyes and markers in recent years has allowed for the development of novel techniques for the in situ assessment of viability. Primarily stains of membrane integrity, fluorescent dyes allow assessment of in situ cellular function while also localizing damage to cells and tissue architecture when used in situ. Again, the most popular use of quantitative fluorescent microscopy utilizes a dual fluorescent staining technique—for instance, combining fluorescein diacetate test (FDA) with ethidium bromide. FDA, as an indicator of membrane integrity and cytoplasmic esterase activity, results in the intracellular accumulation of green fluorescence in cells with intact membranes, while ethidium bromide will enter cells with damaged membranes and form a red complex with nuclear DNA.

The light microscope remains a valuable tool in histologic examination, and routine analysis of nuclear chromatin clumping, mitochondrial degeneration, intracellular edema, and organelle/membrane integrity are highly valued measures of cell viability; however, many new biotechnological modalities are now routinely being applied to the study of cellular viability. They look at a more full range of cellular detail—for example, surface antigen localization by high-powered confocal scanning laser microscopy or ultrastructural analysis by electron microscopy. Scanning electron microscopy and freeze-fracture electron microscopy reveals fine morphological information, and the details revealed have been used in a wide variety of cells (Cerra et al., 1977; Imbert and Cullander, 1999; McCullough et al., 1980; Pegg et al., 1978; Weinstein and Someda, 1967). The primary drawback to the routine use of electron microscopy is the length of processing time. The most valuable use of electron microscopy will be limited to elucidating mechanisms of damage on the cellular level rather than quantification of viability.

Similarly, structural integrity can be measured by a simple biochemical assay of intracellular material in the surrounding fluid or perfusate. The release of ions, such as potassium, or of complex enzymes, such as aminotransferases, can be measured (von Frankenberg et al., 1997; Wang et al., 1998). Lactate dehydrogenase release is a valuable measure of hepatic damage in situ (Bergmeyer and Bernt, 1974), but is also a good measure of renal ischemic damage. Glutamic oxaloacetic transaminase leakage is also commonly measured (McQueen and Williams, 1982). In the quantification of cutaneous toxicity of certain agents, the simple measurement of released intracellular enzymes into skin culture medium (glutamate dehydrogenase, glutamine pyruvate transaminase, glutamic oxaloacetate transaminase, lactate dehydrogenase) has been proposed for an index of toxicity (Kao et al., 1983). Hemoglobin release remains the viability assay of choice for red blood cells. Gross protein measurements have been popular, as they are widely applicable and facile for use, requiring only a cellular homogenate (Peterson, 1977).

Gross pathology can sometimes be useful. In the kidney cryopreservation literature, weight gain (or water uptake) of an organ has been proposed to have many causes and has been suggested as a viability test prior to organ transplantation (Pegg and Farrant, 1969; Sells et al., 1977). Another test of structural integrity is the measurement of vascular resistance in whole kidneys preserved by the Belzer method (Belzer et al., 1970).

Metabolic Assays

Measures of enzyme activity or production, oxygen utilization, or the uptake of synthetic precursers are examples of the quantification of metabolic activity. Additionally, the function of individual subcellular organelles are used to evaluate tissue metabolism. The activity of mitochondria, lysosomes, and microsomes can be quantified through a variety of bioassays. The tissue content of important metabolites, such as adenine nucleotides, Ca and Mg, Na and K, antioxidants (glutathione, vitamin E), and important cofactors (Co Q, pyrimidine and flavin nucleotides), can also be quantified (Southard, 1989). Note that metabolic function is primarily dependent on enzymatic systems and is often independent of the structural integrity of a cell. Indeed, some assays of enzymatic activity require the use of cellular homogenates or call for the disruption of the cell for removal of enzymatic products (Table 32.2; Southard, 1989).

Intact enzyme systems provide perhaps the most simple measurements, as they require artificially introduced substrates in controlled amounts, which react in a quantifiable manner with native enzymes. For instance, the tetrazolium assay is commonly utilized for cell culture, tissue, and whole organ systems. Tetrazolium oxidizes NADH in the intact mitochondria to convert into an intensely colored formazan (Terasaki et al., 1967). This formazan is easily quantified with a spectrophotometer. Similarly, the FDA relies on nonspecific esterases to convert FDA to fluorescein (Rotman and Papermaster, 1966). This viability assay has been widely used on lymphocytes,

TABLE 32.2. *Metabolic assay types*

Mitochondrial enzyme activity
Cytoplasmic enzyme activity
Respiration and energy metabolism
Drug metabolism
Ion Transport

granulocytes, hepatocytes, ova, and cornea (Frim and Mazur, 1980; Leibo et al., 1979; Leibo and Mazur, 1978; Madden, 1987; Nikolai et al., 1991). Acid phosphatase assays simply measure cytosolic acid phosphatase activity (Maghni et al., 1999). In a recent comparison of skin viability assays, Hood and Bronaugh examined the tetrazolium assay, MTT, in comparison to the more common metabolic assay for lactic acid and discovered that lactic acid was an unreliable viability indicator when trying to optimize the lipophilicity of skin perfusion solution medium (Hood and Bronaugh, 1999). This study illustrated the need for multiple approaches to the measurement of viability when optimizing the conditions for metabolism in a cell or organ culture. The living system of multicellular organ culture is affected not only by the "witches' brew" of culture medium, but also by the nature of the demands of a community of cells.

Perhaps the most "vital" of all organelles is the mitochondria, for mitochondrial function has been a common measure of cellular viability. Modulating artificial perfusion solutions to optimize mitochondrial function (through calcium, for example) has been shown to correlate with successful organ preservation (McAnulty et al., 1993). Isolated mitochondria have been used to measure viability in preserved hepatic tissue (Lambotte et al., 1979). Mitochondrial function is often exploited for viability assays, such as the formazan/tetrazolium assays (Imbert and Cullander, 1997; Ishiyama et al., 1997; Supino, 1995) or the AlamarBlue (resazurin) assays (Rasmussen, 1999), which are fluorescent redox indicators of intact mitochondria. ATP levels can be followed with bioluminescence (with luciferin-luciferase) (Bodin and Burnstock, 1998; Brandhorst et al., 1999), and cytochrome P450 activity is routinely measured in hepatocytes (Silva et al., 1999).

Metabolism of energy substrates can be measured in both cell culture as well as in perfused organs. The uptake of oxygen is a commonly studied substrate, and these viability assays have been performed on cell culture, tissue slices, and whole perfused organs (Abbott, 1969; Hardie et al., 1973; Pegg et al., 1986). Cellular respiration is a function of the mitochondria, but the first tests of respiration applied actual respirometers to in vitro whole organs (Cruickshank, 1954). Later, the well-known Warburg manometric technique measured tissue respiration (Umbreit, 1964). The Clark electrode eventually provided measurements of blood-gas analysis as applied to oxygen consumption, and now a miniaturized Clark electrode is used in skin for transcutaneous oxygen consumption determinations (Penneys, 1964).

Viability tests have also been based on the uptake of glucose, pyruvate, lactate, and fatty acids (Fischer et al., 1979; Pegg et al., 1981; Pettersson et al., 1974). Conversion of 14C-glucose to 14CO2 is commonly measured (Collier et al., 1989; May and DeClement, 1981). Pyruvate or lactate can be quantified through a simple colorimetric assay of culture fluid (Bhatt et al., 1997). Fatty acid synthesis can be traced radioactively through 14C-acetate (Bligh and Dyer, 1959). Levels of high energy stores in the form of ADP and ATP have been shown to correlate with the ability of artificially perfused organs to support life (Calman, 1974). Unfortunately, ATP and ADP are labile compounds under normothermic conditions (Sehr et al., 1979). It has been proposed, however, that the ability of an organ to regenerate ATP from substrates and intact enzyme systems is the best correlation with organ transplantation viability (Bore et al., 1979; Buhl, 1976; Collins et al., 1977; Pegg et al., 1989).

Transport functions are also commonly used as viability indicators. For example, sodium pumping can be measured directly or through cell volume regulation during a form of hypotonic stress test. This has·been useful for refining methods for human platelet cryopreservation (Brodthagen et al., 1985; Fahy, 1980; Fuller et al., 1977). Corneal thickness measurements made by specular microscopy reflect the hydration of the corneal stroma, which is ordinarily maintained by the intact endothelial bicarbonate pump (Hodson and Miller, 1976). Also, since potassium and sodium pumping are coupled, measurement of the ratio of intracellular potassium to sodium allows for another measurement of ion pump viability. Para-aminohippurate (PAH) uptake has been extensively studied in the kidney (Dahlager and Bilde, 1976; Pegg et al., 1986; Weinerth and Abbott, 1974), and 5-hydroxytryptamine has been similarly applied to the study of platelets (Armitage, 1982; Brodthagen et al., 1985; Kim and Baldini, 1974).

Many tissues now have well-characterized metabolic pathways which can be established as measures of viability. Studies on cryopreserved bone marrow and cryopreserved rat pancreas have established the use of numerous intact enzymatic pathways for viability measurements (Ashwood-Smith, 1961; Mazur et al., 1976). Similarly, in keratinocytes from postmortem skin, Hirel et al. studied lipid synthesis capacity, keratin production, and transglutaminase activity levels and correlated these measures of viability with drug metabolism capacity (Hirel et al., 1996).

In skin, for example, lipid synthesis is classically analyzed with 14C-acetate incorporation according to the protocols of (Bligh and Dyer, 1959), and (Ponec et al., 1988). Also in skin, keratin production classically looks at the types, maturity, and gross amounts in a skin cell population (Sugimoto et al., 1974). Dermal transglutamase activity has been measured in cell culture homogenates per (Yuspa et al., 1980), and various drug metabolizing enzymatic activities include measurement of glutathione S-transferase activity (Habig and Jacoby, 1981) and glucuronidation activity (Hirel et al., 1996). For murine skin cultures, benzopyrine metabolism has been proven to correlate with the culture viability (Smith and Holland, 1981). In other tissues, particularly hepatocytes, toxin metabolism is commonly studied. Breakdown products of artificially introduced 7-ethoxycoumarin has been measured by HPLC (Coldham et al., 1995), and 7-ethoxyresorufin-O-deethylase activity (Burke and Halmann, 1978) is another drug metabolism pathway which has been studied.

Mechanical Function

When selecting a normal metabolic function as an assay of viability, higher-order functions, such as chemotaxis, are often the appropriate choice. In the analysis of PMN motility, such variables as rate of movement and need for chemotactive stimulus are important study variables (Bank, 1980), but in the study of sperm motility, these factors are not as important. For epidermal cells, migration (epiboly) is also known not to correlate with viability (Mitrani and Marks, 1973; Tammi and Maibach, 1987). (Migrated epidermal cells, in fact, do not incorporate thymidine whereas precursor cells do, suggesting an inverse relationship between migration and epidermal cell proliferation (Gaylard and Sarkany, 1975).) Viability indices have been developed by quantifying motility with the use of time lapse photography (for lymphocytes) or by simply counting the percentage of motile cells (Pulvertaft and Humble, 1956). Motility remains one of the standards in viability assessment for both spermatozoa as well as granulocytes (Bank, 1980; Polge, 1980).

Contraction of smooth muscle is another form of conversion of chemical energy into mechanical force. A common model used in cryobiology to assess the viability of smooth muscle involves the response of smooth muscle to histamine. This dose-response system has been well described with guinea pig taenia coli (Elford and Walter, 1972; Taylor, 1982; Taylor and Pegg, 1983). In this system it has been shown to be important to normalize all tensions as a proportion of that tension developed by each control muscle in response to the largest dose of histamine used because the slopes of the response curves changed with each histamine dose (Elford and Walter, 1972). With cardiac muscle, the spontaneous contraction of cells or entire organ can be observed grossly but is not a reliable prediction of post-transplant viability (Pegg, 1989). Ideally, the cardiac output of an in vivo perfused heart is the best measure of the mechanical work ability, but again this is not a guarantee to transplantation success.

In Vivo Function

In whole organs, quantifiable functions can be the tensile strength in tendons or glomerular filtration rate and protein leakage for kidneys. The examination of the heart can be assessed with a variety of in situ radiolabels and well-established radiographic techniques. The viability of cryopreserved gametes has been extensively studied for reproductive medicine, and sperm motility is a routine measure of sample viability. Similarly, chemotaxis is often the test method of choice for leukocyte (PMN) assessment. Rate of PMN migration and amount of necessary chemotactive stimulus are commonly used study variables.

The "ultimate" test of viability is the function of the cell/tissue/organ under study in its natural environment. Indeed, this is the endpoint of much of the medical applications under study. Transplantation ability as measured in experimental animals, however, is not always perfectly predictive of actual in vivo function. Such predictive assays are better indices of preservation techniques. The exception to this is the highly successful science of gamete preservation. The truly "ultimate" test of viability is the in vivo function of fertilization ability and procession to normal development. Artificial insemination and in vitro fertilization are applications of cryopreservation in medicine where the experimental work has somehow successfully integrated the demands of many complex processes, including cell division, and created a single and reliable viability "test."

In vivo tests of transplantation are not quantitative, except in the case of perhaps bone marrow or tumor cell transplantation, where a single stem cell or tumor cell can propagate (Furth and Kahn, 1937; Hewitt, 1958; Lorenz et al., 1952). Bone marrow has been well studied, and in vitro viability tests have been compared to the in vivo model for cryopreserved cells (Pegg, 1964; Schaefer, 1980). In particular, the trypan blue dye exclusion test was proven to be unreliable on bone marrow in vitro when compared to success of in vivo transplantation (Pegg, 1964). In 1996, Fricker et al. compared embryonic striatal grafts to the use of vital dyes for optimizing the neural cell preparation and culture techniques in animal models of Huntington's disease (Fricker et al., 1996).

In the case of tissues such as skin, actual transplantation after cryopreservation has been studied in animal models (Berggren and Lehr, 1965; Billingham and Medawar, 1952; Bondoc and Burke, 1971). The cornea has also been well studied (Capella and Kaufman, 1972; Fong et al., 1956; Mueller et al., 1967). Cryopreserved cornea has been proven to give reliable viability testing results using colorimetric tests (acridine orange (Smith et al., 1963) or tetrazolium (Capella et al., 1965) assay) when compared to actual in vivo transplantation.

Whole organs have best been studied in "revascularization" models in experimental animals. In the case of the kidney and pancreas, normothermic perfusion assays aided the development of methods that helped to refine whole organ cryopreservation techniques (Bishop and Ross, 1978; Fuller and Pegg, 1976; Idezuki et al., 1968; Klempnauer et al., 1982). The great disadvantage of reperfusion assays, however, is their considerable cost in time and money. Also, these are necessarily short-term assays, as reperfusion causes vascular endothelial damage when prolonged over 4–6 hours (Belzer et al., 1972).

Whole kidney and liver have been studied for the measurement of physiological and secretive functions (as measured by urine composition and production). Kidney and liver viability is limited without a true blood supply to approximately 2 hours, however, and the artificial solutions used to perfuse whole organs have relatively low viscosity and rheologic properties, and perfusion rates are relatively high to provide the required oxygenation (Southard, 1989). More natural conditions can be achieved by using an animal model. It is in this fashion, with whole blood, that the liver and kidney can be successfully transplanted into the major vessels of an animal and studied for at least 4–6 hours (Belzer et al., 1972; Iu et al., 1987; Southard, 1989).

Transplanted pig hearts have been similarly studied, and viability loss has been measured in terms of reperfusion left ventricular systolic or diastolic pressure (Ferrera et al., 1993). Cryopreserved allogeneic heart valves and vessels have now come to be widely used in cardiothoracic surgery (Hawkins et al., 1992). Cryopreserved tracheas are not yet used in humans but have been used in animal transplantation studies (Inutsuka et al., 1996) and in vitro studies (Fricker et al., 1996). (The interest in tracheal tissue is primarily the airway epithelial cell, which is the target of viability testing in vitro.)

   Thin tissue slices are also a well documented source for biochemical or metabolic assessment. Such slices of kidney, liver, heart, and pancreas can be maintained in a physiologically balanced salt solution (with properties similar to plasma) in an oxygenated environment. Studying such tissue's ATP content is a measure of suffi-cient oxygenation (Southard and Belzer, 1980). When such slices are incubated in a respirometer, oxygen utilization can be directly monitored itself to assess mito-chondrial function. Tissue slices have also been used for assessment of tissue edema, intracellular concentrations of added metabolic substrates, and effects of various drugs or toxins (Southard, 1989). This model has been widely used in some laboratories in the creation of techniques for organ preservation, as tissue slices are easily subjected to different preservation media or conditions. The great disadvantage of this model of viability is the lack of vascularity, limiting in vitro methods to approximately 2 hours (similar to whole-organ artificial perfusion. The only organ system which is well suited for tissue slice or whole-organ artificial perfusion is perhaps skin, as up to 70% of whole skin utilizes glucose in an anaerobic fashion (Frienkel, 1960).

CONCLUSION

Intuitively we think of life and death as binary states (i.e., all or nothing). Yet, the viability assay is approached as a condition of gradients or degrees. The term *viability* is used as if it were a quality that could be quantified (Pegg, 1989). Viability is not synonymous with life; it is best approached as a complement of many individual functions. The ideal working definition will necessarily alter for each isolated research objective but should perhaps best be approached from the perspective of multiple concurrent assays of viability in concert with one another.

REFERENCES

Abbott, W. M. (1969): Viability assays as applied to the cryopreservation of hearts and kidneys. *Cryobiology*, 5:454–62.

Abouna, G. J. M. (1974): Viability assays in organ preservation. In: *Organ Preservation for Transplantation*. Edited by A. M. Karow, Jr, G. J. M. Abouna, A. L. Humphreys, 1st ed., pp. 108–26. Little, Brown, Boston.

Armitage, W. J. (1982): Transport of 5-hydroxytryptamine by human platelets incubated in glycerol. *Transfusion*, 22:203–5.

Ashwood-Smith, M. J. (1961): Preservation of mouse bone marrow at −79C with dimethyl sulphoxide. *Nature*, 190:1204–5.

Bank, H. (1980): Granulocyte preservation circa 1980. *Cryobiology*, 17:187–97.

Bank, H. L., Schmehl, M. K. (1989): Parameters for evaluation of viability assays: Accuracy, precision, specificity, sensitivity, and standardization. *Cryobiology*, 26:203–11.

Beckingham Smith, K. (1973): Early effects of vitamin A on protein synthesis in the epidermis of embryonic chick skin cultured in serum-containing medium. *Dev. Biol.*, 30:241–48.

Belzer, F. O., Hoffman R, Huang, J., and Downes, G. (1972): Endothelial damage in perfused dog kidney and cold sensitivity of vascular Na-K-ATPase. *Cryobiology*, 9:457–60.

Belzer, F. O., Reed, T. W., Pryor, J. P., Kountz, S. L., and Dunphy, J. E. (1970): Cause of renal injury in kidneys obtained from cadaver donors. *Surgery*, 130:467–7.

Berggren, R. B., Lehr, H. B. (1965): Clinical use of frozen skin. *J. Amer. Med. Assoc.*, 194:149–151.

Bergmeyer, H. U., Bernt, E. (1974): Lactate dehydrogenase. In: *Methods of Enzymatic Analysis*, Vol II. Edited by H. U. Bergmeyer, pp. 574–9. Academic Press, New York.

Bhatt, R. H., Micali, G, Galinkin, J., Palicharlar, P., Koch, R. L., West, D. P., and Solomon, L. M. (1997): Determination and correlation of in vitro viability for hairless mouse and human neonatal whole skin and stratum corneum/epidermis. *Arch. Dermatol. Res.*, 289:170–3.

Billingham, R. E., Medawar, P. B. (1952): The freezing, drying and storage of mammalian skin. *J. Exp. Biol.*, 29:454–68.

Bishop, M. C., Ross, B. D. (1978): Evaluation of hypertonic citrate flushing solution for kidney preservation using the isolated perfused rat kidney. *Transplantation*, 25:235–9.

Bligh, E. G., Dyer, W. J. (1959): A rapid method of total lipid extraction and purification. *Can. J. Biochem. Physiol.*, 37:911–7.

Bodin, P., Burnstock, G. (1998): Increased release of ATP from endothelial cells during acute inflammation. *Inflammation Research*, 47:351–4.

Bohmer, H., Wohler, W., Wendel, U., Passarge, E., and Rudiger, H. W. (1973): Studies on the optimal cooling rate for freezing human diploid fibroblasts. *Exp. Cell. Res.*, 79:496–8.

Bondoc, C. C., Burke, J. F. (1971): Clinical experience with viable frozen human skin and a frozen skin bank. *Ann. Surg.*, 174:371–81.

Bore, P. J., Papatheofanis, I., Sells, R. A. (1979): Adenosine triphosphate regeneration and function in the rat kidney following warm ischaemia. *Transplantation*, 27:235–7.

Brandhorst, D., Brandhorst, H., Hering, B. J. (1999): Large variability of the intracellular ATP content of human islets isolated from different donors. *Journal of Molecular Medicine*, 77:93–5.

Brodthagen, U. A., Armitage, W. J., Paramar, N. (1985): Platelet cryopreservation with glycerol, dextran and mannitol: Recovery of 5-hydroxytryptamine uptake and hypotonic stress response. *Cryobiology*, 22:1–9.

Buhl, M. R. (1976): The postanozic regeneration of 5′adenine nucleotides in rabbit kidney tissue during in vitro perfusion. *Scand. J. Clin. Lab. Invest.*, 36:175–81.

Burke, M. D., Halmann, H. (1978): Microfluorimetric analysis of the cytochrome P-448 associated ethoxyresorufin-O-deethylase activities of individual isolated rat hepatocytes. *Biochem. Pharmacol.*, 27:2859–63.

Calman, K. C. (1974): Prediction of organ viability. I: An Hypothesis. *Cryobiology*, 11:1–6.

Capella, J. A., Kaufman, H. E., Robbins, J. E. (1965): Preservation of viable corneal tissue. *Cryobiology*, 2:116–21.

Capella, J. A., Kaufman, H. E. (1972): Corneal cryopreservation and its clinical applications. In: *Corneal Grafting*. Edited by T. A. Casey, pp. 219–32. Butterworth, London.

Cerra, F. B., Raza, S., Andres, G. A., and Siegel, J. H. (1977): The endothelial damage of pulsatile renal preservation and its relationship to perfusion pressure and colloid osmotic pressure. *Surgery*, 81:534–41.

Cohen, B. E., Folkman, M. J. (1968): Cell death and the measurement of oxygen consumption in the isolated perfused organ. In: *Organ Perfusion and Preservation*. Edited by J. C. Normal, pp. 471–486. Appleton-Century-Crofts, New York.

Coldham, N. G., Moore, A. S., Dave, M., Graham, P. J., Sivapathasundaram, S., Lake, B. G., and Sauer, M. J. (1995): Imidocarb residues in edible bovine tissues and in vitro assessment of imidocarb metabolism and cytotoxicity. *Drug Metabolism and Disposition*, 23:501–5.

Collier, S. W., Sheikh, N. M., Sakr, A., Lichtin, J. L., Stewart, R. F., and Bronaugh, R. L. (1989): Maintenance of skin viability during in vitro percutaneous absorption/metabolism studies. *Toxicology and Applied Pharmacology*, 99:522–33.

Collins, G. M., Taft, P., Green, R. D., Ruprecht, R., and Halasz, N. A. (1977): Adenine nucleotide levels in preserved and ischaemically injured canine kidneys. *World J. Surg.*, 1:237–42.

Cruickshank, C. N. D. (1954): Continuous observation of the respiration of skin in vitro. *Exp. Cell. Res.*, 7:374

Dahlager, J. I., Bilde, T. (1976): The uptake of hippuran in kidney slices employed as a viability test. *Scand. J. Urol. Nephrol.*, 10:120–5.

Detta, A., Hitchcock, E. (1990): The selective viability of human foetal brain cells. *Brain Research*, 520:277–83.

Elford, B. C., Walter, C. A. (1972): Effects of electrolyte composition and pH on the structure and function of smooth muscle cooled to −79C in unfrozen media. *Cryobiology*, 9:82–100.

Fahy, G. M. (1980): Analysis of "solution effects" injury: Rabbit renal cortex frozen in the presence of dimethyl sulfoxide. *Cryobiology*, 17:371–88.

Fahy, G. M. (1981): Viability assessment. In: *Organ Preservation for Transplantation*. Edited by A. M. Karow, Jr, 0. D. E. Pegg, pp. 53–73. Dekker, New York.

Ferrera, R., Guidollet, J., Larese, A. et al. (1993): Heart graft viability assessed by quantitative reduction of MTT. GRRC Congress, Beaune, France, April, 6–7.

Fischer, J. H., Isselhard, W., Hauer, U., and Menge, M. (1979): Free fatty acid and glucose metabolism during hypothermic perfusion of canine kidneys. *Eur. Surg. Res.*, 11:107–21.

Fong, L. P., Hunt, C. J., Taylor, M. J. et al. (1956): Cryopreservation of rabbit corneas: Assessment by microscopy and transplantation. *Brit. J. Ophthalmol.*, 70:751–60.

Ford, C. H., Richardson, V. J., Tsaltas, G. (1989): Comparison of tetrazolium colorimetric and [3H]-uridine assays for in vitro chemosensitivity testing. *Cancer. Chemother. Pharmacol.*, 24:295–

Foreman, J., Pegg, D. E., Armitage, W. J. (1985): Solutions for preservation of the heart at 0C. *J. Thorac. Cardiovasc. Surg.*, 89:867–71.

Fricker, R. A., Barker, R. A., Fawcett, J. W., and Dunnett, S. B. (1996): A comparative study of preparation techniques for improving the viability of striatal grafts using vital stains, in vitro cultures, and in vivo grafts. *Cell Transplantation*, 5:599–611.

Frienkel, R. J. (1960): Metabolism of glucose C14 by human skin in vitro. *J. Invest. Dermatol.*, 34:37–42.

Frim, J., Mazur, P. (1980): Approaches to the preservation of human granulocytes by freezing. *Cryobiology*, 17:282–6.

Fuller, B. J., Pegg, D. E., Walter, C. A. (1977): An isolated rabbit kidney preparation for use in organ preservation research. *J. Surg. Res.*, 22:128–142.

Fuller, B. J., Pegg, D. E. (1976): The assessment of renal preservation by normothermic bloodless perfusion. *Cryobiology*, 13:177–84.

Furth, J., Kahn, M. C. (1937): The transmission of leukemia of mice with a single cell. *Amer. J. Cancer*, 31:276–82.

Gaylard, P. M., Sarkany, I. (1975): Cell migration and DNA synthesis in organ culture of human skin. *Br. J. Dermatol.*, 92:375–80.

Gillis, S., Ferm, M. M., Ou, W., and Smith, K. A. (1978): T-cell growth factor: Parameters of production and a quantitative microassay for activity. *J. Immunol.*, 120:2027–32.

Green, C. J., Pegg, D. E. (1979): The effects of variations in electrolyte composition and osmolarity of solutions for infusion and hypothermic storage of kidneys. In: *"Organ Preservation II."* Edited by D. E. Pegg and I. A. Jacobsen, pp. 86–97. Churchill Livingstone, Edinburgh/London.

Greene, A. E., Silver, R. K., Krug, M. et al. (1964): Preservation of cell cultures by freezing in liquid nitrogen vapor. *Proc. Soc. Exp. Biol. Med.*, 116:462–7.

Habig, W. H., Jacoby, W. B. (1981): Assays for differentiation of glutathione S-transferases. *Methods Enzymol.*, 77:218–31.

Hardie, I. R., Clunie, G. J. A., Collins, G. M. (1973): Evaluation of simple methods for assessing renal ischemic injury. *Surg. Gynecol. Obstet.*, 136:43–6.

Hawkins, J. A., Bailey, W. W., Dillon, T., and Schwartz, D. C. (1992): Midterm results with cryopreserved allograft valved conduits from the right ventricle to the pulmonary arteries. *J. Thorac. Cardiovasc. Surg.*, 104:910–6.

Hewitt, H. B. (1958): Studies of the dissemination and quantitative transplantation of a lymphocytic leukemia of CBA mice. *Brit. J. Cancer.*, 12:378–401.

Hirel, B., Watier, E., Chesne, C., Patoux-Pibouin, M., and Guillouzo, A. (1996): Culture and drug biotransformation capacity of adult human keratinocytes from post-mortem skin. *Br. J. Dermatol.*, 134:831–6.

Hodson, S., Miller, F. (1976): The bicarbonate ion pump in the endothelium which regulates the hydration of rabbit cornea. *J. Physiol.*, 263:563–77.

Hood, H. L., Bronaugh, R. L. (1999): A comparison of skin viability assays for in vitro skin absorption/metabolism studies. In *Vitro and Molecular Toxicology*, 12:3–9.

Idezuki, Y., Goetz, F. C., Kaufmann, S. E., and Lillehei, R. C. (1968): In vitro insulin productivity of preserved pancreas: A simple test to assess the viability of pancreatic allografts. *Surgery*, 64:940–7.

Imbert, D., Cullander, C. (1999): Buccal mucosa in vitro experiments I: Confocal imaging of vital staining and MTT assays for the determination of tissue viability. *J. Contr. Rel.*, 58:39–50.

Imbert, D., Cullander, C. (1997): Assessment of cornea viability by confocal laser scanning microscopy and MTT assay. *Cornea*, 16:666–74.

Inutsuka, K., Kawahara, K., Takachi, T., and Okabayashi, K. (1996): Reconstruction of trachea and carina with immediate or cryopreserved allografts in dogs. *Ann. Thorac. Surg.*, 62:1480–4.

Ishiyama, M., Miyazono, Y., Sasamoto, K. (1997): A highly water-soluble disulfonated tetrazolium salt as a chromogenic indicator for NADH as well as cell viability. *Talanta*, 44:1299–1305.

Iu, S., Harvey, P. R. C., Makowka, L., Petrunka, C. N., Ilson, R. G., and Strasberg, S. M. (1987): Markers of allograft viability in the rat. Relationship between transplantation viability and liver function in the isolated perfused liver. *Transplantation*, 44:562–9.

Jamieson, N. V., Sundberg, R., Lindell, S., Southard, J. H., and Belzer, F. O. (1988): A comparison of cold storage solutions for hepatic preservation using the isolated perfused rabbit liver. *Cryobiology*, 25:300–10.

Jamieson, N. V., Lindell, S., Sundberg, R., Southard, J. H., and Belzer, F. O. (1988): An analysis of the components in UW Solution using the isolated perfused rabbit liver. *Transplantation*, 46: 512–16.

Kao, J., Hall, J., Holland, J. M. (1983): Quantitation of cutaneous toxicity: An in vitro approach using skin organ culture. *Toxicol. Appl. Pharmacol.*, 68:206–17.

Kearney, J. N., Wheldon, L. A., Gowland, G. (1990): Cryopreservation of skin using a murine model: Validation of a prognostic viability assay. *Cryobiology*, 27:24–30.

Kim, B. K., Baldini, M. G. (1974): Biochemistry, function and hemostatic effectiveness of frozen human platelets. *Proc. Soc. Exp. Biol. Med.*, 145:830–5.

Klempnauer, J., Pegg, D. E., Taylor, M. J. et al. (1982): Hypothermic preservation of the rat pancreas. In: *Organ Preservation Basic and Applied Aspects*. Edited by D. E. Pegg, I. A. Jacobsen, N. A. Halasz. MTP Press, Lancaster.

Knight, S. C., Farrant, J. (1978): The problem of assaying the recovery of lymphocytes and other blood cells after freezing and thawing. *Cryobiology*, 15:230–1.

Lambotte, L., Wojcik, S., Pontegnie-Istace, S. (1979): Investigation on the mechanisms of action of hyperosmolar solutions used in liver preservation. In: *Organ Preservation II*. Edited by D. E. Pegg and I. A. Jacobsen, pp. 292–303. Churchill Livingstone, Edinburgh/London.

Leibo, S. P., von Boehmer, H., Hengartner, H. (1979): Preservation of cloned T-cells cytotoxic for male-antigen- bearing target cells. *Cryobiology*, 16:607.

Leibo, S. P., Mazur, P. (1978): Methods for the preservation of mammalian embryos by freezing. In: *Methods in Mammalian Reproduction*. Edited by J. C. Daniel, pp. 179–201. Academic Press, New York.

Leibo, S. P., Farrant, J., Mazur, P., Hanna, M. G., and Smith, L. H. (1969): Effects of freezing on marrow stem cell suspensions: Interactions of cooling and warming rates in the presence of PVP, sucrose or glycerol. *Cryobiology*, 6:315–32.

Lorenz, E., Congdon, C. C., Uphoff, D. (1952): Modification of acute irradiation injury in mice and guinea-pigs by bone marrow injections. *Radiology*, 58:863–77.

Madden, P. W. (1987): The evaluation of endothelial damage following corneal storage: A comparison of staining methods and the value of scanning electron microscopy. *Curr. Eye Res.*, 6:1441–51.

Maghni, K., Nicolescu, O. M., Martin, J. G. (1999): Suitability of cell metabolic colorimetric assays for assessment of CD4+ T cell proliferation: Comparison to 5-bromo-2-deoxyuridine (BrdU) ELISA. *Journal of Immunological Methods*, 223:185–94.

Malinin, T. I., Perry, V. P. (1967): A review of tissue and organ viability assay. *Cryobiology*, 4:104–15.

May, S. R., DeClement, F. A. (1981): Skin banking. Part III: Cadaveric allograft skin viability. *J. Burn. Care. Rehab.*, 2:128

Mazur, P., Kemp, J. A., Miller, R. H. (1976): Survival of foetal rat pancreas frozen to −78 and −198\6C. *Proc. Natl. Acad. Sci. USA*, 73:4105–9.

McAnulty, J. F., Vreugdenhil, P. K., Lindell, S., Southard, J. H., Belzer, F. O. (1993): Successful 7-day perfusion preservation of the canine kidney. *Transplantation Proceedings*, 25(1 Pt 2):1642–4.

McCullough, J., Weiblen, B. J., Furcht, L. T. (1980): Effect of storage conditions on function of granulocytes. *Cryobiology*, 17:222–9.

McQueen, C. A., Williams, G. M. (1982): Cytotoxicity of xenobiotics in adult rat hepatocytes in primary culture. *Fund. Appl. Toxicol.*, 2:139–44.

Mischell, B. B., Shiigi, S. M. (1980): *Selective Methods in Cellular Immunology*, pp. 21–2. Freeman, San Francisco.

Mitrani, E., Marks, R. (1973): Towards characterization of epidermal cell migration promotion activity in serum. *Br. J. Dermatol.*, 54:673–7.

Moorhead, P. S., Nowell, P. C., Mellman, W. J. et al. (1960): Chromosome preparations of leukocytes cultured from human peripheral blood. *Exp. Cell. Res.*, 20:613–7.

Mueller, F. O., O'Neill, P., Trevor-Roper, P. D. (1967): Full thickness corneal grafts in Addis Ababa, Ethiopia. *Brit. J. Ophthal.*, 51:227–45.

Nakajima, J., Ono, M., Kobayashi, J., Ming-Chung, L., Takeda, M., Kawauchi, M., Takemoto, S., and Takizawa, H. (1998): Viability and allogenicity of airway epithelial cells after cryopreservation. *Transplantation Proceedings*, 20:3395–3396.

Nikolai, T. J., Peshwa, M. V., Goetghebeur, S., and Hu, W. S. (1991): Improved microscopic observation of mammalian cells on microcarriers by fluorescent staining. *Cytotechnology*, 5:141–6.

Pegg, D. E., Foreman, J., Hunt, C. J., and Diaper, M. P. (1989): The mechanism of action of retrograde oxygen persufflation in renal preservation. *Transplantation*, 48:210–7.

Pegg, D. E., Arnaud, F. G. (1988): Equations for obtaining melting points in the quaternary system propane-1,2- diol/glycerol/sodium chloride/water. *Cryo-Lett*, 9:404–17.

Pegg, D. E. (1964): Freezing of bone marrow for clinical use. *Cryobiology*, 1:64–71.

Pegg, D. E., Green, C. J., Walter, C. A. (1978): Attempted canine renal cryopreservation using dimethyl sulfoxide, helium perfusion and microwave thawing. *Cryobiology*, 15:618–26.

Pegg, D. E., Farrant, J. (1969): Vascular resistance and edema in the isolated rabbit kidney perfused with a cell-free solution. *Cryobiology*, 6:200–10.

Pegg, D. E., Jacobsen, I. A., Diaper, M. P., and Foreman, J. (1986): Optimization of a behicle solution for the introduction and removal of glycerol with rabbit kidneys. *Cryobiology*, 23:53–63.

Pegg, D. E., Wusteman, M. C., Foreman, J. (1981): Metabolism of normal and ischemically injured rabbit kidneys during perfusion for 48 hours at 10C. *Transplantation*, 32:437–43.

Pegg, D. E., (1964): In vitro assessment of cell viability in human bone marrow preserved at −79C. *J. Appl. Physiol.*, 19:123–126.

Pegg, P. J. (1965): The preservation of leucocytes for cytogenic and cytochemical studies. *Brit. J. Haematol.*, 11:586–91.

Penneys, R. (1964): Some experiments on the use of the miniaturized Clark electrode and the open-tip electrode, for the measurement of tissue oxygen tension. In: *Polarography*. Edited by G. J. Hills, p. 951. MacMillan, London, Melbourne.

Peterson, G. L. (1977): A simplification of the protein assay method of Lowry et al. which is more generally applicable. *Anal. Biochem.*, 83:346–56.

Pettersson, S., Claes, G., Schersten, T. (1974): Fatty acid and glucose utilization during continuous hypothermic perfusion of dog kidney. *Eur. Surg. Res.*, 6:79–94.

Ploeg, R. J., Vreugdenhil, P., Goossens, D., McAnulty, J. F., Southard, J. H., and Belzer, F. O. (1988): Effect of pharmacologic agents on the function of the hypothermically preserved dog kidney during normothermic reperfusion. *Surgery*, 103:676–83.

Polge, C. (1980): Freezing of spermatozoa. In: *Low Temperature Preservation in Medicine and Biology*. Edited by M. J. Ashwood-Smith and J. Farrant, pp. 45–64. Pitman, Tunbridge Wells.

Ponec, M., Weerheim, A., Kempenaar, J., Mommaas, A. M., and Nugteren, D. (1988): Lipid composition of cultured human keratinocytes in relation to their differentiation. *J. Lipid. Res.*, 29:949–61.

Puck, T. T. (1959): Quantitative studies on mammalian cells in vitro. *Rev. Mod. Phys.*, 31:433–448.

Pulvertaft, R. J. V., Humble, J. G. (1956): Culture de moelle osseuse sur lames tournantes. *Rev. Hematol.*, 11:349–77.

Rasmussen, E. S. (1999): Use of fluorescent redox indicators to evaluate cell proliferation and viability. *In Vitro and Molecular Toxicology*, 12:47–58.

Rotman, B., Papermaster, B. W. (1966): Membrane properties of living mammalian cells as studied by enzymatic hydrolysis of fluorogenic esters. *Proc. Natl. Acad. Sci. USA*, 55:134–41.

Sanfey, H., Bulkley, G. B., Cameron, J. L. (1985): The pathogensis of acute pancreatitis: The source and role of oxygen-derived free radicals in three different experimental models. *Ann. Surg.*, 201:633–8.

Schaefer, U. W. (1980): Bone marrow stem cells. In: *Low Temperature Preservation in Medicine and Biology*. Edited by M. J. Ashwood-Smith and J. Farrant, pp. 139–54. Pitman Medical, Tunbridge Wells.

Sehr, P. A., Bore, P. J., Papatheofanis, J., and Radda, G. K. (1979): Non-destructive measurement of metabolites and tissue pH in the kidney by 31P nuclear magnetic resonance. *Brit. J. Exp. Pathol.*, 60:632–41.

Sells, R. A., Bore, P. H., McLoughlin, G. A., Johnson, J. N., and Tyrrell, I. (1977): A predictive test of renal viability. *Transplant. Proc.*, 9:1557–63.

Silva, J. M., Day, S. H., Nicoll-Griffith, D. A. (1999): Induction of cytochrome-P450 in cryopreserved rat and human hepatocytes. *Chemico-Biological Interactions*, 121:49–63.

Smith, A. U., Ashwood-Smith, M. J., Young, M. R. (1963): Some in vitro studies on rabbit corneal tissue. *Exp. Eye. Res.*, 2:71–87.

Smith, L. H., Holland, J. M. (1981): Interaction between benzo(a)pyrine and mouse skin in organ culture. *Toxicology*, 21:47–57.

Smith, A. U., Ashwood-Smith, M. J., Young, M. R. (1963): Some in-vitro studies on rabbit corneal tissue. *Exp. Eye. Res.*, 2:71–87.

Southard, J. H., Belzer, F. O. (1980): Control of canine cortex slice volume and ion distribution at hypothermia by impermeable anions. *Cryobiology*, 17:540–8.

Southard, J. H. (1989): Viability assays in organ preservation. *Cryobiology*, 26:232–238.

Sugimoto, M., Tajima, K., Kojima, A., and Endo, H. (1974): Differential acceleration by hydrocortisone of the accumulation of epidermal structural proteins in the chick embryonic skin growing in a chemically defined medium. *Dev. Biol.*, 39:295–307.

Sundberg, R., Lindell, S., Jamieson, N. V., Southard, J. H., and Belzer, F. O. (1988): Effects of chlorpromazine and methylprednisolone on perfusion preservation of rabbit livers. *Cryobiology*, 25:417–24.

Supino, R. MTT assays. (1995): *Methods Mol. Biol.*, 43:137–49.

Tammi, R., Maibach, H. (1987): Skin Organ Culture: Why? *Int. J. Dermatol.*, 26:150–60.

Taylor, M. J. (1982): The role of pH and buffer capacity in the recovery of function of smooth muscle cooled to −13C in unfrozen media. *Cryobiology*, 19:585–601.

Taylor, M. J., Pegg, D. E. (1983): The effect of ice formation on the function of smooth muscle tissue stored at −21C or −60C. *Cryobiology*, 20:36–40.

Taylor, R., Adams, G. D. J., Boardman, C. F. B., and Wallis, R. G. (1974): Cryoprotection: Permeant vs nonpermeant additives. *Cryobiology*, 11:430–8.

Terasaki, R. I., Martin, D. C., Smith, R. B. (1967): A rapid metabolism test to screen cadaver kidneys for transplantation. *Transplantation*, 5:76–8.

Thomas, E. D., Lochte, H. L. Jr. (1957): DNA synthesis by bone marrow cells in vitro. *Blood*, 12:1086–95.

Till, J. E., McCulloch, E. A. (1961): A direct measurement of the radiation sensitivity of normal mouse bone marrow cells. *Radiat. Res.*, 14:213–22.

Umbreit, W. W. (1964): The Warburg constant volume respirometer. In: *Manometric Techniques*. Edited by W. W. Umbreit, R. H. Burns, J. F. Stauffer, p. 1. Burgess, Minneapolis.

Uzuka, M., Takeshita, C., Morikawa, F. (1977): In vitro growth of mouse hair root. *Acta. Derm. Venereol.* (Stockh), 57:217–9.

von Frankenberg, M., Forman, D. T., Frey, W., Stachlewitz, R. F., Bunzendahl, H., and Thurman, R. G. (1997): Release of amino acids from fatty livers during organ harvest for transplantation. *Annals of Clinical Laboratory Science*, 27:287–92.

Wang, B. H., Zuzel, K. A., Rahman, K., Billington, D. (1998): Protective effects of aged garlic extract against bromobenzene toxicity to precision cut rat liver slices. *Toxicology*, 126:213–22.

Weinerth, J. L., Abbott, W. M. (1974): Analysis of injury in complex organ preservation. *Ann. Surg.*, 180:840–6.

Weinstein, R. S., Someda, K. (1967): The freeze-cleave approach to the ultrastructure of frozen tissue. *Cryobiology*, 4:116–29.

Yuspa, S. H., Ben, T., Hennings, H., and Lichti, U. (1980): Phorbol ester tumor promoters induce epidermal transglutaminase activity. *Biochem. Biophys. Res. Commun.*, 97:700–8.

Zhang, S. Z., Lipsky, M. M., Trump, B. F., and Hsu, I. C. (1990): Neural red (NR) assay for cell viability and zenobiotic-induced cytotoxicity in primary cultures of human and rat hepatocytes. *Cell. Biol. and Toxicology*, 6(2):219–34.

Pegg, D. E. (1989): Viability assays for preserved cells, tissues and organs. *Cryobiology*, 26:212–31.

Toxicology of Skin
Edited by Howard I. Maibach
Copyright © 2001 Taylor & Francis

33

Dermal Penetration of Calcium Salts and Calcinosis Cutis

Stephen D. Soileau

Gillette Medical Evaluation Laboratories, Needham, Massachusetts, USA

- Introduction
- Clinical Presentation
- Histologic Evaluation
- Calcium Salts that Can Trigger Calcinosis Cutis
- Medical Exposures
- Nonmedical Exposures
- The Calcification Process
- Dermal Penetration of Calcium Salts
- Mechanism of Calcification
- Summary

INTRODUCTION

Calcinosis cutis is described as the deposition of calcium salts in the skin (Johnson et al., 1993). Cutaneous calcification is normally divided into three groups (Bucko and Burgdorf, 1997):

1. *Metastatic calcification*—the deposition of calcium in the skin because of abnormally high serum levels of calcium or phosphate ions, secondary to some metabolic derangement.
2. *Dystrophic calcification*—the deposition of calcium in the skin because of trauma, abnormal connective tissue, the presence of benign or malignant neoplasms, or a variety of other dermal abnormalities.
3. *Idiopathic*—a misnomer because either a metabolic abnormality or an abnormal matrix must be present for calcium deposition. The label *idiopathic* only admits that the likely cause of the calcification remains undiscovered.

Dystrophic calcification is the most common form of cutaneous calcification. As is stated in the definition, one possible component in the dystrophic calcification process is the presence of trauma or the occurrence of abnormal connective tissue. Trauma such as burns (Coskey and Mehregan, 1984; Heim et al., 1986), repeated minor trauma (Withersty and Chillag, 1979; Ellis et al., 1984), repeated needle trauma (Sell et al., 1980; Skidmore et al., 1997), or other types of dermal trauma (Pitt et al., 1990) can lead to the development of calcinosis cutis. Dermal penetration of certain chemicals is also associated with dystrophic calcification. Examination of the scientific literature shows that the majority of the chemicals associated with dystrophic calcinosis cutis are soluble calcium salts. In this chapter, the subclass of dystrophic calcification caused by the dermal penetration of calcium salts will be reviewed.

CLINICAL PRESENTATION

Calcinosis cutis presents in one of two general forms: non-necrotic and necrotic. In the non-necrotic form, the epidermis is either minimally or not involved (Oppenheim, 1935; Heppleston, 1946; Zackheim and Pinkus, 1957; Paradis and Scott, 1989; Jucglà et al., 1995; Moulin, et al., 1995), and subcutaneous calcification is present. This calcification has been described as papular (Zackheim and Pinkus, 1957; Christensen, 1978; Wiley and Eaglstein, 1979; Wheeland and Roundtree, 1985; Sahn et al., 1992; Johnson et al., 1993), nodular (Oppenheim, 1935; Schoenfeld et al., 1965; Goldminz, 1988; Mancuso, 1990; Johnson et al., 1993), linear (Sneddon and Archibald, 1958; Moulin, et al., 1995), or plaque-like (Clendenning and Auerbach, 1964; Christensen, 1978). Redness and inflammation usually occur in conjunction with this condition. In the necrotic form, the epidermis undergoes extensive destruction, either in the form of ulceration (Heppleston, 1946; Edmonds, 1955; Clendenning and Auerbach, 1964; Schick et al., 1987) or the formation of a gray-green to black eschar (Nash, 1955; Zackheim and Pinkus, 1957; Botvinick et al., 1961; Berger et al., 1974; Mancuso, 1990). Calcification is also present in the necrotic form, although it may not be easily visible because of the extensive cutaneous damage. In cases where large quantities of calcium salts are introduced either subcutaneously (Yosowitz et al., 1975), intramuscularly (Lamm, 1945), or orally (Durlacher et al., 1946) in neonates, massive tissue destruction or gangrene can occur, sometimes resulting in limb amputation or even death. Figure 33.1 (see color plate) shows an example of the necrotic form of calcinosis cutis. This reaction was produced by the application of an 8.6% aqueous solution of calcium chloride dihydrate to tape-stripped skin (to the glistening layer) under an occlusive patch for 24 hours. The photograph was taken 25 days after application of the solution.

Although surgical removal of the calcification and damaged skin has been performed (Clendenning and Auerbach, 1964), the most satisfactory cosmetic results have been reported to occur with conservative treatment (Ramamurthy et al., 1975; Wiley and Eaglstein, 1979). Resolution of both forms occurs in a straightforward fashion. In the non-necrotic form, the calcification usually, but not always (Christensen, 1978), is removed via transepidermal elimination. The calcification and any inflammation slowly resolve, with discrete scarring occurring in some cases (Berger et al., 1974; Roberts, 1977; Moulin et al., 1995; Rodriguez-Cano et al., 1996). In the case of the necrotic form, the skin sloughs off and the wound heals by the formation of granulation tissue (Lamm, 1945; Zackheim and Pinkus, 1957; Botvinick et al., 1961; Roberts, 1977; Mancuso, 1990).

HISTOLOGIC EVALUATION

As one would expect, the histologic hallmark of calcinosis cutis is the presence of calcification. Use of the hematoxylin and eosin staining technique has yielded conflicting results. When used, the calcification has stained as a basophilic (Wilson et al., 1967; Wheeland and Roundtree, 1985; Goldminz, 1988; Mancuso, 1990; Jucglà et al., 1995; Rodriguez-Cano et al., 1996), slightly eosinophilic (Johnson et al., 1993), or amphophilic (Johnson et al., 1993) substance, or has stained in an otherwise abnormal manner (Sneddon and Archibald, 1958). Although alizarin red has been used as a stain specific for calcification (Johnson et al., 1993), the primary method used for the detection of calcification has been the von Kossa stain (Nash, 1955; Clendenning and Auerbach, 1964; Schoenfeld et al., 1965; Wilson et al., 1967; Wheeland and Roundtree, 1985; Wang, Lo, and Wong, 1988; Mancuso, 1990; Rodriguez-Cano et al., 1996). von Kossa's stain itself is not specific for calcium but is based on the combination of silver with an anion salt, which may be carbonate, phosphate, oxalate, sulfate, chloride, or sulfocyanide, and reduction to metallic silver by exposure to light. Those salts remaining in tissue after routine processing are almost always calcium phosphate or calcium carbonate (Wilson et al., 1967). Other methods have been used to directly confirm the presence of calcium in these lesions. Analysis of ashed tissue samples (Botvinick et al., 1961) and laser microprobe analysis (Wilson et al., 1967) have confirmed the presence of calcium in the nodules. The calcification associated with calcinosis cutis has been identified as calcium phosphate, or hydroxyapatite, by diffractometer analysis and electron microscopy (Mancuso, 1990).

The calcification appears as a granular (Botvinick et al., 1961; Schoenfeld et al., 1965; Goldminz, 1988), sheet-like (Wheeland and Roundtree, 1985), or bundled (Oppenheim, 1935) deposition in the dermis (Clendenning and Auerbach, 1964; Christensen, 1978; Paradis and Scott, 1989; Johnson et al., 1993) or in the upper portion of the dermis (Sneddon and Archibald, 1958; Schoenfeld et al., 1965; Wiley and Eaglstein, 1979; Wheeland and Roundtree, 1985). The calcification is found in intimate association with collagen or other elastic fibers (Botvinick et al., 1961; Schoenfeld et al.,1965; Christensen, 1978; Schick et al., 1987; Goldminz, 1988; Paradis and Scott, 1989). The calcification is not benign, as the collagen fibers appear distorted (Christensen, 1978; Schick et al., 1987; Paradis and Scott, 1989) or denatured (Heppleston, 1946). In many cases, calcification is noted in blood vessels (Shannon, 1938; Berger et al., 1974; Hironaga et al., 1982; Speer and Rudolf, 1983; Goldminz, 1988; Rodriguez-Cano et al., 1996). In the non-necrotic form, aside from the presence of calcified material (i.e., transdermal elimination of the calcification), the epidermis shows little or no involvement. Very localized or irregular hyperplasia has been reported (Paradis and Scott, 1989; Moulin et al., 1995); one study showed that the hyperplasia was localized in areas damaged by the transdermal elimination process (Moulin et al., 1995). Other case reports have shown that whereas involvement of the dermis was extensive, no epidermal involvement was noted (Heppleston, 1946; Goldminz, 1988; Jucglà et al., 1995). As one would expect, epidermal destruction is noted in the necrotic form (Nash, 1955; Paradis and Scott, 1989).

The inflammatory response associated with calcinosis cutis is variable, depending on the severity of the condition. In general, the reaction can be described as a foreign body reaction. In mild, non-necrotic cases, a mild inflammatory infiltrate (Schoenfeld et al., 1965; Wheeland and Roundtree, 1985) containing macrophages (Wiley and Eaglstein, 1979; Paradis and Scott, 1989), lymphocytes (Wiley and Eaglstein, 1979; Goldminz, 1988; Paradis and Scott, 1989), neutrophils (Schick et al., 1987; Goldminz, 1988; Paradis and Scott, 1989), and eosinophils (Goldminz, 1988), but not foreign body giant cells, was noted in the immediate vicinity of calcified particles. In one case, no inflammatory reaction was noted (Jucglà et al., 1995). In more severe cases, the general inflammatory response is similar except that the presence of giant cells is noted (Kalton and Pedersen, 1974; Paradis and Scott, 1989; Jucglà et al., 1995).

CALCIUM SALTS THAT CAN TRIGGER CALCINOSIS CUTIS

The body of literature on the dermal penetration of calcium salts and the induction of calcinosis cutis indicates that the salt must have some water solubility in order to have the potential to induce a calcification response. In addition, a mechanism must be present to enable the salts to penetrate the stratum corneum. The vast majority of reports of this condition involve exposure to either calcium gluconate or calcium chloride; however, other calcium salts are reported to have a similar potential. These salts are calcium oxide (Clendenning and Auerbach, 1964; Wheeland and Roundtree, 1985; Sukhanov et al., 1990), calcium carbonate (Wheeland and Roundtree, 1985; Paradis and Scott,1989), calcium nitrate (Christensen, 1978; Wheeland and Roundtree, 1985), and calcium levulinate (Tumpeer and Denenholz, 1936). Human dermal exposure to these salts occurs in a variety of situations, including medical, occupational, household, and environmental. A review of these exposure situations is helpful in understanding the mechanism of this subtype of dystrophic calcinosis cutis.

MEDICAL EXPOSURES

Medical exposures to calcium salts associated with the development of calcinosis cutis have predominantly involved parenteral administration. Therefore, by definition, dermal penetration of the salt is known to occur in these cases. A 10% solution of calcium gluconate is commonly administered to neonates for the treatment of hypocalcemia-related tetany (Lee and Gwinn, 1975). In the earliest reports of its usage for this treatment, the solution was administered via intramuscular injection, most commonly in the buttocks. Within a few weeks of administration, the tissues in the vicinity of the injection site became hardened or stony (von Hofe and Jennings, 1936; Shannon, 1938). In the more severe cases, sloughs developed. These were described as pregangrenous, with pus and a chalky material being extruded from the injection sites (Lamm, 1945). In one of the cases reported, gangrene developed, leading to the death of the infant. This chalky material has been identified as calcium phosphate (Tumpeer and Denenholz, 1936). X-rays of the calcified areas showed a diffuse deposition of calcification in the muscles in the vicinity of the injection site, sometimes extending into surrounding tissues (Tumpeer and Denenholz, 1936; von Hofe and Jennings, 1936; Shannon, 1938). The calcification was gradually resorbed, and within 2–6 months, the presence of calcification by x-ray could not be demonstrated (von Hofe and Jennings, 1936; Shannon, 1938).

Given the dangers associated with the intramuscular administration of calcium gluconate, physicians have been advised to change to intravenous administration of calcium gluconate or calcium chloride (Marcus, 1996); however, if the IV fluid is extravasated or if the IV needle remains at the same site for an extended period of time (Weiss et al., 1975), calcinosis cutis commonly develops. Both the non-necrotic and necrotic forms of the conditions have been reported. In the non-necrotic form, calcification appears in the form of papules (Sahn et al., 1992), nodules (Ramamurthy et al., 1975), plaques (Sahn et al., 1992), or large masses (Roberts, 1977) in the vicinity of the venipuncture site. The lesions appear between 2 hours to 24 days after the extravasation occurs (Ramamurthy et al., 1975; Goldminz, 1988). The presentation is similar to that seen with the intramuscular administration of calcium gluconate. Resolution occurs over a period of weeks to months, with resorption of the calcification.

In the necrotic form, involvement of the epidermis occurs. The presentation is generally a black necrosis surrounded by erythema (Berger et al., 1974). The degree of involvement can range from minimal (Weiss et al., 1975; Roberts, 1977; Hironaga et al., 1982; Speer and Rudolf, 1983) to extensive (Yosowitz et al., 1975), with one case requiring limb amputation (Berger et al., 1974). Once again, calcification develops from days to weeks after the extravasation, and the resolution occurs over a period of months with considerable scarring.

There is one report in the literature describing the use of oral calcium chloride for the treatment of tetany in the newborn (Durlacher et al., 1946). Three cases were presented where severe gastric irritation or ulceration developed after the administration of a 16% solution of calcium chloride. Death occurred in one case, and autopsy results showed that the mucosa of the stomach was replaced by calcareous plaques, with similar plaques seen in the duodenum and upper jejunum. Microscopic evaluation revealed ulceration and necrosis of the mucosa and submucosa of the stomach and small intestines, and the presence of numerous von Kossa stain-positive particles, confirming the presence of calcium phosphate. The authors conducted an animal study using 2-day to 6-week-old rabbits. The animals were administered solutions, by gavage, of calcium chloride at concentrations of 20%, 15%, 10%, and 5%. The results indicated a dose-dependent response. At the highest concentration of 20%, severe gastric damage was seen in all animals. At the lower concentrations, similar, but lesser, symptoms were seen in the younger animals, but not in the older animals. Administration of either 10% or 6% calcium lactate, ammonium chloride (0.5– 1.5 g/kg), or mixtures of ammonium chloride and calcium lactate did not produce any gastric lesions. The results of the animal study indicated that the newborn may be particularly susceptible to the irritating or calcifying effects of calcium chloride when administered orally. This is consistent with the experience in humans, where calcium chloride solutions are well tolerated in older infants (Durlacher et al., 1946).

Calcium chloride has been used in electrode pastes to enhance electrode to skin/ tissue conductivity in electroencephalographic and electromyographic studies. There have been a small number of reported cases of calcinosis cutis developing from the use of these pastes. The pastes involved were typically composed of equal parts of (1) a saturated solution of calcium chloride, (2) glycerin, and (3) bentonite clay (Clendenning and Auerbach, 1964; Johnson et al., 1993). The reactions have been associated with either repeated or long-term use of such electrodes, and both necrotic (Clendenning and Auerbach, 1964; Mancuso, 1990) and non-necrotic (Wiley and Eaglstein, 1979; Johnson et al., 1993) lesions have been reported.

Finally, although not strictly a medical use, a psychiatric patient with poison-ivy dermatitis self-treated with a "salve" composed of sulfur, chlorinated lime, and lard, as well as chloride of lime/water compresses. The patient developed numerous subcutaneous nodular and papular calcifications in the region of the poison-ivy dermatitis (Wilson et al., 1967).

NONMEDICAL EXPOSURES

Calcium salts are and have been used in a variety of applications, the bulk of which involve occupational exposure to the salts; however, in some instances persons can be exposed in everyday, household situations. Table 33.1 presents a list of these varied exposures, along with the type of reaction noted and any other relevant information. The concentrations of the salts ranged from very dilute solutions of calcium chloride to solid material. The clinical and histologic presentations of these cases were identical to those seen in medical exposures to calcium salts.

THE CALCIFICATION PROCESS

Intuitively, one can dissect the mechanism of calcium salt–induced calcification into two steps. The first step would be the penetration of the calcium salt to the site or sites of action, and the second step would involve an interaction of the salt with one or more components of the skin, which through one or more steps would lead to the development of visible calcification. There is a large body of information in the literature on each of these steps, both in confirmatory studies in humans and affected animals and in mechanistic studies conducted in experimental animals. A review of this information gives substantial insight into the likely mechanism of action of these salts.

TABLE 33.1. *Occupational and other nonmedical exposures to calcium salts associated with the development of calcinosis cutis*

Salt	Concentration	Use	Type of reaction	Other comments	Reference
CaO, CaCO₃, Ca(NO₂)₂	Unknown	Drilling mud	Non-necrotic		Wheeland and Roundtree, 1985
CaCl₂	3.5%	Mine water	Non-necrotic	Occurred where skin had been scratched	Sneddon and Archibald, 1958
CaCl₂	40%	Cement additive	Necrotic	Occurred where boots rubbed against leg	Nash, 1955
CaCl₂	Powder	Unknown	Non-necrotic	Carried bags on his forearm	Zackheim and Pinkus, 1957
CaCl₂	Powder	Dehumidifier	Necrotic	Placed in pants pocket	Zackheim and Pinkus, 1957
Ca(NO₂)₂	Powder	Fertilizer	Non-necrotic	Occurred where skin was irritated; calcification present decades after incident	Christensen, 1978
CaCl₂	8%	Mine water	Necrotic	No reaction occurred on intact skin	Edmonds, 1955
CaCl₂	Conc. Soln.	Refrigeratant	Non-necrotic		Oppenheim, 1935
CaCl₂	40%	Dust suppression	Both	Occurred where abrasion, friction present	Heppleston, 1946
CaCl₂	Powder	Herbicide	Non-necrotic	Occurred where scratches were present	Moulin et al., 1995
Bone Meal	Powder	Landscaping	Necrotic	Occurred in a dog	Paradis and Scott, 1989
CaCl₂	Flakes	Dust suppression	Necrotic	Placed in pants pocket	Botvinick et al., 1961
CaCl₂	Flakes	Landscaping	Necrotic	Occurred in a dog	Schick et al., 1987

537

DERMAL PENETRATION OF CALCIUM SALTS

Dermal penetration of the calcium salts is a requirement for this type of dystrophic calcinosis cutis to occur. With the exception of solid material and concentrated solutions of the calcium salts, dermal penetration does not occur through an intact stratum corneum. In case reports of calcinosis cutis occurring after exposure to calcium chloride–containing mine water (3.5% calcium chloride, 1% magnesium chloride, 10% sodium chloride), the reaction occurred in those areas prone to friction or scratches (Edmonds, 1955; Sneddon and Archibald, 1958). In order to confirm the hypothesis that damaged skin was a necessary component in this exposure situation for the development of calcinosis cutis, Sneddon and Archibald (1958) scarified the skin of a patient and exposed the site for 72 hours with a gauze pad soaked in mine water. The result was the development of calcification only along the areas of scarification. Conversely, Edmonds (1955) exposed the upper aspect of the arm to a gauze pad soaked with an 8% calcium chloride solution for seven consecutive days. No irritation developed at the exposure sites. Studies conducted with rabbits have shown the same response; acute exposure to either solid calcium chloride (Oppenheim, 1935) or a 50% solution (Sukhanov et al, 1990) to intact skin did not cause any reaction; however, the repeated application of solid calcium chloride did cause a strong reaction (Oppenheim, 1935).

Similar confirmatory studies have been conducted with calcium chloride–containing EEG electrode pastes. Occlusive patching of intact skin for 48 hours with the paste did not produce any reaction; when applied to the tape-stripped skin of one subject, calcinosis cutis developed (Clendenning and Auerbach, 1964). Similarly, Wiley and Eaglstein (1979), using the same protocol, produced calcinosis cutis in three out of four subjects. Chemically induced skin damage is also sufficient to permit the penetration of calcium salts, as the application of a calcium chloride–containing salve to poison ivy–related dermatitis induced calcinosis cutis in one patient (Wilson et al., 1967). As one would expect, the intradermal injection of calcium chloride in experimental animals led to the development of calcinosis cutis in most (Oppenheim, 1935; Padmanbhan et al., 1963; Berger et al., 1974), but not all (Leonard et al., 1972) cases. In a related study, the long-term application of an electrode paste containing 0.33% calcium chloride to drilled skin (i.e., removal of epidermis with a dental burr) did not lead to the development of calcinosis cutis (Montes et al., 1967). The results of these studies suggest that the threshold concentration of calcium chloride to produce calcinosis cutis is somewhere between 3.5% and 0.33%; however, the threshold may be affected by both the salt concentration, the total dose applied, and the vehicle.

The prerequisite for damaged skin clearly applies to lower concentrations of calcium chloride, but at higher concentrations, damaged skin is not a requirement. In a confirmatory study conducted on dogs, application of calcium chloride flakes (presumably calcium chloride dihydrate, as the material is listed as 77%–80% calcium chloride) directly to intact skin for 24–96 hours led to the development of a papular reaction, whereas application to the coat of the animal (without direct skin contact)

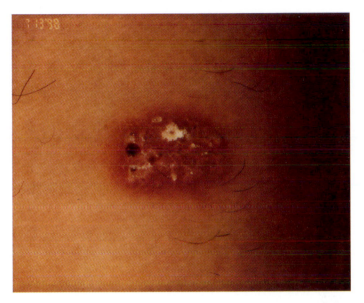

FIG. 33.1. Necrotic form of calcinosis cutis produced by the application of 8.6% calcium chloride dihydrate. The site on the forearm was tape-stripped to the glistening layer, followed by application of the solution under an occlusive patch for 24 hours. The photograph was taken 25 days after application, following the formation and subsequent sloughing of eschar. Note the transdermal elimination of hydroxyapatite, as well as the dark areas of residual necrosis.

did not lead to the development of any dermal reaction (Schick et al., 1987). Also, a one-time application of a 50% calcium chloride solution to intact rabbit skin did not cause any reaction, but the repeated application of the same solution led to ulceration of the skin (Sukhanov et al., 1990).

After the calcium salt has penetrated the initial barrier properties of the skin, it must reach the site where the initial event in calcification occurs. The histological evidence clearly shows that the primary site of action is near the epidermal/dermal interface. Interestingly, there does not appear to be any damage induced, at least initially, in the epidermis. In the non-necrotic lesions, with the exception of minor effects, the epidermis remains essentially intact. In a time course study conducted in dogs, histologic evaluation at 24 hours post-application of solid calcium chloride showed signs of collagen degeneration at the dermal/epidermal interface, whereas the epidermis remained normal; however, epidermal destruction does occur at later time points, indicating that (a) an event subsequent to the initial damage or (b) a delayed toxic reaction is responsible for any epidermal damage that may occur.

MECHANISM OF CALCIFICATION

In the first reports of calcinosis cutis, the mechanisms of necrosis or calcification were described in general terms. For example, (Nash, 1955) postulated that the necrosis was the result of the alkalinity of the concentrated calcium chloride solution, coupled with the general necrosing activity of the salt. The first histologic evaluations also showed the presence of insoluble calcium salts, so it was postulated that soluble calcium was quickly precipitated, and it was this precipitate that started the calcification process. More recently, studies have been conducted to more clearly elucidate the mechanism of calcification, and it appears to be a more complicated process than was originally thought.

The mechanism of calcification is thought to be a two-stage process, with the first step being an initiation event, followed by propagation of the damage into the development of macroscopic calcification (Goldminz, 1988). The initiation stage is believed to be the interaction of the calcium ion with collagen fibers. The mass of histologic evidence shows that when collagen is exposed to soluble calcium salts in vivo, collagen alteration, degeneration, or destruction does occur. In vitro, soluble calcium chloride not only binds to collagen, but it alters the intramolecular structure of collagen (Theis and Jacoby, 1943). When exposed to increasing concentrations of calcium chloride, the amount of calcium that was found bound to collagen increased in a concentration-dependent manner. Calcium chloride–induced alterations in the structure of collagen was evidenced by a shift of approximately one pH unit (in a downward direction) in the isoelectric point of the protein (Theis and Jacoby, 1943). Calcium chloride–treated collagen takes on a shrunken appearance similar to that seen with heat-denatured collagen. Similarly, the presence of calcium chloride lowers the denaturation temperature of collagen, indicating that the intramolecular binding of collagen has been reduced (Theis and Steinhardt, 1942).

If the mechanism of the initiation step is calcium salt–induced alteration in collagen, then one could postulate that calcification can occur as a result of any sort of damage to collagen, and this appears to be the case. Skin damage caused by many different mechanisms has been associated with calcification. Repeated needle sticks (Sell et al., 1980; Skidmore et al., 1997), other acute (Pitt et al., 1990) or chronic (Withersty and Chillag, 1979; Ellis et al., 1984) trauma, burns (Coskey and Mehregan, 1984; Heim et al., 1986), infection-induced necrosis (Rodriguez-Cano et al., 1996), and treatment with certain drugs (Adam et al., 1984; Goldminz, 1988) have been associated with the development of dermal calcification. In animal studies, designed in part to elucidate the mechanism of calcification, collagen damage caused mechanically (Moss and Urist, 1964) or chemically (Padmanbhan et al., 1963; Gabbiani et al., 1969; Bridges and McClure, 1972) induced calcification. In the absence of exogenous calcium, one would require endogenous calcium to initiate the seeding of the damaged collagen. It has been postulated that the hydrolysis of Ca-ATP could provide the initial source of calcium (Leonard and Scullin, 1969; Leonard et al., 1972), and it has also been suggested that mitochondria could provide a ready source of calcium for the initiation step (Goldminz, 1988).

Once the initial alteration in the collagen occurs, what triggers the propagation of this damage into visible calcification? In an elegant time course study in mice and chickens (Bridges and McClure, 1972), the calcification reaction was triggered by the subcutaneous injection of lead acetate. Over a period of 5 minutes to 6 days post-injection, the authors monitored the presence of certain inflammatory cells as well as the presence of lead, calcium and phosphate. At 5 minutes post-injection, lead, but neither calcium nor phosphate, were detected. Mast cells closest to the lead were degranulated. Similar results were obtained at 1, 2, 3, and 4 hours post-injection. At 5 hours post-injection, calcium and phosphate ions were seen around dilated capillaries, near degranulated mast cells. Twenty-four hours after injection, staining for lead was less intense than noted earlier, but stains for calcium (chloranilic acid) and phosphate (von Kossa) were positive. Finally, at 6 days post-injection, no lead was detected, but calcium and phosphate was were abundant. Large giant cells were present in and around the site of calcification and mast cells were intact. Similar results were seen with the chicken, although no mast cells were noted in any of the sections (mast cells are rare and infrequent in the chicken wattle, where the injections were made (Bridges and McClure, 1972)). These experiments indicate that the majority of the calcium found in calcinosis cutis is of endogenous rather than exogenous origin. (Leonard et al., 1972) induced calcification in rats by the subcutaneous injection of Ca-ATP. Chemical analysis of the lesions showed that the amount of calcium found was approximately 100 times the amount injected.

Given this body of information, one can postulate the following general mechanism for calcium salt–induced calcification: Calcium salts penetrate the skin, ultimately binding with collagen at the collagen-rich interface of the epidermis and dermis. If a sufficient amount of calcium ions bind to collagen, a disruption occurs in the intramolecular binding of collagen, which in turn makes available more calcium binding sites. Concurrent with this reaction, an inflammatory response is triggered which results in an influx of inflammatory cells to the region. Either through the degranulation of mast cells or trauma induced by the calcium salts, capillaries dilate and their permeability to calcium and phosphate is increased. The calcium bound to the collagen binds with phosphate and forms hydroxyapatite, with propagation of hydroxyapatite formation occurring by virtue of the increased concentration of calcium and phosphate.

This proposed mechanism of calcification is testable, as there are certain requirements in order for the initial damage of collagen to develop into frank calcinosis cutis. Firstly, exogenous calcium and phosphate must be available at the site of injury. Secondly, crystal growth must be allowed to occur. Several studies have been conducted which address these two points, either directly or indirectly. In a model of calcification in the rat, the subcutaneous injection of the calcifying agent, iron chloride, did not cause the formation of hydroxyapatite, whereas pretreatment with dihydrotachysterol (a substance which mobilizes calcium from bone) (Moss and Urist, 1964) led to higher serum levels of calcium and phosphate and the formation of hydroxyapatite nodules (Gabbiani et al., 1969). Although not directly related to the formation of hydroxyapatite, it has been shown that hydroxyapatite nodules can be reduced in size or completely resorbed by reducing serum levels of phosphate. In a patient with calcinosis cutis induced by juvenile dermatomyositis, oral administration of aluminum hydroxide (which lowered serum phosphate levels) greatly reduced the number and size of hydroxyapatite nodules (Wang et al., 1988). Similar results were noted in treating other cases of metastatic calcinosis cutis (Nassim and Connolly, 1970; Mozaffarian et al., 1972; Kalton and Pedersen, 1974; Verberckmoes et al., 1975). Presumably, calcinosis cutis develops as a result of the hyperphosphatemia, and by creating hypophosphatemia with the administration of aluminum hydroxide, one can reverse the process.

One can also postulate that if the immune system plays a role in increasing the local concentration of calcium and phosphate at the site of collagen damage and in triggering an inflammatory response, then the suppression of the immune system should reduce or inhibit the calcinosis cutis response. Ahn et al. (1997) conducted a study where rabbits received subcutaneous injections of 10% calcium gluconate with or without the coadministration of triamcinolone acetonide. This corticosteroid probably suppresses leukocyte accumulation at the site of inflammation, decreases the accumulation of mast cells, and increases the resorption of foreign materials into the tissue (Ahn et al., 1997). Their hypothesis was that coadministration of this agent with calcium gluconate could prevent the necrotic and calcinosis cutis reaction often associated with the extravasation of calcium chloride in infants. When calcium gluconate was injected alone, a massive necrotic calcinosis cutis-type reaction occurred, with the reaction reaching a maximum at 37 days. When the corticosteroid was coadministered with calcium gluconate, no necrotic reaction developed, and after 1 month the skin of the rabbits appeared normal. Histologic evaluation showed massive calcification in the former group, with minimal calcification in the second group.

There have been no published studies regarding the crystallization phase of the calcification process. Aluminum has been reported to be a crystallization poison

(Fallon and Schwamm, 1990), and preliminary studies in our laboratory have shown that the presence of certain aluminum salts can inhibit calcium chloride–induced calcinosis cutis in compromised human skin. A reasonable interpretation of the results could be that the aluminum salts prevent calcinosis cutis by (1) binding phosphate and lowering cellular phosphate levels sufficiently to prevent crystal growth, or (2) acting as a crystallization poison.

SUMMARY

Calcinosis cutis can occur after dermal exposure to soluble calcium salts. The occurrence of this lesion is dependent upon the concentration of the salt applied to the skin, the duration and frequency of exposure, and the integrity of the stratum corneum. The lesion manifests itself in either a necrotic or non-necrotic form, and in both cases hydroxyapatite nodules develop in the skin and are removed either by resorption or transepidermal elimination. The lesion likely develops through a two-stage process of initiation and propagation, where calcium binds to and disrupts the structure of collagen, forming crystal "seeds" on which endogenous calcium and phosphate can precipitate, leading to hydroxyapatite crystal growth.

REFERENCES

Adam, A., Rakhit, G., Beeton, M. B., and Mitchenere, M. S. (1984): Extensive subcutaneous calcification following injections of pitressin tannate. *Br. J. Radiol.*, 57:921–922.

Ahn, S. K., Kim, K. T., Lee, S. H., Hwang, S. M., Choi, E. H., and Choi, S. (1997): The efficacy of treatment with triamcinolone acetonide in calcinosis cutis following extravasation of calcium gluconate: A preliminary study. *Ped. Dermatol.*, 14:103–109.

Berger, P. E., Heidelberger, K. P., and Poznanski, A. K. (1974): Extravasation of calcium gluconate as a cause of soft tissue calcification in infancy. *Am. J. Roentgenol. Radium Ther. Nucl. Med.*, 121:109–17.

Botvinick, I., Smith, R. G., and Stein, S. C. (1961): Calcium chloride necrosis of the skin. *J. Mich. State Med. Soc.*, 60:743–744.

Bridges, J. B., and McClure, J. (1972): Experimental calcification in a number of species. *Calc. Tiss. Res.*, 10:136–141.

Bucko, A. D., and Burgdorf, W. H. C. (1998): Cutaneous calcification. In: *Clinical Dermatology, Vol. 2.* Edited by Demis, D. J. Lippincott-Raven, Philadelphia-New York.

Christensen, O. B. (1978): An exogenous variety of pseudoxanthoma elasticum in old farmers. *Acta Dermatovener* (Stockholm), 58:319–321.

Clendenning, W. E., and Auerbach, R. (1964): Traumatic calcium deposition in skin. *Arch. Dermatol.*, 89: 360–363.

Coskey, R. J., and Mehregan, A. H. (1984): Calcinosis cutis in a burn scar. *J. Am. Acad. Dermatol.*, 11:666–668.

Durlacher, S. L., Harrison, W., and Darrow, D. C. (1946): The effects of calcium chloride and of calcium lactate administered by gavage. *Yale J. Biol. Med.*, 18:135–143.

Edmonds, O. P. (1955): Dermatitis from solutions of sodium and calcium chloride in the coal-mining industry. *Brit. J. Industr. Med.*, 12:320–321.

Ellis, I. O., Foster, M. C., and Womack, C. (1984): Plumber's knee: Calcinosis cutis after repeated minor trauma in a plumber. *Br. Med. J.*, 288:1723.

Fallon, M. D., and Schwamm, H. A. (1990): Metabolic and other nontumorous disorders of bone. In: *Anderson's Pathology, Vol. 1.* Edited by J. M. Kissane, p. 1946. C.V. Mosby, St. Louis. Company.

Gabbiani, G., Badonnel, M. C., and Baud, C. A. (1969): Relationship between iron, calcium and phosphate during experimental cutaneous calcinosis. *Calc. Tiss. Res.*, 4:224–230.

Goldminz, D., Barnhill, R., McGuire, J., and Stenn, K. S. (1988): Calcinosis cutis following extravasation of calcium chloride. *Arch. Dermatol.*, 124:922–925.

Heim, H., Blankstein, A., Friedman, B., and Horoszowski, H. (1986): Calcinosis cutis: A rare late complication of burns, *Burns*, 12:502–504.

Heppleston, A. G. (1946): Calcium necrosis of the skin. *Br. J. Industr. Med.*, 3:253–254.

Hironaga, M., Fujigaki, T., and Tanaka, S. (1982): Cutaneous calcinosis in a neonate following extravasation of calcium gluconate. *J. Am. Acad. Dermatol.*, 6:392–395.

Johnson, R. C., Fitzpatrick, J. E., and Hahn, D. E. (1993): Calcinosis cutis following electromyographic examination. *Cutis*, 52:161–164.

Jucglà, A., Sais, G., Marcoval, J., Moreno, A., and Peyri, J. (1995): Calcinosis cutis following liver transplantation: A complication of intravenous calcium administration. *Br. J. Dermatol.*, 132:275–278.

Kolton, B., and Pedersen, J. (1974): Calcinosis cutis and renal failure. *Arch. Dermatol.*, 110:256–257.

Lamm, S. S., (1945): The danger of intramuscular injection of calcium gluconate in infancy: Report of three cases, with one death. *JAMA*, 129:347–348.

Lee, F. A., and Gwinn, J. L. (1975): Roentgen patterns of extravasation of calcium gluconate in the tissues of the neonate. *J. Pediatr.*, 86:598–601.

Leonard, F., Boke, J. W., Ruderman, R. J., and Hegyeli, A. F. (1972): Initiation and inhibition of subcutaneous calcification. *Calc. Tiss. Res.*, 10:269–279.

Leonard, F., and Scullin, R. I. (1969): New mechanism for calcification of skeletal tissues. *Nature*, 224:1113–1114.

Mancuso, G., Tosti, A., Fanti, P. A., Berdondini, R. M., Mongiorgi, R., and Morandi, A. (1990): Cutaneous necrosis and calcinosis following electroencephalography. *Dermatologica*, 181:324–326.

Marcus, R. (1996): Agents affecting calcification and bone turnover. In: *Goodman & Gilman's The Pharmacological Basis of Therapeutics*. Edited by Hardman, J. G., and Limbird, L. E., p. 1522. McGraw-Hill, New York.

Montes, L. F., Day, J. L., and Kennedy, L. (1967): The response of human skin to long-term space flight electrodes. *J. Invest. Dermatol.*, 49:100–102.

Moss, M. J., and Urist, M. R. (1964): Experimental cutaneous calcinosis. *Arch. Pathol.*, 78:127–133.

Moulin, G., Balme, B., Musso, M., and Thomas, L. (1995): Le collagénome perforant verruciforme, une dermatose par inclusion exogène? A propos d'un cas induit par le chlorure de calcium. *Ann. Dermatol. Venereol.*, 122:591–594.

Mozaffarian, G., Lafferty, F. W., and Pearson, O. H. (1972): Treatment of tumoral calcinosis with phosphorus deprivation. *Ann. Intern. Med.*, 77:741–745.

Nash, P. H. (1955): Occupational calcium necrosis of the skin. *Br. Med. J.*, 1:586.

Nassim, J. R., and Connolly, C. K. (1970): Treatment of calcinosis universalis with aluminum hydroxide. *Arch. Dis. Child.*, 45:118–121.

Oppenheim, M. (1935): Hautveränderungen durch chlorkalzium bei der herstellung der sogenannten eislutscher. *Wien. Klin. Wschr.*, 48:207–209.

Padmanabhan, N., Tuchweber, B., and Selye, H. (1963): Direct-acting calcifying agents. Arzneimittel-Forsch. 13:429–432.

Paradis, M., and Scott, D. W. (1989): Calcinosis cutis secondary to percutaneous penetration of calcium carbonate in a Dalmation. *Can. Vet. J.*, 30:57–59.

Pitt, E., Ethington, J. E., and Troy, J. L. (1990): Self-healing dystrophic calcinosis following trauma with transepidermal elimination. *Cutis*, 45:28–30.

Ramamurthy, R. S., Harris, V., and Pildes, R. S. (1975): Subcutaneous calcium deposition in the neonate associated with intravenous administration of calcium gluconate. *Pediatrics*, 55:802–806.

Roberts, J. R. (1977): Cutaneous and subcutaneous complications of calcium infusions. *J. Am. Coll. Emerg. Physicians*, 6:16–20.

Rodríguez-Cano, L., Vicente G.-P., Creus, M., Bastida, P., Ortega, J. J., and Castells, A. (1996): Childhood calcinosis cutis. *Ped. Dermatol.*, 13:114–117.

Sahn, E. E., and Smith, D. J. (1992): Annular dystrophic calcinosis cutis in an infant. *J. Am. Acad. Dermatol.*, 26:1015–1017.

Schick, M. P., Schick, R. O., and Richardson, J. A. (1987): Calcinosis cutis secondary to percutaneous penetration of calcium chloride in dogs. *JAVMA*, 191:207–211.

Schoenfeld, R. J., Grekin, J. N., and Mehregan, A. (1965): Calcium deposition in the skin: A report of four cases following electroencephalography. *Neurology*, 15:477–480.

Sell, E. J., Hansen, R. C., and Struck-Pierce, S. (1980): Calcified nodules of the heel: A complication of neonatal intensive care. *J. Pediatr.*, 96:473–475.

Shannon, W. R. 1938. Tetany syndrome in newborn infants: Remote deposition of calcium salts following injection of calcium gluconate. *Am. J. Dis. Child.*, 56:1046–1054.

Skidmore, R. A., Davis, D. A., Woosley, J. T., and McCauliffe, D. P. (1997): Massive dystrophic calcinosis cutis secondary to chronic needle trauma. *Cutis*, 60:259–262.

Sneddon, I. B., and Archibald, R. McL. (1958): Traumatic calcinosis of the skin. *Brit. J. Derm.*, 70:211–214.

Speer, M. E., and Rudolph, A. J. (1983): Calcification of superficial scalp veins secondary to intravenous infusion of sodium bicarbonate and calcium chloride. *Cutis*, 32:65–66.

Sukhanov, V. V., Petul'ko, S. N., Bulonova, L. N., and Yulish, N. R. (1990): Toxicological evaluation of calcium chloride and products containing them. *Gig. Tr. Prof. Zabol.*, 5:51–52.

Theis, E. R., and Jacoby, T. F. (1943): The acid-, base-, and salt-binding capacity of salt-denatured collagen. *J. Biol. Chem.*, 148:603–609.

Theis, E. R., and Steinhardt, R. G. (1942): Studies in animal skin proteins. IV: The theoretical significance of the shrink temperature. *J. Am. Leather Chem. Assn.*, 37:433.

Tumpeer, I. H., and Denenholz, E. J. (1936): Calcium deposit following therapeutic injections in tetany of the new-born. *Arch. Pediatr.*, 53:215–223.

Verberckmoes, R., Bouillon, R., and Krempien, B. (1975): Disappearance of vascular calcifications during treatment of renal osteodystrophy. *Ann. Internal. Med.*, 82:529–533.

von Hofe, F. H., and Jennings, R. E. (1936): Calcium deposition following the intramuscular administration of calcium gluconate: Report of a case in a newborn infant. *J. Pediatr.*, 8:348–351.

Wang, W.-J., Lo, W.-L., and Wong, C.-K. (1988): Calcinosis cutis in juvenile dermatomyositis: Remarkable response to aluminum hydroxide therapy. *Arch. Dermatol.*, 124:1721–1722.

Weiss, Y., Ackerman, C., and Shmilovitz, L. (1975): Localized necrosis of scalp in neonates due to calcium gluconate infusions: A cautionary note. *Pediatrics*, 56:1084–1086.

Wheeland, R. G., and Roundtree, J. M. (1985): Calcinosis cutis resulting from percutaneous penetration and deposition of calcium. *J. Am. Acad. Dermatol.*, 12:172–175.

Wiley, H. E., and Eaglstein, W. E. (1979): Calcinosis cutis in children following electroencephalography. *JAMA*, 242:455–456.

Wilson, R. G., Goldman, L., and Brech, F. (1967): Calcinosis cutis with laser microprobe analysis. *Arch. Dermatol.*, 95:490–495.

Withersty, D. J., and Chillag, S. A. (1979): Seamstress's bottom. *Am. Fam. Physician*, 19:85.

Yosowitz, P., Ekland, D. A., Shaw, R. C., and Parsons, R. W. (1975): Peripheral intravenous infiltration necrosis. *Ann. Surg.*, 182:553–556.

Zackheim, H. S., and Pinkus, H. (1957) Calcium chloride necrosis of the skin. *A.M.A. Arch. Dermatol.*, 76:244–246.

Index